Energy Harvesting Technologies

Shashank Priya · Daniel J. Inman

Editors

Energy Harvesting Technologies

 Springer

Editors

Shashank Priya
Virginia Tech
Center for Intelligent Material Systems
    and Structures
Department of Mechanical Engineering
304A Holden Hall
Blacksburg, VA 24061
spriya@mse.vt.edu

Daniel J. Inman
Virginia Tech
Center for Intelligent Material Systems
    and Structures
Department of Mechanical Engineering
310 Durham Hall
Blacksburg, VA 24061
dinman@vt.edu

ISBN 978-0-387-76463-4   e-ISBN 978-0-387-76464-1
DOI 10.1007/978-0-387-76464-1

Library of Congress Control Number: 2008934452

springer.com

Preface

Energy harvesting materials and systems have emerged as a prominent research area and continues to grow at rapid pace. A wide range of applications are targeted for the harvesters, including distributed wireless sensor nodes for structural health monitoring, embedded and implanted sensor nodes for medical applications, recharging the batteries of large systems, monitoring tire pressure in automobiles, powering unmanned vehicles, and running security systems in household conditions. Recent development includes the components and devices at micro–macro scales covering materials, electronics, and integration. The growing demand for energy harvesters has motivated the publication of this book to present the current state of knowledge in this field.

The book is addressed to students, researchers, application engineers, educators, developers, and producers of energy harvesting materials and systems. The chapters mainly consist of technical reviews, discussions, and basic knowledge in the design and fabrication of energy harvesting systems. It brings the leading researchers in the world in the field of energy harvesting and associated fields on to one platform to provide a comprehensive overview of the fundamentals and developments. The book has good mix of researchers from academics, industry, and national laboratories. All the important energy harvesting technologies including piezoelectric, inductive, thermoelectric, and microbatteries are addressed by the leading authors. Furthermore, the book covers the principles and design rules of the energy harvesting circuits in depth. The chapters on demonstrated applications of the energy harvesting-based technologies will allow readers to conceptualize the promise of the field.

The first section in the book provides discussions on background, theoretical models, equivalent circuit models, lumped models, distributed models, and basic principles for design and fabrication of bulk and MEMS-based vibration-based energy harvesting systems. The second section addresses the theory and design rules required for the fabrication of the efficient electronics. The third section discusses the progress in the field of thermoelectric energy harvesting systems. The fourth section addresses the important subject of storage systems. The fifth section describes some of the prototype demonstrations reported so far utilizing energy harvesting. The sixth section reports some initial standards for vibration energy harvesting being formalized by a nationwide committee consisting of researchers from academia and industry. This standard will lay the basic rules for conducting and reporting the research on vibration energy harvesting. The publication of this standard follows the annual energy harvesting workshop. Fourth workshop in this series will be held at Virginia Tech on January 28–29, 2009. It is worthwhile to mention here that this workshop in the last 3 years has grown in size and numbers with growing participation from academia and industry.

The chapters published here are mostly the invited technical submissions from the authors. The editors did not make any judgment on the quality and organization of the text in the chapters and it was mostly left to the decision of the authors. In this regard, the editors do not accept the responsibility for any technical errors present in the chapters and those should be directly discussed with the authors of the relevant chapter.

It was an honor editing this book consisting of contributions from knowledgeable and generous colleagues. Thanks to all the authors for their timely assistance and cooperation during the course of this book. Without their continual support, this work would not have been possible. We hope that readers will find the book informative and instructive and provide suggestions and comments to further improve the text in eventual second edition.

Blacksburg, VA   Shashank Priya and Dan J. Inman

Contents

Part III   Thermoelectrics

11   Thermoelectric Energy Harvesting 325
G. Jeffrey Snyder

12   Optimized Thermoelectrics For Energy Harvesting Applications 337
James L. Bierschenk

Contributors

S.W. Arms

MicroStrain, Inc., 459 Hurricane Lane, Williston, Vermont 05495, USA,
swarms@microstrain.com

M. Augustin

MicroStrain, Inc., 310 Hurricane Lane, Williston, Vermont 05495, USA,
MAugustin@bellhelicopter.textron.com

Adrien Badel

Ferroelectricity and Electrical Engineering Laboratory (LGEF), National
Institute of Applied Science Lyon (INSA de Lyon), 69621 Villeurbanne, France,
adrien.badel@insa-lyon.fr

Stephen P Beeby

University of Southampton, Highfield, Southampton, SO17 1BJ, UK,
spb@ecs.soton.ac.uk

Dinesh Bhatia

Electrical Engineering Department, University of Texas at Dallas, 800 W Campbell
Road, Richardson, TX 75080, USA, dinesh@utdallas.edu

James L. Bierschenk

Marlow Industries, Inc., 10451 Vista Park Road, Dallas TX 75238, USA,
JBIERSCHENK@marlow.com

Raj Bridgelall

Axcess International Inc, 3208 Commander Drive, Carrollton, TX 75006, USA,
r.bridgelall@axcessinc.com

D.L. Churchill

MicroStrain, Inc., 310 Hurricane Lane, Williston, Vermont 05495, USA,
tlbarrows@microstrain.com

William W. Clark

University of Pittsburgh, Pittsburgh, PA 15261, 412-624-9794, USA,
wclark@pitt.edu, bluetick@pitt.edu

Mohammed F. Daqaq

Department of Mechanical Engineering, Clemson University, daqaq@clemson.edu

Nancy J. Dudney

Material Science and Technology Division, Oak Ridge National Laboratory, Oak Ridge, TN, USA, dudneynj@ornl.gov

Alper Erturk

Center for Intelligent Material Systems and Structures, Department of Engineering Science and Mechanics, Virginia Polytechnic Institute and State University, Blacksburg, VA 24061, USA, erturk@vt.edu

Kevin M. Farinholt

The Engineering Institute, Los Alamos National Laboratory, Los Alamos, New Mexico 87545, USA, farinholt@lanl.gov

Charles R. Farrar

The Engineering Institute, Los Alamos National Laboratory, Los Alamos, New Mexico 87545, USA, farrar@lanl.gov

K. J. Gustafson

Case Western Reserve University, Department of Biomedical Engineering, Cleveland, OH 44106, USA; Louis Stokes Cleveland Department of Veterans Affairs Medical Center, Cleveland, OH 44106, USA

Daniel Guyomar

Laboratoire de Génie Electrique et de Ferroélectricité, INSA Lyon, France, daniel.guyomar@insa-lyon.fr

M.J. Hamel

MicroStrain, Inc., 310 Hurricane Lane, Williston, Vermont 05495, USA

Abhiman Hande

Electrical Engineering Department, University of Texas at Dallas, 800 W Campbell Road, Richardson, TX 75080, USA; Texas MicroPower Inc., 18803 Fortson Ave, Dallas Texas, 75252, USA, ahande@texasmicropower.com

Daniel J. Inman

Center for Intelligent Material Systems and Structures, Department of Mechanical Engineering, Virginia Polytechnic Institute and State University, Blacksburg, VA 24061, USA, dinman@vt.edu

K. L. Kilgore

Case Western Reserve University, Department of Biomedical Engineering, Cleveland, OH 44106, USA; Metro Health Medical Center, Cleveland, OH 44109, USA; Louis Stokes Cleveland Department of Veterans Affairs Medical Center, Cleveland, OH 44106, USA

Hyunuk Kim
Center for Intelligent Material Systems and Structures, Center for Energy
Harvesting Materials and Systems, Virginia Tech, Blacksburg, VA 24061, USA.

Mickaël Lallart
Laboratoire de Génie Electrique et de Ferroélectricité, INSA Lyon, France,
mickael.lallart@insa-lyon.fr

Elie Lefeuvre
Laboratoire de Génie Electrique et de Ferroélectricité, INSA Lyon, France

B. E. Lewandowski
NASA Glenn Research Center, Bioscience and Technology Branch, Cleveland,
OH 44135, USA; Case Western Reserve University, Department of Biomedical
Engineering, Cleveland, OH 44106, USA, beth.e.lewandowski@nasa.gov

Arumugam Manthiram
Electrochemical Energy Laboratory, Materials Science and Engineering
Program, The University of Texas at Austin, Austin, TX 78712, USA,
rmanth@mail.utexas.edu

Changki Mo
University of Pittsburgh, Pittsburgh, PA 15261, 412-624-9794, USA

Dr Terence O'Donnell
Tyndall National Institute, Lee Maltings, Prospect Row, Cork, Ireland

Gyuhae Park
The Engineering Institute, Los Alamos National Laboratory, Los Alamos,
New Mexico 87545, USA, gpark@lanl.gov

N. Phan
MicroStrain, Inc., 310 Hurricane Lane, Williston, Vermont 05495, USA

Shashank Priya
Center for Intelligent Material Systems and Structures, Center for Energy
Harvesting Materials and Systems, Virginia Tech, Blacksburg, VA 24061, USA,
spriya@mse.vt.edu

Jamil M. Renno
Center for Intelligent Material Systems and Structures, Virginia Polytechnic
Institute and State University, VA, USA, renno@vt.edu

Claude Richard
INSA Lyon, France

Björn Richter
Heinz Nixdorf Institute, University of Paderborn, 33102 Paderborn, Germany

Gabriel A. Rincón-Mora
School of Electrical and Computer Engineering, Georgia Institute of Technology,
Atlanta, GA 30332-0250, USA, rincon-mora@ece.gatech.edu

Tajana Rosing
Jacobs of School of Engineering, University of California, San Diego, La Jolla, CA
92093, USA

Yi-Chung Shu
Institute of Applied Mechanics, National Taiwan University, Taipei 106, Taiwan,
ROC, yichung@spring.iam.ntu.edu.tw

G. Jeffrey Snyder
Materials Science, California Institute of Technology, 1200 East California,
Boulevard, Pasadena, California 91125, USA, jsnyder@caltech.edu

Henry A. Sodano
Department of Mechanical Engineering – Engineering Mechanics, Michigan
Technological University, Houghton, MI 49931-1295, USA,
Henry.Sodano@asu.edu

Dan Steingart
Department of Chemical Engineering, City College of New York, 140th Street at
Convent Avenue, New York, NY 10031, USA, dan.steingart@gmail.com

Yonas Tadesse
Center for Intelligent Material Systems and Structures, Center for Energy
Harvesting Materials and Systems, Virginia Tech, Blacksburg, VA 24061, USA,
yonas@vt.edu

Michael D. Todd
Jacobs of School of Engineering, University of California, San Diego, La Jolla, CA
92093, USA, mdt@ucsd.edu

C.P. Townsend
MicroStrain, Inc., 310 Hurricane Lane, Williston, Vermont 05495, USA

Jens Twiefel
Institute of Dynamics and Vibration Research, Leibniz University Hannover, 30167
Hannover, Germany, twiefel@ids.uni-hannover.de

Jörg Wallaschek
Institute of Dynamics and Vibration Research, Leibniz University Hannover, 30167
Hannover, Germany, wallaschek@ids.uni-hannover.de

D. Yeary
MicroStrain, Inc., 310 Hurricane Lane, Williston, Vermont 05495, USA,
Ryeary@bellhelicopter.textron.com

Part I
Piezoelectric and Electromagnetic Energy Harvesting

Chapter 1
Piezoelectric Energy Harvesting

Hyunuk Kim, Yonas Tadesse, and Shashank Priya

Abstract This chapter provides the introductory information on piezoelectric energy harvesting covering various aspects such as modeling, selection of materials, vibration harvesting device design using bulk and MEMS approach, and energy harvesting circuits. All these characteristics are illustrated through selective examples. A simple step-by-step procedure is presented to design the cantilever beam based energy harvester by incorporating piezoelectric material at maximum stress points in first and second resonance modes. Suitable piezoelectric material for vibration energy harvesting is characterized by the large magnitude of product of the piezoelectric voltage constant (g) and the piezoelectric strain constant (d) given as ($d \cdot g$). The condition for obtaining large magnitude of $d \cdot g$ has been shown to be as $|d| = \varepsilon^n$, where ε is the permittivity of the material and n is a material parameter having lower limit of 0.5. The material can be in the form of polycrystalline ceramics, textured ceramics, thin films, and polymers. A brief coverage of various material systems is provided in all these categories. Using these materials different transducer structures can be fabricated depending upon the desired frequency and vibration amplitude such as multilayer, MFC, bimorph, amplified piezoelectric actuator, QuickPack, rainbow, cymbal, and moonie. The concept of multimodal energy harvesting is introduced at the end of the chapter. This concept provides the opportunity for further enhancement of power density by combining two different energy-harvesting schemes in one system such that one assists the other.

In last decade, the field of energy harvesting has increasingly become important as evident from the rising number of publications and product prototypes. Several excellent review articles have been published on this topic covering wide variety of mechanisms and techniques (Priya 2007, Anton and Sodano 2007, Beeby et al. 2006, Roundy and Wright 2004, Sodano et al. 2004). At the same time, several applications have been projected for the energy harvesters covering wide range of civilian and defense components. Out of these different applications, the prominent

H. Kim (✉)
Center for Intelligent Material Systems and Structures, Center for Energy Harvesting Materials and Systems, Virginia Tech, Blacksburg, VA 24061.

S. Priya, D.J. Inman (eds.), *Energy Harvesting Technologies*,
DOI 10.1007/978-0-387-76464-1_1 © Springer Science+Business Media, LLC 2009

use of harvester is to power the wireless sensor node. A major challenge in the implementation of multi-hop sensor networks is supplying power to the nodes (Gonzalez et al. 2002). Powering of the densely populated nodes in a network is a critical problem due to the high cost of wiring or replacing batteries. In many cases, these operations may be prohibited by the infrastructure (Raghunathan et al. 2005, Paradiso and Starner 2005).

Outdoor solar energy has the capability of providing power density of $15,000\,\mu W/cm^3$ which is about two orders of magnitudes higher than other sources. However, solar energy is not an attractive source of energy for indoor environments as the power density drops down to as low as $10–20\,\mu W/cm^3$. Mechanical vibrations ($300\,\mu W/cm^3$) and air flow ($360\,\mu W/cm^3$) are the other most attractive alternatives (Roundy et al. 2005, Roundy et al. 2003, Starner and Paradiso 2004). In addition to mechanical vibrations, stray magnetic fields that are generated by AC devices and propagate through earth, concrete, and most metals, including lead, can be the source of electric energy. The actual AC magnetic field strengths encountered within a given commercial building typically range from under $0.2\,mG$ in open areas to several hundred near electrical equipment such as cardiac pace makers, CRT displays, oscilloscopes, motor vehicles (approximately up to $5\,G$ max); computers, magnetic storage media, credit card readers, watches (approximately up to $10\,G$ max); magnetic power supply, liquid helium monitor (approximately up to $50\,G$ max); magnetic wrenches, magnetic hardware, and other machinery (approximately up to $500\,G$ max). AC magnetic fields decrease naturally in intensity as a function of distance (d) from the source. The rate of decrease, however, can vary dramatically depending on the source. For example, magnetic fields from motors, transformers, and so on, decrease very quickly ($1/d^3$), while circuits in a typical multi-conductor circuit decay more slowly ($1/d^2$). Magnetic fields from "stray" current on water pipes, building steel, and so on, tend to decay much more slowly ($1/d$). The other important sources of energy around us are radio frequency waves and acoustic waves.

This chapter provides the introductory information on piezoelectric energy harvesting covering various aspects such as modeling, materials, device design, circuits, and example applications. All of these aspects have been discussed in much detail in the subsequent chapters.

1.1 Energy Harvesting Basics

Vibration of a rigid body can be caused by several factors such as unbalanced mass in a system, tear and wear of materials and can occur in almost all dynamical systems. The characteristic behavior is unique to each system and can be simply described by two parameters: damping constant and natural frequency. Most commonly, a single degree of freedom lumped spring mass system is utilized to study the dynamic characteristics of a vibrating body associated with energy harvesting (Laura et al. 1974). The single degree of freedom helps to study unidirectional

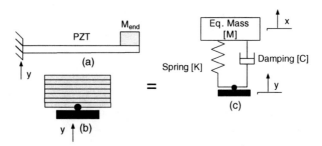

Fig. 1.1 (a) Cantilever beam with tip mass, (b) multilayer PZT subjected to transverse vibration excited at the base, and (c) equivalent lumped spring mass system of a vibrating rigid body

response of the system. Figure 1.1 shows a diagram of a cantilever beam with piezo-electric plates bonded on a substrate and a proof mass at the end; multilayer piezo-electric plates and equivalent lumped spring mass with external excitation. Can-tilever structure with tip mass is the most widely used configuration for piezoelectric energy harvesting device. The source of vibration is shown with an arrow at the base of the contact point. The stiffness of the structure depends on the loading condition, material, and cross-sectional area perpendicular to the direction of vibration. The governing equation of motion for the system shown in Fig. 1.1(c) can be obtained from energy balance equation or D'Alembert's principle. This configuration applies to both the energy harvesting mechanisms shown in Fig. 1.1(a) and (b).

The governing equation of motion of a lumped spring mass system can be written as:

$$M\ddot{z} + C\dot{z} + Kz = -M\ddot{y} \tag{1.1}$$

where $z = x - y$ is the net displacement of mass. Equation (1.1) can also be written in terms of damping constant and natural frequency. A damping factor, ζ, is a dimensionless number defined as the ratio of system damping to critical damping as:

$$\zeta = \frac{c}{c_c} = \frac{c}{2\sqrt{mK}} \tag{1.2a}$$

The natural frequency of a spring mass system is defined by Eq. (1.2b) as:

$$\omega_n = \sqrt{\frac{K}{M}} \tag{1.2b}$$

where the stiffness K for each loading condition should be initially calculated. For example, in case of a cantilever beam, the stiffness K is given by $K = 3EI/L^3$, where E is the modulus of elasticity, I is the moment of inertia, and L is the length of beam. The moment of inertia for a rectangular cross-sectional can be obtained from expression, $I = (1/12)bh^3$, where b and h are the width and thickness of the beam in transverse direction, respectively. For the other cross-sectional area and stiffness, formulas are available in standard mechanical engineering handbook (Blevins, 1979). The power output of piezoelectric system will be higher if system is operating at natural frequency which dictates the selection of material and dimensions. The terms "natural frequency" and "resonant frequency" are used alternatively in literature, where natural frequency of piezoelectric system should not be confused with natural frequency of mechanical system.

The ratio of output $z(t)$ and input $y(t)$ can be obtained by applying Laplace transform with zero initial condition on Eq. (1.1) as:

$$\left| \frac{Z(s)}{Y(s)} \right| = \frac{s^2}{s^2 + 2\zeta\omega_n S + \omega_n^2} \tag{1.3}$$

The time domain of the response can be obtained by applying inverse Laplace transform on Eq. (1.3) and assuming that the external base excitation is sinusoidal given as: $y = Y \sin(\omega t)$:

$$z(t) = \frac{\left(\frac{\omega}{\omega_n}\right)^2}{\sqrt{\left(1 - \left(\frac{\omega}{\omega_n}\right)^2\right)^2 + \left(2\zeta\frac{\omega}{\omega_n}\right)^2}} Y \sin(\omega t - \phi) \tag{1.4}$$

The phase angle between output and input can be expressed as $\Phi = \arctan\left(\frac{C\omega}{K - \omega^2 M}\right)$. The approximate mechanical power of a piezoelectric transducer vibrating under the above-mentioned condition can be obtained from the product of velocity and force on the mass as:

$$P(t) = \frac{m\zeta Y^2 \left(\frac{\omega}{\omega_n}\right)^3 \omega^3}{\left(1 - \left(\frac{\omega}{\omega_n}\right)^2\right)^2 + \left(2\zeta\frac{\omega}{\omega_n}\right)^2} \tag{1.5}$$

The maximum power can be obtained by setting the operating frequency as natural frequency in Eq. (1.5):

$$P_{max} = \frac{m Y^2 \omega_n^3}{4\zeta} \tag{1.6}$$

Using Eq. (1.6), it can be seen that power can be maximized by lowering damping, increasing natural frequency, mass and amplitude of excitation.

There are two common modes utilized for piezoelectric energy harvesting: 33-mode (stack actuators) and 31-mode (bimorphs). In 33-mode, the direction of applied stress (force) and generated voltage is the same, while in 31-mode the stress is applied in axial direction but the voltage is obtained from perpendicular direction

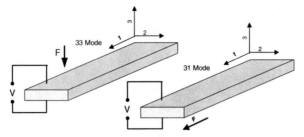

Fig. 1.2 Operating modes of piezoelectric transducer

as shown in Fig. 1.2. For a cantilever beam with long length, the lumped parameter model may not provide reasonable estimate of the output. Contrary to the single degree-of-freedom model (lumped spring mass system), the continuous system has infinite number of natural frequencies and is a logical extension of discrete mass systems where infinite numbers of masses are connected to each other, each having their own degree of freedom.

1.2 Case Study: Piezoelectric Plates Bonded to Long Cantilever Beam with tip mass

Sometimes, small size piezoelectric plates are bonded to a long cantilever beam and need arises to find the stress distribution along the length as a function of excitation frequency. We outline here a simple step-by-step procedure as a starting guideline to find the stress distribution along the continuous beam that can be used to locate the position of piezoelectric plates.

1. Using the governing equation of motion, find the relative displacement which is a function of position and time. The curvature and transverse displacement of a beam can be obtained from the fundamental Euler–Bernoulli beam equation for the given boundary condition expressed as:

$$EI\frac{\partial^4 w(x,t)}{\partial x^4} = -\lambda_m \frac{\partial^2 w(x,t)}{\partial t^2} \tag{1.7}$$

where $\lambda_m = \rho A$ is the linear mass density of the beam.

2. Apply the boundary condition and solve the differential equation. For the cantilever beam of mass M_v and loaded with tip mass M, the boundary conditions are given as:

$$w(0, t) = \frac{\partial w}{\partial x}(0, t) = 0,$$

$$\frac{\partial^2 w}{\partial x^2}(L, t) = 0, \tag{1.8}$$

$$EI\frac{\partial^3 w}{\partial x^3}(L, t) = M\frac{\partial^2 w}{\partial x^2}(L, t)$$

3. Obtain the solution for governing equation using separation of variables method. The general solution for Eq. (1.7) is given as:

$$w_i(x, t) = \phi(x)q(t) \tag{1.9}$$

$$\phi(x) = C_1 \cos \lambda \frac{x}{L} + C_2 \sin \lambda \frac{x}{L} + C_3 \cos \lambda \frac{x}{L} + C_4 \sin \lambda \frac{x}{L}$$

4. Apply the boundary condition and solve for unknown C's. The natural frequency of transversal vibration of a continuous cantilever beam can be obtained analytically from the decoupled equation of Euler–Bernoulli beam and is given by Eq. (1.10) as:

$$f_i = \frac{1}{2\pi}\left(\frac{\lambda}{L}\right)^2 \sqrt{\frac{EI}{\rho A}} \tag{1.10}$$

where i is the mode index, ρ is the mass density, A is the cross-sectional area of beam, and L is the length of the beam.

5. Obtain the solution for forcing term $q(t)$ from equation of motion. The solution of Eq. (1.7) for a cantilever beam of mass $M_v = \rho A$, with a tip mass (M) and boundary condition (Eq. (1.8)), was derived by Erturk and Inman (2007) as follows:

$$w_i(x, t) = \phi(x)q(t) = \omega^2 \sum_{r=1}^{\infty} \frac{\phi(x)(\psi)}{\omega_r^2 - \omega^2 + i2\zeta\omega_r\omega}$$

where

$$\Phi(x) = C_r\left\{\cos\left(\frac{\lambda x}{L}\right) - \cosh\frac{\lambda x}{L} - \beta\left[\sin\frac{\lambda x}{L} - \sinh\frac{y_i x}{L}\right]\right\}$$

$$\beta = \frac{mL(\sin\lambda - \sinh\lambda) + \lambda M(\cos\lambda - \cosh\lambda)}{mL(\sin\lambda + \cosh\lambda) - \lambda M(\sin\lambda - \sinh\lambda)}$$

$$q(t) = \frac{\psi\omega^2}{\omega_{r^2} - \omega^2 + i2\zeta\omega\omega_r}y_o e^{j\omega t}$$

$$\psi = -m\int_0^L \Phi(x)dx + M_z\Phi(L) \tag{1.11}$$

6. The strain ε on surface of beam at a distance y from the neutral axis can be obtained by taking second partial derivative of transverse displacement $w_i(x, t)$:

$$\varepsilon(x) = -y\frac{\partial^2 w}{\partial x^2} \tag{1.12}$$

7. The approximate stress as a function of ratio of distance from the fixed end to the specific location on beam can be obtained from Hooke's law as follows:

$$\sigma(x) = E\varepsilon(x) \tag{1.13}$$

where E is the modulus of elasticity of beam material. If piezoelectric plates are bonded at certain section of a beam, the output voltage from PZT can be estimated by just multiplying the stress at that location with the piezoelectric voltage constant, g. Assuming that the material is linear, elastic, and isotropic with an average stress applied along the 1–1 direction, the output voltage can be determined as follows:

$$V\left(\frac{x}{L}\right) = g_{31} E\varepsilon\left(\frac{x}{L}\right) L_b \tag{1.14}$$

The output power of a PZT at location x from the clamped end and connected to a resistive load can be expressed as:

$$P = \frac{v^2}{R_L} == \frac{1}{R_L}\left\{g_{31} E\varepsilon\left(\frac{x}{L}\right) L_b\right\}^2 \tag{1.15}$$

where R_L is the load resistance and L_b is the length of piezoelectric crystal bonded to substrate beam.

1.3 Piezoelectric Materials

There are two extreme cases of the high-energy density material, PVDF piezoelectric polymer ($d_{33} = 33\,\text{pC/N}$, $\varepsilon_{33}/\varepsilon_0 = 13$, $g_{33} = 286.7\times10^{-3}\,\text{m}^2/\text{C}$), and relaxor piezoelectric single crystals such as PZN – 7%PT ($d_{33} = 2500\,\text{pC/N}$, $\varepsilon_{33}/\varepsilon_0 = 6700$, $g_{33} = 42.1 \times 10^{-3}\,\text{m}^2/\text{C}$). It can be seen from this data that piezoelectric polymer has the highest piezoelectric voltage constant, g_{33}, of $286.7 \times 10^{-3}\,\text{m}^2/\text{C}$ and relaxor-based single crystals have the highest product ($d_{33}.g_{33}$) of the order of $105,250 \times 10^{-15}\,\text{m}^2/\text{N}$. However, the synthesis of both single crystal materials and polymers in large volume is challenging and expensive. Thus, for mass applications, current focus is on improving the properties of polycrystalline ceramics. In this section, we will review some of the developments in the synthesis of high-energy density materials covering ceramics, single crystals, polymers, and thin films.

1.3.1 Piezoelectric Polycrystalline Ceramics

A high-energy density material is characterized by the large magnitude of product of the piezoelectric voltage constant (g) and the piezoelectric strain constant (d) given as ($d.g$). The condition for obtaining large magnitude of $d.g$ has been shown to be as $|d| = \varepsilon^n$, where ε is the permittivity of the material and n is a material parameter having lower limit of 0.5. Table 1.1 shows the relationship between magnitude of n and g_{33} for various commercial compositions. It can be clearly seen from this data that as the magnitude of n decreases the magnitude of g_{33} increases. Islam and Priya (2006a, 2006b) have shown that high-energy density piezoelectric polycrystalline ceramic composition can be realized in the system $Pb(Zr_{1-x}Ti_x)O_3 - Pb[(Zn_{1-y}Ni_y)_{1/3}Nb_{2/3}]O_3$ (PZT – PZNN). The compositions investigated in their study can be represented as: $0.9\,Pb(Zr_{0.52}Ti_{0.48})O_3 - 0.1\,Pb(Zn_{1/3}Nb_{2/3})O_3$ [0.9PZT(52:48) – 0.1PZN] + y wt% $MnCO_3$, where y varies from 0 wt% to 0.9 wt% and $0.9\,Pb(Zr_{0.56}Ti_{0.44})O_3 - 0.1\,Pb[(Zn_{0.8}Ni_{0.2})_{1/3}Nb_{2/3}]O_3$ [0.9PZT (56:44) – 0.1PZNN] + y mol% MnO_2, where y varies from 1 mol% to 3 mol%. The $d_{33}.g_{33}$ values of the samples having composition 0.9PZT (56:44) – 0.1PZNN + 2 mol% MnO_2 (sintered in two steps at 1100–1000 °C) was found to be as $18,456.2 \times 10^{-15}$ m^2/N. This composition was also found to exhibit a high magnitude of g_{33} as 83.1 V m/N, corresponding to the magnitude of n as 1.126.

Table 1.1 Piezoelectric properties and energy harvesting parameter of various commercially available piezoelectric ceramic materials (Copyright: Blackwell Publishing)

Composition	$\varepsilon_{33}/\varepsilon_o$	d_{33} (pC/N)	g_{33} (V m/N)	$d_{33}.g_{33}$ (m^2/N)	n
Morgan electroceramics					
PZT 701	425	153	41×10^{-3}	6273×10^{-15}	1.165
PZT 703	1100	340	30×10^{-3}	10200×10^{-15}	1.181
PZT 502	1950	450	25×10^{-3}	11250×10^{-15}	1.204
PZT 507	3900	700	20×10^{-3}	14000×10^{-15}	1.227
American Piezoelectric Ceramics International					
APC 880	1000	215	25×10^{-3}	5375×10^{-15}	1.20
APC 840	1250	290	26.5×10^{-3}	7685×10^{-15}	1.198
APC 841	1350	300	25.5×10^{-3}	7650×10^{-15}	1.202
APC 850	1750	400	26×10^{-3}	10400×10^{-15}	1.203
APC 855	3400	620	21×10^{-3}	12600×10^{-15}	1.224
Ferroperm Piezoceramics					
Pz 24	400	190	54×10^{-3}	10260×10^{-15}	1.150
Pz 26	1300	300	28×10^{-3}	8400×10^{-15}	1.199
Pz 39	1780	480	30×10^{-3}	14400×10^{-15}	1.194
Pz 52	1900	420	25×10^{-3}	10500×10^{-15}	1.206
Pz 29	2900	575	23×10^{-3}	13225×10^{-15}	1.217
Edo Corporation					
EC-63	1300	295	24.1×10^{-3}	7109.5×10^{-15}	1.20
EC-65	1725	380	25×10^{-3}	9500×10^{-15}	1.205
EC-70	2750	490	20.9×10^{-3}	10241×10^{-15}	1.222
EC-76	3450	583	19.1×10^{-3}	11135.3×10^{-15}	1.228

The selection of piezoelectric ceramic composition for a particular application is dependent on parameters such as operating temperature range ($-20 \leq T \leq 80\,°C$), operating frequency range (10–200 Hz), external force amplitude (0.1–3N), and lifetime ($> 10^6$ cycles). The operating temperature range is determined by the Curie temperature of material which for most of the $Pb(Zr, Ti)O_3$ ceramics is greater than $200\,°C$.

Recently, there has been emphasis on utilizing lead-free materials in domestic and medical applications. Out of all the possible choices for lead-free ceramics, $(Na, K)NbO_3$ (KNN)-based ceramics such as $KNN-LiNbO_3$, $KNN-LiTaO_3$, $KNN-LiSbO_3$, $KNN-Li(Nb, Ta, Sb)O_3$, $KNN-BaTiO_3$ (BT), $KNN-SrTiO_3$, and $KNN-CaTiO_3$ have gained prominence mainly for two reasons: (i) piezoelectric properties exist over a wide range of temperature and (ii) there are several possibilities for substitution and additions. Table 1.2 lists some of the prominent lead-free compositions based on KNN and $(Na_{1/2}Bi_{1/2})TiO_3$ (NBT) – $(K_{1/2}Bi_{1/2})TiO_3$ (KBT) (Shrout and Zhang 2007, Guo et al. 2004, Yuan et al. 2006, Takenaka and Nagata 2005, Zhao et al. 2007, Zang et al. 2006, Ming et al. 2007, Park et al. 2006, Ahn et al. 2007, Cho et al. 2007). NBT-KBT-based ceramics suffer from drawback that there is anti-ferroelectric phase transition at low temperatures that limits the operating range of transducers. Alkali niobate-based ceramics are currently being commercialized by several companies in Europe and Japan and are expected to be available in large quantities in near future.

1.3.2 Piezoelectric Single Crystal Materials

Oriented single crystals of $(1-x)Pb(Zn_{1/3}Nb_{2/3})O_3 - xPbTiO_3$ (PZN-PT) and $(1-x)Pb(Mg_{1/3}Nb_{2/3})O_3 - xPbTiO_3$ (PMN-PT) have been reported to have exceptional properties, such as longitudinal electromechanical coupling factors of 0.95 (Kuwata et al. 1981, 1982; Park and Shrout 1997a, 1997b, 1997c), longitudinal piezoelectric coefficients between 1500 and 2500 pC/N (Kuwata et al. 1981, 1982; Park and Shrout 1997a, 1997b, 1997c;), and electrically induced strains of up to 1.7% (Park and Shrout 1997a, 1997b, 1997c). Single crystals of PZN-PT are grown

Table 1.2 Summary of the lead-free compositions based on KNN system

System	d_{33} (pC/N)	$\varepsilon_3^T/\varepsilon_o$	$tan\,\delta$	k_p	$T_d/T_c(°C)$
NBT-KBT-BT	183	770	0.03	0.37	100/290
NBT-KBT-LBT	216	1550	0.03	0.40	160/350
KNN–LiNbO$_3$	235	500	0.04	0.42	$-/460$
KNN–LiTaO$_3$	268	570	0.01	0.46	$-/430$
KNN–LiSbO$_3$	283	1288	0.02	0.50	$-/392$
KNN–Li(Nb,Ta,Sb)O$_3$	308	1009	0.02	0.51	$-/339$
KNN–BaTiO$_3$	225	1058	0.03	0.36	$-/304$
KNN–SrTiO$_3$	220	1447	0.02	0.40	$-$
KNN–CaTiO$_3$	241	1316	0.09	0.40	$-/306$

T_d: depolarization temperature; T_c: Curie temperature.

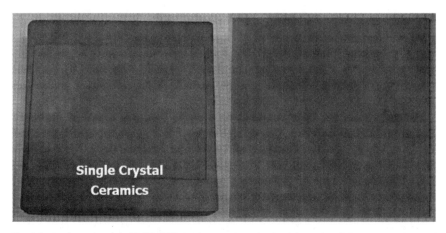

Fig. 1.3 A schematic of the PMN-PZT single crystal synthesized through solid-state crystal growth technique (Copyright: Ho-Yong Lee, Ceracomp Co., Ltd.)

widely using self-flux method (Koyabashi et al. 1997; Mulvihill et al. 1996) while that of PMN-PT by Bridgman's technique (Commercial suppliers such as H. C. Materials Corporation, TRS Technologies, and IBULE Photonics).

Recently, Lee et al. have shown the possibility of synthesizing large size crystals through solid state conversion process as shown in Fig. 1.3 (Ceracomp Co. Ltd.). In this method, a seed crystal is bonded to the surface of the ceramic compact or embedded in the powder compact and the composite sample is carefully sintered at high temperatures. Table 1.3 lists the properties of $Pb(Mg_{1/3}Nb_{2/3})O_3 - PbTiO_3$ (PMN-PT) crystals grown by Bridgman's technique and PMN-PZT by solid-state conversion. Clearly, the system PMN-PZT offers higher rhombohedral–tetragonal transition temperature (T_{R-T}) extending the operating range of the transducer.

The piezoelectric coefficients of single crystal can be enhanced by special cuts and poling as shown in Fig. 1.4 (Zhang et al. 2004). This is quite useful for designing the bimorph-type transducer structures, which mainly utilize d_{31} or d_{32} coefficients. A simple vibration energy harvesting device using d_{32}-mode piezoelectric single crystals can be designed as shown in Fig. 1.5. The structure consists of unimorph- or bimorph-type transducers with single crystal plates bonded on one side or both sides of the metal plates (e.g., brass and aluminum). The transducers are rigidly

Table 1.3 Properties of <001> oriented piezoelectric single crystals available through commercial sources

Material	T_c (°C)	T_{R-T} (°C)	d_{33} (pC/N)	$\varepsilon_{33}/\varepsilon_o$	tan[TM] (%)	k_{33}
PMN-PT-B (HC Material)	–	~75	2000–3500	5500–6500	0.8	0.90–0.94
TRS-X2C (TRS Tech.)	160	75	2200–2700	6500–8500	1	0.92
Type IB (Ibule Ph.)	–	88	1871	6502	< 1	0.91
70PMN-30PT (Ceracomp)	130	90	1500	5000	< 1	0.9
CPSC20-130 (Ceracomp)	195	130	1450	4200	< 1	0.9

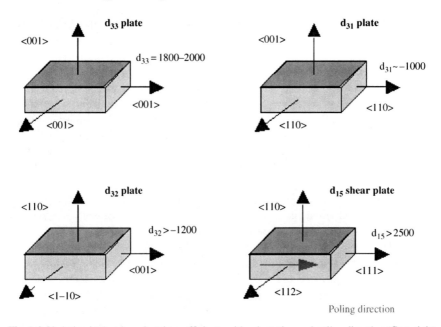

Fig. 1.4 Variation in the piezoelectric coefficients with orientation and poling direction (Copyright: TRS Technologies, State College, PA 16801)

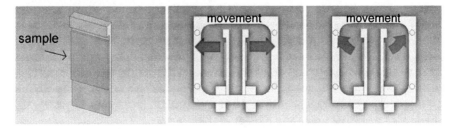

Fig. 1.5 Schematic representation of a bimorph transducer-based uniaxial energy harvester. The operating frequency is tuned by the material and dimensions of the beam and tip mass

fixed in a small cubical box with freedom to oscillate in a specific direction. The tip mass, material for beam, and dimensions of beam determine the operating frequency which could be matched with the resonance frequency of the system. The advantage of bimorph-based devices is that they are simple to fabricate and cheap. The disadvantage is that they are uniaxial systems and limited in power density.

1.3.3 Piezoelectric and Electrostrictive Polymers

Polyvinylidene fluoride (PVDF) is a semi-crystalline high-molecular weight polymer with repeat unit (CH_2-CF_2), whose structure is essentially head-to-tail, i.e., CH2-CF2-(CH2-CF2)n-CH2-CF2. PVDF is approximately half crystalline and half amorphous. In the semi-crystalline polymers such as PVDF, there are regions where the chains exhibit a short- and long-term ordering (crystalline regions). A net dipole moment (polar phase) is obtained by applying a strong electric field at or above T_g and then is frozen in by cooling the material resulting in a piezoelectric-like effect. PVDF sensors have been successfully used as NDE transducers in pulse-echo, through-transmission, and acousto-ultrasonic techniques to monitor curing, and to detect porosity and crack propagation in different model joint geometries (Smolorz and Grill 1995, Chen and Payne 1995, Lee and Moon 1990, Lee 1990, Ohigashi 1988, Harsanji 1995). The curie temperature of PVDF is nearly 110°C, which also makes it useful for some elevated temperature applications. Some of the other piezoelectric polymers that are known today include: polyparaxylene, poly-bischloromethyloxetane (Penton), aromatic polyamides, polysulfone, synthetic polypeptide, and cyanoethyl cellulose (Fukuda and Yasuda 1964).

PVDF has the advantage that it is mechanically strong, resistant to a wide variety of chemicals including acids and can be manufactured on a continuous reel basis. Another important property of PVDF is that it shows a strong piezoelectric response even at microwave frequencies (Smolorz and Grill 1995). Furthermore, the very high g constant of PVDF at the level of -339×10^{-3} V m/N, when compared with the other commonly used piezoelectrics, makes this polymer an ideal transduction material. Table 1.4 compares the piezoelectric properties of the PVDF with other piezoelectric materials.

Recently, research has focused on electrostrictive polymers where it is possible to induce large piezoelectric effect by applying high-DC bias field due to Maxwell stress. Electrostrictive polymers have been discovered that generate large strain (above 5%) under moderate electric field intensity (400–800 V on a 20 μm film). Poly(vinylidene fluoride-trifluoroethylene) or P[VDF:TrFe] copolymers have been shown to exhibit d_{31} of 8.0×10^{-11} pC/N, and polyurethane has d_{31} of 1.7×10^{-10} pC/N. A theoretical investigation conducted by Liu et al. (2005) has shown that energy densities of the order of 0.221 J/cm^3 is possible in the constant field condition using polyurethane material.

1.3.4 Piezoelectric Thin Films

Piezoelectric thin films from material systems such as Pb(Zr, Ti)O$_3$ (PZT), ZnO, AlN, and BaTiO$_3$ (BT) are commonly employed for applications such as sensors,

Table 1.4 Comparison between commonly used piezoelectric materials and PVDF

Material	Relative dielectric constant, $\varepsilon/\varepsilon_o$	Piezoelectric constant d_{33} (pC/N)	Piezoelectric voltage constant, g_{33} (10^{-3} V m/N)
BaTiO$_3$	1700	191	12.6
Quartz	4.5	2.3 (d 11)	50.0 (g 31)
PVDF	13	−33	−339.0
PZT-4	1300	289	25.1
BaTiO$_3$	1700	191	12.6

Table 1.5 Piezoelectric and dielectric properties of thin films (taken from Troiler-Mckinstry and Muralt 2004)

Coefficients/figures of merit	ZnO	AlN	PZT
$e_{31,f}(C/m^2)$	-1.0	-1.05	$-8\ldots-12$
$d_{33,f}(pm/V)$	5.9	3.9	$60\ldots130$
ε_{33}	10.9	10.5	$300\ldots1300$
$\tan\delta$ (at 1–10 kHz, 10^5 V/m)	$0.01\ldots0.1$	0.003	$0.01\ldots0.03$

switches, and actuation diaphragms. In energy harvesting application, same deposition techniques have been used to synthesize films on $Pt/Ti/SiO_2/Si$ wafer. PZT polycrystalline thick films are widely deployed in order to fabricate MEMS scale energy harvesting devices owing to their superior piezoelectric constant as compared to other materials. Table 1.5 lists the properties of various thin film materials (taken from Troiler-Mckinstry and Muralt 2004).

There is a difference in the magnitude of piezoelectric constant between thin/thick film and bulk material due to the constrained imposed by substrate on the film. The film is clamped in-plane but free to move in off-plane direction. As a consequence, piezoelectric constant, $d_{33,f}$, is smaller for film than that of the bulk. The relationships for the effective film piezoelectric constant can be expressed as below (Troiler-Mckinstry and Muralt 2004):

$$
\begin{aligned}
e_{31,f} &= \frac{d_{31}}{s_{11}^E + s_{12}^E} = e_{31} - \frac{c_{13}^E}{c_{33}^E}e_{33} \\
d_{33,f} &= \frac{e_{33}}{c_{33}^E} = d_{33} - \frac{2s_{13}^E}{s_{11}^E + s_{12}^E}d_{31}
\end{aligned}
\tag{1.16}
$$

Chemical solution deposition or sol–gel method is most commonly employed for synthesizing thick films due to various advantages such as cost effectiveness, texturing, and good control on stoichiometry. The compositions of interest are around the morphotropic phase boundary (MPB) which in PZT system is near the Zr/Ti ratio of 52/48. The important variables affecting the piezoelectric properties are grain size, tensile stress, film thickness, and texture. It is well known that (100) oriented PZT thin films show higher piezoelectric response. The texture can be controlled by selecting the suitable substrate or electrode, deposition conditions, and heating rate. The texture can also be controlled using buffer layer, for example, $PbTiO_3$ adopts (100) orientation on Pt and thin layer of TiO_2 adopts (111) orientation (Muralt 2000). The dielectric and piezoelectric properties increase with film thickness and become equivalent to that of bulk ceramic data in the vicinity of 10μm. Piezoelectric properties also increase with grain size. Thus, by controlling the nucleation phenomenon occurring at the film–electrode interface, better properties can be obtained.

There are several requirements for deposition of piezoelectric thin film such as processing temperature range, stoichiometry, and surface energy. Higher processing temperatures induce stress in the film leading to defects. Stress in the film can be expressed by Stoney equation (Glang et al. 1965):

$$\sigma_f = \frac{1}{3} \frac{E_s}{1 - v_s} \frac{t_s^2}{t_f} \frac{\delta}{r^2} \tag{1.17}$$

where σ_f is film stress, E_s, v_s, and t_s are Young's modulus, Poisson's ratio, and the thickness of substrate, respectively, t_f is the thickness of film and δ is the deflection of substrate. The deflection of the substrate is proportional to the temperature that increases with processing temperature. In chemical solution deposition, the crystallization temperatures are in the range of 600–750 °C which is high. This step is repeated many times to obtain thick films. For example, a 2 μm film required 32 spinning and pyrolysis steps, and eight crystallization anneals (Muralt et al. 2002). Thus, heating and cooling rates are selected such that it can reduce the magnitude of built-in stress.

1.4 Piezoelectric Transducers

There has been significant progress in the design and fabrication of piezoelectric transducer structures. Table 1.6 lists the designs which can be easily obtained from commercial sources and are promising for energy harvesting application.

1.5 Meso-macro-scale Energy Harvesters

Once a suitable piezoelectric material is selected, then meso-macro-scale devices can be fabricated using two methods: (i) laser micromachining followed by die bonding and (ii) fiber synthesis followed by macro-fiber composite fabrication.

1.5.1 Mechanical Energy Harvester Using Laser Micromachining

Kim et al. (2008) have demonstrated the micromachining technique using pulsed laser to machine piezoelectric wafers in a desired geometry and pattern with precision of 50–75 μm. The precision can be further improved by selecting the proper laser spot diameter, scanning mirror, and f-theta lens. This technique provides freedom for selecting any desired piezoelectric composition (ceramics, single crystal, and polymer). It reduces the difficulty associated with prior approaches in achieving meso-scale structures such as complex synthesis technique which involves multiple deposition steps, clean room conditions, and limits the piezoelectric composition. The machining process consists of following steps: (i) sintered ceramic wafer is grinded and polished to have a flat surface, (ii) wafer is electroded and poled, and (iii) poled wafer is kept on platform with laser beam and machined. The movement of laser is automated and guided by CAD drawing of pattern. In the work of Kim

Table 1.6 Promising bulk transducer structures for energy harvesting.

Transducer Products	Company/Characteristics
Multilayer	Supplier example: Morgan electroceramics, APC International, Tokin, PI. Characteristics: low frequency (~10 Hz), suitable under large uniaxial stress condition, easy mounting.
Macro Fiber Composite (MFC)	Supplier example: Smart Material. Characteristics: flexible, both d_{33} and d_{31} mode possible, low strain high frequency application, large area coverage, can be used as a bimorph element.
Thunder	Supplier example: Face International Characteristics: Various curvatures and heights possible providing wide range of stress amplification, suitable for very low frequencies (~1 Hz).
Bimorphs	Supplier example: APC International. Characteristics: resonance frequency can be tuned in the range of 5–100 Hz, used in various configuration such as cantilever, end–end clamped, etc.
Amplified Piezoelectric Actuator	Supplier example: Cedrat Characteristics: higher efficiency under large stress, resonance frequency can be tuned to lower ranges (~100 Hz).
QuickPack	Supplier example: Mide Characteristics: similar to bimorphs but easier mounting, wide bandwidth, widely used in cantilever configuration
Rainbow	Characteristics: curved surface resulting in higher charge under a given stress level, can be stacked to amplify charge.
Moonie Cymbal	Supplier: Micromechatronics Characteristics: metal caps protect ceramic allowing application under high stress levels, higher charge due to stress amplification, resonance frequency can be tuned by changing cap dimensions and material.

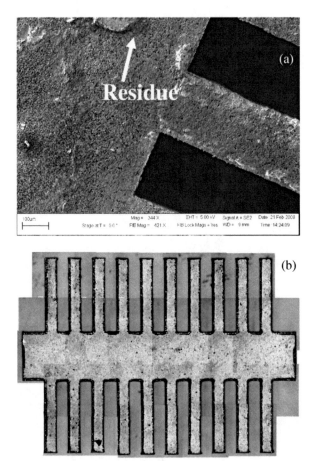

Fig. 1.6 Laser-machined cantilever beam. (**a**) SEM picture and (**b**) optical picture

et al. (2008), YAG laser with wavelength of 1064 nm and pulse width on the order of ∼1 ns with high power was used for machining. The time for whole machining process was about 3 hr 10 min. The smallest feature that can be machined using the nanosecond pulse was on the order of 75 μm. Figure 1.6(a) and (b) shows the pictures of laser-machined cantilever beam.

Conventional microelectronics packaging process was adopted to mount the cantilever beams. For the bottom die bonding, nonconductive thermal curing epoxy was used as the attachment between top PZT and pads. Gold wire (1.0 mil diameter) was bridged for the electrical connection between top silver electrodes and bonding pad of CERDIP (Ceramic Dual In-line Package). Temperature of wire-bonding chuck was maintained at 130 °C and instant glue was used for attaching the tip mass. Sn/Pb solder ball was used as tip mass with diameter and mass of 0.5 mm and 1.53 mg. There were 10 cantilevers on both side of the bridge: five of them with tip mass and five without tip mass. The tip masses were used alternately so that the

Fig. 1.7 Assembled energy harvesting device using laser micromachining (Copyright: IEEE)

cantilevers vibrate with different resonance frequency providing voltage response over a wider frequency range. The laser-machined cantilever beam and assembled energy harvesting device are shown in Fig. 1.7.

The maximum V_{rms} voltage from this device was measured to be 655 mV at 870 Hz with tip mass of 3.06 mg. Figure 1.8 shows the variation of output voltage and power as a function of load for two different frequencies. A maximum power of 1.13 µW was measured across the matching load of 288.5 kΩ at 870 Hz with a power density of 301.3 µW/cm³.

The device shown in Fig. 1.7 can find application in various scenarios such as wireless sensor node for aircraft health monitoring, self-powered sensors in automobile applications, and wireless sensor nodes for industrial machine health mon-

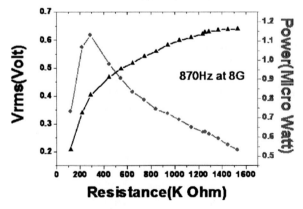

Fig. 1.8 Variation of the output power and voltage as a function of load at frequency of 870 Hz under 8 g acceleration (Copyright: IEEE)

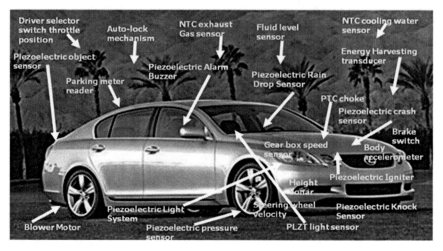

Fig. 1.9 Sensor placement in future automobiles

itoring. Figure 1.9 shows the futuristic view of automobile consisting of multiple sensors based on smart materials. It will be cost-effective and less tedious if many of these sensors have built-in power-harvesting mechanism eliminating the need for wiring. The device shown in Fig. 1.7 can provide the architecture for "self-powered sensors".

1.5.2 Mechanical Energy Harvester Using Piezoelectric Fibers

Macro-fiber composite (MFC), shown in Fig. 1.10 inset, was developed at the NASA Langley Research Center. It consists of uniaxially aligned fibers with rectangular cross-section surrounded by a polymer matrix. MFCs are marketed by Smart Materials Corporation, FL. The fibers in MFCs are machined from low-cost piezoelectric ceramic wafers using a computer controlled dicing saw. Similar to laser micromachining, this process also allows utilizing various forms of piezoelectric materials such as ceramics, composites, and single crystals. It does not restrict any particular ceramic composition. MFCs can be operated in either d_{33}- or d_{31}-modes by designing two different electrode patterns. MFC operating in d_{33}-mode has higher energy conversion rate but lower electrical current when compared with d_{31}-mode. These transducers have been shown to have reliability of above 10^9 cycles operating at maximum strain (Fig. 1.10). The energy harvesting tests conducted on two kinds of MFCs have shown that:

- d_{33} effect is less suitable for energy harvesting due to lower charge output
- electric charge generated is proportional to strain and frequency, and
- low strain, high frequency (>20 Hz) is suitable for continuous charge generation

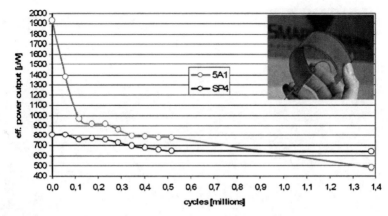

Fig. 1.10 Reliability test on MFCs conducted over 1.4 million cycles (Copyright: T. Daue, Smart Material Corp., Sarasota, FL 34236)

1.6 Piezoelectric Microgenerator

There are five components of energy harvesting device based on MEMS process, namely thin film composition and deposition technique, device design, fabrication process, including etching procedure, electrical connections, and packaging. As mentioned earlier, PZT composition close to MPB are commonly selected and deposited using chemical solution deposition technique. Wolf and Troiler-McKinstry (2004) have proposed the chemical deposition process for PZT films. In this process, lead acetate trihydrate is added to 2-methoxyethanol (2-MOE) under argon atmosphere in the flask maintained in an oil bath at 70 °C on a hot plate and the temperature of the bath is gradually increased to 120 °C. The solution is then dehydrated under vacuum until a semi-dry powder remains. A mixture of zirconium *n*-propoxide and titanium *iso*-propoxide in 2-MOE at room temperature is added and the entire solution is refluxed for 2 h under Ar at 120 °C. After refluxing, the solution is vacuum distilled and 2-MOE is added until the desired solution molarity is achieved. Acetylacetonate and acetic acid are used to adjust the molarity of the solution. Desired modifiers can be added before the refluxing stage.

1.6.1 Piezoelectric Microcantilevers

Kim (2008) has demonstrated piezoelectric mirco-cantilevers-based energy harvester utilizing lead-free barium titanate thin film. This device has interdigital electrode pattern to access the d_{33} response. The microcantilevers were connected in series to increase the output voltage. Figure 1.11 shows the design of mircogenerator and cantilever beam layouts. The fabrication of microcantilever beams requires three masks: electrode mask, cantilever beam mask, and backside mask. The electrode mask is used to pattern metal electrode (red color), the cantilever beam mask is used

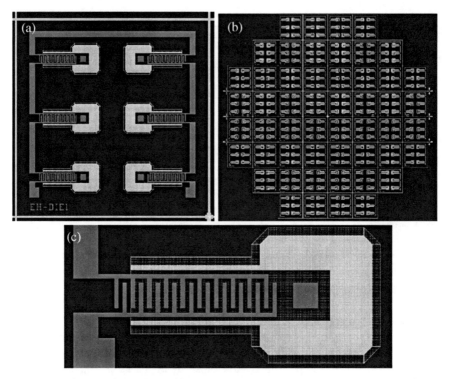

Fig. 1.11 (a) Layout of six microcantilever energy harvesting beams, (b) layout of mask for 4" wafer, and (c) close up of a d_{33}-mode cantilever beam

to pattern microcantilever beam structures (green color), and the backside mask is used to etch out backside of silicon layer (yellow color).

The process starts with 4" <100> SOI (silicon on insulator) wafer as shown in Fig. 1.12(a). The first step was to deposit thick barium titanate layer by sol–gel method. After PR deposition (Futurrex PR1-2000A) for patterning top electrode, Cr/Au layer was deposited as a top electrode. Chrome layer was used for good adhesion between silicon and gold. The 2% HF solution and DRIE (Trion) were used for etching the BT and Si layers, respectively. Before backside Si-etching process, a 3 μm PECVD SiO_2 was deposited on the backside of SOI wafer. PECVD oxide was selected for the hard mask instead of other materials. The SEM picture of the fabricated MEMS microcantilever beam is shown in Fig. 1.13.

It should be pointed out here that there is relatively large variance of etching rate for the BT thin film using 2% HF solution when compared with the dry etching. The products of surface reaction between the barium titanate thin film and 2% HF can be expressed as:

$$BaTiO_3 + 2HF \rightarrow BaF_2 + TiO_2 + H_2O \qquad\qquad \text{Reaction 1}$$
$$2BaTiO_3 + 2HF \rightarrow BaF_2 + BaTi_2O_5 + H_2O \qquad\qquad \text{Reaction 2}$$

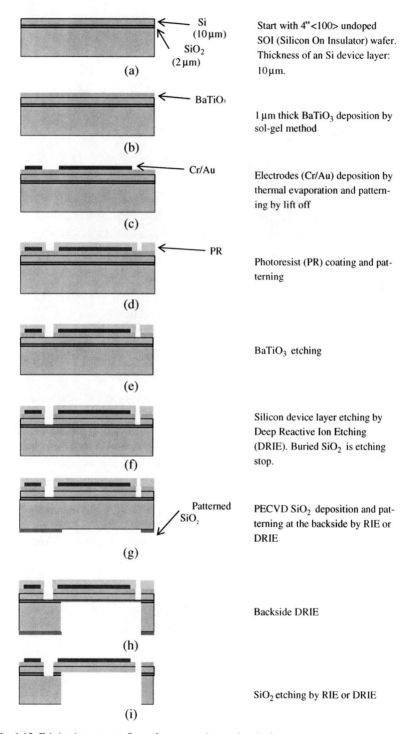

(a) — Si (10 μm), SiO₂ (2 μm)

Start with 4"<100> undoped SOI (Silicon On Insulator) wafer. Thickness of an Si device layer: 10 μm.

(b) — BaTiO₃

1 μm thick BaTiO₃ deposition by sol-gel method

(c) — Cr/Au

Electrodes (Cr/Au) deposition by thermal evaporation and patterning by lift off

(d) — PR

Photoresist (PR) coating and patterning

(e)

BaTiO₃ etching

(f)

Silicon device layer etching by Deep Reactive Ion Etching (DRIE). Buried SiO₂ is etching stop.

(g) — Patterned SiO₂

PECVD SiO₂ deposition and patterning at the backside by RIE or DRIE

(h)

Backside DRIE

(i)

SiO₂ etching by RIE or DRIE

Fig. 1.12 Fabrication process flow of an energy harvesting device

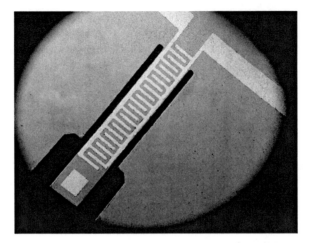

Fig. 1.13 SEM image of cantilever beam after laser wafer saw process

In the second reaction, a Ba-polytitanate phase $BaTi_2O_5$ is produced. The control of the time is essential in order to avoid the second reaction.

1.7 Energy Harvesting Circuits

A simple energy harvesting circuit consists of a diode rectifier (AC/DC) and a DC–DC converter, as shown in Fig. 1.14. The addition of DC–DC converter has been shown to improve energy harvesting by a factor of 7. The efficiency of the step-down converter was between 74 and 88%. Through the exploitation of discontinuous conduction mode operation of the DC–DC converter, a stand-alone energy harvesting system with significantly simplified control circuitry has been proposed (Kim et al. 2007).

Guyomar et al. (2005) have presented a nonlinear processing technique "Synchronized Switch Harvesting on Inductor" (SSHI) for harvesting energy, which consists of a switching device in parallel with piezoelectric element. The device

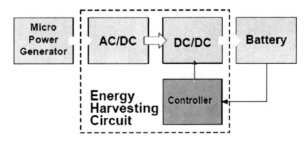

Fig. 1.14 Block diagram for the energy harvesting circuit

(a)   (b)

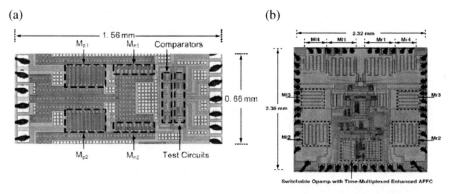

Fig. 1.15 (a) A high-efficiency CMOS rectifier and (b) a SC voltage doubler. (Copyright: IEEE; Permission granted by H. Lee.)

is composed of a switch and an inductor connected in series. The switch is in open state except when the maximum displacement occurs in the transducer. At that instant, the switch is closed and the capacitance of the piezoelectric element and inductance together constitute an oscillator. The switch is kept closed until the voltage on the piezoelectric element has been reversed. In case of nonlinear AC device, a resistive load is directly related on the piezoelectric element in parallel with the switching device. This nonlinear technique has been shown to significantly enhance the performance of the energy harvesting circuit and will be well suited for the resonating structures (Guyomar et al. 2005).

Piezoelectric microgenerators require development of CMOS-based components for efficient energy harvesting and storage. Typically, piezoelectric generator's internal impedance is high in the range of few tens of $k\Omega$. This high-internal impedance restricts the amount of output current that can be driven by the source to microampere range. Therefore, it is important to have low dissipation of the quiescent current for both rectifier and controller of the SC-regulated doubler in order to enhance the efficiency of the power conversion system. Recently, Guo and Lee have proposed a high-efficiency rectifier for transcutaneous power transmission in biomedical application as shown in Fig. 1.15(a) (Guo and Lee 2007). The rectifier was implemented in a standard 0.35 μm CMOS and was found to achieve peak conversion ratio of 95% with output current handling capability of up to 20 mA. The rectifier adopts two comparator-controlled switches to allow unidirectional current flow with only $2V_{ds}$ drop in the conducting path, where V_{ds} is the drain–source voltage across a transistor in linear region. With the use of comparator-controlled switches, this design was not only functional under the lowest input amplitude, but also has good power efficiency. In the subsequent work, Lee and Mok have developed a switched-capacitor (SC) voltage doubler with pseudo-continuous control as shown in Fig. 1.15(b) (Lee and Mok 2005). The SC-regulated voltage doubler provides a constant DC voltage through a step-up DC–DC conversion. These developments are important for full implementation of microgenerator.

In the case of wireless sensor node, an energy storage mechanism with significant energy density is required between the energy harvesting circuitry and the wireless transmission unit. Possible solutions are lithium ion rechargeable batteries and double-layer ultracapacitors. The appropriate choice is based on the required energy density and power density of the application, with lithium ion batteries being used if high "energy" density is required, and ultracapacitors if high "power" density is required.

1.8 Strategies for Enhancing the Performance of Energy Harvester

1.8.1 Multi-modal Energy Harvesting

In order to effectively harvest energy from the available environmental resources, the concept of multi-modal energy harvesting is being pursued. This concept improves upon the existing systems by integrating the following ideas: (i) in a given scenario, two or more energy sources may be available for harvesting energy such as solar, wind, and vibration and (ii) two different energy harvesting schemes can be combined in one system such that one assists the other.

A suitable example of multi-modal energy harvesting device is a system that combines electromagnetic and piezoelectric energy harvesting mechanisms (Poulin et al. 2004). Such a device consists of piezoelectric transducer and electromagnetic system mutually contributing towards power generation (Tadesse et al. 2008). The design consists of cantilever beam with piezoelectric plates bonded at maximum stress locations as shown in Fig. 1.16(a) and (b). At the tip of cantilever beam, a permanent magnet is attached which oscillates within a stationary coil fixed to the top of package. The permanent magnet serves two functions: acts as a proof mass for the cantilever beam and acts as a core which oscillates between the inductive coils resulting in electric current generation through Faraday's effect.

Figure 1.17 shows the diagram of the prototype device. The piezoelectric single crystal plates were bonded on both top and bottom of the beam. The plates were isolated from beam using an insulating epoxy. The output voltages of top and bottom row of piezoelectric were connected in parallel. The beam has tapered geometry with linear variation in dimension along the length such that its moment of inertia along the axis perpendicular to the direction of vibration also varies linearly. This allows the beam to exhibit higher sensitivity. When a cantilevered beam is subjected to external excitation, various modes occur at respective resonance frequencies depending on the geometry, modulus of elasticity, density, and boundary conditions. The first two vibration modes were considered for experimentation as they cover the desired low-operating range of 20–400 Hz for the dimension of cantilever ($25 \times 30 \times 125$ mm^3).

The advantage of this design is that at constant acceleration, the output power from electromagnetic is much higher at lower frequencies (first transversal resonance

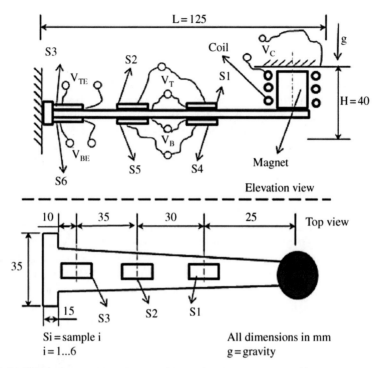

Fig. 1.16 Schematic structure of the multi-modal energy harvester, side-view, and top-view (Copyright: SAGE Publications)

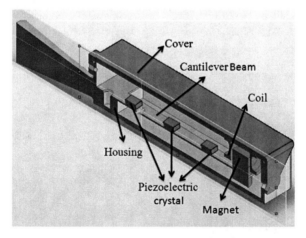

Fig. 1.17 Isometric view of the hand held multi-modal prototype mechanical energy harvester

mode), whereas the output from piezoelectric is higher at higher frequencies (second transversal resonance mode). Thus, this combination allows obtaining significant magnitude of output power from the same device over a wide operating frequency range. Since the frequency content of a random vibration is wide, such a device can utilize multiple spectrum. The bonded piezoelectric plates were located at maximum stress position for the cantilever beam operating in the second resonance mode.

The first resonance frequency of the beam was found to be 20 Hz. The output power of electromagnetic coil at the first resonance frequency across a resistive load of 120 Ω was found to be 8.3 mW. The output power from the top (S2 and S1) and bottom piezoelectric plates (S4 and S5) across a resistive load of 3.3MΩ was found to be 19 and 15 μW, respectively. Piezoelectric plates attached at the end (S3 and S6) generated the higher power of the order of 120 μW in the first resonance mode. The second resonance frequency was found to be 300 Hz. At this frequency, the output power from electromagnetic coil reduced to 3.1 μW. The output power from top piezoelectric plates (S2 and S1) increased to 60 μW, whereas from the bottom plates (S4 and S5) reduced to 14 μW. Figure 1.18 shows the frequency dependence of output power from the fabricated prototype. Typical output power of both the electromagnetic coil and the piezoelectric crystals for various acceleration magnitudes are shown in Fig. 1.19 operating at first natural frequency (20 Hz). It can be seen from this figure that the maximum power was obtained at 35 g acceleration and 20 Hz frequency with a magnitude of 0.25 mW from all piezoelectric plates.

Figure 1.20 shows the photograph of EVA 100/105 wireless transceiver and the fabricated multi-modal harvester. By connecting the on board RS232 DB9 connector through a serial PC port, the received data can be saved on a PC through the application software. The push button transmitter module, PTM 200, works based on electro-dynamic energy transducer actuated by a bow that can be pushed from outside the module on the left or right. The transmission frequency and power were 868.3 MHz and 10 mW max, respectively. The electrical power generated from

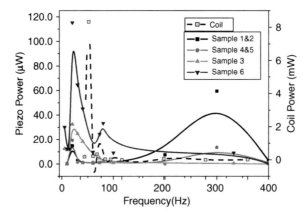

Fig. 1.18 Output power as a function of frequency from the electromagnetic and piezoelectric system (Copyright: SAGE Publications)

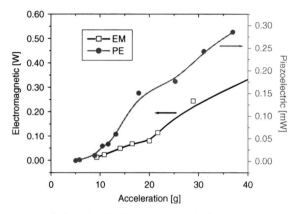

Fig. 1.19 Power output of piezoelectric (*right* y-axis) and electromagnetic coil (*left* y-axis) for various acceleration

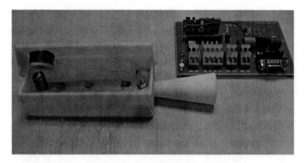

Fig. 1.20 EVA 100/105 evaluation kits and the multi-modal harvester

the fabricated prototype is able to power up the processor and HF transmitter on the board. We are working on developing an integrated unit that combines the wireless transceiver and multi-modal energy harvester in one package.

1.8.2 Magnetoelectric Composites

Magnetoelectric composites combine piezoelectric and magnetostrictive phases in one material thus providing the opportunity to convert magnetic field into stress. Furthermore, it can improve the efficiency of vibration energy harvesting system. The voltage generated across the piezoelectric device under an applied stress X can be simply expressed by the relation, $V = gXt$, where t is the thickness. The voltage can be significantly enhanced using the magnetic stress, through the expression, $V + \Delta V^{\text{magnetic}} = g.(X + \Delta X^{\text{magnetic}})t$, generated through magnetostriction. Exact calculations by Dong et al. (2006) show the voltage generated through the magnetoelectric composite (piezoelectric – magnetostrictive (fiber – matrix)) structure is given as:

$$V = \frac{N d_{33,m} g_{33,p}}{N s_{33}^{E}(1 - k_{33}^{2}) + (1 - N)s_{33}^{H}} H.t \quad \text{(open circuit voltage)} \quad (1.18)$$

where s_{33}^{E} and s_{33}^{H} are the elastic compliances for the piezoelectric and magnetostrictive layers, respectively, k_{33} is the electromechanical coupling coefficient of the piezoelectric layer, $d_{33,m}$ and $g_{33,p}$ are the longitudinal piezomagnetic and piezoelectric voltage coefficients, respectively, and n is the thickness fraction of magnetostrictive layers. The magnetoelectric coefficients as high as 20 V/cm Ωe at 1 Hz have been reported as predicted by the above expression (after using the correction factor). This result indicates that by utilizing magnetoelectric laminate composites, a higher power density can be obtained from the vibration energy harvester where vibration is used to create magnetic field by oscillating the magnets.

Figure 1.21 shows some examples of the structures that can be designed using the two phases in thin film and bulk forms. It can be seen from this figure that the possibilities are numerous and there can be several phases (amorphous, ceramic, metal, polymer, and so on), shapes (disk, cylinder, plate, toroid, sphere, and so on), and sizes (number of layers, layer thicknesses, length and width can be varied differently) for obtaining the magnetoelectric properties. All of these combinations will exhibit ME response with varying degrees of magnitude. For n phases, the number of connectivity patterns is given as $(n + 3)!/3!n!$, which for two phase composites comes out to be 10, for three phase as 20, and 35 for four phase patterns (Newnham 1986). Furthermore, in each of these shapes, there is a possibility of orienting the polarization along different axes and applying the electric (E) and magnetic (H) fields along different axes. In addition, there are several choices for materials depending on magnetostriction constant, resistivity, permeability, permittivity, piezoelectric strain and voltage constant, sintering temperature, and chemical reac-

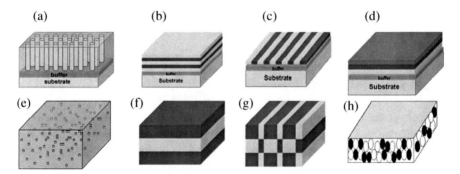

Fig. 1.21 Schematics of the magnetoelectric composites illustrating the multiple possibilities of (**a**) self-assembled nanopillar structures, (**b**) layer-by-layer deposition, (**c**) deposition accompanied by masking and patterning, (**d**) three phase depositions for gradient materials, (**e**) particles dispersed in a polymer matrix, (**f**) lamination, (**g**) checkerboard arrangement, and (**h**) sintered particulates

tivity (magnetostrictive: $MnFe_2O_4$, $CoFe_2O_4$, $NiFe_2O_4$, $ZnFe_2O_4$, YFe_5O_{12}, $SmFe_5O_{12}$, YIG, Terfenol-D, Metglas 2605SC, Ni, Co, and so on; piezoelectric: PZT, $BaTiO_3$, PMN – PT, PVDF, $SrBi_4Ti_4O_{15}$, $(Na_{0.5}K_{0.5})NbO_3)$, and so on). Several reports have been published on the magnetoelectric composites with different materials and geometry and these have contributed in understanding the effect of some of the variables associated with design. Recently, the focus of much research has been on the laminated magnetoelectric composites made by using piezoelectric and magnetostrictive materials (Ryu et al. 2002, Srinivasan et al. 2003, Srinivasan et al. 2002, Dong et al. 2006b, 2003, 2004, 2005, Jia et al. 2006, Wan et al. 2006). We have reported results on laminate composites (MLCs) made from the giant magnetostrictive material, Terfenol-D, and relaxor-based piezocrystals $Pb(Mg_{1/3}Nb_{2/3})$ O_3 – $PbTiO_3$ (PMN-PT) showing high ME coupling (Ryu et al. 2002).

1.8.3 Self-Tuning

One element of efficiency is the magnitude of power harvested when the mechanical input frequency is different from the resonance of piezoelectric harvester. It is highly desired to have an energy harvester that is able to self-tune to optimize its power output in virtually any vibration environment. One common approach is using the shunt. In this case, the non-dimensional mechanical impedance (Z) is given by (Muriuki 2004):

$$\bar{Z}_{jj} = \frac{Z^{\text{SHUNTED}}}{Z^{\text{OPEN}}} = \frac{1 - k_{ij}^2}{1 - k_{ij}^2 \left(\frac{Z^{\text{ELECT}}}{Z^{\text{PIEZO}}}\right)} \tag{1.19}$$

Muriuki (2004) has demonstrated the concept of shunt capacitor to alter the natural frequency of the beam illustrated by expression, $\omega = \sqrt{k_{eff} + C^{-1}d^2/m_{eff}}$, where k is the electromechanical coupling factor. This method has drawback that it requires a complex electronic with programmable microcontroller to connect or disconnect a series of capacitive loads. Another method which can be conceived to alter the frequency of the harvester is by cascading electrical connections, i.e., changing the number of transducers connected in series or parallel by an active switch. The switch can regulate the connections by comparing the power levels over a frequency band. This method has drawback that it requires several transducers to be available and arranged in a systematic fashion for implementing the cascading. The device is bulky and has low power density.

Another approach can be developed based on the nonlinearity of piezoelectric material. It is known that the applied electrical voltage has an effect on the resonance which is attributed to the changes in stiffness. As the applied voltage increases the resonance frequency of the piezoelectric material shifts to lower frequencies. Previously, we have shown that such phenomenon is related to the elastic nonlinearities and the shift in the resonance frequency with increasing excitation can be represented by an empirical response relationship given as (Priya 2001, Nayfeh

1979):

$$f_r(\text{eff}) = f_r^{\text{lin}} + \frac{3}{8}\left(\frac{\alpha\varepsilon^2}{f_r^{\text{lin}}}\right) \pm \sqrt{\left(\frac{K}{f_r^{\text{lin}}\varepsilon}\right)^2 - \delta^2}, \qquad (1.20)$$

where $f_r(\text{eff})$ is the effective resonance frequency, f_r^{lin} is the amplitude independent linear resonance frequency, α is the nonlinear elastic constant, δ is the linear anelastic constant, K is the external excitation, and ε is the rms strain amplitude of the vibration of the sample. The value for the nonlinear constant α is material dependent and determines the shift in the resonance frequency under an external field. For example, it has been found that α/f_r^{lin} is about 40 times higher in soft PZT than in hard PZT. In an energy harvester, the shift in operating frequency can be achieved using the bias voltage or stress.

1.8.4 Frequency Pumping

A method to increase the power harvested by piezoelectric transducers is to achieve more deflections in the same time period ($P \propto f$). In other words, enhance the ratio, $f_{\text{piezo}}/f_{\text{source}} > 1$. Typical vibration frequency of the machines, aircraft, ship hulls, and deck is in the range of 5–100 Hz, which is much lower than the operating frequency of the microgenerator. If the source vibration frequency can be amplified before being applied to the piezoelectric transducer than an improved efficiency can be obtained. One of the simple ways to implement this is using mechanical gears or springs. We have used this technique before to design piezoelectric windmills as shown in Fig. 1.22, where each bimorph goes through 5 Hz oscillation for source frequency of 1 Hz.

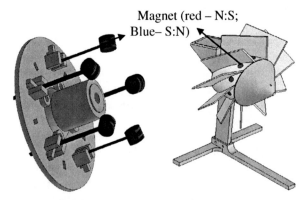

Fig. 1.22 Two magnets sandwich the tip of the bimorphs. These interact with 10 magnets mounted on the vanes in alternating polarities which causes the bimorphs to go through five cycles for every rotation of the vanes

1.8.5 Wide-Bandwidth Transducers

An efficient scheme to increase overall efficiency of an energy harvester is to ensure that it harvests power across varying vibration inputs. The primary problem of the current energy harvester's is its narrow bandwidth. For example, the conventional piezoelectric harvester comprises a single piezoceramic cantilever beam, where the operating frequency is given by: $\omega = \sqrt{k/m}$ (k is the equivalent beam stiffness and m is the weighted sum of cantilever and tip mass). Unfortunately, for minute departures from the beams resonance frequency, there is a significant drop in power levels produced. It should be pointed here that several simple schemes can be designed to achieve the wide-bandwidth operation, but most of these have fundamental drawbacks that power density decreases. Figure 1.23 is an example of some of the common approaches. Recent research is focused on developing transducer structures that exhibit multi-pleresonance which are closely spaced.

1.9 Selected Applications

1.9.1 Border Security Sensors

Border intrusion monitoring requires sensors to be powered continuously for a long period of time. This puts limit on the usage of microbatteries until they can be frequently recharged using environmental resources such as sunlight and wind. The sensors dispersed in rough terrains may be constantly under shade or covered requiring an alternative to photovoltaic's. In these circumstances, wind flow becomes an attractive source for generating small magnitudes of electrical energy and recharging the batteries (Priya 2005). Figure 1.24 shows the schematic representation of the mini-windmill integrated with a section of the sensor nodes through control circuits.

For a flow of air with velocity v and density ρ through unit area A perpendicular to the wind direction, the kinetic energy per unit time is given by P:

$$P = \frac{1}{2}mv^2 = \frac{1}{2}(Av\rho)v^2 = \frac{1}{2}A\rho v^3$$

Fig. 1.23 Structures for achieving wide-bandwidth at the cost of power density

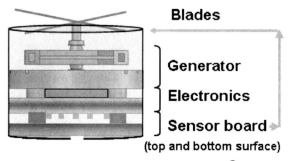

Blades

Generator

Electronics

Sensor board

(top and bottom surface)

Fig. 1.24 Schematic diagram of miniature Piezoelectric Windmill® with integrated electronics and sensor nodes

The air density varies with altitude, atmospheric pressure, and temperature. The power density function $p(v)$ of a windmill may be expressed as:

$$p(v) = C_p \frac{1}{2} A_r \rho v^3 f(v)$$

and

$$f(v) = \left(\frac{k}{c}\right) \left(\frac{v}{c}\right)^{k-1} e^{-(v/c)^k}$$

where C_p is the power coefficient of the windmill defined as the ratio of the power output from the windmill to energy available in the wind, $f(v)$ is the wind speed probability density function, $\underline{A_r}$ ($= \pi R^2$, where R is the length of blade arm) is the rotor area, k is the shape parameter, and c is the scale parameter. The coefficient C_p is limited to 0.59 by the Betz limit. The power generated with 80% blade efficiency and 80% generator efficiency can be determined from the expression:

$$P = (2R)^2 v^3 \times 0.0012$$

Using the above expression, the power available from the blade diameter of 5 in. is 210 mW at 10 mph wind speed and 50 mW for 2.5 in. blade diameter. Clearly, a power of 25 mW can be easily generated if one can design the efficient generator and blades.

Figure 1.25 shows the picture of small-scale Piezoelectric Windmill® which uses three fan blades to enhance the AC stress and effectively capture the wind flow (American Windmills, Diamond Springs, CA). The inner structure of the windmill consists of a vertical shaft connected to a lever arm that converts the rotational motion into translation motion. All three fans are connected into the single vertical shaft through an adjustable gear ratio. The windmill consists of two rows of piezoelectric bimorphs, mentioned here as front and back rows, where each row has nine bimorphs. The number of the bimorphs was selected from the force–

 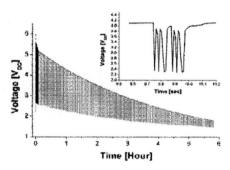

Fig. 1.25 Picture of the windmill and real time transmission of the sensor data

displacement load line curve where the optimum point approximately corresponds to 50% reduction in available force. The operating frequency of the windmill can be easily adjusted by changing the gear ratio. The translation distance of the bimorphs can be adjusted by swapping the crank at the end of the piezoelectric pulling arm. This windmill was able to provide continuous power of 5–6 mW when the wind speeds are in the range of 9–12 mph.

1.9.2 Biomedical Applications

The design of human body-based vibration energy scavenging device incorporates mechanism that can respond to high acceleration and low frequency. The acceleration magnitude from human motion is in the range of $\sim$100 m/s^2 ($\sim$10 g) at a frequency of 2 Hz while jogging (Mathuna et al. 2008). This frequency range is very low for implementing piezoelectric transduction mechanism without gears. However, incorporating complex gear boxes comes at the cost of comfort in motion and defeats the goal of harvesting. Most of the research on harvesting mechanical energy available during human motion has been on modifying the shoe structure. MIT Media lab has reported results on two different heel inserts, one made from PVDF (polyvinylidone fluoride) and the other from PZT. Power levels of up to 9.7 mW were obtained from these shoe inserts (Paradiso and Starner 2005).

1.10 Summary

Self-powering can be achieved by developing smart architecture which utilizes all the environmental resources available for generating electrical power at any given location as exemplified in Fig. 1.26 (Priya et al. 2006). These resources can be vibrations, wind, magnetic fields, light, sound, temperature gradients, and water currents. The generated electric energy can be directly used or stored in the matching media selected by the microprocessor depending on the power magnitude and

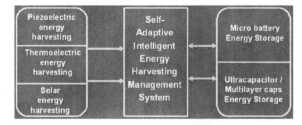

Fig. 1.26 Schematic of the architecture required for self-powered wireless sensor network to achieve desired reliability for long period of time

output impedance. In order to implement this scheme for powering sensors, it will be crucial to design and fabricate them with proper interface. In addition, a special ASIC for matching the widely different output impedance of the harvesters is needed for combined storage.

Acknowledgments The authors gratefully acknowledge the financial support from Texas Advanced Research Program, Pratt & Whitney, Office of Naval Research, and Army Research Office. The authors would also like to thank W.H. Lee for helping with MEMS fabrication.

References

Ahn, CW, Park, HY, Nahm, S, et al. (2007) Structural variation and piezoelectric properties of $0.95(Na_{0.5}K_{0.5})NbO_3-0.05BaTiO_3$ ceramics. *Sens. Actuators A* 136: 255

Anton, SR and Sodano, HA (2007) A review of power harvesting using piezoelectric materials (2003–2006). *Smart Mater. Struct.* 16: R1–R21.

Beeby, SP, Tudor, MJ, and White, NM (2006) Energy harvesting vibration sources for microsystems applications. *Meas. Sci. Technol.* 17: R175–R195.

Blevins, RD (1979) *Formulas for natural frequencies and mode shape.* Robert E. Krieger Publishing, Malabar, Florida, USA.

Chen, QX and Payne, DA (1995) Industrial applications of piezoelectric polymer transducers. *Meas. Sci. Technol.* 6: 249–267.

Cho, KH, Park, HY, Ahn, CW, Nahm, S, Lee, HG, and Lee, HJ (2007) Microstructure and piezoelectric properties of $0.95(Na_{0.5}K_{0.5})NbO_3-0.05SrTiO_3$ ceramics. *J. Am. Ceram. Soc.* 90 [6]: 1946–1949.

Dong, S, Zhai, J, Li, J, and Viehland, D (2006) Near-ideal magnetoelectricity in high-permeability magnetostrictive/piezofiber laminates with a (2-1) connectivity. *Appl. Phys. Lett.* 89: 252904.

Dong, SX, Li, JF, and Viehland, D (2003) Longitudinal and transverse magnetoelectric voltage coefficients of magnetostrictive/piezoelectric laminate composite: theory. *IEEE Trans. Ultrason. Ferroelec. Freq. Control* 50 [10]: 1253–1261.

Dong, SX, Li, JF, and Viehland, D (2004) Longitudinal and transverse magnetoelectric voltage coefficients of magnetostrictive/piezoelectric laminate composite: experiments. *IEEE Trans. Ultrason. Ferroelec. Freq. Control* 51 [7]: 794–799.

Dong, SX, Zhai, J, Li, JF, and Viehland, D (2006) Small dc magnetic field response of magnetoelectric laminate composites. *J. Appl. Phys.* 88: 082907.

Dong, SX, Zhai, J, Wang, N, Bai, F et al. (2005) Fe-Ga/Pb$(Mg_{1/3}Nb_{2/3})O_3$-PbTiO$_3$ magnetoelectric laminate composites. *Appl. Phys. Lett.* 87: 222504.

Erturk, A and Inman, DJ (2007) Mechanical considerations for modeling of vibration-based energy harvester. *Proc.IDETC/CIE 2007.*

Fukuda, E and Yasuda, J (1964) *Jpn. J. Appl. Phys.* 3: 117.

Glang, R, Holmwood, RA, and Rosenfeld, RL (1965) *Rev. Sci. Instrum.* 36, 7.

Gonzalez, JL, Rubio, A, and Moll, F (2002) Human powered piezoelectric batteries to supply power of wereables electronic devices. *Int. J. Soc. Mater. Eng. Resour.* 10 [1]: 34–40.

Guo, S and Lee, H (2007) An efficiency-enhanced integrated CMOS rectifier with comparator-controlled switches for transcutaneous powered implants. Proc. *IEEE Custom Integrated Circuits Conference,* Sep. 2007, 385–388.

Guo, Y, Kakimoto, K, and Ohsato, H (2004) Phase transitional behavior and piezoelectric properties of $(Na_{0.5}K_{0.5})NbO_3$–$LiNbO_3$ ceramics. *Appl. Phys. Lett.* 85: 4121–4123.

Guyomar, D, Badel, A, Lefeuvre, E, and Richard, C (2005) *IEEE Trans. Ultrason. Ferroelectr. Freq. Control* 52: 584–595.

Harsanji, G (1995) *Polymer films in sensor applications,* Technomic Publishing Co., Lancaster, PA.

Islam, RA, Priya, S (2006a) Realization of high-energy density polycrystalline piezoelectric ceramics. Appl. Phys. Lett. 88: 032903.

Islam, RA, Priya, S (2006b) High energy density composition in the system PZT – PZNN. *J. Amer. Ceram. Soc.* 89: 3147–3156.

Jia, Y, Or, SW, et al. (2006) Converse magnetoelectric effect in laminated composites of PMN–PT single crystal and Terfenol-D alloy. *Appl. Phys. Lett.* 88: 242902.

Kim, H, Priya, S, Stephanou, H, and Uchino, K (2007) Consideration of impedance matching techniques for efficient piezoelectric energy harvesting. *IEEE Trans. Ultrason. Ferroelectr. Freq. Control.* 54: 1851–1859.

Kim,H (2008) Design and fabrication of piezoelectric microgenerator using laser micromachining and MEMS techniques, MS Thesis, UT Arlington.

Kim, H, Bedekar, V, Islam, R, Lee, WH, Leo, D, and Priya, S. (2008) Laser micro-machined piezoelectric cantilevers for mechanical energy harvesting. *IEEE Ultrason. Freq. Ferroelect. Cntrl.* 55: 1900–1905.

Koyabashi, T, Shimanuki, S, Saitoh, S, and Yamashita, Y, (1997) *Jpn. J. Appl. Phys.* 36: 6035.

Kuwata, J, Uchino, K, and Nomura, S (1981) *Ferroelectrics* 37: 579.

Kuwata, J, Uchino, K, and Nomura, S (1982) *Jpn. J. Appl. Phys.* 21: 1298.

Laura, PAA, Pombo, JL, and Susemihl, EA (1974) A note on the vibration of clamped-free beam with a mass at the free end. *J. Sound and Vib.* 37(2): 161–168.

Lee, CK and Moon, FC (1990) Modal sensors/actuators. Trans. ASME, *J. Appl. Mech.* 57: 434–441.

Lee, CK (1990) Theory of laminated piezoelectric plates for the design of distributed sensors/actuators, Part I and II. *J. Acoust. Soc. Am.* 89: 1144–1158.

Lee, H and Mok, PKT A SC voltage doubler with pseudo-continuous output regulation using a three-stage switchable opamp. *IEEE International Solid-State Circuits Conference Dig. Tech. Papers,* Feb. 2005, 288–289.

Liu, Y, Ren, KL, Hofmann, HF, and Zhang, Q (2005) *IEEE Trans. Ultrason. Ferroelec. Freq. Control.* 52: 2411–2417.

Mathuna, CO, O'Donnell, T, Matrtinez-Catala, RV, et al. (2008) *Talanta* 75: 613–624.

Ming, BQ, Wang, JF, Qi, P, and Zang, GZ (2007) Piezoelectric properties of (Li, Sb, Ta) modified $(Na,K)NbO_3$ lead-free Ceramics. *J. Appl. Phys.* 101: 054103

Mulvihill, ML, Park, SE, Risch, G, Li, Z, Uchino, K, Shrout, TR, (1996) *Jpn. J. Appl. Phys.* 35: 3984.

Muralt, P (2000) *J. Micromech. Microeng.* 10: 136.

Muralt, P, Baborowski, J, and Ledermann, N (2002) Piezoelectric micro-electromechanical systems with $PbZr_xTi_{1-x}O_3$ Thin Films: Integration and Application Issues, in Piezoelectric Materials in Devices, Ed. N. Setter, EPFL Swiss Federal Institute of Technology, Lausanne, Switzerland.

Muriuki, MG (2004) *An Investigation into the Design and Control of Tunable Piezoelectric Resonators* Ph.D. thesis, University of Pittsburgh, PA.

Nayfeh, AH and Mook, D *Nonlinear Oscillations* (Wiley, New York,1979).

Newnham, RE (1986) Composite Electroceramics. *Ferroelectrics* 68: 1–32.

Ohigashi, H (1988) Ultrasonic Transducers in the Megahertz Range, *The Application of Ferroelectric Polymers*, Ed. T. T. Wang, J. M. Herbert and A. M. Glass, Chapman and Hall, NY.

Paradiso, J and Starner, T (2005) Energy Scavenging for Mobile and Wireless Electronics. *Pervasive Computing*, Jan–March: 18–27.

Park, S and Shrout, TR (1997a) *IEEE Trans. Ultrason. Ferroelectr. Freq. Control* 44: 1140.

Park, S and Shrout, TR (1997b) *J. Appl. Phys.* 82: 1804.

Park, S and Shrout, TR (1997c) *Mater. Res. Innovations* 1: 20.

Park, HY, Ahn, CW, Song, HC, Lee, JH, Nahm, S, et al. (2006) Microstructure and piezoelectric properties of $0.95(Na_{0.5}K_{0.5})NbO_3$–$0.05BaTiO_3$ ceramics. *Appl. Phys. Lett.* 89: 062906

Poulin, G, Sarraute, E and Costa, F (2004) Generation of electrical energy for portable devices Comparative study of an electromagnetic and piezoelectric system. *J. Sens. Actuators A* 116: 461–471.

Priya, S (2005) Modeling of electric energy harvesting using piezoelectric windmill. *Appl. Phys. Lett.* 87: 184101.

Priya, S (2007) Advances in energy harvesting using low profile piezoelectric transducers. *J. Electroceram.* 19:165–182.

Priya, S, Popa, D, and Lewis, F (2006) Energy efficient mobile wireless sensor networks. *ASME Congress 2006*, Nov. 5–10, Chicago, Illinois, IMECE2006-14078.

Priya, S, Viehland, D, Carazo, AV, Ryu, J, and Uchino, K, (2001) *J. Appl. Phys.* 90: 1469.

Raghunathan, V, Kansal, A, Hsu, J, Friedman, J, Srivastava, MB (2005) Design considerations for solar energy harvesting wireless embedded systems. IEEE International Symp. on Information Processing in Sensor Networks (IPSN), April 2005 (TR-UCLA-NESL-200503-10).

Roundy, S and Wright, PK (2004) A piezoelectric vibration based generator for wireless electronics. *Smart Mater. Struct.* 13: 1131–1142.

Roundy, S, Leland, ES, Baker, J, Carleton, E, Reilly, E, Lai, E, Otis, B, Rabaey, JM, Wright, PK, and Sundararajan, V (2005) Improving power output for vibration-based energy scavengers. *Pervasive Comput.* 4 [1] January–March: 28–36.

Roundy, S, Wright, PK, and Rabaey, J (2003) A study of low level vibrations as a power source for wireless sensor nodes. *Comput. Commun.* 26 [11]: 1131–1144.

Ryu, J, Priya, S, and Uchino, K (2002) Magnetoelectric laminate composites of piezoelectric and magnetostrictive materials. *J. Electroceram.* 8: 107–119.

Shrout, TR, and Zhang, SJ, (2007) Lead-free piezoelectric ceramics: Alternatives for PZT?. *J. Electroceram.* 19: 111–124.

Smolorz, S and Grill, W (1995) Focusing PVDF transducers for acoustic microscopy. *Research in Nondestructive Evaluation*, Springer-Verlag, NY, Vol. 7: 195–201.

Sodano, H, Inman, DJ and Park, G (2004) A review of power harvesting from vibration using piezoelectric materials. *Shock Vib. Digest.* 36: 197–205.

Srinivasan, G, Rasmussen, E, Levin, B, and Hayes, R (2002) Magnetoelectric effects in bilayers and multilayers of magnetostrictive and piezoelectric perovskite oxides. *Phys. Rev. B* 65: 134402.

Srinivasan, G, Rasmussen, ET, and Hayes, R (2003) Magnetoelectric effects in ferrite-lead zirconate titanate layered composites: The influence of zinc substitution in ferrites. *Phys. Rev. B* 67 [1]: 014418.

Starner, T and Paradiso, JA (2004) Human-Generated Power for Mobile Electronics. *Low-power electronics design*, C. Piguet, ed., CRC Press, Chapter 45, 1–35.

Tadesse, Y, Zhang, S, and Priya, S (May 2008 submitted) Multimodal energy harvesting system: piezoelectric and electromagnetic. *J. Intell. Mater. Syst. Struct.* (In press)

Takenaka, T and Nagata, H (2005) Current status and prospects of lead-free. piezoelectric ceramics. *J. Eur. Ceram. Soc.* 25: 2693–2700.

Troiler-Mckinstry, S and Muralt, P (2004) Thin film piezoelectrics for MEMS. *J. Electroceram.* 12: 7–17.

Wan, JG, Liu, JM, Wang, GH, and Nan, CW (2006) Electric-field-induced magnetization in $Pb(Zr, Ti)O_3$/Terfenol-D composite structures. *Appl. Phys. Lett.* 88: 182502.

Williams, CB and Yates, RB (1996) Analysis of micro electric generator for micro-electric generator for microsystems. *J. Sens. Actuator* A 52:8–11.

Wolf, RA and Troiler-McKinstry, S (2004) Temperature dependence of the piezoelectric response in lead zirconate titanate films. *J. Appl. Phys.* 95, 1397–406.

Yuan, Y, Zhang, S, Zhou, X, and Liu, J (2006) Phase transition and temperature dependences of electrical properties of $[Bi_{0.5}(Na_{1-x-y}K_xLi_y)_{0.5}]TiO_3$ ceramics. *Jpn. J. Appl. Phys.* 45: 831–834.

Zang, GZ, Wang, JF et al. (2006) Perovskite $(Na_{0.5}K_{0.5})_{1-x}(LiSb)_xNb_{1-x}O_3$ lead-free piezoceramics. *Appl. Phys. Lett.* 88: 212908.

Zhang, S, Lebrun, L, Randall, CA, and Shrout, TR (2004) Growth and electrical properties of (Mn,F) co-doped $0.92Pb(Zn_{1/3}Nb_{2/3})O_3 - 0.08PbTiO_3$ single crystal. *J. Crystal Growth* 267: 204–212.

Zhao, P, Zhang, BP, and Li, JF (2007) Enhancing piezoelectric d_{33} coefficient in Li/Ta-codoped lead-free $(Na, K)NbO_3$ ceramics by compensating Na and K at a fixed ratio. *Appl. Phys. Lett.* 91: 172901.

Chapter 2
Electromechanical Modeling of Cantilevered Piezoelectric Energy Harvesters for Persistent Base Motions

Alper Erturk and Daniel J. Inman

Abstract This chapter investigates electromechanical modeling of cantilevered piezoelectric energy harvesters excited by persistent base motions. The modeling approaches are divided here into two sections as lumped parameter modeling and distributed parameter modeling. The first section discusses the amplitude-wise correction of the existing lumped parameter piezoelectric energy harvester model for base excitation. For cantilevers operating in the transverse and longitudinal vibration modes, it is shown that the conventional base excitation expression used in the existing lumped parameter models may yield highly inaccurate results in predicting the vibration response of the structure. Dimensionless correction factors are derived to improve the predictions of the coupled lumped parameter piezoelectric energy harvester model. The second section of this chapter presents coupled distributed parameter modeling of unimorph and bimorph cantilevers under persistent base excitations for piezoelectric energy harvesting. Closed-form solutions are obtained by considering all vibration modes and the formal representation of the direct and converse piezoelectric effects. Steady state electrical and mechanical response expressions are derived for arbitrary frequency excitations. These multi-mode solutions are then reduced to single-mode solutions for excitations around the modal frequencies. Finally, the analytical expressions derived here are validated experimentally for a cantilevered bimorph with a proof mass.

2.1 Introduction

Vibration-based energy harvesting has received a great attention in the past decade. Research motivation in this field is due to the reduced power requirement of small electronic components, such as the wireless sensors used in structural health monitoring applications. The ultimate goal is to power such small electronic devices using the vibration energy available in their ambient so the requirement of an

A. Erturk (✉)
Center for Intelligent Material Systems and Structures, Department of Engineering Science and Mechanics, Virginia Polytechnic Institute and State University, Blacksburg, VA 24061, USA
e-mail: erturk@vt.edu

S. Priya, D.J. Inman (eds.), *Energy Harvesting Technologies*,
DOI 10.1007/978-0-387-76464-1_2 © Springer Science+Business Media, LLC 2009

external power source or periodic battery replacement can be removed or at least minimized. Research in this area involves understanding the mechanics of vibrating structures, the constitutive behavior of piezoelectric materials and the electrical circuit theory. This promising way of powering small electronic components and remote sensors has attracted researchers from different disciplines of engineering, including mechanical, electrical, and civil as well as researchers from the field of material science.

As described by Williams and Yates (1996), the three basic vibration-to-electric energy conversion mechanisms are the electromagnetic (Williams and Yates, 1996, Glynne-Jones et al., 2004, Arnold, 2007), electrostatic (Roundy et al., 2002, Mitcheson, 2004), and piezoelectric (Roundy et al., 2003, Sodano et al., 2005, Jeon et al., 2005) transductions. In the last decade, these transduction mechanisms have been investigated by numerous researchers for vibration-based energy harvesting and extensive discussions can be found in the existing review articles (e.g., Beeby et al., 2006). The literature of the last five years shows that piezoelectric transduction has received the most attention for vibration-to-electricity conversion and four review articles focusing on piezoelectric energy harvesting have been published in the last four years (Sodano et al., 2004a, Priya, 2007, Anton and Sodano, 2007, Cook-Chennault et al., 2008). The relevant experimental research and possible applications of piezoelectric energy harvesting can be found in the aforementioned review articles.

Typically, a piezoelectric energy harvester is a cantilevered beam with one or two piezoceramic layers (a unimorph or a bimorph). The harvester beam is located on a vibrating host structure and the dynamic strain induced in the piezoceramic layer(s) generates an alternating voltage output across the electrodes covering the piezoceramic layer(s). In addition to the experimental research on the practical applications of such energy harvesters, researchers have proposed various mathematical models. Developing a reliable mathematical model can allow predicting the electrical power output of a given energy harvester under the prescribed base excitation conditions. Moreover, it can allow designing and optimizing the energy harvester for a given set of electrical and mechanical variables. Although the implementation of piezoelectric energy harvesting for charging a real battery in an efficient way is more sophisticated (Ottman et al., 2002), researchers have considered a resistive load in the electrical circuit to come up with simple models for predicting the electrical response of the harvester for a given base motion input.

The coupled problem of predicting the voltage across the resistive load connected to the electrodes of a vibrating energy harvester (under base excitation) has been investigated by many authors. In the early mathematical modeling treatments, researchers (Roundy et al., 2003, duToit et al., 2005) employed lumped parameter (single-degree-of-freedom) type solutions. Lumped parameter modeling is a convenient modeling approach since the electrical domain already consists of lumped parameters: a capacitor due to the internal (or inherent) capacitance of the piezoceramic and a resistor due to an external load resistance. Hence, the only thing required is to obtain the lumped parameters representing the mechanical domain so that the mechanical equilibrium and electrical loop equations can be coupled through the piezoelectric constitutive relations (IEEE, 1987) and a transformer relation can be established. This was the main procedure followed by Roundy et al. (2003) and duToit et al. (2005) in their lumped parameter model derivations. Although lumped parameter modeling gives initial insight into the problem by allowing simple expressions, it is an approximation limited to a single vibration mode and lacks important aspects of the coupled physical system, such as the information of dynamic mode shape and accurate strain distribution as well as their effects on the electrical response.

Since cantilevered energy harvesters are basically excited due to the motion of their base, the well-known lumped parameter harmonic base excitation relation taken from the elementary vibration texts has been used in the energy harvesting literature for both modeling (duToit et al., 2005) and studying the maximum power generation and parameter optimization (Stephen, 2006, Daqaq et al., 2007). It was shown (Erturk and Inman, 2008a) that the conventional form of the lumped parameter harmonic base excitation relation may yield highly inaccurate results both for the transverse and longitudinal vibrations of cantilevered structures depending on the tip (proof) mass to beam/bar mass ratio. The correction is due to the contribution of the distributed mass of the cantilevered structure to the excitation amplitude, which is not modeled in Roundy et al. (2003) and underestimated due to using the lumped parameter base excitation model in duToit et al. (2005). The contribution of the distributed mass to the excitation amplitude is important especially if the harvester does not have a very large proof mass. Correction factors were derived (Erturk and Inman, 2008a) to improve the predictions of the lumped parameter electromechanical relations (duToit et al., 2005) for cantilevered energy harvesters under base excitation. Amplitude-wise correction of the lumped parameter electromechanical relations is summarized in Section 2.2 of this chapter.

As an improved modeling approach, the Rayleigh–Ritz-type discrete formulation originally derived by Hagood et al. (1990) for piezoelectric actuation (based on the generalized Hamilton's principle for electromechanical systems given by Crandall et al. (1968)) was employed by Sodano et al. (2004b) and duToit et al. (2005) for modeling of cantilevered piezoelectric energy harvesters (based on the Euler–Bernoulli beam theory). The Rayleigh–Ritz model gives a discrete model of the distributed parameter system and it is a more accurate approximation when compared with lumped parameter modeling. In order to represent the electrical outputs analytically, Lu et al. (2004) used the vibration mode shapes obtained from the Euler–Bernoulli beam theory and the piezoelectric-constitutive relation (IEEE, 1987) that gives the electric displacement to relate the electrical outputs to the vibration mode shape. Similar models were given by Chen et al. (2006) and Lin et al. (2007) where the electrical response is expressed in terms of the beam vibration response. The deficiencies in these analytical-modeling attempts include lack of consideration of the resonance phenomenon, ignorance of modal expansion, and oversimplified modeling of piezoelectric coupling in the beam equation as viscous damping (e.g., Lu et al., 2004, Chen et al., 2006, Lin et al., 2007). As shown in this work, representing the effect of piezoelectric coupling in the beam equation as viscous damping fails in predicting the coupled system dynamics of a piezoelectric energy harvester, although this simplified model (viscous damping representation of electromechanical coupling) is a reasonable approximation for certain electromagnetic energy harvesters (Williams and Yates, 1996). One particular consequence of misrepresenting or ignoring piezoelectric coupling in the mechanical equation is highly inaccurate prediction of the optimum load resistance that gives the maximum electrical power (Erturk and Inman, 2008b). Moreover, simplified modeling of piezoelectric coupling cannot predict the variation of the resonance frequencies with changing load resistance. In terms of analytical modeling, Ajitsaria et al. (2007) presented a bimorph cantilever model, where they attempted to combine the static sensing/actuation equations (with constant radius of curvature and a static tip force) with the dynamic Euler–Bernoulli beam equation (where the radius of curvature varies) under base excitation (where there is no tip force). Thus, highly different modeling approaches have appeared in the literature during the last five years and some of them might be misleading due to the simplified modeling approaches presented.

Erturk and Inman (2008c) presented the analytical solution to the coupled problem of a unimorph piezoelectric energy harvester configuration based on the Euler–Bernoulli beam assumptions. They obtained the coupled voltage response across the resistive load and the coupled vibration response of the harvester explicitly for harmonic base excitations in the form of translation with small rotation. The short-circuit and open-circuit trends and the effect of piezoelectric coupling were investigated extensively. Later, Elvin and Elvin (2008) observed the convergence of the Rayleigh–Ritz-type solution formerly introduced by Hagood et al. (1990) to the analytical solution given by Erturk and Inman (2008c) when sufficient number of vibration modes is used with appropriate admissible functions. Erturk and Inman (2008d) extended their closed-form distributed parameter solution to bimorph configurations and presented experimental validations. Section 2.3 of this chapter summarizes the distributed parameter models for the unimorph and bimorph cantilevers and their closed-form solutions (Erturk and Inman, 2008c, 2008d) along with experimental validations. The mathematical derivations, theoretical, and experimental demonstrations presented in Section 2.3 of this chapter aim to provide a basic understanding of piezoelectric energy harvesting and also to clarify the modeling issues observed and repeated in the literature (Erturk and Inman, 2008b).

2.2 Amplitude-Wise Correction of the Lumped Parameter Model

This section discusses the amplitude-wise correction of the coupled lumped parameter piezoelectric energy harvester model for base excitation. The details of the following subsections can be found in Erturk and Inman (2008a). First, the uncoupled lumped parameter base excitation model is reviewed as it is commonly used for representing the mechanical equation of motion in vibration-based energy harvesting. Then the uncoupled distributed parameter solution is presented for cantilevered beams in transverse vibrations and cantilevered bars in longitudinal vibrations and deficiency of the lumped parameter model is shown. Correction factors are derived for improving the excitation amplitude of the respective lumped parameter models for the transverse and longitudinal vibrations. For the presence of a tip mass (or proof mass), variation of the correction factors tip mass-to-beam/bar mass ratio are presented graphically along with curve-fit relations obtained from numerical solutions. Finally, the amplitude-wise correction factor for longitudinal vibrations is used in the lumped parameter electromechanical equations for improving the coupled lumped parameter predictions.

2.2.1 Uncoupled Lumped Parameter Base Excitation Model

Since vibration-based energy harvesters are excited due to the motion of their base, the lumped parameter representation of base excitation model shown in Fig. 2.1 has been used by many authors. The governing equation of motion due to the vibrating base is

$$m_{eq}\ddot{z} + c_{eq}\dot{z} + k_{eq}z = -m_{eq}\ddot{y}, \tag{2.1}$$

where m_{eq}, k_{eq}, and c_{eq} are the equivalent (or effective) mass, stiffness, and viscous damping terms, respectively, y is the base displacement and z is the displacement of the lumped mass relative to the vibrating (or moving) base. Hereafter, an over dot represents differentiation with respect to time and therefore, $\ddot{y}$ is simply the base acceleration. If the base displacement is harmonic in the form of $y(t) = Y_0 \, e^{j\omega t}$ (where Y_0 is the displacement amplitude, ω is the frequency, and j is the unit imaginary number), then the steady-state displacement response of the lumped mass relative to the base becomes

$$z(t) = \frac{\omega^2}{\omega_n^2 - \omega^2 + j2\zeta\omega_n\omega} Y_0 \, e^{j\omega t}, \tag{2.2}$$

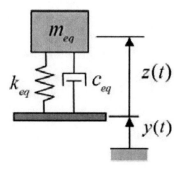

Fig. 2.1 Lumped parameter base excitation model

where ω_n is the undamped natural frequency and ζ is the mechanical damping ratio such that $\omega_n = \sqrt{k_{eq}/m_{eq}}$ and $\zeta = c_{eq}/\sqrt{4k_{eq}m_{eq}}$.

In the literature of vibration-based energy harvesting, Eq. (2.1) was first used by Williams and Yates (1996) for modeling the dynamics of their magnet-coil-type electromagnetic generator. The viscous damping coefficient was defined to have an electromagnetically induced component, which allowed expressing the electrical power output of the electromagnetic generator in a convenient way. This simple approach has been employed by several others for modeling of electromagnetic energy harvesters (El-hami et al., 2001, Beeby et al., 2007).

Although modeling of electromechanical coupling in piezoelectric energy harvesting is more complicated than adding an electrically induced viscous damping, the form of Eq. (2.1) was used for the mathematical representation of the piezoelectric energy harvesting problem by Jeon et al. (2005) and Fang et al. (2006) in their experimental papers investigating cantilevered piezoelectric energy harvesters under base excitation. Ajitsaria et al. (2007) used this uncoupled form of lumped parameter base excitation relation for predicting the voltage response of a cantilever, where they did not model piezoelectric coupling in the mechanical equation. As a coupled lumped parameter modeling approach, duToit et al. (2005) introduced piezoelectric coupling to Eq. (2.1) in a formal way (i.e., based on the piezoelectric constitutive laws), rather than introducing an electrically induced viscous damping. However, the origin of their (duToit et al., 2005) mechanical equation was still the conventional lumped parameter base excitation relation given by Eq. (2.1). Since the lumped parameter base excitation model has been widely used in the literature of vibration-based energy harvesting, it is a useful practice to compare its accuracy with the distributed parameter model. The uncoupled distributed parameter solution of the base excitation problem is reviewed in the next section.

2.2.2 Uncoupled Distributed Parameter Base Excitation Model

In the following, the uncoupled distributed parameter solution of the harmonic base excitation problem is reviewed for cantilevered beams in transverse vibrations and for cantilevered bars in longitudinal vibrations.

2.2.2.1 Cantilevered Beams in Transverse Vibrations

Consider the uniform cantilevered beam shown in Fig. 2.2. Based on the Euler–Bernoulli beam assumptions (i.e., shear deformations and rotary inertias are negligible), free vibrations of the beam are governed by the following partial differential equation:

$$YI\frac{\partial^4 w(x,t)}{\partial x^4} + c_s I\frac{\partial^5 w(x,t)}{\partial x^4 \partial t} + c_a \frac{\partial w(x,t)}{\partial t} + m\frac{\partial^2 w(x,t)}{\partial t^2} = 0, \qquad (2.3)$$

where YI is the bending stiffness, m is the mass per unit length, c_a is the external viscous damping term (due to air or the respective surrounding fluid), $c_s I$ is the internal strain rate (or Kelvin–Voigt) damping term, and $w(x, t)$ is the absolute transverse displacement of the beam at point x and time t. Note that both of the damping mechanisms considered here are assumed to satisfy the proportional damping criterion and they are mathematically convenient for the modal analysis solution procedure (Clough and Penzien, 1975, Banks and Inman, 1991). Strain-rate damping is assumed to be stiffness proportional, whereas air damping is assumed to be mass proportional and this type of combined proportional damping is also known as Rayleigh damping (Clough and Penzien, 1975).

As proposed by Timoshenko et al. (1974), the absolute displacement response $w(x, t)$ can be given in terms of the base displacement $w_b(t)$ and the transverse displacement response $w_{rel}(x, t)$ relative to the base as

$$w(x, t) = w_b(t) + w_{rel}(x, t), \qquad (2.4)$$

which can be substituted into Eq. (2.3) to yield the following forced vibration equation for the transverse vibrations of the beam relative to its moving base:

$$YI \frac{\partial^4 w_{rel}(x, t)}{\partial x^4} + c_s I \frac{\partial^5 w_{rel}(x, t)}{\partial x^4 \partial t} + c_a \frac{\partial w_{rel}(x, t)}{\partial t} + m \frac{\partial^2 w_{rel}(x, t)}{\partial t^2}$$
$$= -m \frac{d^2 w_b(t)}{dt^2} - c_a \frac{d w_b(t)}{dt}. \qquad (2.5)$$

A more general form of the base excitation (with small rotation) is given in the relevant work (Erturk and Inman, 2008a). It is important to note from Eq. (2.5) that the right-hand side forcing function has two components. One component is expectedly due to the distributed inertia of the beam, whereas the other component is due to external viscous damping. The excitation component induced by the external damping component is peculiar to support motion problems. If the surrounding fluid is air, then the excitation term due to air damping is typically negligible when compared with the inertial excitation term. In the absence of a tip mass, it was shown (Erturk and Inman, 2008a) in a dimensionless basis that the contribution from air damping to the excitation force at resonance is less than 5% of the total (inertial and damping) excitation, if the damping ratio due to air damping (defined as ζ_r^a in the following) is less than 2.5%. However, one should be careful with this external damping related excitation term for energy harvesters excited by base motions in fluids with larger damping effect.

Based on the expansion theorem, $w_{rel}(x, t)$ can be represented by an absolutely and uniformly convergent series of the eigenfunctions as

$$w_{rel}(x, t) = \sum_{r=1}^{\infty} \phi_r(x) \eta_r(t), \qquad (2.6)$$

Fig. 2.2 Cantilevered beam excited by the motion of its base in the transverse direction

where $\phi_r(x)$ and $\eta_r(t)$ are the mass normalized eigenfunction and the modal response of the clamped-free beam for the rth vibration mode, respectively. In the absence of a tip mass (which is the case in Fig. 2.2), the mass normalized eigenfunctions are

$$\phi_r(x) = \sqrt{\frac{1}{mL}} \left[\cosh \frac{\lambda_r}{L}x - \cos \frac{\lambda_r}{L}x - \sigma_r \left(\sinh \frac{\lambda_r}{L}x - \sin \frac{\lambda_r}{L}x \right) \right], \qquad (2.7)$$

where the λ_r s are the dimensionless frequency numbers obtained from the characteristic equation given by

$$1 + \cos \lambda \cosh \lambda = 0, \qquad (2.8)$$

and σ_r is expressed as

$$\sigma_r = \frac{\sinh \lambda_r - \sin \lambda_r}{\cosh \lambda_r + \cos \lambda_r}. \qquad (2.9)$$

Using Eq. (2.6) in Eq. (2.5) along with the orthogonality conditions of the eigenfunctions (Erturk and Inman, 2008a) yield

$$\ddot{\eta}_r(t) + 2\zeta_r\omega_r\dot{\eta}_r(t) + \omega_r^2\eta_r(t) = [-m\ddot{w}_b(t) - c_a\dot{w}_b(t)] \int_0^L \phi_r(x)dx, \qquad (2.10)$$

where ω_r is the undamped natural frequency of the rth mode given by

$$\omega_r = \lambda_r^2 \sqrt{\frac{YI}{mL^4}}, \qquad (2.11)$$

and ζ_r is the modal mechanical damping ratio and it satisfies the following relation since the strain rate and air damping components are assumed to be stiffness and mass proportional, respectively:

$$\zeta_r = \frac{c_s I \omega_r}{2YI} + \frac{c_a}{2m\omega_r}. \qquad (2.12)$$

If the base displacement is assumed to be harmonic in the form of $w_b(t) = Y_0 \, e^{j\omega t}$, the steady-state modal response can be obtained from Eq. (2.10) as

$$\eta_r(t) = \frac{m\omega^2 - j\omega c_a}{\omega_r{}^2 - \omega^2 + j2\zeta_r\omega_r\omega} Y_0 \, e^{j\omega t} \int_0^L \phi_r(x)\,dx, \qquad (2.13)$$

which can then be used in Eq. (2.6) along with Eq. (2.7) to obtain $w_{rel}(x, t)$. Then, the transverse displacement at the tip of the beam $(x = L)$ relative to the moving base is

$$w_{rel}(L, t) = Y_0 \, e^{j\omega t} \sum_{r=1}^{\infty} \frac{2\sigma_r \, [\cosh\lambda_r - \cos\lambda_r - \sigma_r \, (\sinh\lambda_r - \sin\lambda_r)] \, (\omega^2 - j2\zeta_r^a\omega_r\omega)}{\lambda_r \, (\omega_r{}^2 - \omega^2 + j2\zeta_r\omega_r\omega)},$$
$$(2.14)$$

where ζ_r^a is the viscous air damping component of the modal mechanical damping ratio $(\zeta_r^a = c_a/2m\omega_r)$.

Note that the above analysis is given for a uniform cantilevered beam without a tip mass. Presence of a tip mass changes not only the eigenfunctions and the characteristic equation given by Eqs. (2.7) and (2.8), respectively, but also the excitation term since the inertia of the tip mass also contributes to the forcing function in that case (Erturk and Inman, 2008a). The effect of a tip mass on the ultimate results of the analysis given here (which are the correction factors) is addressed in Section 2.2.3.1.

2.2.2.2 Cantilevered Bars in Longitudinal Vibrations

A similar procedure is applied in this section to obtain the distributed parameter solution of the base excitation problem for the longitudinal vibrations of the uniform cantilevered bar shown in Fig. 2.3. The partial differential equation of motion for free vibrations of the bar can be given by

$$Y A \frac{\partial^2 u(x, t)}{\partial x^2} + c_s A \frac{\partial^3 u(x, t)}{\partial x^2 \partial t} - c_a \frac{\partial u(x, t)}{\partial t} - m \frac{\partial^2 u(x, t)}{\partial t^2} = 0, \qquad (2.15)$$

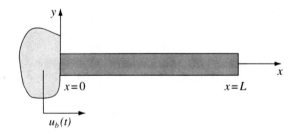

Fig. 2.3 Cantilevered bar excited by the motion of its base in the longitudinal direction

where YA is the axial stiffness, m is the mass per unit length, c_a is the external viscous damping term, $c_s A$ is the internal strain rate damping term, and $u(x, t)$ is the absolute longitudinal displacement of the bar at point x and time t.[1] As in the case of transverse vibrations, strain rate damping is assumed to be stiffness proportional, whereas viscous air damping is assumed to be mass proportional.

The absolute longitudinal displacement response is then represented in terms of the base displacement and the displacement response relative to the base as

$$u(x, t) = u_b(t) + u_{rel}(x, t). \tag{2.16}$$

One can then substitute Eq. (2.16) into Eq. (2.15) to obtain a forced partial differential equation for the longitudinal vibrations of the bar relative to its moving base. After following similar steps given for the transverse vibrations case, the displacement response at the free end of the bar ($x = L$) relative to its moving base can be expressed for harmonic base displacement $u_b(t) = X_0 \, e^{j\omega t}$ as

$$u_{rel}(L, t) = X_0 \, e^{j\omega t} \sum_{r=1}^{\infty} \frac{2 \sin \alpha_r \, (1 - \cos \alpha_r) \left(\omega^2 - j2\zeta_r^a \omega_r \omega \right)}{\alpha_r \left(\omega_r^2 - \omega^2 + j2\zeta_r \omega_r \omega \right)}, \tag{2.17}$$

where X_0 is the harmonic base displacement amplitude (at frequency ω), ω_r is the undamped natural frequency, ζ_r is the modal mechanical damping ratio, ζ_r^a is the viscous air damping component of the modal mechanical damping ratio, and the eigenvalues denoted by α_r are obtained from the following characteristic equation,

$$\cos \alpha_r = 0, \tag{2.18}$$

and the undamped natural frequencies are given by

$$\omega_r = \alpha_r \sqrt{\frac{YA}{mL^2}}. \tag{2.19}$$

Note that the foregoing analysis is valid for a uniform cantilevered bar without a tip mass and details of the derivation in the presence of a tip mass is given by Erturk and Inman (2008a). The effect of a tip mass on the results of the analysis given here is discussed in Section 2.2.3.2.

2.2.3 Correction Factors for the Lumped Parameter Model

In this section, the distributed parameter solutions are reduced to single-mode expressions and these relations are compared with the lumped parameter solution

---

[1] Although the same notation is used for the mass per length and the damping terms in Eqs. (2.3) and (2.15) for convenience, these terms are not necessarily identical.

given by Eq. (2.2). Deficiency of the lumped parameter solution is shown and correction factors are introduced to improve its predictions. For the presence of a tip mass, variation of the correction factors with tip mass-to-beam/bar mass ratio are presented.

2.2.3.1 Cantilevered Beams in Transverse Vibrations

For excitations around the first natural frequency (i.e., for $\omega \cong \omega_1$), the distributed parameter solution given by Eq. (2.14) can be reduced to

$$w_{\text{rel}}(L, t) \cong \frac{\mu_1 \omega^2}{\omega_1{}^2 - \omega^2 + j2\zeta_1\omega_1\omega} Y_0 \, e^{j\omega t},$$ (2.20)

which is in the form of the lumped parameter solution given by Eq. (2.2).[2] Thus, μ_1 is the amplitude-wise correction factor for the lumped parameter model (for transverse vibrations) and it can be given by

$$\mu_1 = \frac{2\sigma_1 \left[\cosh \lambda_1 - \cos \lambda_1 - \sigma_1 (\sinh \lambda_1 - \sin \lambda_1) \right]}{\lambda_1},$$ (2.21)

where subscript 1 stands for the first vibration mode. It is then straightforward to obtain from Eqs. (2.8), (2.9) and (2.21) that $\mu_1 = 1.566$. Here, in correcting the amplitude-wise prediction of Eq. (2.2), it is assumed that the lumped parameter estimate of the undamped natural frequency (ω_n in Eq. (2.2)) is sufficiently accurate (i.e., $\omega_n \cong \omega_1$ where ω_1 is obtained from Eq. (2.11) for $r = 1$). Typically, the following lumped parameter relation is used to predict the natural frequency of a uniform cantilevered beam in transverse vibrations (Lord Rayleigh, 1894)

$$\omega_n = \sqrt{\frac{k_{\text{eq}}}{m_{\text{eq}}}} = \sqrt{\frac{3YI/L^3}{(33/140)mL + M_t}},$$ (2.22)

where M_t is the tip mass (if exists) and it is zero in the above discussion. It can be shown that the error due to using Eq. (2.22) in predicting the fundamental natural frequency is about 0.5% in the absence of a tip mass (relative to the Euler–Bernoulli model fundamental natural frequency ω_1). The prediction of Eq. (2.22) is improved in the presence of a tip mass.

In order to compare the lumped parameter and the distributed parameter solutions, the *relative displacement transmissibility FRF* (frequency response function) forms a convenient basis and it is simply $z(t)/y(t)$ for the lumped parameter model and $w_{\text{rel}}(L, t)/w_b(t)$ for the distributed parameter model. Figure 2.4a shows the amplitude-wise error in the lumped parameter displacement transmissibility FRF as a function of dimensionless frequency $\Omega = \omega/\omega_1$ where the correct solution is taken as the distributed parameter solution given by Eq. (2.14). It is clear from

---

[2] Note that the excitation component due to viscous air damping is neglected in Eq. (2.20).

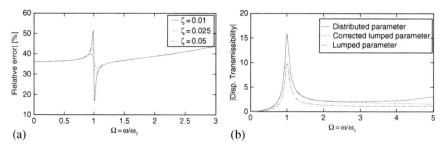

Fig. 2.4 (**a**) Relative error in the lumped parameter displacement transmissibility FRF and the (**b**) displacement transmissibility FRFs obtained from the distributed parameter, corrected lumped parameter, and lumped parameter models

Fig. 2.4a that the error due to using the lumped parameter approach for predicting the relative motion at the tip of the beam is very large. In the vicinity of the first natural frequency (i.e., at $\Omega \cong 1$), the error of the lumped parameter model can be greater than 35% regardless of the damping ratio. The interesting behavior in the relative error plot at resonance is due to the slight inaccuracy of the lumped parameter natural frequency prediction by Eq. (2.22), which was mentioned before. If the lumped parameter natural frequency was taken to be identical to the first natural frequency of the Euler–Bernoulli beam model, one would obtain a smooth behavior in the error. At higher frequencies, the error in the lumped parameter model increases drastically as one approaches to the region of the second vibration mode, which cannot be captured by the lumped parameter model.

Figure 2.4b shows the displacement transmissibility FRFs for the distributed parameter solution given by Eq. (2.14), corrected lumped parameter solution given by Eq. (2.20) and the original lumped parameter solution given by Eq. (2.2). The corrected lumped parameter model obtained by Eq. (2.20) or by multiplying Eq. (2.2) with the correction factor μ_1 represents the distributed parameter model perfectly in a wide range of frequencies around the fundamental vibration mode. Note that the corrected lumped parameter model deviates from the distributed parameter model at high frequencies, since it does not include the information of the higher vibration modes.

Based on the above discussion, it can be concluded that the correct form of the lumped parameter equation of motion given by Eq. (2.1) is

$$m_{eq}\ddot{z} + c_{eq}\dot{z} + k_{eq}z = -\mu_1 m_{eq}\ddot{y}, \qquad (2.23)$$

Thus, Eq. (2.2) becomes

$$z(t) = \frac{\mu_1 \omega^2}{\omega_n^2 - \omega^2 + j2\zeta\omega_n\omega} Y_0 \, e^{j\omega t}. \qquad (2.24)$$

Here, μ_1 corrects the excitation amplitude and therefore the response amplitude of the lumped parameter model. Otherwise, in the absence of a tip mass, the original lumped parameter model underestimates the response amplitude with an error of about 35%.

So far, the case without a tip mass has been discussed. In many cases, a tip mass (or proof mass) is attached rigidly at the free end of the cantilever to tune its fundamental natural frequency to a desired value or to increase its flexibility especially in microscale applications. The presence of a tip mass changes the forcing function due to base excitation as well as the eigenfunctions and the eigenvalues. After working out the problem for the presence of a tip mass, one can obtain the variation of the correction factor with tip mass-to-beam mass ratio as shown in Fig. 2.5a for transverse vibrations. As can be seen from this figure, if the tip mass of the cantilevered beam is very large, the correction factor tends to unity (i.e., as $M_t/mL \rightarrow \infty$, $\mu_1 \rightarrow 1$). Therefore, there is no need to use the correction factor if a large tip mass is attached to the cantilevered beam. This makes perfect physical sense, since the contribution of the distributed mass of the beam to the excitation amplitude becomes negligible in the presence of a large tip mass. In such a case, there is no need to correct the original form of the lumped parameter solution given by Eq. (2.2). The curve shown in Fig. 2.5a is obtained by numerical solution and the following curve fit relation represents it with an error of less than $9 \times 10^{-3}\%$ for all values of M_t/mL:

$$\mu_1 = \frac{(M_t/mL)^2 + 0.603(M_t/mL) + 0.08955}{(M_t/mL)^2 + 0.4637(M_t/mL) + 0.05718}. \tag{2.25}$$

2.2.3.2 Cantilevered Bars in Longitudinal Vibrations

In a similar way, the distributed parameter solution given by Eq. (2.17) can be reduced to the following expression with the assumption of modal excitation around

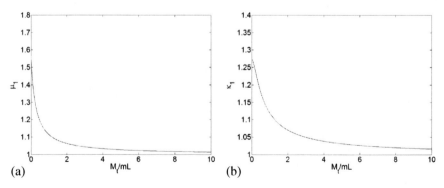

Fig. 2.5 Variation of the correction factor (**a**) with tip mass-to-beam mass ratio for transverse vibrations and (**b**) with tip mass-to-bar mass ratio for longitudinal vibrations

the first natural frequency (i.e., for $\omega \cong \omega_1$):

$$u_{\text{rel}}(L, t) \cong \frac{\kappa_1 \omega^2}{\omega_1^2 - \omega^2 + j2\zeta_1\omega_1\omega} X_0\, e^{j\omega t}, \tag{2.26}$$

where the excitation due to air damping is neglected. Equation (2.26) is in the form of the lumped parameter solution given by Eq. (2.2) and therefore κ_1 is the amplitude-wise correction factor for the lumped parameter model (for longitudinal vibrations) and it is simply

$$\kappa_1 = \frac{2\sin\alpha_1\,(1 - \cos\alpha_1)}{\alpha_1} = \frac{4}{\pi} \cong 1.273, \tag{2.27}$$

where subscript 1 stands for the first vibration mode. Once again, in correcting the amplitude-wise prediction of Eq. (2.2), it is assumed that the lumped parameter estimate of the undamped natural frequency (ω_n in Eq. (2.2)) is sufficiently accurate (i.e., $\omega_n \cong \omega_1$ where ω_1 is obtained from Eq. (2.19) for $r = 1$). The following lumped parameter relation is commonly used (e.g., duToit et al., 2005) to predict the natural frequency of a uniform cantilevered bar in longitudinal vibrations (Lord Rayleigh, 1894)

$$\omega_n = \sqrt{\frac{k_{\text{eq}}}{m_{\text{eq}}}} = \sqrt{\frac{YA/L}{mL/3 + M_t}}, \tag{2.28}$$

where M_t is the tip mass (if exists) and it is zero in the above discussion. In the absence of a tip mass, this expression gives an error of about 10.3% (compared with the distributed parameter model fundamental natural frequency ω_1 obtained from Eq. (19)) and the prediction of Eq. (2.28) is improved in the presence of a tip mass. However, it is recommended that one should use Eq. (2.19) with $r = 1$ to predict the fundamental natural frequency for longitudinal vibrations rather than using Rayleigh's relation (Eq. (2.28)) especially if the cantilevered bar does not have a large tip mass. Then, from Eqs. (2.2) and (2.26), it can be shown that the error in the predicted motion transmissibility using the former expression can be as high as 21.4% in the absence of a tip mass even if the accurate natural frequency is used in Eq. (2.2).

As in the case of transverse vibrations, the base excitation problem in the presence of a tip mass can be studied and the variation of the correction factor κ_1 with tip mass-to-bar mass ratio can be obtained as shown in Fig. 2.5b. Equation (2.17) takes a more complicated form in the presence of a tip mass (Erturk and Inman, 2008a). Expectedly, the correction factor tends to unity for large values of tip mass due to the reasoning given in the transverse vibrations case (Section 2.2.3.1). Therefore, there is no need to correct the original lumped parameter model if the cantilevered bar has a sufficiently large tip mass. Otherwise, in order not to underestimate the vibration response of the bar, the lumped parameter relations given by Eqs. (2.1) and (2.2) must be corrected to

$$m_{eq}\ddot{z} + c_{eq}\dot{z} + k_{eq}z = -\kappa_1 m_{eq}\ddot{y}, \tag{2.29}$$

and

$$z(t) = \frac{\kappa_1\omega^2}{\omega_n^2 - \omega^2 + j2\zeta\omega_n\omega}Y_0\,e^{j\omega t}, \tag{2.30}$$

respectively. Note that the curve shown in Fig. 2.5b is obtained by numerical solution and the following curve fit relation represents it with an error of less than $4.5 \times 10^{-2}\%$ for all values of M_t/mL:

$$\kappa_1 = \frac{(M_t/mL)^2 + 0.7664(M_t/mL) + 0.2049}{(M_t/mL)^2 + 0.6005(M_t/mL) + 0.161}. \tag{2.31}$$

2.2.4 Correction Factor in the Piezoelectrically Coupled Lumped Parameter Equations

We introduce the amplitude-wise correction factor to the piezoelectrically coupled lumped parameter equations. The "general 1D model of piezoelectric vibration energy harvester" shown in Fig. 2.6 as well as the sample numerical values shown in the same figure are from a paper by duToit et al. (2005) and these data are used here for the purpose of demonstration.

The electromechanically coupled equations of the lumped parameter model shown in Fig. 2.6 were given by (duToit et al., 2005).[3]

$$\ddot{u}_{rel}(t) + 2\zeta\omega_n\dot{u}_{rel}(t) + \omega_n^2 u_{rel}(t) - \omega_n^2 d_{33}v(t) = -\ddot{u}_b(t), \tag{2.32}$$

$$R_{eq}C_p\dot{v}(t) + v(t) + m_{eq}R_{eq}d_{33}\omega_n^2\dot{u}_{rel}(t) = 0, \tag{2.33}$$

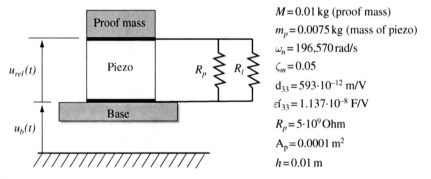

$M = 0.01\,\text{kg (proof mass)}$

$m_p = 0.0075\,\text{kg (mass of piezo)}$

$\omega_n = 196{,}570\,\text{rad/s}$

$\zeta_m = 0.05$

$d_{33} = 593 \cdot 10^{-12}\,\text{m/V}$

$\epsilon^{\bar{s}}_{33} = 1.137 \cdot 10^{-8}\,\text{F/V}$

$R_p = 5 \cdot 10^9\,\text{Ohm}$

$A_p = 0.0001\,\text{m}^2$

$h = 0.01\,\text{m}$

Fig. 2.6 Lumped parameter piezoelectric energy harvester model with sample numerical values by duToit et al. (2005)

---

[3] Some of the variables defined in the relevant work have been adapted to the notation in our text.

where m_{eq} is the equivalent mass of the bar, ζ is the mechanical damping ratio, ω_n is the undamped natural frequency, d_{33} is the piezoelectric constant, R_{eq} is the equivalent resistance (due to the external load resistance R_l and the piezoelectric leakage resistance R_p), C_p is the internal capacitance of the piezoceramic, $u_b(t)$ is the harmonic base displacement, $u_{rel}(t)$ is the displacement of the proof mass relative to the base and $v(t)$ is the voltage output. The lumped parameter model introduced by duToit et al. (2005) is an improved approach with the formal treatment of piezoelectric coupling (which results in the voltage term in the mechanical equation) when compared with the models which considered the effect of piezo-electric coupling as viscous damping. However, based on our previous discussion, we know that the mechanical equilibrium equation given by Eq. (2.32) relies on the lumped parameter base excitation relation given by Eq. (2.1) and it may need a correction factor depending on the proof mass-to-bar mass ratio. From the numerical values given in Fig. 2.6, the proof mass-to-bar mass ratio of this device is $M_t/mL \cong 1.33$. From Fig. 2.5b or Eq. (2.31), the correction factor for the fundamental mode can be obtained as $\kappa_1 \cong 1.0968$. Therefore, the corrected form of Eq. (2.32) is

$$\ddot{u}_{rel}(t) + 2\zeta\omega_n\dot{u}_{rel}(t) + \omega_n^2 u_{rel}(t) - \omega_n^2 d_{33}v(t) = -\kappa_1\ddot{u}_b(t). \tag{2.34}$$

Note that the resulting vibration amplitude of the proof mass relative to the moving base and the voltage amplitude are linearly proportional to the amplitude of the right-hand side of Eq. (2.34), whereas the power output is proportional to the square of this term:

$$\left|\frac{u_{rel}(t)}{\ddot{u}_b(t)}\right| = \frac{\kappa_1/\omega_n^2\sqrt{1+(r\Omega)^2}}{\sqrt{\left[1-(1+2\zeta r)\Omega^2\right]^2 + \left[(1+k_e^2)r\Omega + 2\zeta\Omega - r\Omega^3\right]^2}}, \tag{2.35}$$

$$\left|\frac{v(t)}{\ddot{u}_b(t)}\right| = \frac{\kappa_1 m_{eq}R_{eq}d_{33}\omega_n\Omega}{\sqrt{\left[1-(1+2\zeta r)\Omega^2\right]^2 + \left[(1+k_e^2)r\Omega + 2\zeta\Omega - r\Omega^3\right]^2}}, \tag{2.36}$$

$$\left|\frac{P(t)}{\ddot{u}_b^2(t)}\right| = \frac{\kappa_1^2 m_{eq}/\omega_n r k_e^2 R_{eq}/R_l\Omega^2}{\left[1-(1+2\zeta r)\Omega^2\right]^2 + \left[(1+k_e^2)r\Omega + 2\zeta\Omega - r\Omega^3\right]^2}. \tag{2.37}$$

Here, $\Omega = \omega/\omega_n$ is the dimensionless frequency (where ω is the excitation frequency), k_e^2 is the coupling coefficient and $r = \omega_n R_{eq}C_p$ (duToit et al., 2005). If the correction factor is not used for the energy harvester device described by the data provided in Fig. 2.6, all of these terms ($u_{rel}(t)$, $v(t)$, and $P(t)$) are underestimated. In such a case, the error in the relative tip motion and the voltage amplitudes is about 8.83%, whereas the error in the electrical power amplitude is about 16.9% (since $\kappa_1 \cong 1.0968$) relative to the distributed parameter solution.

2.3 Coupled Distributed Parameter Models and Closed-Form Solutions

This section presents the distributed parameter models for unimorph and bimorph cantilevers and their closed-form solutions for harmonic base excitation input. The mathematical derivation steps are summarized here and the details of the following subsections can be found in the relevant papers by Erturk and Inman (2008c, 2008d). Experimental validation of the coupled distributed parameter relations is presented for a cantilevered bimorph with a tip mass.

2.3.1 Modeling Assumptions

The three basic cantilevered piezoelectric energy harvester configurations are shown in Fig. 2.7. The first cantilever shown in Fig. 2.7a is a unimorph configuration (with a single piezoceramic layer), whereas the other two cantilevers (Fig 2.7b and c) are bimorph configurations (with two piezoceramic layers). These configurations are modeled here as uniform composite beams for linearly elastic deformations and

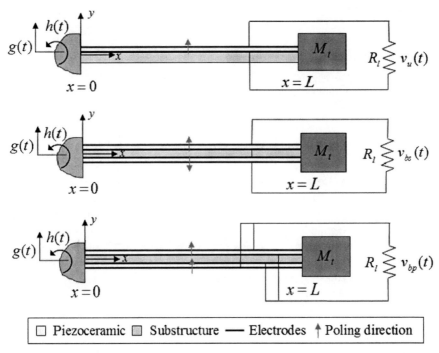

Fig. 2.7 Cantilevered piezoelectric energy harvester configurations under base excitation; (**a**) unimorph configuration, (**b**) bimorph configuration (series connection), and (**c**) bimorph configuration (parallel connection)

geometrically small oscillations based on the Euler–Bernoulli beam assumptions. Therefore, the effects of shear deformation and rotary inertia are neglected and this is a reasonable assumption since typical piezoceramic benders are designed and manufactured as thin beams. The mechanical losses are represented by the internal (strain rate) and external (viscous air) damping mechanisms. The piezoceramic and substructure layers are assumed to be perfectly bonded to each other. The electrodes covering the opposite faces of the piezoceramic layers are assumed to be very thin so that their contribution to the thickness dimension is negligible.

The piezoceramic layers are poled in the thickness direction as depicted in Fig. 2.7. Since the bimorph configurations have two piezoceramic layers, one can connect the electrical outputs of the piezoceramic layers either in series (Fig. 2.7b) or in parallel (Fig. 2.7c). The continuous electrode pairs covering the top and the bottom faces of the piezoceramic layers are assumed to be perfectly conductive so that a single electric potential difference can be defined between them. Therefore, the instantaneous electric fields induced in the piezoceramic layers are assumed to be uniform throughout the length of the beam. A resistive electrical load (R_l) is considered in the circuit along with the internal capacitances of the piezoceramic layers. As mentioned in the introduction, considering a resistive load in the electrical domain is a common practice in modeling of vibration-based energy harvesters. As a consequence, it is assumed that the base motion input is persistent so that continuous electrical outputs can be extracted from the electromechanical system.

2.3.2 Mathematical Background

For each of the configurations shown in Fig. 2.7, free vibrations of the cantilevered beam are governed by

$$\frac{\partial^2 M(x,t)}{\partial x^2} + c_s I \frac{\partial^5 w(x,t)}{\partial x^4 \partial t} + c_a \frac{\partial w(x,t)}{\partial t} + m \frac{\partial^2 w(x,t)}{\partial t^2} = 0, \tag{2.38}$$

which is similar to Eq. (2.3) given in the uncoupled analysis and definition of the relevant terms can be found in Section 2.2.2.1. However, in the coupled treatment given here, the internal bending moment term, $M(x,t)$, which is the first moment of the axial stress integrated over the beam cross-section, yields not only the first term of Eq. (2.3) but also a piezoelectrically induced voltage term. In order to introduce the piezoelectric coupling to the beam equation, the following axial stress expression (IEEE, 1987) must be used in expressing $M(x,t)$ over the thickness of the beam occupied by the piezoceramic layer(s):

$$T_1 = c_{11}^E S_1 - e_{31} E_3. \tag{2.39}$$

Here, T_1 is the axial stress component, S_1 is the axial strain component, c_{11}^E is the elastic stiffness (i.e., Young's modulus) of the piezoceramic layer at constant electric field, e_{31} is the piezoelectric constant, E_3 is the electric field component, and 1- and

3-directions coincide with the longitudinal and thickness directions, respectively (where $c_{11}^E = 1/s_{11}^E$ in terms of the compliance s_{11}^E and $e_{31} = d_{31}/s_{11}^E$ in terms of the piezoelectric constant d_{31} with the plane-stress assumption). Note that the axial strain component at a certain level (y) from the neutral axis of the composite beam is proportional to the curvature of the beam at that position (x):

$$S_1(x, y, t) = -y \frac{\partial^2 w_{\text{rel}}(x, t)}{\partial x^2}. \tag{2.40}$$

The base excitation $w_b(x, t)$ is then introduced to the free vibration equation such that

$$w(x, t) = w_b(x, t) + w_{\text{rel}}(x, t), \tag{2.41}$$

where the base displacement is

$$w_b(x, t) = g(t) + x h(t). \tag{2.42}$$

Here, as depicted in Fig. 2.7, the base excitation consists of translation in the transverse direction (denoted by $g(t)$) and superimposed small rotation (denoted by $h(t)$). Equation (2.41) can then be substituted into Eq. (2.38) to obtain a forced vibration equation for the transverse displacement response of the beam relative to its base ($w_{\text{rel}}(x, t)$):

$$YI \frac{\partial^4 w_{\text{rel}}(x, t)}{\partial x^4} + c_s I \frac{\partial^5 w_{\text{rel}}(x, t)}{\partial x^4 \partial t} + c_a \frac{\partial w_{\text{rel}}(x, t)}{\partial t} + m \frac{\partial^2 w_{\text{rel}}(x, t)}{\partial t^2} + \vartheta v(t)$$
$$\times \left[\frac{d\delta(x)}{dx} - \frac{d\delta(x - L)}{dx} \right] = -[m + M_t \delta(x - L)] \frac{\partial^2 w_b(x, t)}{\partial t^2}, \tag{2.43}$$

where the bending moment term is also expanded (to give a term related to bending stiffness and an electrical term due to piezoelectric coupling) and the excitation force due to air damping is neglected. In Eq. (2.43), $\delta(x)$ is the Dirac delta function, ϑ is the piezoelectric coupling term (given in the following sections), and $v(t)$ is the voltage across the resistive load for the given cantilever configuration (Fig. 2.7). Equation (2.43) is the expanded form of the coupled mechanical equation of motion in the physical coordinates. Note that the electrical and the mechanical terms as well as the piezoelectric coupling term ϑ (Erturk and Inman, 2008c, 2008d) differ for the three configurations shown in Fig. 2.7 as summarized in the following sections. Since a tip mass (M_t) is present in the configurations shown in Fig. 2.7, the right-hand side inertial forcing term of Eq. (2.43) has an additional term compared with Eq. (2.5).

The presence of a tip mass changes the eigenfunctions and the characteristic equation of the free vibration problem. For the modal expansion of the eigenfunctions given by Eq. (2.6), the eigenfunctions to be used in the presence of a tip mass are

$$\phi_r(x) = C_r \left[\cos \frac{\lambda_r}{L} x - \cosh \frac{\lambda_r}{L} x + \varsigma_r \left(\sin \frac{\lambda_r}{L} x - \sinh \frac{\lambda_r}{L} x \right) \right], \qquad (2.44)$$

where ς_r is obtained from

$$\varsigma_r = \frac{\sin \lambda_r - \sinh \lambda_r + \lambda_r (M_t/mL)(\cos \lambda_r - \cosh \lambda_r)}{\cos \lambda_r + \cosh \lambda_r - \lambda_r (M_t/mL)(\sin \lambda_r - \sinh \lambda_r)}, \qquad (2.45)$$

and C_r is a modal constant which should be evaluated by normalizing the eigenfunctions according to the following orthogonality conditions:

$$\int_0^L \phi_s(x) m \phi_r(x) dx + \phi_s(L) M_t \phi_r(L) + \left[\frac{d\phi_s(x)}{dx} I_t \frac{d\phi_r(x)}{dx} \right]_{x=L} = \delta_{rs},$$

$$\int_0^L \phi_s(x) YI \frac{d^4 \phi_r(x)}{dx^4} dx - \left[\phi_s(x) YI \frac{d^3 \phi_r(x)}{dx^3} \right]_{x=L} + \left[\frac{d\phi_s(x)}{dx} YI \frac{d^2 \phi_r(x)}{dx^2} \right]_{x=L} = \omega_r^2 \delta_{rs}.$$

$$(2.46)$$

Here, I_t is the rotary inertia of the tip mass M_t and δ_{rs} is Kronecker delta, defined as being equal to unity for $s = r$ and equal to zero for $s \neq r$.

The undamped natural frequencies are obtained using Eq. (2.11); however, in the presence of a tip mass, the eigenvalues of the system are obtained from

$$1 + \cos \lambda \cosh \lambda + \lambda \frac{M_t}{mL} (\cos \lambda \sinh \lambda - \sin \lambda \cosh \lambda) - \frac{\lambda^3 I_t}{mL^3}$$

$$\times (\cosh \lambda \sin \lambda + \sinh \lambda \cos \lambda) + \frac{\lambda^4 M_t I_t}{m^2 L^4} (1 - \cos \lambda \cosh \lambda) = 0. \qquad (2.47)$$

For a single piezoceramic layer operating into a circuit of admittance $1/R_1$ (as in the unimorph configuration, Fig. 2.7a) the coupled electrical circuit equation is obtained from the following relation (IEEE, 1987):

$$\frac{d}{dt} \left(\int_A \mathbf{D} \cdot \mathbf{n} \, dA \right) = \frac{v(t)}{R_1}, \qquad (2.48)$$

where $v(t)$ is the voltage across the electrodes of the piezoceramic, $\mathbf{D}$ is the vector of electric displacement components, $\mathbf{n}$ is the unit outward normal, and the integration is performed over the electrode area A. The relevant component of the electric displacement to be used in the inner product of the integrand of Eq. (2.48) is

$$D_3 = e_{31} S_1 + \varepsilon_{33}^S E_3, \qquad (2.49)$$

where D_3 is the electric displacement component and ε_{33}^S is the permittivity component at constant strain (IEEE, 1987) and $\varepsilon_{33}^S = \varepsilon_{33}^T - d_{31}^2/s_{11}^E$ in terms of the permittivity at constant stress. Note that the axial strain component S_1 introduces the beam deflection $w_{rel}(x, t)$ to the circuit equation due to Eq. (2.40), so that Eq. (2.48) yields a coupled electrical circuit equation.

The coupled distributed parameter equations for the cantilevered piezoelectric energy harvester configurations shown in Fig. 2.7 and their closed-form solutions are derived in the following sections based on the foregoing introduction.

2.3.3 Unimorph Configuration

After expressing the coupled beam equation in the physical coordinates (Erturk and Inman, 2008c), the modal expansion and the orthogonality conditions of the vibration modes (given by Eqs. (2.6) and (2.46)) can be used to obtain

$$\ddot{\eta}_r^u(t) + 2\zeta_r \omega_r \dot{\eta}_r^u(t) + \omega_r^2 \eta_r^u(t) + \chi_r^u v_u(t) = f_r(t), \tag{2.50}$$

where $\eta_r^u(t)$ is the modal mechanical response of the unimorph cantilever, $v_u(t)$ is the voltage across the resistive load as depicted in Fig. 2.7a, and χ_r^u is the *backward* modal coupling term given by

$$\chi_r^u = \vartheta_u \left. \frac{d\phi_r(x)}{dx} \right|_L, \tag{2.51}$$

and

$$\vartheta_u = \frac{e_{31} b}{2 h_{pu}} \left(h_b^2 - h_c^2 \right) \tag{2.52}$$

is the coupling term in the physical equation, where b is the width of the beam,[4] h_{pu} is the thickness of the piezoceramic layer, and the subscript (or superscript) u stands for the *unimorph* configuration. The positions of the bottom and the top of the piezoceramic layer from the neutral axis are denoted by h_b and h_c, respectively, and these terms are given in terms of the thicknesses and the Young's moduli ratio of the piezoceramic and the substructure layers in Erturk and Inman (2008c). In Eq. (2.50), the modal mechanical forcing function due to base excitation is

$$f_r(t) = -m \left(\ddot{g}(t) \int_0^L \phi_r(x) dx + \ddot{h}(t) \int_0^L x \phi_r(x) dx \right) - M_t \phi_r(L) \left(\ddot{g}(t) + L \ddot{h}(t) \right). \tag{2.53}$$

---

[4] The width of the piezoceramic layer(s) (and the electrodes) is assumed to be identical to the width of the substructure layer in the entire analysis.

The electrical circuit equation can be obtained from Eq. (2.48) as

$$\frac{\varepsilon_{33}^S bL}{h_{pu}} \dot{v}_u(t) + \frac{v_u(t)}{R_1} = \sum_{r=1}^{\infty} \varphi_r^u \dot{\eta}_r^u(t), \tag{2.54}$$

where φ_r^u is the *forward* modal coupling term of the unimorph configuration and can be given by

$$\varphi_r^u = -e_{31} h_{pc}^u b \int_0^L \frac{d^2\phi_r(x)}{dx^2} dx = -e_{31} h_{pc}^u b \left. \frac{d\phi_r(x)}{dx} \right|_L, \tag{2.55}$$

where h_{pc}^u is the distance between the neutral axis and the center of the piezoceramic layer.

The forward coupling term φ_r^u has important consequences as extensively discussed by Erturk et al. (2008c, 2008e). According to Eq. (2.54), which originates from Eq. (2.48), excitation of the simple electrical circuit considered here as well as that of more sophisticated harvesting circuit topologies (Ottman et al., 2002) is proportional to the integral of the dynamic strain distribution over the electrode area. For vibration modes of a cantilevered beam other than the fundamental mode, the dynamic strain distribution over the beam length changes sign at the *strain nodes*. It is known from Eq. (2.40) that the curvature at a point is a direct measure of the bending strain. Hence, for modal excitations, strain nodes are the *inflection points* of the eigenfunctions and the integrand in Eq. (2.55) is the curvature eigenfunction. If the electric charge developed at the opposite sides of a strain node is collected by continuous electrodes for vibrations with a certain mode shape, cancellation occurs due to the phase difference in the mechanical strain distribution. Mathematically, the partial areas under the integrand function of the integral in Eq. (2.55) cancel each other over the domain of integration. As an undesired consequence, the excitation of the electrical circuit, and therefore the electrical outputs may diminish drastically. In order to avoid cancellations, segmented electrodes can be used in harvesting energy from the modes higher than the fundamental mode. The leads of the segmented electrodes can be combined in the circuit in an appropriate manner (Erturk et al., 2008e). Note that the rth vibration mode of a clamped-free beam has $r - 1$ strain nodes, and consequently, the first mode of a cantilevered beam has no cancellation problem. Some boundary conditions are more prone to strong cancellations. For instance, a beam with clamped–clamped boundary conditions has $r + 1$ strain nodes for the rth vibration mode and even the first vibration mode may yield strong cancellations if continuous electrodes are used. This discussion regarding the forward coupling term is valid for the bimorph configurations discussed in the following sections as well as for more sophisticated geometric arrangements. Attention should be given to the mode shape-dependent optimization of the electrode locations to avoid cancellations.

The circuit equation given by Eq. (2.54) is in the form of an RC electrical circuit excited by a current source (all three elements being connected in parallel to each other). Thus, Eq. (2.54) can be rewritten as

$$C_{pu} \dot{v}_u(t) + \frac{v_u(t)}{R_1} = i_p^u(t), \tag{2.56}$$

where the internal capacitance and the current source terms can be extracted by matching equations (2.54) and (2.56) as

$$C_{pu} = \frac{\varepsilon_{33}^S b L}{h_{pu}} \qquad i_p^u(t) = \sum_{r=1}^{\infty} \varphi_r^u \dot{\eta}_r^u(t) \tag{2.57}$$

Equations (2.50) and (2.56) constitute the coupled equations for the modal mechanical response $\eta_r^u(t)$ of the unimorph and the voltage response $v_u(t)$ across the resistive load. If the translational and rotational components of the base displacement given by Eq. (2.42) are harmonic in the forms of $g(t) = Y_0\, e^{j\omega t}$ and $h(t) = \theta_0\, e^{j\omega t}$, then the modal mechanical forcing function given by Eq. (2.53) can be expressed as $f_r(t) = F_r\, e^{j\omega t}$, where the amplitude F_r is

$$F_r = \omega^2 \left[m \left(Y_0 \int_0^L \phi_r(x)dx + \theta_0 \int_0^L x\phi_r(x)dx \right) + M_t \phi_r(L)(Y_0 + L\theta_0) \right]. \tag{2.58}$$

Based on the linear electromechanical system assumption, the steady-state modal mechanical response of the beam and the steady-state voltage response across the resistive load are assumed to be harmonic at the same frequency as $\eta_r^u(t) = H_r^u\, e^{j\omega t}$ and $v_u(t) = V_u\, e^{j\omega t}$, respectively, where the amplitudes H_r^u and V_u are complex valued. Then, Eqs. (2.50) and (2.56) yield the following two equations for H_r^u and V_u:

$$\left(\omega_r^2 - \omega^2 + j2\zeta_r \omega_r \omega \right) H_r^u + \chi_r^u V_u = F_r, \tag{2.59}$$

$$\left(\frac{1}{R_1} + j\omega C_{pu} \right) V_u - j\omega \sum_{r=1}^{\infty} \varphi_r^u H_r^u = 0. \tag{2.60}$$

The complex modal mechanical response amplitude H_r^u can be extracted from Eq. (2.59) and can be substituted into Eq. (2.60) to obtain the complex voltage response amplitude V_u explicitly. The resulting complex voltage response amplitude can then be used in $v_u(t) = V_u\, e^{j\omega t}$ to express the steady-state voltage response as

$$v_u(t) = \frac{\displaystyle\sum_{r=1}^{\infty} \frac{j\omega\varphi_r^u F_r}{\omega_r^2 - \omega^2 + j2\zeta_r \omega_r \omega}}{\frac{1}{R_1} + j\omega C_{pu} + \displaystyle\sum_{r=1}^{\infty} \frac{j\omega\varphi_r^u \chi_r^u}{\omega_r^2 - \omega^2 + j2\zeta_r \omega_r \omega}} e^{j\omega t}. \tag{2.61}$$

The complex voltage amplitude V_u can be substituted into Eq. (2.59) to obtain the steady-state modal mechanical response of the beam as

$$\eta_r^u(t) = \left(F_r - \chi_r^u \frac{\sum\limits_{r=1}^{\infty} \frac{j\omega\varphi_r^u F_r}{\omega_r^2 - \omega^2 + j2\zeta_r\omega_r\omega}}{\frac{1}{R_l} + j\omega C_{pu} + \sum\limits_{r=1}^{\infty} \frac{j\omega\varphi_r^u \chi_r^u}{\omega_r^2 - \omega^2 + j2\zeta_r\omega_r\omega}} \right) \frac{e^{j\omega t}}{\omega_r^2 - \omega^2 + j2\zeta_r\omega_r\omega}.$$

$$(2.62)$$

The transverse displacement response of the unimorph relative to the base can be obtained in the physical coordinates by substituting Eq. (2.62) into the modal expansion (Eq. (6)) as

$$w_{rel}^u(x, t) = \sum\limits_{r=1}^{\infty} \left(F_r - \chi_r^u \frac{\sum\limits_{r=1}^{\infty} \frac{j\omega\varphi_r^u F_r}{\omega_r^2 - \omega^2 + j2\zeta_r\omega_r\omega}}{\frac{1}{R_l} + j\omega C_{pu} + \sum\limits_{r=1}^{\infty} \frac{j\omega\varphi_r^u \chi_r^u}{\omega_r^2 - \omega^2 + j2\zeta_r\omega_r\omega}} \right) \frac{\phi_r(x)e^{j\omega t}}{\omega_r^2 - \omega^2 + j2\zeta_r\omega_r\omega}.$$

$$(2.63)$$

2.3.4 Bimorph Configurations

As shown in Fig. 2.7, depending on the poling directions of the piezoceramic layers and the connection of the electrode leads, the electrical outputs of the piezoceramic layers can be combined in series or in parallel for bimorph cantilevers. Although the bimorph configurations shown in Fig. 2.7b and 2.7c have the same geometric and material properties, the different combinations of the piezoceramic layers in the electrical circuit (in series or in parallel) changes not only the voltage response across the resistive load, but also the coupled vibration response. The coupled models of these two bimorph configurations are derived in the following.

2.3.4.1 Series Connection of the Piezoceramic Layers

The partial differential equation governing the forced vibrations of the bimorph configuration shown in Fig. 2.7b can be reduced to

$$\ddot{\eta}_r^{bs}(t) + 2\zeta_r\omega_r\dot{\eta}_r^{bs}(t) + \omega_r^2\eta_r^{bs}(t) + \chi_r^{bs}v_{bs}(t) = f_r(t),$$

$$(2.64)$$

where $\eta_r^{bs}(t)$ is the modal mechanical response of the bimorph cantilever for series connection of the piezoceramic layers (b stands for *bimorph configuration* and s stands for *series connection*), $v_{bs}(t)$ is the voltage across the resistive load, $f_r(t)$ is the modal mechanical forcing function given by Eq. (2.53), and χ_r^{bs} is the backward modal coupling term given by

$$\chi_r^{bs} = \vartheta_{bs} \left. \frac{d\phi_r(x)}{dx} \right|_L,$$

$$(2.65)$$

where

$$
\vartheta_{bs} = \frac{e_{31}b}{2h_{pb}}\left[\frac{h_{sb}^2}{4} - \left(h_{pb} + \frac{h_{sb}}{2}\right)^2\right] \tag{2.66}
$$

is the coupling term in the physical equation. Here, b is the width of the beam, h_{pb} and h_{sb} are the thicknesses of the piezoceramic and the substructure layers of the bimorph configuration shown in Fig. 2.7b, respectively.

The electrical circuit equation for the series connection case is

$$
\frac{C_{pb}}{2}\dot{v}_{bs}(t) + \frac{v_{bs}(t)}{R_l} = i_p^{bs}(t), \tag{2.67}
$$

where the internal capacitance and the current source terms are

$$
C_{pb} = \frac{\varepsilon_{33}^S bL}{h_{pb}} \qquad i_p^{bs}(t) = \sum_{r=1}^{\infty}\varphi_r^b\dot{\eta}_r^{bs}(t), \tag{2.68}
$$

respectively, and the forward modal coupling term for the bimorph configuration is

$$
\varphi_r^b = -\frac{e_{31}b(h_{pb}+h_{sb})}{2}\int_0^L\frac{d^2\phi_r(x)}{dx^2}dx = -\frac{e_{31}(h_{pb}+h_{sb})b}{2}\left.\frac{d\phi_r(x)}{dx}\right|_L. \tag{2.69}
$$

Equations (2.64) and (2.67) are the coupled equations for the modal mechanical response $\eta_r^{bs}(t)$ of the bimorph and the voltage response $v_{bs}(t)$ across the resistive load. For harmonic base motions ($g(t) = Y_0\,e^{j\omega t}$ and $h(t) = \theta_0\,e^{j\omega t}$), the modal mechanical forcing $f_r(t)$ function becomes $f_r(t) = F_r\,e^{j\omega t}$, where F_r is given by Eq. (2.58).

The steady-state voltage response across the resistive load can then be obtained as

$$
v_{bs}(t) = \frac{\displaystyle\sum_{r=1}^{\infty}\frac{j\omega\varphi_r^b F_r}{\omega_r^2-\omega^2+j2\zeta_r\omega_r\omega}}{\displaystyle\frac{1}{R_l}+j\omega\frac{C_{pb}}{2}+\sum_{r=1}^{\infty}\frac{j\omega\varphi_r^b\chi_r^{bs}}{\omega_r^2-\omega^2+j2\zeta_r\omega_r\omega}}e^{j\omega t}, \tag{2.70}
$$

and the transverse displacement response of the bimorph relative to its base is

$$
w_{rel}^{bs}(x,t) = \sum_{r=1}^{\infty}\left(F_r - \chi_r^{bs}\frac{\displaystyle\sum_{r=1}^{\infty}\frac{j\omega\varphi_r^b F_r}{\omega_r^2-\omega^2+j2\zeta_r\omega_r\omega}}{\displaystyle\frac{1}{R_l}+j\omega\frac{C_{pb}}{2}+\sum_{r=1}^{\infty}\frac{j\omega\varphi_r^b\chi_r^{bs}}{\omega_r^2-\omega^2+j2\zeta_r\omega_r\omega}}\right)\frac{\phi_r(x)e^{j\omega t}}{\omega_r^2-\omega^2+j2\zeta_r\omega_r\omega}.
$$
$$
\tag{2.71}
$$

2.3.4.2 Parallel Connection of the Piezoceramic Layers

The coupled equation of motion for the modal mechanical response can be obtained as

$$\ddot{\eta}_r^{\text{bp}}(t) + 2\zeta_r\omega_r\dot{\eta}_r^{\text{bp}}(t) + \omega_r^2\eta_r^{\text{bp}}(t) + \chi_r^{\text{bp}}v_{\text{bp}}(t) = f_r(t), \qquad (2.72)$$

where $\eta_r^{\text{bp}}(t)$ is the modal mechanical response of the bimorph cantilever for parallel connection of the piezoceramic layers and $v_{\text{bp}}(t)$ is the voltage across the resistive load (b stands for *bimorph configuration* and p stands for *parallel connection*). The modal mechanical forcing function $f_r(t)$ is given by Eq. (2.53) and the backward modal coupling term in Eq. (2.72) is

$$\chi_r^{\text{bp}} = \vartheta_{\text{bp}} \left. \frac{\mathrm{d}\phi_r(x)}{\mathrm{d}x} \right|_L, \qquad (2.73)$$

where the coupling term in the physical coordinates can be given by

$$\vartheta_{\text{bp}} = \frac{e_{31}b}{h_{\text{pb}}} \left[\frac{h_{\text{sb}}^2}{4} - \left(h_{\text{pb}} + \frac{h_{\text{sb}}}{2} \right)^2 \right]. \qquad (2.74)$$

Here, b is the width of the beam, h_{pb} and h_{sb} are the thicknesses of the piezoceramic and the substructure layers of the bimorph configuration displayed in Fig. 2.7c, respectively.

The electrical circuit equation for the parallel connection case is

$$C_{\text{pb}}\dot{v}_{\text{bp}}(t) + \frac{v_{\text{bp}}(t)}{2R_1} = i_{\text{p}}^{\text{bp}}(t), \qquad (2.75)$$

where the internal capacitance and the current source terms are

$$C_{\text{pb}} = \frac{\varepsilon_{33}^S bL}{h_{\text{pb}}} \qquad i_{\text{p}}^{\text{bp}}(t) = \sum_{r=1}^{\infty} \varphi_r^b \dot{\eta}_r^{\text{bp}}(t), \qquad (2.76)$$

respectively, and the forward modal coupling term φ_r^b is given by Eq. (2.69).

Equations (2.72) and (2.75) constitute the coupled equations for the modal mechanical response $\eta_r^{\text{bp}}(t)$ of the bimorph and the voltage response $v_{\text{bp}}(t)$ across the resistive load. For harmonic base motions, one can obtain the steady-state voltage response across the resistive load as

$$v_{\text{bp}}(t) = \frac{\displaystyle\sum_{r=1}^{\infty} \frac{j\omega\varphi_r^b F_r}{\omega_r^2 - \omega^2 + j2\zeta_r\omega_r\omega}}{\displaystyle\frac{1}{2R_1} + j\omega C_{\text{pb}} + \sum_{r=1}^{\infty} \frac{j\omega\varphi_r^b \chi_r^{\text{bp}}}{\omega_r^2 - \omega^2 + j2\zeta_r\omega_r\omega}} e^{j\omega t}, \qquad (2.77)$$

and the transverse displacement response of the bimorph relative to its base can be expressed as

$$
w_{\text{rel}}^{\text{bp}}(x, t) = \sum_{r=1}^{\infty} \left(F_r - \chi_r^{\text{bp}} \frac{\sum_{r=1}^{\infty} \frac{j\omega\varphi_r^{\text{b}} F_r}{\omega_r{}^2 - \omega^2 + j2\zeta_r\omega_r\omega}}{\frac{1}{2R_1} + j\omega C_{\text{pb}} + \sum_{r=1}^{\infty} \frac{j\omega\varphi_r^{\text{b}}\chi_r^{\text{bp}}}{\omega_r{}^2 - \omega^2 + j2\zeta_r\omega_r\omega}} \right) \frac{\phi_r(x) e^{j\omega t}}{\omega_r{}^2 - \omega^2 + j2\zeta_r\omega_r\omega}.
$$

$$(2.78)$$

2.3.5 Single-Mode Electromechanical Equations

The steady-state voltage response and vibration response expressions obtained in Sections 2.3.3 and 2.3.4 are valid for harmonic excitations at any arbitrary frequency ω. That is, these expressions are the *multi-mode* solutions as they include all vibration modes of the respective cantilevered piezoelectric energy harvester beams. Hence, the resulting equations can predict the coupled systems dynamics not only for resonance excitations but also for excitations at the off-resonance frequencies of the configurations shown in Fig. 2.7.

Resonance excitation is a special case of the derivation given here and is commonly used in the literature in order to investigate the maximum performance of the harvester in electrical power generation. Therefore, excitation of a unimorph/bimorph at or very close to one of its natural frequencies is a very useful problem to investigate through the resulting equations derived in Sections 2.3.3 and 2.3.4. This is the *modal excitation* condition and mathematically it corresponds to $\omega \cong \omega_r$. With this assumption on the excitation frequency, the major contribution in the summation terms of Eqs. (2.61), (2.63), (2.70), (2.71), (2.77) and (2.78) are from the rth vibration mode, which allows drastic simplifications in the coupled voltage response and vibration response expressions. In the following, the reduced *single-mode* expressions are given for excitations at or very close to the r-th natural frequency. However, it should be noted that the fundamental vibration mode is the main concern in the energy harvesting problem (which corresponds to $r = 1$).

2.3.5.1 Unimorph Configuration

If the unimorph configuration shown in Fig. 2.7a is excited at $\omega \cong \omega_r$, the contribution of all the vibration modes other than the rth mode can be ignored in the summation terms. Then, the steady-state voltage response given by Eq. (2.61) can be reduced to

$$
\hat{v}_{\text{u}}(t) = \frac{j\omega R_1 \varphi_r^{\text{u}} F_r \, e^{j\omega t}}{(1 + j\omega R_1 C_{\text{pu}})(\omega_r^2 - \omega^2 + j2\zeta_r\omega_r\omega) + j\omega R_1 \varphi_r^{\text{u}} \chi_r^{\text{u}}},
$$

$$(2.79)$$

and the transverse displacement relative to the moving base is obtained from equation Eq. (2.63) as

$$\hat{w}^{u}_{rel}(x,t) = \frac{\left(1 + j\omega R_1 C_{pu}\right) F_r \phi_r(x) e^{j\omega t}}{(1 + j\omega R_1 C_{pu})(\omega_r^2 - \omega^2 + j2\zeta_r \omega_r \omega) + j\omega R_1 \varphi_r^u \chi_r^u}. \tag{2.80}$$

Here and below, a hat ($^\wedge$) denotes that the respective term is reduced from the full (multi-mode) solution for excitations very close to a natural frequency. The relevant terms in Eqs. (2.79) and (2.80) can be found in Section 2.3.3.

2.3.5.2 Bimorph Configuration (Series Connection)

For excitation of the bimorph shown in Fig. 2.7b around its rth natural frequency, the single-mode steady-state voltage response given by Eq. (2.70) can be reduced to

$$\hat{v}_{bs}(t) = \frac{j2\omega R_1 \varphi_r^b F_r e^{j\omega t}}{(2 + j\omega R_1 C_{pb})(\omega_r^2 - \omega^2 + j2\zeta_r \omega_r \omega) + j2\omega R_1 \varphi_r^b \chi_r^{bs}}. \tag{2.81}$$

Similarly, the transverse displacement relative to the moving base is reduced from Eq. (2.71) as

$$\hat{w}^{bs}_{rel}(x,t) = \frac{(2 + j\omega R_1 C_{pb}) F_r \phi_r(x) e^{j\omega t}}{(2 + j\omega R_1 C_{pb})(\omega_r^2 - \omega^2 + j2\zeta_r \omega_r \omega) + j2\omega R_1 \varphi_r^b \chi_r^{bs}}, \tag{2.82}$$

where the relevant terms can be found in Section 2.3.4.1.

2.3.5.3 Bimorph Configuration (Parallel Connection)

If the bimorph configuration displayed in Fig. 2.7c is excited at $\omega \cong \omega_r$, the steady-state voltage response given by Eq. (2.77) can be reduced to

$$\hat{v}_{bp}(t) = \frac{j2\omega R_1 \varphi_r^b F_r e^{j\omega t}}{(1 + j2\omega R_1 C_{pb})(\omega_r^2 - \omega^2 + j2\zeta_r \omega_r \omega) + j2\omega R_1 \varphi_r^b \chi_r^{bp}}, \tag{2.83}$$

and the transverse displacement relative to the base is obtained from Eq. (2.78) as

$$\hat{w}^{bp}_{rel}(x,t) = \frac{(1 + j2\omega R_1 C_{pb}) F_r \phi_r(x) e^{j\omega t}}{(1 + j2\omega R_1 C_{pb})(\omega_r^2 - \omega^2 + j2\zeta_r \omega_r \omega) + j2\omega R_1 \varphi_r^b \chi_r^{bp}}, \tag{2.84}$$

where the relevant terms can be found in Section 2.3.4.2.

2.3.6 Experimental Validation

The aim of this section is to validate the distributed parameter relations for a bimorph cantilever with a tip mass experimentally. The experimentally measured voltage response-to-base acceleration FRFs and the vibration response-to-base acceleration FRFs are compared with the closed-form relations derived in this chapter. Variations of the voltage output and the tip velocity response of the bimorph with changing load resistance are also investigated and predicted using the analytical relations. Optimum resistive loads of the bimorph cantilever are identified for excitations at the short-circuit and open-circuit resonance frequencies.

2.3.6.1 Experimental Setup

The experimental setup used for measuring the voltage-to-base acceleration and tip velocity-to-base acceleration FRFs of the bimorph cantilever is shown in Fig. 2.8a. The bimorph cantilever analyzed in this experiment is displayed in Fig. 2.8b and is manufactured by Piezo Systems, Inc. (T226-A4-503X). The same type of bimorph was used by duToit et al. (2007) for verification of their Rayleigh–Ritz type of approximate model. In that work, duToit et al. (2007) underestimated the vibration response and the power outputs especially around the resonance frequency and they attributed this inaccuracy in the results to the "unmodeled nonlinear piezoelectric response."

Here, a tip mass is attached to the bender to make the problem relatively sophisticated in terms of modeling (Fig. 2.8b). The bimorph consists of two oppositely poled PZT-5A piezoceramic layers bracketing a brass substructure layer. Therefore, the piezoelectric elements are connected in series as given in Fig. 2.7b. Table 2.1 shows

(a)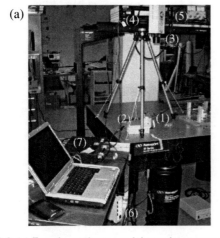

(b)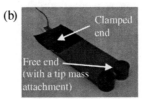

(1) Shaker, bimorph cantilever and a low mass accelerometer
(2) Circuit with a resistive load
(3) Laser vibrometer head
(4) Laser vibrometer controller
(5) Power amplifier
(6) Charge amplifier
(7) Data acquisition system

Fig. 2.8 (a) Experimental setup and the equipments used for analyzing the bimorph cantilever and (b) a detailed view of the bimorph cantilever with a tip mass

Table 2.1 Geometric and material parameters of the bimorph cantilever used for the experimental validation

Geometric Parameters	Piezoceramic		Substructure
Length, L (mm)	50.8		50.8
Width, b (mm)	31.8		31.8
Thickness, h (mm)	0.26 (each)		0.14
Tip mass, M_t (kg)		0.012	
Material Parameters	**Piezoceramic (PZT-5A)**		**Substructure (Brass)**
Mass density, ρ (kg/m^3)	7800		9000
Young's modulus, Y (GPa)	66		105
Piezo. constant, d_{31} (pm/V)	−190		−
Permittivity, ε_{33}^S (nF/m)	$1500\varepsilon_0$		−

the geometric and material properties of the piezoceramic and the substructure layers, respectively. Note that the length described by L is the overhang length of the harvester in the clamped condition, i.e., it is not the total free length (63.5 mm) of the bender as acquired from the manufacturer. In addition, permittivity component at constant strain is given in Table 2.1 in terms of the permittivity of free space, $\varepsilon_0 = 8.854$ pF/m (IEEE, 1987).

The bimorph cantilever shown in Fig. 2.8b is excited from its base with a sine sweep generated by an electromagnetic LDS shaker. The base acceleration of the harvester is measured by a low mass accelerometer (PCB U352C22) on the shaker and the velocity response of the harvester at the free end is measured by a laser vibrometer (Polytec OFV303 laser sensor head, OFV3001 controller). The experimental voltage FRF (in V/g) and tip velocity FRF (in (m/s)/g) obtained for a resistive load of 1 kΩ are shown in Fig. 2.9a (where g is the gravitational acceleration). The coherence functions of these measurements are given by Fig. 2.9b. The coherence is considerably low for frequencies less than 30 Hz but it is good around the first resonance frequency (which is approximately 45.6 Hz for a 1 kΩ resistive load).

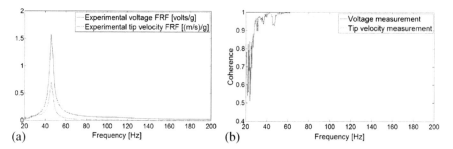

Fig. 2.9 (a) Experimental voltage and tip velocity FRFs of the bimorph cantilever and (b) their coherence functions (for a resistive load of 1 kΩ)

2.3.6.2 Validation of the Distributed Parameter Model

Since the vibration mode seen around 45.6 Hz (for 1 kΩ) is the only mode in a wide frequency range and as this is the mode of primary concern (i.e., the most flexible vibration mode), the single-mode relations given by Eqs. (2.81) and (2.82) are used for predicting the electromechanical FRFs. Note that the base of the cantilever is not rotating, i.e., $\theta_0 = 0$ in Eq. (2.58). In addition, Eq. (2.82) must be rearranged in order to predict the experimental velocity measurement since the laser vibrometer measures the *absolute* velocity response at the tip of the beam (rather than the velocity *relative* to the moving base). Thus, the electromechanical FRFs used for predicting the experimental measurements can be extracted from Eqs. (2.81) and (2.82) as

$$\frac{\hat{v}_{bs}(t)}{-\omega^2 Y_0\, e^{j\omega t}} = \frac{-j2\omega R_l \varphi_r^b \left(m \int_0^L \phi_r(x)dx + M_t\phi_r(L)\right)}{(2 + j\omega R_l C_{pb})(\omega_r^2 - \omega^2 + j2\zeta_r\omega_r\omega) + j2\omega R_l \varphi_r^b \chi_r^{bs}}, \quad (2.85)$$

$$\frac{\frac{d\hat{w}^{bs}(L,t)}{dt}}{-\omega^2 Y_0\, e^{j\omega t}} = \frac{1}{j\omega} + \frac{-j\omega(2 + j\omega R_l C_{pb})\left(m \int_0^L \phi_r(x)dx + M_t\phi_r(L)\right)\phi_r(L)}{(2 + j\omega R_l C_{pb})(\omega_r^2 - \omega^2 + j2\zeta_r\omega_r\omega) + j2\omega R_l \varphi_r^b \chi_r^{bs}}, \quad (2.86)$$

where $r = 1$ since the vibration mode of interest is the fundamental vibration mode.

In order to identify the modal mechanical damping ratio, one way is to force the system to short-circuit conditions (by shorting the electrodes of the bimorph) as done by duToit et al. (2007). However, if the electromechanical model is *self-consistent*, one must be able to identify the mechanical damping ratio for any value of load resistance. Furthermore, either the voltage FRF or the tip velocity FRF can be used for identifying modal mechanical damping ratio, since the bimorph energy harvester itself is a *transducer*. In other words, theoretically, the coupled tip velocity information is included in the voltage output information of the harvester, and the voltage and tip motion predictions for the same mechanical damping ratio must be in agreement.

The mechanical damping ratio of the first vibration mode is identified as $\zeta_1 = 0.027$ using the voltage FRF as shown in Fig. 2.10a (for 1 kΩ load resistance). For this identified damping ratio, the voltage FRF of the model (obtained from Eq. (2.85)) is in perfect agreement with the experimental FRF as shown in Fig. 2.10a. As discussed in the previous paragraph, for the same damping ratio (2.7%), the tip velocity FRF obtained from the model should predict the experimental tip velocity FRF accurately. The tip velocity FRF obtained from Eq. (2.86) for 2.7% mechanical damping is plotted with the laser vibrometer measurement in Fig. 2.10b. The agreement between the theoretical and the experimental tip velocity FRFs is very good, which clearly shows the consistency of the linear electromechanical model proposed here.

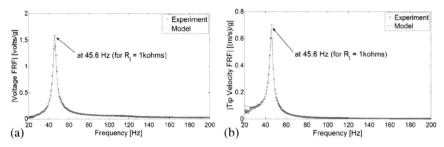

Fig. 2.10 Comparison of the model predictions and the experimental measurements; (**a**) voltage FRF and (**b**) tip velocity FRF (for 1 kΩ)

The experimental measurements are repeated for eight different values of load resistance: 1, 6.7, 11.8, 22, 33, 47, 100, and 470 kΩ. Each of the resistive loads results in a different voltage FRF and a tip velocity FRF. Figure 2.11a and 2.11b, respectively, display enlarged views of the voltage output and tip velocity FRFs around the first vibration mode for these eight different values of load resistance. The direction of increasing load resistance is depicted with an arrow and it is clear from Fig. 2.11a that the voltage across the resistive load increases monotonically with increasing load resistance at every excitation frequency. For the extreme values of the load resistance, the frequency of maximum voltage output moves from the short-circuit resonance frequency (for $R_1 \rightarrow 0$) to the open-circuit resonance frequency (for $R_1 \rightarrow \infty$). The experimentally obtained short-circuit and open-circuit resonance frequencies for the first mode of this cantilever are approximately 45.6 and 48.4 Hz, respectively. The analytical model predicts these two frequencies as 45.7 and 48.2 Hz, respectively. For a moderate value of load resistance, the frequency of maximum voltage has a value in between these two extreme frequencies. Even for more sophisticated energy harvesting circuits, the resonance frequency of the beam usually takes a value between these two extreme frequencies since the impedance across the electrodes can change at most from zero to infinity regardless of the circuit elements. This can be observed in the experimental vibration FRF (Lesieutre et al., 2004) of a cantilevered piezoelectric energy harvester beam con-

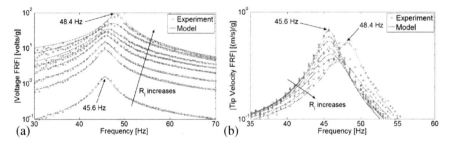

Fig. 2.11 Enlarged views of the (**a**) voltage FRFs and the (**b**) tip velocity FRFs for eight different values of load resistance (model predictions and the experimental measurements)

nected to an electrical circuit with more complicated elements (such as a full-wave rectifier).

The shift in the frequencies of the maximum response amplitude is also the case in the tip velocity FRF (Fig. 2.11b). However, the variation of tip velocity with load resistance is not necessarily monotonic at every frequency. For excitation at 45.6 Hz, the tip motion is suppressed as the resistive load is increased up to a certain value. It is very important to note that this suppression in the motion amplitude is more sophisticated than viscous damping. With increasing load resistance (up to a certain value), the motion is *attenuated* at 45.6 Hz, whereas it is *amplified* at 48.4 Hz. Hence, if one focuses on the open-circuit resonance frequency (48.4 Hz), *both the voltage output and the vibration amplitude* at the tip of the beam *increase* with increasing load resistance. Therefore, modeling the effect of piezoelectric coupling in the beam equation as viscous damping clearly *fails* in predicting this phenomenon (as it cannot predict the frequency shift due to changing load resistance). Note that, for eight different resistive loads, the model successfully predicts the frequency response of the voltage output and tip velocity (where the mechanical damping ratio is kept constant at 2.7%).

The electric current FRF (obtained from $I = V/R_1$) exhibits the opposite behavior of the voltage FRF with changing load resistance as shown in Fig. 2.12a. Hence, the electric current decreases monotonically with increasing load resistance at every excitation frequency. Figure 2.12b displays the electrical power FRF for three different resistive loads.[5] The trend in the electrical power FRF with changing load resistance is more interesting as it is the multiplication of two FRFs (voltage and current) with the opposite trends. As can be seen in Fig. 2.12b, the electrical power FRFs for different resistive loads intersect each other just like the tip velocity FRFs (Fig. 2.11b). For a given excitation frequency, there exists a certain value of load resistance that gives the maximum electrical power. This value is called the *optimum*

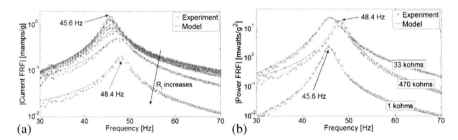

Fig. 2.12 Enlarged views of the (**a**) current FRFs for eight different values of load resistance and the (**b**) power FRFs for three different values of load resistance (model predictions and experimental measurements)

---

[5] In order to avoid confusion with 8 intersecting curves, the electrical power FRFs are given for 3 resistive loads only.

load resistance and can be observed more easily if the frequency of interest is kept constant and the power amplitude is plotted against load resistance as addressed next.[6]

The short-circuit and open-circuit resonance frequencies of the first mode are defined for the extreme cases of load resistance (45.6 Hz as $R_l \to 0$ and 48.4 Hz as $R_l \to \infty$) and these frequencies are of practical interest. The variation of the voltage output with changing load resistance for excitations at these two frequencies are shown in Fig. 2.13a. In both cases, voltage increases monotonically with load resistance. The voltage output for excitation at the short-circuit resonance frequency is larger when the system is close to short-circuit conditions and vice versa. The maximum voltage amplitude in the limit $R_l \to \infty$ is about 54.5 V/*g* for excitation at 45.6 Hz and is about 108.8 V/*g* for excitation at 48.4 Hz. Figure 2.13b displays the variation of the electric current with changing load resistance for excitations at these two frequencies. The trend of the current amplitude with changing load resistance is opposite to that of the voltage amplitude. That is, the current amplitude decreases monotonically with increasing load resistance. The current output for excitation at the short-circuit resonance frequency is larger when the system is close to short-circuit conditions and vice versa. In the limit $R_l \to 0$, the maximum current

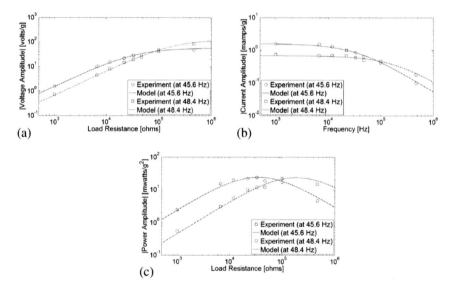

Fig. 2.13 Variations of the **(a)** voltage amplitude, **(b)** current amplitude, and the **(c)** electrical power amplitude with load resistance for excitations at the short-circuit and open-circuit resonance frequencies of the first vibration mode

---

[6] The amplitudes given in the voltage, current and tip velocity FRFs are the peak amplitudes (not the root mean square values). Therefore, the electrical power amplitude is the peak power amplitude (which is twice the average power).

amplitude is about 1.57 mA/g for excitation at 45.6 Hz and is about 0.68 mA/g for excitation at 48.4 Hz.

The variation of the electrical power with changing load resistance is given in Fig. 2.13c for the short-circuit and open-circuit resonance frequency excitations. These two excitation frequencies have different optimum resistive loads which yield the maximum electrical power. The optimum load resistance for excitation at 45.6 Hz is about 35 kΩ, yielding a maximum electrical power of about 23.9 mW/g^2, whereas the optimum resistive load for excitation at 48.4 Hz is 186 kΩ, yielding approximately the same power output. As in the case of the voltage and current outputs, the electrical power output for excitation at the short-circuit resonance frequency is larger when the system is close to short-circuit conditions and vice versa. The respective trends in the electrical outputs at the short-circuit and open-circuit resonance frequencies of the first mode are successfully predicted by the single-mode analytical relations derived in this chapter.

A useful practice to obtain some additional information regarding the performance of the energy harvester device results from dividing the electrical power by the volume and by the mass of the device. The total overhang volume of the device (including the volume of the tip mass) is about 3.52 cm^3 and the total overhang mass of the device is about 20.6 g. The electrical power versus load resistance graph given by Fig. 2.13c can therefore be used to obtain the maximum *power density* (power per device volume) and *specific power* (power per device mass) values. The variations of the power density and the specific power with load resistance are given by Fig. 2.14a and 2.14b, respectively (for the short-circuit and open-circuit resonance excitations). For instance, for excitation at 45.6 Hz, the maximum power density is about 6.8 (mW/g^2)/cm^3 and the maximum specific power is about 1.15 (W/g^2)/kg (for a 35 kΩ resistive load). It is very important to note that the power density and the specific power concepts are *not* complete non-dimensional representations. For instance, the same device volume can be occupied by the same amount of material (piezoceramic, substructure, and tip mass) for a different aspect ratio of the beam, yielding a larger or smaller electrical power with totally different natural frequencies. However, these representations have been found useful for comparison of the energy harvester devices in the literature.

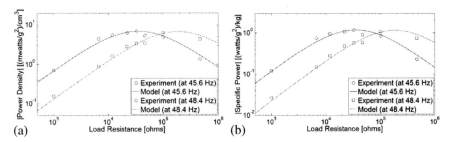

(a)　　　　　　　　　　　　　　　(b)

Fig. 2.14 Variations of the (**a**) power density and the (**b**) specific power amplitudes with load resistance for excitations at the short-circuit and open-circuit resonance frequencies of the first vibration mode

References

Ajitsaria J, Choe S Y, Shen D, and Kim D J 2007 Modeling and analysis of a bimorph piezoelectric cantilever beam for voltage generation *Smart Materials and Structures* **16**:447–454

Anton S R and Sodano H A 2007 A review of power harvesting using piezoelectric materials (2003–2006) *Smart Materials and Structures* **16**:R1–R21

Arnold D 2007 Review of microscale magnetic power generation *IEEE Transactions on Magnetics* **43**:3940–3951

Banks T L and Inman D J 1991 On damping mechanisms in beams *ASME Journal of Applied Mechanics* **58**:716–723

Beeby S P, Tudor M J, and White N M 2006 Energy harvesting vibration sources for microsystems applications, *Measurement Science and Technology* **13**:175–195

Chen S -N, Wang G –J, and Chien M -C 2006 Analytical modeling of piezoelectric vibration-induced micro power generator, *Mechatronics* **16**:387–397

Clough R W and Penzien J 1975 *Dynamics of Structures* John Wiley and Sons, New York

Cook-Chennault K A, Thambi N, and Sastry A M 2008 Powering MEMS portable devices – a review of non-regenerative and regenerative power supply systems with emphasis on piezo-electric energy harvesting systems, *Smart Materials and Structures* **17**:043001:1–33

Crandall S H, Karnopp D C, Kurtz Jr E F, and Pridmore-Brown D C 1968 *Dynamics of Mechanical and Electromechanical Systems* McGraw-Hill, New York

Daqaq M, Renno J M, Farmer J R, and Inman D J 2007 Effects of system parameters and damping on an optimal vibration-based energy harvester *Proceedings of the 48th AIAA/ASME/ASCE/AHS/ASC Structures, Structural Dynamics, and Materials Conference*

duToit N E, Wardle B L, and Kim S-G 2005 Design considerations for MEMS-scale piezoelectric mechanical vibration energy harvesters, *Integrated Ferroelectrics* **71**:121–160

duToit N E and Wardle B L 2007 Experimental verification of models for microfabricated piezoelectric vibration energy harvesters, *AIAA Journal* **45**:1126–1137

Elvin N and Elvin A 2008 A general equivalent circuit model for piezoelectric generators, *Journal of Intelligent Material Systems and Structures* **19** in press (DOI: 10.1177/1045389X08089957)

Erturk A and Inman D J 2008a On mechanical modeling of cantilevered piezoelectric vibration energy harvesters, *Journal of Intelligent Material Systems and Structures* **19**:1311–1325

Erturk A and Inman D J 2008b Issues in mathematical modeling of piezoelectric energy harvesters, *Smart Materials and Structures* in press

Erturk A and Inman D J 2008c A distributed parameter electromechanical model for cantilevered piezoelectric energy harvesters, *ASME Journal of Vibration and Acoustics* **130**:041002-1-15

Erturk A and Inman D J 2008d An experimentally validated bimorph cantilever model for piezoelectric energy harvesting from base excitations, *Smart Materials and Structures* accepted

Erturk A and Tarazaga P A, Farmer J R, and Inman D J 2008e Effect of strain nodes and electrode configuration on piezoelectric energy harvesting from cantilevered beams, *ASME Journal of Vibration and Acoustics* in press (DOI: 10.1115/1.2981094)

Fang H-B, Liu J-Q, Xu Z-Y, Dong L, Chen D, Cai B-C, and Liu Y 2006 A MEMS-based piezoelectric power generator for low frequency vibration energy harvesting, *Chinese Physics Letters* **23**:732–734

Glynne-Jones P, Tudor M J, Beeby S P, and White N M 2004 An electromagnetic, vibration-powered generator for intelligent sensor systems, *Sensors and Actuators A* **110**:344–349

Hagood N W, Chung W H, and Von Flotow A 1990 Modelling of piezoelectric actuator dynamics for active structural control, *Journal of Intelligent Material Systems and Structures* **1**:327–354

IEEE Standard on Piezoelectricity 1987 IEEE, New York.

Jeon Y B, Sood R, Jeong J H, and Kim S 2005 MEMS power generator with transverse mode thin film PZT, *Sensors & Actuators A* **122**:16–22

Lesieutre G A, Ottman G K, and Hofmann H F 2004 Damping as a result of piezoelectric energy harvesting, *Journal of Sound and Vibration* **269**:991–1001

Lin J H, Wu X M, Ren T L, and Liu L T 2007 Modeling and simulation of piezoelectric MEMS energy harvesting device, *Integrated Ferroelectrics* **95**:128–141.

Lu F, Lee H, and Lim S 2004 Modeling and analysis of micro piezoelectric power generators for micro-electromechanical-systems applications, *Smart Materials and Structures* **13**:57–63

Mitcheson P, Miao P, Start B, Yeatman E, Holmes A, and Green T 2004 MEMS electrostatic micro-power generator for low frequency operation, *Sensors and Actuators A* **115**:523–529

Ottman G K, Hofmann H F, Bhatt A C, and Lesieutre G A 2002 Adaptive piezoelectric energy harvesting circuit for wireless remote power supply, *IEEE Transactions on Power Electronics* **17**:669–676.

Priya S 2007 Advances in energy harvesting using low profile piezoelectric transducers, *Journal of Electroceramics* **19**:167–184

Roundy S, Wright P K, and Rabaey J 2002 Micro-electrostatic vibration-to-electricity converters *Proceedings of the ASME 2002 International Mechanical Engineering Congress and Exposition*

Roundy S, Wright P K, and Rabaey J 2003 A study of low level vibrations as a power source for wireless sensor nodes, *Computer Communications* **26**:1131–1144

Sodano H A, Inman D J, and Park G 2004a A review of power harvesting from vibration using piezoelectric materials, *The Shock and Vibration Digest* **36**:197–205

Sodano H A, Park G, and Inman D J 2004b Estimation of electric charge output for piezoelectric energy harvesting, *Strain* **40**:49–58

Sodano H, Inman D, and Park G 2005 Generation and storage of electricity from power harvesting devices, *Journal of Intelligent Material Systems and Structures* **16**:67–75

Stephen N G 2006 On energy harvesting from ambient vibration, *Journal of Sound and Vibration* **293**:409–425

Strutt J W (Lord Rayleigh) 1894 *The Theory of Sound* MacMillan Company, London

Timoshenko S, Young D H, and Weaver W 1974 *Vibration Problems in Engineering* John Wiley and Sons, New York

Williams C B and Yates R B 1996 Analysis of a micro-electric generator for microsystems, *Sensors and Actuators A* **52**:8–11

Chapter 3
Performance Evaluation of Vibration-Based Piezoelectric Energy Scavengers

Yi-Chung Shu

Abstract This chapter summarizes several recent activities for fundamental understanding of piezoelectric vibration-based energy harvesting. The developed framework is able to predict the electrical behavior of piezoelectric power harvesting systems using either the standard or the synchronized switch harvesting on inductor (SSHI) electronic interface. In addition, some opportunities for new devices and improvements in existing ones are also pointed here.

3.1 Introduction

The development of wireless sensor and communication node networks has received great interests in research communities over the past few years (Rabaey et al. 2000). Applications envisioned from these node networks include building the health monitoring for civil infrastructures, environmental control systems, hazardous materials detection, smart homes, and homeland security applications. However, as the networks increase in the number and the devices decrease in the size, the proliferation of these autonomous microsensors raises the problem of effective power supply. As a result, the foremost challenge for such dense networks to achieve their full potential is to manage power consumption for a large number of nodes.

Unlike cellphones and laptops, whose users can periodically recharge, embedded devices must operate in their initial batteries. However, batteries cannot only increase the size and weight of microsensors but also suffer from the limitations of a brief service life. For example, at an average power consumption of 100 μW (an order of magnitude smaller than any existing node), a sensor node would last only 1 yr if a 1 cm^3 of lithium battery (at the maximum energy density of 800 W hr per L) was used to supply power (Kansal & Srivastava 2005, Roundy et al. 2005). A lifetime of approximately 1 yr is obviously not practical for many applications. In addition, the need for constant battery replacement can be very tedious and expensive task. In many other cases, these operations may be prohibited by the

Y.C. Shu (✉)

Institute of Applied Mechanics, National Taiwan University, Taipei 106, Taiwan, ROC
email: yichung@spring.iam.ntu.edu.tw

S. Priya, D.J. Inman (eds.), *Energy Harvesting Technologies*,
DOI 10.1007/978-0-387-76464-1_3 © Springer Science+Business Media, LLC 2009

infrastructure. Therefore, energy supply using batteries is currently a major bottleneck for system lifetime, and harvesting energy from the deployment environment can help to relieve it.

Simultaneous advances in low-power electronic design and fabrication have reduced power requirements for individual nodes, and therefore, allow the feasibility of self-powering these autonomous electronic devices. This stems from the fact that power consumption in integrated circuits (IC) will continue to decrease as IC processing moves towards smaller feature sizes (Chandrakasan et al. 1998, Yoon et al. 2005). This opens the possibility for completely self-powered sensor nodes, and the notion of a small smart material generator-producing enough power is not far fetched. However, these smart material generators need to be optimally designed, which is the central topic discussed in this chapter. The goal here is to develop a smart architecture which utilizes the environmental resources available for generating electrical power. These resources include solar power, thermal gradients, acoustic, and mechanical vibration (Roundy et al. 2004a, Sebald et al. 2008, Whalen et al. 2003). Among these energy scavenging sources, we are particularly interested in mechanical vibration since it is a potential power source that is abundant enough to be of use, is easily accessible through microelectromechanical systems (MEMS) technology for conversion to electrical energy, and is ubiquitous in applications from small household appliances to large infrastructures (Roundy et al. 2004b, Sodano et al. 2004).

Vibration energy can be converted into electrical energy through piezoelectric, electromagnetic, and capacitive transducers (Beeby et al. 2007, Cheng et al. 2007, Lee et al. 2004, Nakano et al. 2007, Poulin et al. 2004, Roundy et al. 2003, Stephen 2006a,b, Williams & Yates 1996, Zhao & Lord 2006). Among them, piezoelectric vibration-to-electricity converters have received much attention, as they have high-electromechanical coupling and no external voltage source requirement, and they are particularly attractive for use in MEMS (Choi et al. 2006, Fang et al. 2006, Horowitz et al. 2006, Jeon et al. 2005, Lu et al. 2004). As a result, the use of piezoelectric materials for scavenging

energy from ambient vibration sources has recently seen a dramatic rise for power harvesting (duToit et al. 2005, Elvin et al. 2006, Guyomar et al. 2005, Hu et al 2007a, Kim et al. 2005a, Liao et al. 2001, Ng & Liao 2005, Ottman et al. 2002, Richards et al. 2004, Richter et al. 2006, Roundy & Wright 2004, Shu & Lien 2006a, Sodano et al. 2006). The optimum design and setup of an energy harvesting system using piezoelectric generators depends on the kind of the surrounding kinetic energy to be exploited (amplitude and frequency) as well as on the electrical application to be powered. Thus, it is necessary to use the model-based design methods instead of using try-and-error schemes. To realize such design models, a good understanding of the piezoelectric energy harvesting system is inevitable.

To be precise, an energy conversion device considered here includes a vibrating piezoelectric structure together with an energy storage system, as shown in Fig. 3.1. The piezoelectric generator is modeled as a mass+spring+damper+piezostructure and is connected to a storage circuit system. An AC–DC rectifier followed by a filtering capacitance C_e is added to smooth the DC voltage. A controller placed between the rectifier output and the battery is included to regulate the

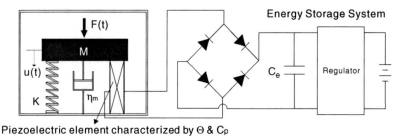

Piezoelectric element characterized by Θ & C$_p$

Fig. 3.1 An equivalent model for a piezoelectric vibration energy harvesting system

output voltage. From Fig. 3.1, it is clear to see that the left-hand side is related to the design of piezoelectric power generators and the right-hand side is associated with the design of power electronics. Indeed, current piezoelectric-harvesting research falls mainly into two key areas: developing optimal energy harvesting structures and highly efficient electrical circuits to store the generated charges (Johnson & Clark 2005). The former includes the works by duToit et al. (2005), Lu et al. (2004), Richards et al. (2004), Roundy & Wright (2004), Sodano et al. (2004), and the latter contains the works by Guan & Liao (2007), Lefeuvre et al. (2005b), Ngo et al. (2006), Ottman et al. (2002), Ottman et al. (2003). However, the linkage between these two has not been exploited in detail until recently by Ottman et al. (2002) and Guyomar et al. (2005). They have provided different estimations of AC–DC power output. The former has assumed that the vibration amplitude is not affected by the load resistance, while the latter has hypothesized that the periodic external excitation and the speed of mass are in-phase. In contrast with the estimates based on these two approaches, Shu & Lien (2006a) have provided an analysis of AC–DC power output for a rectified piezoelectric harvester. They have proposed a new method to determine AC–DC power flow without the uncoupled and in-phase assumptions and concluded that their estimation is more accurate than the other two. In addition, Shu & Lien (2006b) have also shown that the conversion efficiency and optimization criteria vary according to the relative strength of the electromechanical coupling and mechanical damping ratio. These results are crucial for the choice of the optimal power converter.

Interest in the application of piezoelectric vibration-based energy scavengers for converting mechanical energy to electrical energy has increased significantly in the very recent years. An overview of research in this field has recently been provided by Sodano et al. (2004) and Anton & Sodano (2007) for review articles as well as by Roundy et al. (2004b) for an advanced book. The following summarizes recent activities in energy harvesting using piezoelectric materials.

3.1.1 Piezoelectric Bulk Power Generators

An early work at MIT Media Lab has investigated the feasibility of harnessing energy parasitically from various human activities (Starner 1996). It was later confirmed that energy generated by walking can be collected using piezoelectric ceramics (Shenck & Paradiso 2001). Since then, piezoelectric elements used for power harvesting in various forms of structures have been proposed to serve specific purposes. Subsequent studies for generating electricity from walking with loads were discussed by Rome et al. (2005) and Kuo (2005). Granstrom et al. (2007) investigated energy harvesting from a backpack instrumented with piezoelectric shoulder straps. (Elvin et al. 2001, 2003) and Ng & Liao (2005) used the piezo-electric element simultaneously as a power generator and a sensor. They evaluated the performance of the piezoelectric sensor to power wireless transmission and val-idated the feasibility of the self-powered sensor system. Elvin et al. (2006) further provided the evidence of ability of the harvesting electrical energy generated from the vibration of typical civil structures such as bridges and buildings. Roundy & Wright (2004) analyzed and developed a piezoelectric generator based on a two-layer bending element and used it as a basis for generator design optimization. Renaud et al. (2007) investigated the performances of a piezoelectric bender for impact or shock energy harvesting. Similar ideas based on cantilever-based devices using piezoelectric materials to scavenge vibration energy included the works by Ajitsaria1 et al. (2007), Cornwell et al. (2005), Hu et al. (2007), Jiang et al. (2005), Mateu & Moll (2005), Mossi et al. (2005), and Yoon et al. (2005).

Instead of 1D design, Kim et al. (2005a,b), and Ericka et al. (2005) have modeled and designed piezoelectric plates (membranes) to harvest energy from pulsing pres-sure sources. Yang et al. (2007) analyzed a rectangular plate piezoelectric generator. Guigon et al. (2008a) and Guigon et al. (2008b) studied the feasibility of scav-enging vibration energy from a piezoelectric plate impacted by water drop. Other harvesting schemes included the use of long strips of piezoelectric polymers (Energy Harvesting Eel) in ocean or river-water flows (Allen & Smits 2001, Taylor et al. 2001), the use of ionic polymer metal composites (IPMCs) as generating materials (Brufau-Penella et al. 2008), the use of piezoelectric "cymbal" transducers operated in the $\{3–3\}$ mode (Kim et al. 2004, 2005), the use of drum transducer (Wang et al. 2007), and the use of a piezoelectric windmill for generating electric power from wind energy (Priya 2005, Priya et al. 2005).

3.1.2 Piezoelectric Micro Power Generators

Jeon et al. (2005) and Choi et al. (2006) at MIT have successfully developed the first MEMS-based microscale power generator using a $\{3–3\}$ mode of PZT transducer. A 170 μm × 260 μm PZT beam has been fabricated and a maximum DC voltage of 3 V across the load 10.1 MΩ has been observed. In addition, the energy density of the power generator has been estimated at around 0.74 mW h/cm^2, which compares favorably to the use of lithium–ion batteries. Fang et al. (2006) subsequently fabri-cated another MEMS-based microscale power generator utilizing a PZT thick film as the transducer to harvest ambient vibration energy. The dimension of their beam is around 2000 μm × 500 μm (length × width) with 500 μm × 500 μm (length × height) metal mass. Different from the previous group, a $\{3–1\}$ piezo-mode is operated in their design. The natural frequency is amazingly reduced to only 609 Hz which is two orders of magnitude smaller than that observed by the previous group. However, the maximum AC voltage is only around 0.6 V which may be too low to overcome the forward bias of the rectifying bridge in order to convert AC to DC voltage.

In addition, Roundy et al. (2005) created prototyes of thin PZT structures with target volume power density of $80\,\mu\text{W/cm}^3$. Recently, duToit et al. (2005) and duToit & Wardle (2006) provided in-depth design principles for MEMS-scale piezoelectric energy harvesters and proposed a prototype of $30\,\mu\text{W/cm}^3$ from low-level vibration. Related issues on the modeling of miniaturized piezoelectric power harvesting devices include the works by duToit & Wardle (2007), Feng (2007), Horowitz et al. (2006), Lu et al. (2004), Prabhakar & Vengallatore (2007), Ramsay & Clark (2001), Trolier-Mckinstry & Muralt (2004), White et al. (2001), Xu et al. (2003), and Yeatman (2007).

3.1.3 Conversion Efficiency and Electrically Induced Damping

The efficiency of mechanical to electrical energy conversion is a fundamental parameter for the development and optimization of a power generation device. Umeda et al. (1996, 1997) have studied the efficiency of mechanical impact energy to electrical energy using a piezoelectric vibrator. Goldfarb & Jones (1999) subsequently investigated the efficiency of the piezoelectric material in a stack configuration for converting mechanical harmonic excitation into electrical energy. Roundy (2005) provided an expression for effectiveness that can be used to compare various approaches and designs for vibration-based energy harvesting devices (see also the work by Wang et al. (1999)). Recently, in contrast to efforts where the conversion efficiency was examined numerically (Umeda et al.1996), Richards et al. (2004) and Cho et al. (2006) derived an analytic formula to predict the energy conversion efficiency of piezoelectric energy harvesters in the case of AC power output. Since the electronic load requires a stabilized DC voltage while a vibrating piezoelectric element generates an AC voltage, the desired output needs to be rectified, filtered, and regulated to ensure the electric compatibility. Thus, Shu & Lien (2006b) investigated the conversion efficiency for a rectified piezoelectric power-harvesting system. They have shown that the conversion efficiency is dependent on the frequency ratio, the normalized resistance and, in particular, the ratio of electromechanical coupling coefficient to mechanical damping. In general, the conversion efficiency can be improved with a larger coupling coefficient and smaller damping. Recently, Cho et al. (2005a) and Cho et al. (2005b) performed a series of experiments and proposed a set of design guidelines for the performance optimization of micromachined piezoelectric membrane generators by enhancing the electromechanical coupling coefficient.

When an energy harvester is applied to a system, energy is removed from the vibrating structure and supplied to the desired electronic components, resulting in additional damping of the structure (Sodano et al. 2004). Because the efficiency is defined as the ratio of the time-averaged power dissipated across the load to that done by the external force, electrically-induced damping can be defined explicitly. Lesieutre et al. (2004) have investigated the damping added to a structure due to the removal of electrical energy from the system during power harvesting. They have shown that the maximum induced electric damping corresponds to the optimal power transfer in the case of weak electromechanical coupling. However, unlike the work by Lesieutre et al. (2004), Shu & Lien (2006b) provided a new finding showing that the optimal electric load maximizing the conversion efficiency and induced electric damping is very different from that maximizing the harvested power in strongly coupled electromechanical systems. This shows that optimization criteria vary according to the relative strength of the coupling.

3.1.4 Power Storage Circuits

The research works cited above focus mainly on developing optimal energy-harvesting structures. However, the electrical outputs of these devices in many cases are too small to power electric devices directly. Thus, the methods of accumulating and storing parasitic energy are also the key to develop self-powered systems. Sodano et al. (2005a,b) have investigated several piezoelectric power-harvesting devices and the methods of accumulating energy by utilizing either a capacitor or a rechargeable battery. Ottman et al. (2002, 2003) have developed highly efficient electric circuits to store the generated charge or present it to the load circuit. They have claimed that at high levels of excitation, the power output can be increased by as much as 400%. In contrast to the linear load impedance adaptation by Ottman et al. (2002, 2003), Guyomar et al. (2005), Lefeuvre et al. (2005a,b) and Badel et al. (2006b) developed a new power flow optimization principle based on the technique, called *synchronized switch harvesting on inductor* (SSHI), for increasing the converted energy. They claimed that the electric harvested power may be increased by as much as 900% over the standard technique. Badel et al. (2005) subsequently extended to the case of pulsed excitation and Makihara et al. (2006) improved the SSHI technique by proposing a low-energy dissipation circuit. Recently, Shu et al. (2007) provided an improved analysis for the performance evaluation of a piezoelectric energy harvesting system using the SSHI electronic interface. They found that the best use of the SSHI harvesting circuit is for systems in the mid-range of electromechanical coupling. The degradation in harvested power due to the non-perfect voltage inversion is not pronounced in this case, and the reduction in power is much less sensitive to frequency deviations than that using the standard technique.

3.2 Approach

3.2.1 Standard AC–DC Harvesting Circuit

Consider an energy conversion device which includes a vibrating piezoelectric structure together with an energy storage system. If the modal density of such a

device is widely separated and the structure is vibrating at around its resonance frequency, we may model the power generator as a mass+spring+damper+piezo structure, as shown in Fig. 3.1 (Guyomar et al. 2005, Richards et al. 2004). It consists of a
piezoelectric element coupled to a mechanical structure. In this approach, a forcing
function $F(t)$ is applied to the system and an effective mass M is bounded on a
spring of effective stiffness K, on a damper of coefficient η_m, and on a piezoelectric
element characterized by effective piezoelectric coefficient Θ and capacitance C_p.
These effective coefficients are dependent on the material constants and the design
of energy harvesters and can be derived using the standard modal analysis (Hagood
et al. 1990, Wang & Cross 1999).

For example, consider a triple-layer bender mounted as a cantilever beam with
polarization poled along the thickness direction, as shown in Fig. 3.2. The electric field is generated through the direction of thickness of the piezoelectric layers,
while strain is in the axial direction; consequently, the transverse, or {3–1}, mode
is utilized. The effective coefficients related with material constants and structural
geometry can be derived using the modal analysis (Shu & Lien 2006a).

$$M = \beta_M(m_p + m_b) + m_a,$$

$$K = \beta_K \, S \left\{ \left(\frac{2}{3} \frac{t^3}{L^3} + \frac{ht^2}{L^3} + \frac{1}{2} \frac{th^2}{L^3} \right) c^E_{p_{11}} + \frac{1}{12} \frac{h^3}{L^3} c^E_{b_{11}} \right\},$$

$$\Theta = \beta_\Theta \frac{S(h+t)}{2L} e_{31},$$

$$C_p = \frac{SL}{2t} \varepsilon^S_{33},$$

where β_M, β_K, and β_Θ are constants derived from the Rayleigh–Ritz approximation,
e_{31} and ε^S_{33} are the piezoelectric and clamped dielectric constants, respectively, S and

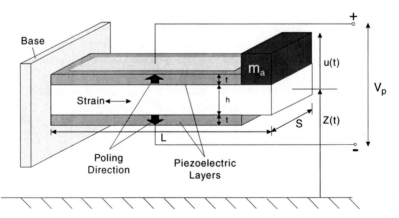

Fig. 3.2 A common piezoelectric-based power generator: a cantilever triple-layer bender operated
in the {3–1} mode. The base is excited with acceleration $\ddot{z}(t)$

L are the width and axial length of the cantilever beam, respectively, t and h, $C^E_{p_{11}}$ and $C^E_{b_{11}}$, m_p and m_b are the thicknesses, elastic moduli, and masses of the piezo-electric and central passive layers, and m_a is the attached mass. Another common piezoelectric power generator operated in the longitudinal or $\{3\text{--}3\}$ mode has been developed recently by Jeon et al. (2005) using interdigitated electrode configuration, as shown in Fig. 3.7. The advantage of utilizing this mode is that the longitudinal piezoelectric effect is usually much larger than the transverse effect ($d_{33} > d_{31}$).

Let u be the displacement of the mass M and V_p be the voltage across the piezoelectric element. The governing equations of the piezoelectric vibrator can be described by (Guyomar et al. 2005, Richards et al. 2004)

$$M\ddot{u}(t) + \eta_m\dot{u}(t) + Ku(t) + \Theta V_p(t) = F(t), \tag{3.1}$$

$$-\Theta\dot{u}(t) + C_p\dot{V}_p(t) = -I(t), \tag{3.2}$$

where $I(t)$ is the current flowing into the specified circuit as shown in Fig. 3.3. Since most applications of piezoelectric materials for power generation involve the use of periodic straining of piezoelectric elements, the vibrating generator is assumed to be driven at around resonance by the harmonic excitation

$$F(t) = F_0 \sin wt, \tag{3.3}$$

where F_0 is the constant magnitude and w (in radians per second) is the angular frequency of vibration.

The power generator considered here is connected to a storage circuit system, as shown in Fig. 3.1. Since the electrochemical battery needs a stabilized DC voltage while a vibrating piezoelectric element generates an AC voltage, this requires a suitable circuit to ensure the electric compatibility. Typically, an AC–DC rectifier followed by a filtering capacitance C_e is added to smooth the DC voltage, as shown in Fig. 3.1. A controller placed between the rectifier output and the battery is included to regulate the output voltage. Figure 3.3(a) is a simplified energy harvesting circuit commonly adopted for design analysis. It can be used to estimate an upper bound of the real power that the piezoelectric generator is able to deliver at a given excitation. Note that the regulation circuit and battery are replaced with an equivalent resistor R and V_c is the rectified voltage across it.

The common approach to have the stable output DC voltage is to assume that the filter capacitor C_e is large enough so that the rectified voltage V_c is essentially constant (Ottman et al. 2002). Specifically, $V_c(t) = \langle V_c(t)\rangle + V_{\text{ripple}}$, where $\langle V_c(t)\rangle$ and

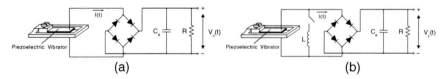

(a)   (b)

Fig. 3.3 (a) A standard energy harvesting circuit. (b) An SSHI energy harvesting circuit

V_{ripple} are the average and ripple of $V_c(t)$, respectively. This average $\langle V_c(t) \rangle$ is independent of C_e provided that the time constant RC_e is much larger than the oscillating period of the generator (Guyomar et al. 2005). The magnitude of V_{ripple}, however, depends on C_e and is negligible for large C_e. Under this hypothesis, $V_c(t) \approx \langle V_c(t) \rangle$, and therefore in the following, we use V_c, instead of $\langle V_c(t) \rangle$, to represent the average of $V_c(t)$ for notation simplicity.

The rectifying bridge shown in Fig. 3.3 is assumed to be perfect here. Thus, it is open circuited if the piezovoltage $|V_p|$ is smaller than the rectified voltage V_c. As a result, the current flowing into the circuit vanishes, and this implies $\dot{V}_p(t)$ varies proportionally with respect to $\dot{u}(t)$ as seen from (3.2). On the other hand, when $|V_p|$ reaches V_c, the bridge conducts and the piezovoltage is kept equal to the rectified voltage; i.e., $|V_p| = V_c$. Finally, the conduction in the rectifier diodes is blocked again when the absolute value of the piezovoltage $|V_p(t)|$ starts decreasing. Hence, the piezoelectric voltage $V_p(t)$ either varies proportionally with the displacement $u(t)$ when the rectifying bridge is blocking, or is kept equal to V_c when the bridge conducts.

The model equations (3.1, 3.2 and 3.3) are developed at the resonance mode of the device, and therefore, a single-mode vibration of the structure at the steady-state operation is expected with

$$u(t) = u_0 \sin (wt - \theta), \tag{3.4}$$

where u_0 is the magnitude and θ is the phase shift. This assumption of choosing the sinusoidal form for displacement has been made by Guyomar et al. (2005) excluding the effect of the phase shift θ. Shu & Lien (2006a) have included this effect and validated it both numerically and experimentally for the standard interface. The corresponding waveforms of $u(t)$ and $V_p(t)$ are shown in Fig. 3.4(a). Let $T = \frac{2\pi}{w}$ be the period of vibration, and t_i and t_f be two time instants ($t_f - t_i = \frac{T}{2}$) such that the displacement u undergoes from the minimum $-u_0$ to the maximum u_0, as shown in Fig. 3.4(a). Assumed that $\dot{V}_p \geq 0$ during the semi-period from t_i to t_f. It follows that $\int_{t_i}^{t_f} \dot{V}_p(t)dt = V_c - (-V_c) = 2V_c$. Note that $C_e \dot{V}_c(t) + \frac{V_c}{R} = 0$ for $t_i < t < t^*$ during which the piezovoltage $|V_p| < V_c$ and $I(t) = C_e \dot{V}_c(t) + \frac{V_c}{R}$ for $t^* \leq t < t_f$ during which the rectifier conducts. This gives

$$-\int_{t_i}^{t_f} I(t)dt = -\frac{T}{2}\frac{V_c}{R}$$

since the average current flowing through the capacitance C_e is zero; i.e., $\int_{t_i}^{t_f} C_e \dot{V}_c(t) dt = 0$ at the steady-state operation. The integration of (3.2) from time t_i to t_f is therefore

$$-2\Theta u_0 + 2C_p V_c = -\frac{T}{2}\frac{V_c}{R},$$

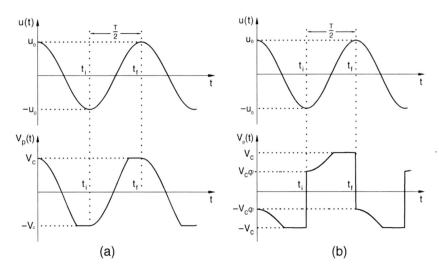

Fig. 3.4 Typical waveforms of displacement and piezoelectric voltage for (**a**) the standard and (**b**) the SSHI electronic interfaces

or

$$V_c = \frac{w\Theta R}{wC_p R + \frac{\pi}{2}} u_0. \tag{3.5}$$

Thus, from (3.5), u_0 has to be determined to decide V_c. There are three approaches in the recent literature for estimating it (Guyomar et al. 2005, Ottman et al. 2002, Shu & Lien 2006a). The first one models the piezoelectric device as the current source in parallel with its internal parasitic capacitance C_p (Jeon et al. 2005, Ng & Liao 2005, Ottman et al. 2002). It is based on the assumption that the internal current source of the generator is independent of the external load impedance. However, the amplitude of the current source is closely related to that of displacement which depends not only on the mechanical damping but also on the electrical damping at the resonant vibration (Lesieutre et al. 2004, Shu & Lien 2006b). This assumption is, therefore, not suitable when the effect of the electrical damping is significant. As a result, Guyomar et al. (2005) have proposed another estimation accounting for the effect of electromechanical coupling. Their estimation is based on the assumption that the external forcing function and the velocity of the mass are in-phase, or in other words, the phase shift effect is neglected in (3.4). Instead, Shu & Lien (2006a) have included this phase factor in their improved analysis, and derived the analytic expressions of displacement magnitude u_0, rectified voltage V_c, and harvested average power P. Their results are summarized as follows:

$$\bar{u}_0 = \frac{u_0}{\frac{F_0}{K}} = \frac{1}{\left\{ \left(2\zeta_m + \frac{2k_c^2 r}{(r\Omega + \frac{\pi}{2})^2} \right)^2 \Omega^2 + \left(1 - \Omega^2 + \frac{k_c^2 r\Omega}{r\Omega + \frac{\pi}{2}} \right)^2 \right\}^{\frac{1}{2}}}, \tag{3.6}$$

$$\overline{V}_c = \frac{V_c}{\frac{F_0}{\Theta}} = \left(\frac{r\Omega}{r\Omega + \frac{\pi}{2}}\right) \frac{k_e^2}{\left\{\left(2\zeta_m + \frac{2k_e^2 r}{(r\Omega+\frac{\pi}{2})^2}\right)^2 \Omega^2 + \left(1 - \Omega^2 + \frac{k_e^2 r\Omega}{r\Omega+\frac{\pi}{2}}\right)^2\right\}^{\frac{1}{2}}}, \quad (3.7)$$

$$\overline{P} = \frac{P}{\frac{F_0^2}{w_{sc}M}} = \frac{1}{\left(r\Omega + \frac{\pi}{2}\right)^2} \frac{k_e^2 \Omega^2 r}{\left\{\left(2\zeta_m + \frac{2k_e^2 r}{(r\Omega+\frac{\pi}{2})^2}\right)^2 \Omega^2 + \left(1 - \Omega^2 + \frac{k_e^2 r\Omega}{r\Omega+\frac{\pi}{2}}\right)^2\right\}}, \quad (3.8)$$

where several dimensionless variables are introduced by

$$k_e^2 = \frac{\Theta^2}{KC_p}, \quad \zeta_m = \frac{\eta_m}{2\sqrt{KM}}, \quad w_{sc} = \sqrt{\frac{K}{M}}, \quad \Omega = \frac{w}{w_{sc}}, \quad r = C_p w_{sc} R. \quad (3.9)$$

Above k_e^2 is the alternative electromechanical coupling coefficient, ζ_m is the mechanical damping ratio, w_{sc} is the natural oscillation frequency (of the piezo-electric vibrator under the short circuit condition), Ω and r are the normalized frequency and electric resistance, respectively. Note that there are two resonances for the system since the piezoelectric structure exhibits both short-circuit and open-circuit stiffness. They are defined by

$$\Omega_{sc} = 1, \quad \Omega_{oc} = \sqrt{1 + k_e^2}, \quad (3.10)$$

where Ω_{sc} and Ω_{oc} are the frequency ratios of short- and open-circuits, respectively. The shift in device natural frequency is pronounced if the coupling factor k_e^2 is large.

3.2.2 SSHI-Harvesting Circuit

An SSHI electronic interface consists of adding up a switch and an inductance L connected in series and is in parallel with the piezoelectric element, as shown in Fig. 3.3(b). The electronic switch is triggered according to the maximum and minimum of the displacement of the mass, causing the processing of piezoelectric voltage to be synchronized with the extreme values of displacement.

To illustrate the electrical behavior of this nonlinear processing circuit, consider the harmonic excitation given by (3.3). In view of the single-mode excitation, the mechanical displacement $u(t)$ is assumed to be sinusoidal as in (3.4) in steady-state operation. The validation of this assumption has been examined by considering the output voltage (Shu et al. 2007). The waveform of the piezoelectric voltage $V_p(t)$, however, may not be sinusoidal and is dependent on the specific type of the interface circuit connected to the piezoelectric element. To see it, let $T = \frac{2\pi}{w}$ be the period of mechanical excitation and t_i and t_f be two time instants such that the displacement $u(t)$ undergoes from the minimum $-u_0$ to the maximum u_0, as illustrated in Fig. 3.4(b). The switch is turned off most of time during this semi-period (t_i^+, t_f). When it is turned on at the time instant t_i, $|V_p(t)|$ remains lower than the rectified voltage V_c. So the rectifying bridge is open circuited, and an oscillating electrical circuit composed by the inductance L and the piezoelectric capacitance C_p is established, giving rise to an inversion process for the piezoelectric voltage V_p. Specifically, let Δt be the half electric period of this oscillating L-C_p circuit. It is equal to (Guyomar et al. 2005).

$$\Delta t = \pi \sqrt{LC_p}.$$

We assume that the inversion process is quasi-instantaneous in the sense that the inversion time is chosen to be much smaller than the period of mechanical vibration; i.e., $\Delta t = t_i^+ - t_i \ll T$. The switch is kept closed during this small time period Δt, resulting in the reverse of voltage on the piezoelectric element; i.e.,

$$V_p(t_i^+) = -V_p(t_i)e^{\frac{-\pi}{2Q_I}} = V_c q_1, \qquad q_1 = e^{\frac{-\pi}{2Q_I}}, \tag{3.11}$$

as shown in Fig. 3.4(b). Above Q_I is the inversion quality factor due to the energy loss mainly from the inductor in series with the switch. As a result, the current outgoing from the piezoelectric element through the rectifier during a half vibration period can be obtained by integrating (3.2) from time t_i^+ to t_f

$$\int_{t_i^+}^{t_f} \left\{ -\Theta \dot{u}(t) + C_p \dot{V}_p(t) \right\} dt = -2\Theta u_0 + C_p \left(1 - e^{-\frac{\pi}{2Q_I}} \right) V_c = -\frac{T}{2} \frac{V_c}{R},$$

since the rectifier bridge is blocking during the inversion process and the inversion time $\Delta t \ll T$. The relation between the magnitude of displacement u_0 and the rectified voltage V_c is therefore obtained by

$$V_c = \frac{2R\Theta w}{(1 - q_1)C_p Rw + \pi} u_0. \tag{3.12}$$

The rest of the derivation is to estimate the magnitude of displacement u_0 and the phase shift θ, and we refer to the work by Shu et al. (2007) for details. The results for normalized displacement magnitude $\overline{u}_0^{\text{SSHI}}$, rectified voltage $\overline{V}_c^{\text{SSHI}}$, and average-harvested power $\overline{P}^{\text{SSHI}}$ are given, respectively, by

$$\overline{u}_0^{\text{SSHI}} = \frac{u_0^{\text{SSHI}}}{\frac{F_0}{K}}$$

$$= \frac{1}{\left\{ \left(2\zeta_m + \frac{2\left[1 + \frac{r\Omega}{2\pi}(1-q_i^2)\right]k_e^2 r}{\left(\frac{(1-q_1)}{2}r\Omega + \frac{\pi}{2}\right)^2} \right)^2 \Omega^2 + \left(1 - \Omega^2 + \frac{\frac{(1-q_1)}{2}k_e^2 r\Omega}{\frac{(1-q_1)}{2}r\Omega + \frac{\pi}{2}} \right)^2 \right\}^{\frac{1}{2}}}, \tag{3.13}$$

$$\overline{V}_c^{SSHI} = \frac{V_c^{SSHI}}{\frac{F_0}{\Theta}}$$

$$= \left(\frac{r\Omega}{\frac{(1-q_1)}{2}r\Omega + \frac{\pi}{2}} \right)$$

$$\times \frac{k_e^2}{\left\{ \left(2\zeta_m + \frac{2\left[1 + \frac{r\Omega}{2\pi}(1-q_1^2)\right]k_e^2 r}{\left(\frac{(1-q_1)}{2}r\Omega + \frac{\pi}{2}\right)^2} \right)^2 \Omega^2 + \left(1 - \Omega^2 + \frac{\frac{(1-q_1)}{2}k_e^2 r\Omega}{\frac{(1-q_1)}{2}r\Omega + \frac{\pi}{2}} \right)^2 \right\}^{\frac{1}{2}}}, \quad (3.14)$$

$$\overline{P}^{SSHI} = \frac{P^{SSHI}}{\frac{F_0^2}{w_{sc}M}}$$

$$= \left(\frac{1}{\frac{(1-q_1)}{2}r\Omega + \frac{\pi}{2}} \right)^2$$

$$\times \frac{k_e^2\Omega^2 r}{\left(2\zeta_m + \frac{2\left[1 + \frac{r\Omega}{2\pi}(1-q_1^2)\right]k_e^2 r}{\left(\frac{(1-q_1)}{2}r\Omega + \frac{\pi}{2}\right)^2} \right)^2 \Omega^2 + \left(1 - \Omega^2 + \frac{\frac{(1-q_1)}{2}k_e^2 r\Omega}{\frac{(1-q_1)}{2}r\Omega + \frac{\pi}{2}} \right)^2}. \quad (3.15)$$

Above all are expressed in terms of dimensionless parameters defined in (3.9) and (3.11).

Finally, Guyomar et al. (2005) have used the in-phase assumption to analyze the electrical performance of the power generator using the SSHI interface. To be precise, they have assumed that the external forcing function and the velocity of the mass are in-phase, giving rise to no phase shift effect in their formulation. The following summaries their results for comparisons:

$$\overline{u}_{in\text{-}phase}^{SSHI} = \frac{u_{in\text{-}phase}^{SSHI}}{\frac{F_0}{K}} = \frac{1}{\left\{ 2\zeta_m + \frac{2\left[1 + \frac{r\Omega}{2\pi}(1-q_1^2)\right]k_e^2 r}{\left(\frac{(1-q_1)}{2}r\Omega + \frac{\pi}{2}\right)^2} \right\}\Omega}, \quad (3.16)$$

$$\overline{V}_{in\text{-}phase}^{SSHI} = \frac{V_{in\text{-}phase}^{SSHI}}{\frac{F_0}{\Theta}} = \left(\frac{r}{\frac{(1-q_1)}{2}r\Omega + \frac{\pi}{2}} \right) \frac{k_e^2}{\left\{ 2\zeta_m + \frac{2\left[1 + \frac{r\Omega}{2\pi}(1-q_1^2)\right]k_e^2 r}{\left(\frac{(1-q_1)}{2}r\Omega + \frac{\pi}{2}\right)^2} \right\}}, \quad (3.17)$$

$$\overline{P}_{in\text{-}phase}^{SSHI} = \frac{P_{in\text{-}phase}^{SSHI}}{\frac{F_0^2}{w_{sc}M}} = \frac{1}{\left(\frac{(1-q_1)}{2}r\Omega + \frac{\pi}{2}\right)^2} \frac{k_e^2 r}{\left\{ 2\zeta_m + \frac{2\left[1 + \frac{r\Omega}{2\pi}(1-q_1^2)\right]k_e^2 r}{\left(\frac{(1-q_1)}{2}r\Omega + \frac{\pi}{2}\right)^2} \right\}^2}. \quad (3.18)$$

3.3 Results

3.3.1 Standard Interface

The improved estimates in (3.6, 3.7 and 3.8) for the standard AC–DC interface have been found to agree well with experimental observations and numerical simulations of (3.1) and (3.2) under (3.3) (Shu & Lien 2006a). Therefore, these estimates are suitable for the electrical performance evaluation of the piezoelectric energy harvesting system embedded with the standard electronic interface. Basically from (3.8), the harvested average power increases significantly for smaller mechanical damping ratio ζ_m or larger electromechanical coupling coefficient k_e^2. It is consistent with that found by (Badel et al. 2006a), who have performed an interesting experiment by comparing the performances of vibration-based piezoelectric power generators using a piezoelectric ceramic and a single crystal. Under the same operating condition, the power generated using the single crystal is much higher than that using the ceramic, since according to their measurements the coupling factor k_e^2 of the former is 20 times larger than that of the latter. However, one has to be cautious that the average-harvested power approaches to its saturation value for much larger k_e^2, as shown in Fig. 3.5.

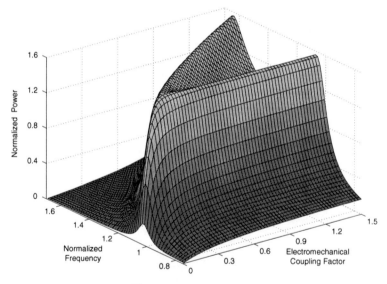

Fig. 3.5 The normalized power $\overline{P}$ against the normalized frequency Ω and the electromechanical coupling factor k_e^2 at the optimal condition in the sense that $\overline{P}^{\text{opt}}(\Omega, k_e^2, \zeta_m) = \overline{P}(r^{\text{opt}}(\Omega), \Omega, k_e^2, \zeta_m)$ and $r^{\text{opt}}(\Omega)$ is determined by solving $\frac{\partial}{\partial r}\overline{P}(r, \Omega, k_e^2, \zeta_m) = 0$. We use $\zeta_m = 0.04$ here. Note that for large k_e^2 there are two identical peaks of power evaluated at the frequency ratio close to $\Omega_{\text{sc}} = 1$ and $\Omega_{\text{oc}} = \sqrt{1 + k_e^2}$. These peaks are saturated for much higher coupling factor $k_e^2 >> 1$ (Shu et al. 2007)

The improved estimates in (3.6, 3.7 and 3.8) have also been compared with the uncoupled and in-phase estimates according to the relative magnitudes of electromechanical coupling coefficient and mechanical damping ratio. The results given by Shu & Lien (2006a) show that the conventional uncoupled solutions and in-phase estimates are suitable, provided that the ratio $\frac{k_e^2}{\zeta_m} << 1$, while the discrepancies among these distinct approaches become significant when $\frac{k_e^2}{\zeta_m}$ increases. If the shift in device natural frequency is pronounced and the mechanical damping ratio of the system is small; i.e. $\frac{k_e^2}{\zeta_m} >> 1$, the harvested power is shown to have two optimums evaluated at $(r_1^{opt}, \Omega_1^{opt})$ and $(r_2^{opt}, \Omega_2^{opt})$, where Ω_1^{opt} is close to Ω_{sc} and the electric load r_1^{opt} is very small, while Ω_2^{opt} is close to Ω_{oc} and r_2^{opt} is large. Indeed, Table 3.1 summarizes the relationship between the system parameters k_e^2 and ζ_m and the normalized load, displacement, voltage, and power at these two optimal conditions. The first optimal pair is designed at the short-circuit resonance Ω_{sc} with the optimal

Table 3.1 The relation between the system parameters k_e^2 and ζ_m and the normalized electric resistance, displacement, voltage, and power operated at the short-circuit (Ω_{sc}) and open-circuit (Ω_{oc}) resonances (Shu & Lien 2006a). Note that the condition $\frac{k_e^2}{\zeta_m} >> 1$ is implied in the analysis

Optimal conditions	Ω_{sc}		Ω_{oc}
Resistance	$r_{sc}^{opt} \propto \frac{1}{\frac{k_e^2}{\zeta_m}}$	$<$	$r_{oc}^{opt} \propto \frac{1}{(1+k_e^2)} \frac{k_e^2}{\zeta_m}$
Displacement	$\bar{u}_0^{opt} \propto \frac{1}{\zeta_m}$	$>$	$\bar{u}_0^{opt} \propto \frac{1}{\zeta_m(\sqrt{1+k_e^2})}$
Voltage	$\bar{V}_c^{opt} \propto 1$	$<$	$\bar{V}_c^{opt} \propto \frac{1}{\sqrt{1+k_e^2}} \frac{k_e^2}{\zeta_m}$
Power	$\bar{P}^{opt} \propto \frac{1}{\zeta_m}$	$=$	$\bar{P}^{opt} \propto \frac{1}{\zeta_m}$

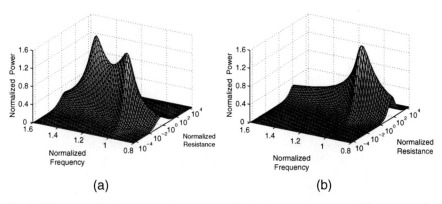

(a)   (b)

Fig. 3.6 The normalized power against the normalized electric resistance and frequency ratio. (a) A strongly coupled electromechanical system using the standard AC/DC electronic interface ($k_e^2 = 1.0$, $\zeta_m = 0.04$ $\frac{k_e^2}{\zeta_m} = 25$). (b) A weakly coupled electromechanical system using the ideal SSHI electronic interface ($k_e^2 = 0.01$, $\zeta_m = 0.04$, $\frac{k_e^2}{\zeta_m} = 0.25$, $Q_I = \infty$). Notice that both (a) and (b) provide the identical peaks of harvested power evaluated at different conditions (Shu et al. 2007)

load $r_{sc}^{opt} \propto \frac{1}{\frac{k_e^2}{\zeta_m}}$, while the second one is designed at the open-circuit resonance Ω_{oc}

with the optimal load $r_{oc}^{opt} \propto \frac{1}{(1+k_e^2)} \frac{k_e^2}{\zeta_m}$. They give the identical value of maximum harvested power which depends only on the mechanical damping ratio ζ_m. Unlike the power, the displacement is higher at Ω_{sc} than at Ω_{oc}, while the voltage operating at the first peak is one order of magnitude smaller than that operating at the second peak.

Figure 3.6(a) gives the dependence of the normalized harvested power on the normalized resistance and frequency ratio for the case of strong electromechanical coupling. While such a strong coupling is not commonly observed in the conventional piezoelectric power generators, we particularly emphasize it here since it has been shown by Shu et al. (2007) that the behavior of an ideal SSHI system is similar to that of a strongly coupled electromechanical standard system excited at around the short-circuit resonance. This finding is generally valid no matter whether the real electromechanical system is weakly or strongly coupled.

Discussions

If the vibration source is due to the periodic excitation of some base, this gives $F_0 = MA$ where A is the magnitude of acceleration of the exciting base. From (3.8), the harvested average power per unit mass becomes

$$\frac{P}{M} = \frac{A^2}{w_{sc}} \overline{P}(r, \Omega, k_e^2, \zeta_m). \tag{3.19}$$

As (3.19) is expressed in terms of a number of dimensionless parameters, an effective power normalization scheme is provided and can be used to compare power-harvesting devices of various sizes and with different vibration inputs to estimate efficiencies. Conceptually, the formula (3.19) provides a design guideline to optimize AC–DC power output either by tuning the electric resistance, selecting suitable operation points, or by adjusting the coupling coefficient by careful structural design. However, it needs much more efforts to make this scheme feasible due to the following various reasons.

(a) It may not be an easy task to adjust one parameter with other parameters fixed. For example, adjusting the dimensions of the device may result in the simultaneous changes of the whole dimensionless parameters r, Ω, k_e^2, and ζ_m.

(b) Current design requires that the natural frequency of the device is below 300 Hz since a number of common ambient sources have significant vibration components in the frequency range of 100–300 Hz (duToit et al. 2005). This adds a constraint for optimizing (3.19). Moreover, the most common geometry for piezoelectric power generators is the cantilever beam configuration. It is well known that its first resonance is proportional to the inverse of the beam length, causing the pronounced increase of natural frequency at the microscale. For example, the resonance frequency of the MEMS-based piezoelectric micro power generator developed by MIT has been measured as high as 13.9 kHz (Jeon et al. 2005). However, this order of magnitude of 10 kHz frequency is not low enough to meet the frequency range of common ambient sources, neither high enough to match that of supersonic transducers, for example.

(c) Operating the piezoelectric element in the {3–3} actuation mode is advantageous since better coupling between the mechanical and the electrical domain is possible ($d_{33} > d_{31}$ in general). Conventionally, longitudinal mode operation occurs through the use of interdigitated electrodes, as shown in Fig. 3.7, since a large component of the electric field can be produced in the axial direction. Based on this design, Jeon et al. (2005) have successfully developed the first MEMS-based piezoelectric power generator and found a maximum DC voltage of 3 V across the load 10.1 MΩ. However, another very recent paper in 2006 by Sodano et al. (2006) have observed the poor performance of bulk energy harvester with interdigitated electrode configuration. The reasons for these two seemingly contradictory experiments are not clear. One possible explanation is that the generated electric field decays significantly far away from the interdigitated electrode surface, causing the degradation of performance for bulk generators.

(d) In general, the harvested average power arises as the electromechanical coupling coefficient ascends at the early stage, as can be seen from Fig. 3.5. One proposed method to increase the coupling coefficient k_e^2 is to apply destabilizing axial loads, as shown in Fig. 3.8 (Lesieutre & Davis 1997). The idea behind it is that the beam's apparent stiffness is the function of the axial compressive preload; it theoretically reduces to zero as the axial preload approaches the critical bucking load. Using this idea, Leland & Wright (2006) have observed the coupling coefficient rising as much as 25%. However, the device damping also rises 67% which is not favorable for improving harvested power extraction. Hence, it needs more quantitative efforts to investigate this idea and therefore, to find out the suitable trade-off relationship (see also a recent investigation by Hu et al. (2007b)).

(e) Vibration-based power generators achieve the maximum power when their resonance frequency matches the driving frequency. However, the scavengeable power decreases significantly and almost goes away if the frequency deviation is more than 5% from the resonant frequency (Charnegie et al. 2006, Muriuki & Clark 2007, Shahruz 2006). Due to inconsistencies in the fabrication of the

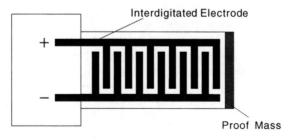

Fig. 3.7 An interdigitated electrode configuration

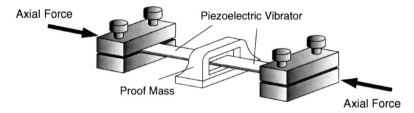

Fig. 3.8 The beam stiffness is able to be reduced by applying destabilizing compressive preloads

harvester or variations in the source, frequency matching can be difficult to achieve. Thus, it is obviously advantageous to have a single design operating effectively over a range of vibration frequencies, and Roundy et al. (2005) have suggested to use multi-mass and multi-mode resonators to enlarge the bandwidth of generators. In addition, Challa et al. (2008) have used a magnetic force technique to develop a resonance frequency tunable energy harvesting device.

(f) Note that (3.19) is an estimation of an upper bound of the real power that a piezoelectric energy harvester is able to deliver at a given excitation. Thus, research on power circuit designs can be viewed as efforts to improve the actual power extraction (but cannot increase its maximum power). The most common circuit design is to use the principle of load impedance adaptation by tuning the load impedance to achieve the higher power flow (Ottman et al. 2003). Guyomar and co-workers (Badel et al. 2006a, Badel et al. 2005, Guyomar et al. 2005, Lefeuvre et al. 2005b, 2006) have developed another new technique (SSHI) for increasing the converted energy, which has been discussed in Section 3.2.2. Recently, Liu et al. (2007) have investigated the electromechanical conversion capacity of a piezoelectric power generator by considering a quasi-static work cycle, and pointed out theoretically that their proposed method yields more power than SSHI.

3.3.2 SSHI Interface

The in-phase estimates in (3.16), (3.17) and (3.18) provided by Guyomar et al. (2005) are lack of frequency dependence. Thus, they are unable to predict the system behavior when the applied driving frequency deviates from the system resonance frequency. As the reduction in power is significant due to frequency deviation, such an effect cannot be ignored in practical design. The improved estimates in (3.13), (3.14) and (3.15) for the SSHI interface, on the other hand, exhibit the frequency dependence, and have also been validated numerically by Shu et al. (2007). Therefore, these estimates are suitable for the electrical performance evaluation of the piezoelectric energy harvesting system embedded with an SSHI interface circuit.

To see how the SSHI electronic interface boosts power extraction, consider an ideal case where the inversion of the piezoelectric voltage V_p is complete; i.e., $Q_I = \infty$. From (3.11) this gives $q_I = 1$ and the normalized harvested power from (3.15) becomes

$$\overline{P}^{\text{SSHI}} = \frac{4}{\pi^2} \frac{r k_e^2 \Omega^2}{\left\{ 4 \left(\zeta_m + \frac{4 k_e^2 r}{\pi^2} \right)^2 \Omega^2 + (1 - \Omega^2)^2 \right\}}. \tag{3.20}$$

The optimal electric load resistance and the normalized power operated at Ω_{sc} are therefore

$$r^{\text{opt}} = \frac{\pi^2}{4} \frac{1}{\frac{k_e^2}{\zeta_m}}, \qquad \overline{P}^{\text{SSHI}} \Big|_{r=r^{\text{opt}}, \Omega=1} = \frac{1}{16 \zeta_m}. \tag{3.21}$$

From (3.21), the optimal load resistance is inversely proportional to the ratio $\frac{k_e^2}{\zeta_m}$, while the corresponding optimal power depends only on the mechanical damping ratio ζ_m and is independent of the electromechanical coupling coefficient k_e^2. Comparing all of these features with Table 3.1 suggests that the behavior of the power-harvesting system using the SSHI interface is similar to that of a strongly coupled electromechanical system using the standard interface and operated at the short-circuit resonance Ω_{sc}. In addition, according to Table 3.1, there exists another identical peak of power operated at the open-circuit resonance. But Shu et al. (2007) have also shown that the second peak of power is moved to the infinite point in the (r, Ω) space, and therefore, there is only one peak of power for the SSHI electronic interface no matter whether the real electromechanical system is weakly or strongly coupled, as schematically shown in Fig. 3.6(b). Note that we particularly take $k_e^2 = 0.01$ and $\zeta_m = 0.04$ in Fig. 3.6(b) so that the electromechanical generator itself is weakly coupled $\left(\frac{k_e^2}{\zeta_m} = 0.25 \right)$. The harvested power obtained using the standard harvesting circuit is pretty small in this case, since it has been shown that (Shu & Lien 2006a)

$$\overline{P} \left(r^{\text{opt}} = \frac{\pi}{2}, \Omega = 1, k_e^2, \zeta_m \right) \approx \left(\frac{2}{\pi} \frac{k_e^2}{\zeta_m} \right) \frac{1}{16 \zeta_m} \ll \frac{1}{16 \zeta_m} = \left(\overline{P}^{\text{SSHI}} \right)_{\text{max}}$$

if $\frac{k_e^2}{\zeta_m} \ll 1$. But the inclusion of SSHI circuit boosts the average harvested power whose maximum is the same as that using a strongly coupled electromechanical generator connected to the standard interface, as illustrated in Fig. 3.6(a) ($k_e^2 = 1.0$, $\zeta_m = 0.04$ and $\frac{k_e^2}{\zeta_m} = 25$). Therefore, the harvested power increases tremendously for any weak coupling SSHI system at the cost of using a much larger optimal electric load which is proportional to $\frac{1}{\frac{k_e^2}{\zeta_m}}$ according to (3.21).

As in many practical situations, the inversion of the piezoelectric voltage V_p is not perfect ($Q_1 \neq \infty$), which accounts for a certain amount of the performance degradation using the SSHI electronic interface. We take $Q_1 = 2.6$ for the comparisons of the electrical performance of a vibration-based piezoelectric power generator using the standard and SSHI electronic interfaces according to the different ratios of $\frac{k_e^2}{\zeta_m}$. The results are shown in Fig. 3.9 (Shu et al. 2007). Note that it is possible to have a larger value of quality factor Q_1 by requiring the use of the low losses inductor (Lefeuvre et al. 2006).

To explain Fig. 3.9, first consider a weakly coupled electromechanical system; i.e., the ratio $\frac{k_e^2}{\zeta_m} \ll 1$. We take $k_e^2 = 0.01$ and $\zeta_m = 0.04$ for demonstration. This gives $\frac{k_e^2}{\zeta_m} = 0.25$. Comparing Fig. 3.9(a) with Fig. 3.9(d) gives the achieved optimal power using SSHI is three times larger than that using the standard interface ($\overline{P}^{SSHI}|_{Q_1=2.6} = 0.67$ and $\overline{P} = 0.23$). However, there is a significant performance degradation in this case since the maximum normalized power generated for the ideal voltage inversion is around $\overline{P}^{SSHI}|_{Q_1=\infty} = 1.56$.

Next, suppose the electromechanical coupling is in the medium range; i.e., the ratio of $\frac{k_e^2}{\zeta_m}$ is of order 1. We take $k_e^2 = 0.09$ and $\zeta_m = 0.04$. This gives $\frac{k_e^2}{\zeta_m} = 2.25$. Comparing Fig. 3.9(b) with Fig. 3.9(e) gives $\overline{P}^{SSHI}|_{Q_1=2.6} = 1.38$, the maximum normalized power for the non-ideal voltage inversion, and $\overline{P} = 1.20$, the maximum normalized power for the standard interface. While there is no significant increase of power output using the SSHI electronic interface in this case, Fig. 3.9(e) demonstrates that the harvested power evaluated at around the optimal load is less sensitive to frequency deviated from the resonant vibration. For example, the amount of normalized harvested power $\overline{P}$ evaluated at $r = \frac{\pi}{2}$ in the standard case drops from 1.2 to 0.6 for about 5% frequency deviation, and from 1.2 to 0.2 for about 10% frequency deviation. However, under the same conditions, the normalized harvested power $\overline{P}^{SSHI}$ in the SSHI circuit drops from 1.3 to only 1.0 for about 5% frequency deviation, and from 1.3 to 0.5 for about 10% frequency deviation. It has also been shown that this frequency-insensitive feature is much more pronounced if the quality factor Q_1 is further improved (Shu et al. 2007).

Finally, we turn to a strongly coupled electromechanical system $\left(\frac{k_e^2}{\zeta_m} \gg 1 \right)$. We take $k_e^2 = 1.0$ and $\zeta_m = 0.04$, and this gives $\frac{k_e^2}{\zeta_m} = 25$. The results are shown in Fig. 3.9(c) based on the standard interface and in Fig. 3.9(f) based on the SSHI interface. In the standard case, the harvested power has two identical optimal peaks, and the switching between these two peaks can be achieved by varying the electric loads. The envelope of these peaks has a local minimum, which is closely related to the minimum proof mass displacement. On the other hand, there is only one peak of power in the SSHI circuit, as explained previously. Unlike the standard case, as illustrated in Fig. 3.9(c), the peaks of the average harvested power decrease significantly as the load resistances increase, as shown in Fig. 3.9(f). In addition, it can be seen from (3.21) that the optimal electric load for the SSHI system is very small, since $\frac{k_e^2}{\zeta_m} \gg 1$. Thus, Fig. 3.9(f) indicates that any deviation in the load resistance will cause a significant power drop in the SSHI case. Such an effect cannot be ignored in practical design, since there may exist other inherent electrical damping in the whole circuit system; for example, the diode loss is not taken into account in the present analysis. As a result, there seems to be no obvious advantage of using the SSHI electronic interface from the comparison between Fig. 3.9(c) and Fig. 3.9(f).

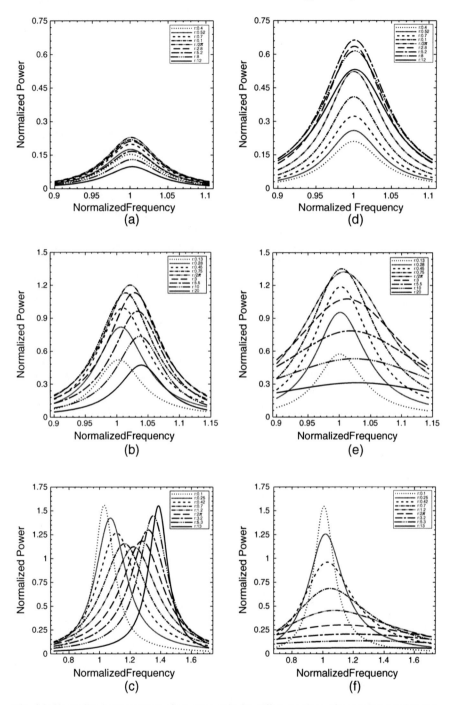

Fig. 3.9 Normalized power versus frequency ratio for different values of normalized resistances. Notice that (**a**)–(**c**) are obtained using the standard electronic interface, while (**d**)–(**f**) are obtained using the SSHI electronic interface (Shu et al. 2007)

3.4 Conclusion

This chapter presents a theory of piezoelectric vibration-based energy harvesting following the works by Shu & Lien (2006a), Shu & Lien (2006b) and Shu et al. (2007). The theory is able to predict the electrical behavior of piezoelectric power-harvesting systems using either the standard or the SSHI electronic interface. It shows that power extraction depends on the input vibration characteristics (frequency and acceleration), the mass of the generator, the electrical load, the natural frequency, the mechanical damping ratio, the electromechanical coupling coefficient of the system, and/or the inversion quality factor of an SSHI circuit. An expression of average harvested power incorporating all of these factors is analytically provided by (3.8) or (3.15) for the standard or SSHI interface. As the formula is expressed in terms of a number of dimensionless parameters, an effective power normalization scheme is provided and can be used to compare power harvesting devices of various sizes and with different vibration inputs to estimate efficiencies. Further, it is also highly recommended to provide all these parameters in all future publications to facilitate the relative comparison of various devices. Finally, the developed theory also points to opportunities for new devices and improvements in existing ones. For example, it shows that optimization criteria vary according to the relative strength of coupling, and scavenger bandwidth is improved by SSHI technique.

Acknowledgments The author is grateful to W. J. Wu and I. C. Lien for many pleasant collaborations and very helpful comments on this chapter. The author is glad to acknowledge the partial supports from National Science Council under Grant No. 96-2628-E-002-119-MY3, and from Ministry of Economic Affair under Grant No. 96-EC-17-A-05-S1-017 (WHAM-BioS).

References

Ajitsaria1, J., Choe, S. Y., Shen, D. and Kim, D. J. (2007). Modeling and Analysis of a Bimorph Piezoelectric Cantilever Beam for Voltage Generation, *Smart Materials and Structures* **16**: 447–454.

Allen, J. J. and Smits, A. J. (2001). Energy Harvesting EEL, *Journal of Fluids and Structures* **15**: 629–640.

Anton, S. R. and Sodano, H. A. (2007). A Review of Power Harvesting using Piezoelectric Materials (2003-2006), *Smart Materials and Structures* **16**: R1–R21.

Badel, A., Benayad, A., Lefeuvre, E., Lebrun, L., Richard, C. and Guyomar, D. (2006a). Single Crystals and Nonlinear Process for Outstanding Vibration-Powered Electrical Generators, *IEEE Transaction on Ultrasonics, Ferroelectrics, and Frequency Control* **53**: 673–684.

Badel, A., Guyomar, D., Lefeuvre, E. and Richard, C. (2005). Efficiency Enhancement of a Piezoelectric Energy Harvesting Device in Pulsed Operation by Synchronous Charge Inversion, *Journal of Intelligent Material Systems and Structures* **16**: 889–901.

Badel, A., Guyomar, D., Lefeuvre, E. and Richard, C. (2006b). Piezoelectric Energy Harvesting Using a Synchronized Switch Technique, *Journal of Intelligent Material Systems and Structures* **17**: 831–839.

Beeby, S. P., Torah, R. N., Tudor, M. J., Glynne-Jones, P., O'Donnell, T., Saha, C. R. and Roy, S. (2007). A Micro Electromagnetic Generator for Vibration Energy Harvesting, *Journal of Micromechanics and Microengineering* **17**: 1257–1265.

Brufau-Penella, J., Puig-Vidal, M., Giannone, P., Graziani, S. and Strazzeri, S. (2008). Characterization of the Harvesting Capabilities of an Ionic Polymer Metal Composite Device, *Smart Materials and Structures* **17**: 015009.

Challa, V. R., Prasad, M. G., Shi, Y. and Fisher, F. T. (2008). A Vibration Energy Harvesting Device with Bidirectional Resonance Frequency Tunability, *Smart Materials and Structures* **17**: 015035.

Chandrakasan, A., Amirtharajah, R., Goodman, J. and Rabiner, W. (1998). Trends in Low Power Digital Signal Processing, *International Symposium on Circuits and Systems* **4**: 604–607.

Charnegie, D., Mo, C., Frederick, A. A. and Clark, W. W. (2006). Tunable Piezoelectric Cantilever for Energy Harvesting, *Proceedings of 2006 ASME International Mechanical Engineering Congress and Exposition*, pp. IMECE2006–14431.

Cheng, S., Wang, N. and Arnold, D. P. (2007). Modeling of Magnetic Vibrational Energy Harvesters using Equivalent Circuit Representations, *Journal of Micromechanics and Microengineering* **17**: 2328–2335.

Cho, J., Anderson, M., Richards, R., Bahr, D. and Richards, C. (2005a). Optimization of Electromechanical Coupling for a Thin-Film PZT Membrane: I. Modeling, *Journal of Micromechanics and Microengineering* **15**: 1797–1803.

Cho, J., Anderson, M., Richards, R., Bahr, D. and Richards, C. (2005b). Optimization of Electromechanical Coupling for a Thin-Film PZT Membrane: II. Experiment, *Journal of Micromechanics and Microengineering* **15**: 1804–1809.

Cho, J. H., Richards, R. F., Bahr, D. F., Richards, C. D. and Anderson, M. J. (2006). Efficiency of Energy Conversion by Piezoelectrics, *Applied Physics Letter* **89**: 104107.

Choi, W. J., Jeon, Y., Jeong, J. H., Sood, R. and Kim, S. G. (2006). Energy Harvesting MEMS Device Based on Thin Film Piezoelectric Cantilevers, *Journal of Electroceramics* **17**: 543–548.

Cornwell, P. J., Goethal, J., Kowko, J. and Damianakis, M. (2005). Enhancing Power Harvesting Using a Tuned Auxiliary Structure, *Journal of Intelligent Material Systems and Structures* **16**: 825–834.

duToit, N. E. and Wardle, B. L. (2006). Performance of Microfabricated Piezoelectric Vibration Energy Harvesters, *Integrated Ferroelectrics* **83**: 13–32.

duToit, N. E. and Wardle, B. L. (2007). Experimental Verification of Models for Microfabricated Piezoelectric Vibration Energy Harvesters, *AIAA Journal* **45**: 1126–1137.

duToit, N. E., Wardle, B. L. and Kim, S. G. (2005). Design Considerations for MEMS-Scale Piezoelectric Mechanical Vibration Energy Harvesters, *Integrated Ferroelectrics* **71**: 121–160.

Elvin, N., Elvin, A. and Choi, D. H. (2003). A Self-Powered Damage Detection Sensor, *Journal of Strain Analysis* **38**: 115–124.

Elvin, N. G., Elvin, A. A. and Spector, M. (2001). A Self-Powered Mechanical Strain Energy Sensor, *Smart Materials and Structures* **10**: 293–299.

Elvin, N. G., Lajnef, N. and Elvin, A. A. (2006). Feasibility of Structural Monitoring with Vibration Powered Sensors, *Smart Materials and Structures* **15**: 977–986.

Ericka, M., Vasic, D., Costa, F., Poulin, G. and Tliba, S. (2005). Energy Harvesting from Vibration Using a Piezoelectric Membrane, *J. Phys. IV France* **128**: 187–193.

Fang, H. B., Liu, J. Q., Xu, Z. Y., Dong, L., Chen, D., Cai, B. C. and Liu, Y. (2006). A MEMS-Based Piezoelectric Power Generator for Low Frequency Vibration Energy Harvesting, *Chinese Physics Letters* **23**: 732–734.

Feng, G. H. (2007). A Piezoelectric Dome-shaped-diaphragm Transducer for Microgenerator Applications, *Smart Materials and Structures* **16**: 2636–2644.

Goldfarb, M. and Jones, L. D. (1999). On the Efficiency of Electric Power Generation with Piezoelectric Ceramic, *Trans. ASME, Journal of Dynamic Systems, Measurement, and Control* **121**: 566–571.

Granstrom, J., Feenstra1, J., Sodano, H. A. and Farinholt, K. (2007). Energy Harvesting from a Backpack Instrumented with Piezoelectric Shoulder Straps, *Smart Materials and Structures* **16**: 1810–1820.

Guan, M. J. and Liao, W. H. (2007). On the Efficiencies of Piezoelectric Energy Harvesting Circuits towards Storage Device Voltages, *Smart Materials and Structures* **16**: 498–505.

Guigon, R., Chaillout, J. J., Jager, T. and Despesse, G. (2008a). Harvesting Raindrop Energy: Experimental Study, *Smart Materials and Structures* **17**: 015039.

Guigon, R., Chaillout, J. J., Jager, T. and Despesse, G. (2008b). Harvesting Raindrop Energy: Theory, *Smart Materials and Structures* **17**: 015038.

Guyomar, D., Badel, A., Lefeuvre, E. and Richard, C. (2005). Toward Energy Harvesting Using Active Materials and Conversion Improvement by Nonlinear Processing, *IEEE Transaction on Ultrasonics, Ferroelectrics, and Frequency Control* **52**: 584–595.

Hagood, N. W., Chung, W. H. and Flotow, A. V. (1990). Modelling of Piezoelectric Actuator Dynamics for Active Structural Control, *Journal of Intelligent Material Systems and Structures* **1**: 327–354.

Horowitz, S. B., Sheplak, M., III, L. N. C. and Nishida, T. (2006). A MEMS Acoustic Energy Harvester, *Journal of Micromechanics and Microengineering* **16**: S174–S181.

Hu, H. P., Cao, J. G. and Cui, Z. J. (2007). Performance of a Piezoelectric Bimorph Harvester with Variable Width, *Journal of Mechanics* **23**: 197–202.

Hu, H. P., Xue, H. and Hu, Y. T. (2007a). A Spiral-Shaped Harvester with an Improved Harvesting Element and an Adaptive Storage Circuit, *IEEE Transaction on Ultrasonics, Ferroelectrics, and Frequency Control* **54**: 1177–1187.

Hu, Y. T., Xue, H. and Hu, H. P. (2007b). A Piezoelectric Power Harvester with Adjustable Frequency through Axial Preloads, *Smart Materials and Structures* **16**: 1961–1966.

Jeon, Y. B., Sood, R., Jeong, J. H. and Kim, S. G. (2005). MEMS Power Generator with Transverse Mode Thin Film PZT, *Sensors and Actuators* A **122**: 16–22.

Jiang, S., Li, X., Guo, S., Hu, Y., Yang, J. and Jiang, Q. (2005). Performance of a Piezoelectric Bimorph for Scavenging Vibration Energy, *Smart Materials and Structures* **14**: 769–774.

Johnson, T. J. and Clark, W. W. (2005). Harvesting Energy from Piezoelectric Material, *IEEE Pervasive Computing* **4**: 70–71.

Kansal, A. and Srivastava, M. B. (2005). Distributed Energy Harvesting for Energy-Neutral Sensor Networks, *IEEE Pervasive Computing* **4**: 69–70.

Kim, H. W., Batra, A., Priya, S., Uchino, K., Markley, D., Newnham, R. E. and Hofmann, H. F. (2004). Energy Harvesting Using a Piezoelectric "Cymbal" Transducer in Dynamic Environment, *Japanese Journal of Applied Physics* **43**: 6178–6183.

Kim, H. W., Priya, S., Uchino, K. and Newnham, R. E. (2005). Piezoelectric Energy Harvesting under High Pre-Stressed Cyclic Vibrations, *Journal of Electroceramics* **15**: 27–34.

Kim, S., Clark, W. W. and Wang, Q. M. (2005a). Piezoelectric Energy Harvesting with a Clamped Circular Plate: Analysis, *Journal of Intelligent Material Systems and Structures* **16**: 847–854.

Kim, S., Clark, W. W. and Wang, Q. M. (2005b). Piezoelectric Energy Harvesting with a Clamped Circular Plate: Experimental Study, *Journal of Intelligent Material Systems and Structures* **16**: 855–863.

Kuo, A. D. (2005). Harvesting Energy by Improving the Economy of Human Walking, *Science* **309**: 1686–1687.

Lee, C. K., Hsu, Y. H., Hsiao, W. H. and Wu, J. W. J. (2004). Electrical and Mechanical Field Interactions of Piezoelectric Systems: foundation of smart structures-based piezoelectric sensors and actuators, and free-fall sensors, *Smart Materials and Structures* **13**: 1090–1109.

Lefeuvre, E., Badel, A., Benayad, A., Lebrun, L., Richard, C. and Guyomar, D. (2005a). A Comparison between Several Approaches of Piezoelectric Energy Harvesting, *Journal de Physique IV. France* **128**: 177–186.

Lefeuvre, E., Badel, A., Richard, C. and Guyomar, D. (2005b). Piezoelectric Energy Harvesting Device Optimization by Synchronous Electric Charge Extraction, *Journal of Intelligent Material Systems and Structures* **16**: 865–876.

Lefeuvre, E., Badel, A., Richard, C., Petit, L. and Guyomar, D. (2006). A Comparison between Several Vibration-Powered Piezoelectric Generators for Standalone Systems, *Sensors and Actuators* A **126**: 405–416.

Leland, E. S. and Wright, P. K. (2006). Resonance Tuning of Piezoelectric Vibration Energy Scavenging Generators Using Compressive Axial Preload, *Smart Materials and Structures* **15**: 1413–1420.

Lesieutre, G. A. and Davis, C. L. (1997). Can a Coupling Coefficient of a Piezoelectric Device be Higher than Those of its Active Material?, *Journal of Intelligent Material Systems and Structures* **8**: 859–867.

Lesieutre, G. A., Ottman, G. K. and Hofmann, H. F. (2004). Damping as a Result of Piezoelectric Energy Harvesting, *Journal of Sound and Vibration* **269**: 991–1001.

Liao, W. H., Wang, D. H. and Huang, S. L. (2001). Wireless Monitoring of Cable Tension of Cable-Stayed Bridges Using PVDF Piezoelectric Films, *Journal of Intelligent Material Systems and Structures* **12**: 331–339.

Liu, W. Q., Feng, Z. H., He, J. and Liu, R. B. (2007). Maximum Mechanical Energy Harvesting Strategy for a Piezoelement, *Smart Materials and Structures* **16**: 2130–2136.

Lu, F., Lee, H. P. and Lim, S. P. (2004). Modeling and Analysis of Micro Piezoelectric Power Generators for Micro-Electro-Mechanical-Systems Applications, *Smart Materials and Structures* **13**: 57–63.

Makihara, K., Onoda, J. and Miyakawa, T. (2006). Low Energy Dissipation Electric Circuit for Energy Harvesting, *Smart Materials and Structures* **15**: 1493–1498.

Mateu, L. and Moll, F. (2005). Optimum Piezoelectric Bending Beam Structures for Energy Harvesting Using Shoe Inserts, *Journal of Intelligent Material Systems and Structures* **16**: 835–845.

Mossi, K., Green, C., Ounaies, Z. and Hughes, E. (2005). Harvesting Energy Using a Thin Unimorph Prestressed Bender: Geometrical Effects, *Journal of Intelligent Material Systems and Structures* **16**: 249–261.

Muriuki, M. G. and Clark, W. W. (2007). Analysis of a Technique for Tuning a Cantiliver Beam Resonator Using Shunt Switching, *Smart Materials and Structures* **16**: 1527–1533.

Nakano, K., Elliott, S. J. and Rustighi, E. (2007). A Unified Approach to Optimal Conditions of Power Harvesting using Electromagnetic and Piezoelectric Transducers, *Smart Materials and Structures* **16**: 948–958.

Ng, T. H. and Liao, W. H. (2005). Sensitivity Analysis and Energy Harvesting for a Self-Powered Piezoelectric Sensor, *Journal of Intelligent Material Systems and Structures* **16**: 785–797.

Ngo, K. D., Phipps, A., Nishida, T., Lin, J. and Xu, S. (2006). Power Converters for Piezoelectric Energy Extraction, *Proceedings of 2006 ASME International Mechanical Engineering Congress and Exposition*, pp. IMECE2006–14343.

Ottman, G. K., Hofmann, H. F., Bhatt, A. C. and Lesieutre, G. A. (2002). Adaptive Piezoelectric Energy Harvesting Circuit for Wireless Remote Power Supply, *IEEE Transactions on Power Electronics* **17**: 669–676.

Ottman, G. K., Hofmann, H. F. and Lesieutre, G. A. (2003). Optimized Piezoelectric Energy Harvesting Circuit Using Step-Down Converter in Discontinuous Conduction Mode, *IEEE Transactions on Power Electronics* **18**: 696–703.

Poulin, G., Sarraute, E. and Costa, F. (2004). Generation of Electric Energy for Portable Devices: Comparative Study of an Electromagnetic and a Piezoelectric system, *Sensors and Actuators* A **116**: 461–471.

Prabhakar, S. and Vengallatore, S. (2007). Thermoelastic Damping in Bilayered Micromechanical Beam Resonators, *Journal of Micromechanics and Microengineering* **17**: 532–538.

Priya, S. (2005). Modeling of Electric Energy Harvesting Using Piezoelectric Windmill, *Applied Physics Letters* **87**: 184101.

Priya, S., Chen, C. T., Fye, D. and Zahnd, J. (2005). Piezoelectric Windmill: A Novel Solution to Remote Sensing, *Japanese Journal of Applied Physics* **44**: L104–L107.

Rabaey, J. M., Ammer, M. J., da Silva Jr., J. L., Patel, D. and Roundy, S. (2000). PicoRadio Supports Ad Hoc Ultra-Low Power Wireless Networking, *Computer* **33**: 42–48.

Ramsay, M. J. and Clark, W. W. (2001). Piezoelectric Energy Harvesting for Bio MEMS Applications, *Proceedings of the SPIE*, Vol. **4332**, pp. 429–438.

Renaud, M., Fiorini, P. and Hoof, C. V. (2007). Optimization of a Piezoelectric Unimorph for Shock and Impact Energy Harvesting, *Smart Materials and Structures* **16**: 1125–1135.

Richards, C. D., Anderson, M. J., Bahr, D. F. and Richards, R. F. (2004). Efficiency of Energy Conversion for Devices Containing a Piezoelectric Component, *Journal of Micromechanics and Microengineering* **14**: 717–721.

Richter, B., Twiefel, J., Hemsel, T. and Wallaschek, J. (2006). Model based Design of Piezoelectric Generators Utilizing Geometrical and Material Properties, *Proceedings of 2006 ASME International Mechanical Engineering Congress and Exposition*, pp. IMECE2006–14862.

Rome, L. C., Flynn, L., Goldman, E. M. and Yoo, T. D. (2005). Generating Electricity while Walking with Loads, *Science* **309**: 1725–1728.

Roundy, S. (2005). On the Effectiveness of Vibration-Based Energy Harvesting, *Journal of Intelligent Material Systems and Structures* **16**: 809–823.

Roundy, S., Leland, E. S., Baker, J., Carleton, E., Reilly, E., Lai, E., Otis, B., Rabaey, J. M., Wright, P. K. and Sundararajan, V. (2005). Improving Power Output for Vibration-Based Energy Scavengers, *IEEE Pervasive Computing* **4**: 28–36.

Roundy, S., Steingart, D., Frechette, L., Wright, P. and Rabaey, J. (2004a). Power Sources for Wireless Sensor Networks, *Lecture Notes in Computer Science* **2920**: 1–17.

Roundy, S. and Wright, P. K. (2004). A Piezoelectric Vibration Based Generator for Wireless Electronics, *Smart Materials and Structures* **13**: 1131–1142.

Roundy, S., Wright, P. K. and Rabaey, J. (2003). A Study of Low Level Vibrations as Power Source for Wireless Sensor Nodes, *Computer Communications* **26**: 1131–1144.

Roundy, S., Wright, P. K. and Rabaey, J. M. (2004b). *Energy Scavenging for Wireless Sensor Networks with Special Focus on Vibrations*, Kluwer Academic Publishers, Boston.

Sebald, G., Pruvost, S. and Guyomar, D. (2008). Energy Harvesting based on Ericsson Pyroelectric Cycles in a Relaxor Ferroelectric Ceramic, *Smart Materials and Structures* **15**: 015012.

Shahruz, S. M. (2006). Design of Mechanical Band-Pass Filters with Large Frequency Bands for Energy Scavenging, *Mechatronics* **16**: 523–531.

Shenck, N. S. and Paradiso, J. A. (2001). Energy Scavenging with Shoe-Mounted Piezoelectrics, *IEEE Micro* **21**: 30–42.

Shu, Y. C. and Lien, I. C. (2006a). Analysis of Power Output for Piezoelectric Energy Harvesting Systems, *Smart Materials and Structures* **15**: 1499–1512.

Shu, Y. C. and Lien, I. C. (2006b). Efficiency of Energy Conversion for a Piezoelectric Power Harvesting System, *Journal of Micromechanics and Microengineering* **16**: 2429–2438.

Shu, Y. C., Lien, I. C. and Wu, W. J. (2007). An Improved Analysis of the SSHI Interface in Piezoelectric Energy Harvesting, *Smart Materials and Structures* **16**: 2253–2264.

Sodano, H. A., Inman, D. J. and Park, G. (2004). A Review of Power Harvesting from Vibration Using Piezoelectric Materials, *The Shock and Vibration Digest* **36**: 197–205.

Sodano, H. A., Inman, D. J. and Park, G. (2005a). Comparison of Piezoelectric Energy Harvesting Devices for Recharging Batteries, *Journal of Intelligent Material Systems and Structures* **16**: 799–807.

Sodano, H. A., Inman, D. J. and Park, G. (2005b). Generation and Storage of Electricity from Power Harvesting Devices, *Journal of Intelligent Material Systems and Structures* **16**: 67–75.

Sodano, H. A., Lloyd, J. and Inman, D. J. (2006). An Experimental Comparison between Several Active Composite Actuators for Power Generation, *Smart Materials and Structures* **15**: 1211–1216.

Sodano, H. A., Park, G. and Inman, D. J. (2004). Estimation of Electric Charge Output for Piezoelectric Energy Harvesting, *Journal of Strain* **40**: 49–58.

Starner, T. (1996). Human-Powered Wearable Computing, *IBM Systems Journal* **35**: 618–629.

Stephen, N. G. (2006a). On Energy Harvesting from Ambient Vibration, *Journal of Sound and Vibration* **293**: 409–425.

Stephen, N. G. (2006b). On Energy Harvesting from Ambient Vibration, *Proceedings of the Institution of Mechanical Engineers Part C - Journal of Mechanical Engineering Science* **220**: 1261–1267.

Taylor, G. W., Burns, J. R., Kammann, S. M., Powers, W. B. and Welsh, T. R. (2001). The Energy Harvesting Eel: A Small Subsurface Ocean/River Power Generator, *IEEE Journal of Oceanic Engineering* **26**: 539–547.

Trolier-Mckinstry, S. and Muralt, P. (2004). Thin Film Piezoelectrics for MEMS, *Journal of Electroceramics* **12**: 7–17.

Umeda, M., Nakamura, K. and Ueha, S. (1996). Analysis of the Transformation of Mechanical Impact Energy to Electric Energy Using Piezoelectric Vibrator, *Japanese Journal of Applied Physics* **35**: 3267–3273.

Umeda, M., Nakamura, K. and Ueha, S. (1997). Energy Storage Characteristics of a Piezo-Generator Using Impact Induced Vibration, *Japanese Journal of Applied Physics* **36**: 3146–3151.

Wang, Q. M. and Cross, L. E. (1999). Constitutive Equations of Symmetrical Triple Layer Piezoelectric Benders, *IEEE Transaction on Ultrasonics, Ferroelectrics, and Frequency Control* **46**: 1343–1351.

Wang, Q. M., Du, X. H., Xu, B. and Cross, L. E. (1999). Electromechanical Coupling and Output Efficiency of Piezoelectric Bending Actuators, *IEEE Transaction on Ultrasonics Ferroelectrics and Frequency Control* **46**: 638–646.

Wang, S., Lam, K. H., Sun, C. L., Kwok, K. W., Chan, H. L. W., Guo, M. S. and Zhao, X. Z. (2007). Energy Harvesting with Piezoelectric Drum Transducer, *Applied Physics Letters* **90**: 113506.

Whalen, S., Thompson, M., Bahr, D., Richards, C. and Richards, R. (2003). Design, Fabrication and Testing of the P^3 Micro Heat Engine, *Sensors and Actuators* A **104**: 290–298.

White, N. M., Glynne-Jones, P. and Beeby, S. P. (2001). A Novel Thick-Film Piezoelectric Micro-Generator, *Smart Materials and Structures* **10**: 850–852.

Williams, C. B. and Yates, R. B. (1996). Analysis of a Micro-Electric Generator for Microsystems, *Sensors and Actuators* A **52**: 8–11.

Xu, C. G., Fiez, T. S. and Mayaram, K. (2003). Nonlinear Finite Element Analysis of a Thin Piezoelectric Laminate for Micro Power Generation, *Journal of Microelectromechanical Systems* **12**: 649–655.

Yang, J., Chen, Z. and Hu, Y. T. (2007). An Exact Analysis of a Rectangular Plate Piezoelectric Generator, *IEEE Transaction on Ultrasonics Ferroelectrics and Frequency Control* **54**: 190–195.

Yeatman, E. M. (2007). Applications of MEMS in Power Sources and Circuits, *Journal of Micromechanics and Microengineering* **17**: S184–S188.

Yoon, H. S., Washington, G. and Danak, A. (2005). Modeling, Optimization, and Design of Efficient Initially Curved Piezoceramic Unimorphs for Energy Harvesting Applications, *Journal of Intelligent Material Systems and Structures* **16**: 877–888.

Zhao, X. and Lord, D. G. (2006). Application of the Villari Effect to Electric Power Harvesting, *Journal of Applied Physics* **99**: 08M703.

Chapter 4
Piezoelectric Equivalent Circuit Models

Björn Richter, Jens Twiefel and Jörg Wallaschek

Abstract Electromechanical equivalent circuits can be used to model the dynamics of piezoelectric systems. In the following, they will be applied for the modeling of piezoelectric bending generators for energy harvesting. Therefore, the basic analogies between electrical and mechanical systems will be discussed and a simple piezoelectric equivalent circuit model for a system which can be described by a single mechanical modal coordinate will be derived. In a next step, an experimentally based method for the determination of the model parameters will be presented. The modeling of additional mechanical degrees of freedom as well as the modeling of force and kinematic base excitation will also be addressed.

4.1 Model Based Design

Models are used in different stages of the design process. In the early stages, the emphasis is on conceptual modeling and system representations of low complexity are preferred. Later, more detailed models, which take all relevant design parameters into account, will be needed. Piezoelectric equivalent circuit models are typically employed in overall system studies in the early design phases.

Equivalent circuit models can be used for analysis and design of piezoelectric systems. The parameters which enter into these models can be obtained by various methods. Among these methods, the 'experimental' parameter identification which is based on the systems' transfer functions. It can be applied whenever a physical prototype is available. Of course, it is also applicable if the data of the transfer function are gained by numerical calculations. In the early design phases, when the overall system is synthesized, prototypes are in general not available. In this situation, the model parameters can be estimated using certain analytical solutions of the Euler–Bernoulli beam equation. The experimental parameter identification as

J. Wallaschek (✉)
Gottfried Wilhelm Leibniz Universität Hannover, D-30167 Hannover, Germany

S. Priya, D.J. Inman (eds.), *Energy Harvesting Technologies*,
DOI 10.1007/978-0-387-76464-1_4 © Springer Science+Business Media, LLC 2009

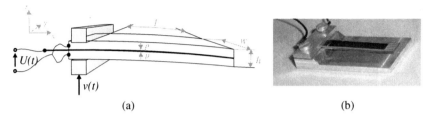

(a)   (b)

Fig. 4.1 Piezoelectric parallel bimorph cantilever used as generator element. (**a**) General configuration. (**b**) Demonstrator system

well as the parameter estimation based on the analytical model will be described in the subsequent sections of this chapter.

4.1.1 Basic Configurations of Piezoelectric Generators

Depending on the application, different types of piezoelectric generators (PEG) can be used for energy harvesting, e.g., stack, bimorph, membrane or spiral rotational springs. Each of these configurations has specific advantages and design limits. Most often, resonance is used in the mechanical–electrical energy conversion and the systems are typically operated at vibration frequencies up to a few hundred Hertz.

The most widespread PEGs are bimorph cantilever structures which have eigenfrequencies in the range of a few hundred Hertz and are easy to handle. Piezoelectric bimorphs consist of two or more piezoelectric layers mounted on a passive metallic or ceramic layer (Tokin Corp. 1987). The piezoelectric layers are connected in series for high-output impedance and they are connected in parallel for low-output impedance, as generally desired for driving electric circuits. Figure 4.1 shows a generator based on a cantilever element.

4.2 Linear Constitutive Equations for Piezoelectric Material

According to IEEE Std 176 (1978), the linear constitutive equations for piezoelectric materials are given by Eq. 4.1.

$$\begin{pmatrix} \mathbf{S} \\ \mathbf{D} \end{pmatrix} = \begin{pmatrix} \mathbf{s}^{\mathbf{E}} & \mathbf{d} \\ \mathbf{d_t} & \boldsymbol{\varepsilon}^{\mathbf{T}} \end{pmatrix} \begin{pmatrix} \mathbf{T} \\ \mathbf{E} \end{pmatrix} \tag{4.1}$$

$\mathbf{S}$ is the mechanical strain, $\mathbf{T}$ is the mechanical stress, $\mathbf{D}$ is the dielectric charge displacement, and $\mathbf{E}$ is the electrical field strength. These quantities are tensors of order 2 and 1, respectively. $\mathbf{s}^{\mathbf{E}}$ is the compliance tensor under the condition of a constant electric field defined as strain generated per unit stress. The tensor $\mathbf{d}$ contains the piezoelectric charge constants; it gives the relationship between electric charge and mechanical stress. The representation can be used for both the direct and the inverse piezoelectric effects. Further, $\boldsymbol{\varepsilon}^T$ gives the absolute permittivity defined as dielectric displacement per unit electric field for constant stress. For most practical applications, there exists a dominant deformation mode and these equations can be reduced to scalar form. If, for example, the piezoelectric material is polarized in 3-direction and the electrical field is applied in the same direction, while the predominant mechanical stress and strain is in the 1-direction, the constitutive equations can be reduced to Eq. 4.2.

$$
\begin{aligned}
S_1 &= s_{11}^E T_1 + d_{31} E_3 \\
D_3 &= d_{31} T_1 + \varepsilon_{33}^T E_3
\end{aligned}
\tag{4.2}
$$

The constitutive Eq. 4.2 can be transformed into a form where the mechanical strain and the dielectric charge displacement are being considered as independent variables.

$$
\begin{pmatrix} \mathbf{T} \\ \mathbf{E} \end{pmatrix} = \begin{pmatrix} \mathbf{c}^D & \mathbf{h} \\ \mathbf{h}_t & \boldsymbol{\beta}^S \end{pmatrix} \begin{pmatrix} \mathbf{S} \\ \mathbf{D} \end{pmatrix}
\tag{4.3}
$$

The coefficients (cp. IEEE Std 176 (1978)) in this representation can, of course, be expressed in terms of the coefficients in Eq. 4.1. In the special case of Eq. 4.2, the corresponding form of Eq. 4.3 is

$$
\begin{aligned}
T_1 &= \frac{1}{s_{11}^E(1 - k_{31}^2)} S_1 - \frac{1}{d_{31}} \left(\frac{k_{31}^2}{1 - k_{31}^2} \right) D_3 \\
E_3 &= -\frac{1}{d_{31}} \left(\frac{k_{31}^2}{1 - k_{31}^2} \right) S_1 + \frac{1}{\varepsilon_{33}^T(1 - k_{31}^2)} D_3
\end{aligned}
\tag{4.4}
$$

with the piezoelectric coupling factor

$$
k_{31}^2 = \frac{d_{31}^2}{\varepsilon_{33}^T s_{11}^E}
\tag{4.5}
$$

This particular representation of the material law will be used in a later section.

4.3 Piezoelectric Equivalent Circuit Models for Systems with Fixed Mechanical Boundary

In this section, we directly introduce a piezoelectric equivalent circuit model by considering a simple relationship between local and global state variables of the system. The equivalent circuit model is based on an analogy between mechanical and electrical state variables. Only piezoelectric systems which have a fixed mechanical boundary are studied in this chapter, e.g., a cantilever or simply supported beam. The case of base excitation will be studied in a subsequent section.

4.3.1 Quasi-Static Regime

The linear constitutive equations describe the behavior of the piezoelectric material on a "local" scale in terms of mechanical stress and strain and dielectric charge displacement and electrical field strength. The corresponding "global" state variables of the system are force, displacement, charge flow, and voltage. Local and global state variables are interrelated by linear mathematical operators, involving differentiation and integration, respectively. The mechanical strain, e.g., is determined by the derivatives of the displacement field which itself can in most cases be characterized by a single generalized coordinate x. If the piezoelectric element is fixed in space at one of its mechanical boundaries, x can be directly related to the displacement of a characteristic point of the system, e.g., the tip displacement of an actuator. The charge flow is determined by the integral of the dielectric charge displacement on the electrode surfaces of the piezoelectric system. The electric field is determined by the gradient of the electrical potential which itself depends on the voltage difference between the electrical ports (electrodes) of the element.

As a consequence, the quasi-static electro-mechanical behavior of a piezoelectric system can, on a global scale, be described by

$$F = c\,x - \alpha U$$
$$Q = \alpha\,x + CU \tag{4.6}$$

where c is the mechanical stiffness of the system with short-circuited electrodes and C is the electrical capacitance of the mechanically unconstrained system. U is the applied electric voltage, Q is the charge, and F is the force. The coordinate x represents the mechanical displacement and the electro-mechanical coupling is expressed by the coupling factor α which has the dimension N/V or As/m. Note that dynamical effects (acceleration and damping terms in the equations of motion) have been neglected and that Eq. 4.6 is only valid for quasi-static conditions, i.e., frequencies which are far below the first eigenfrequency of the piezoelectric system.

Using mechanical (or electrical) standard elements, Eq. 4.6 can be displayed as a system of springs and ideal levers or as a system of capacitors and transformers. There are two common analogies between the electrical and the mechanical domains. In the first one, the mechanical force is considered to be analog to the electric voltage and the velocity is equivalent to the electric current. In the second analogy, the mechanical force is considered to be analog to the electric current and the velocity is equivalent to the electric voltage. The second analogy is used mainly for electromagnetic systems, while the first one is typically used for piezoelectric systems. The corresponding analogies for the circuit parameters are shown in Table 4.1. In the following only the first analogy will be used.

Using the first analogy, the mechanical and electrical representations of Eq. 4.6 can be sketched easily, cp. Fig. 4.2. The system contains two springs (one represents the structural stiffness and the other one is the capacitive behavior of the piezoelectric material) and an ideal lever with the ratio $1 : \alpha$. This representation is

Table 4.1 Electro-mechanical analogies

	Force	Velocity	Mass	Spring	Damper
First Analogy	Voltage	Current	Inductance	1/Capacitance	Resistance
Second Analogy	Current	Voltage	Capacitance	1/Inductance	1/Resistance

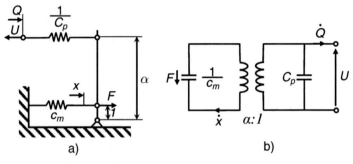

a)   b)

Fig. 4.2 Quasi-static equivalent model of piezoelectric systems: (**a**) mechanical representation and (**b**) electrical representation

independent of the geometrical shape and material parameters of the piezoelectric system. It can be used for stacked or multilayer longitudinal elements as well as for beam or plate benders. In any case, Eq. 4.2 and Fig. 4.2 give a valid representation of the systems' behavior as long as dynamic effects can be neglected and the deformation of the system can be represented by a single mechanical degree of freedom.

4.3.2 Single Degree of Freedom Model for Dynamic Regime

Structural damping and dielectric losses can be taken into account by introducing damper elements in parallel to the springs. And inertia effects finally can be modeled by introducing the equivalent mass m. Thus the system shown in Fig. 4.3 results; its equations of motion are

$$m\ddot{x} + d\dot{x} + cx = F + \alpha U$$
$$\frac{1}{C}[Q - \alpha x] + R[\dot{Q} - \alpha \dot{x}] = \dot{U} \tag{4.7}$$

Depending on the boundary conditions of the system and the choice of the generalized coordinate x, the equivalent mass m represents the modal mass associated with the deformation mode of the system. This can be a static deformation mode if frequencies below the first eigenfrequency are considered. If the equivalent circuit model is used to represent the behavior of the piezoelectric system in a certain neighborhood of one of its resonance frequencies, it can be one of the eigenfunctions of the system, either.

The particular topology of Fig. 4.3 is valid for the case of fixed mechanical boundary conditions of the piezoelectric system, i.e., for a cantilever beam actuator

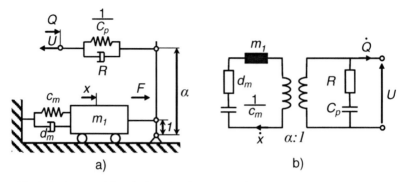

Fig. 4.3 Equivalent models of piezoelectric systems in mechanical (**a**) and electrical (**b**) representation

which is fixed at one end. In this case, the modal coordinate x can be chosen as the tip displacement. We will later also discuss the case of a base-excited system.

The parameters of the mechanical model are the modal stiffness c_m, modal damping d_m, and modal mass m_1. The parameters describing the electrical properties are the shunted capacitance C_p and resistor R, which represents dielectric losses.

Equivalent circuits as model for the systems behavior at the electrical ports have already been used by Cady (1922) for modeling a lossless resonator. He represented the mechanical properties of the piezoelectric system by an inductance and a capacitor in parallel to an electric capacitance representing the electrical properties of the system. Cady's model was later expanded by Van Dyke (1925), who included mechanic losses – modeled by a resistor – resulting in an RLC-series network in parallel to the capacitor of the electrical system.

The models of Cady and Van Dyke describe the electromechanical coupling in an implicit way. In order to give an explicit description of the coupling, Mason (1935) introduced an ideal transformer, cp. Fig 4.3. In the mechanical circuit, this transformer can be represented by an ideal lever with a coupling factor α of dimension [N/V].

The models of Fig. 4.3 can also be described as a two-port network. If the velocity $v = dx/dt$ and force F as well as the electric voltage U and current $I = dQ/dt$ are chosen as port variables for the system, the dynamic behavior can be fully characterized by the admittance matrix of the system:

$$\begin{pmatrix} \dot{Q} \\ \dot{x} \end{pmatrix} = \begin{pmatrix} Y_{11} & Y_{12} \\ Y_{21} & Y_{22} \end{pmatrix} \begin{pmatrix} U \\ F \end{pmatrix} \tag{4.8}$$

Y_{11} is the so-called short-circuit input impedance of the system. It describes the ratio between the (complex) amplitudes of the electrical current and the voltage for harmonic vibrations of the system in the absence of a mechanical force F (short-circuit condition for the mechanical port, $F = 0$). In the following, Y_{11} will be renamed to Y_{el} for simplicity. Y_{22} is the short-circuit mechanical output impedance of the system. It describes the ratio between the (complex) amplitudes of the mechanical force and the vibration for harmonic vibrations of the system in the absence of an electrical voltage (short-circuit condition for the electrical port, $U = 0$).

Y_{12} and Y_{21}, respectively, describe the electro-mechanical conversion. Y_{12} is the ratio between the (complex) amplitudes of the electrical current and the mechanical force under force excitation and short-circuit conditions for the electrical port ($U = 0$). Y_{21} is the ratio between the (complex) amplitudes of the velocity and the voltage under voltage excitation and short-circuit conditions for the mechanical port ($F = 0$). The admittances Y_{12} and Y_{21} are identical and will be called Y_{mech} in the following.

$$Y_{11} = Y_{\mathrm{el}} = \left.\frac{\mathrm{j}\,\omega\, Q}{U}\right|_{F=0} = \frac{\alpha^2}{\mathrm{j}\omega\, m_1 + d_m + c_m/(\mathrm{j}\,\omega)} + \frac{1}{R + 1/(\mathrm{j}\omega\, C_p)} \qquad (4.9)$$

$$Y_{12} = Y_{21} = Y_{\mathrm{mech}} = \frac{\mathrm{j}\,\omega\, x}{U} = \frac{\alpha}{\mathrm{j}\,\omega\, m_1 + d_m + c_m/(\mathrm{j}\,\omega)} \qquad (4.10)$$

$$Y_{22} = \left.\frac{\mathrm{j}\,\omega\, Q}{F}\right|_{U=0} = \frac{1}{\mathrm{j}\,\omega\, m_1 + d_m + c_m/(\mathrm{j}\,\omega)} \qquad (4.11)$$

4.3.3 Multi-Degree of Freedom Model for Dynamic Regime

It should be mentioned once more, that the models of Fig. 4.3 and the respective admittances of Eqs. 4.9ff are only valid in a neighborhood of the eigenmode under investigation. If the deformation of the piezoelectric system is more complex – and more than one eigenfunction have to be superimposed in order to represent the motion of the system – additional degrees of freedom must be taken into account. They can be represented by additional mass–spring–damper systems as shown in Fig. 4.4.

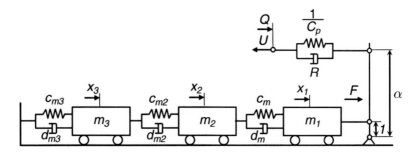

Fig. 4.4 Equivalent model for a system described by three vibration modes

4.3.4 Experimental Parameter Identification

The equivalent circuit models are mainly used for two main purposes: in analysis of existing systems, and as simplified model for the design of new systems. The parameters which enter the equivalent circuit model must be determined before the model can be applied in the design process. If a prototype exists, the parameters can be identified experimentally. Although the energy flow in piezoelectric energy-harvesting systems is from the mechanical port to the electrical port, the simplest way to determine the parameters of the equivalent circuit model is to excite the system electrically and measure its reaction. This will be explained in the following. The case of a single degree of freedom is studied here, see Fig. 4.3.

The system is excited by a harmonic voltage with amplitude U. The complex amplitudes of current and velocity are measured. The system is mechanically unloaded, i.e., $F = 0$.

The identification of the model parameters is performed into two steps. First the electric circuit of Fig. 4.4 is transformed to the form of Fig. 4.5b. Here, the mechanical branch has been converted to the electrical side resulting in the well-known Butterworth–Van Dyke topology, with parameters:

$$L_m = \frac{m_1}{\alpha^2}; \quad R_m = \frac{d_m}{\alpha^2}; \quad \frac{1}{C_m} = \frac{c_m}{\alpha^2}, \tag{4.12}$$

The electric admittance of the system is given by Eq. 4.9. In practice, the dielectric losses are small and R can be neglected. The electric admittance is then given by

$$Y_{el}(j\,\omega) = j\,\omega \left(C_p + \frac{C_m}{1 + \omega\,C_m\,(jR_m - \omega\,L_m)} \right) \tag{4.13}$$

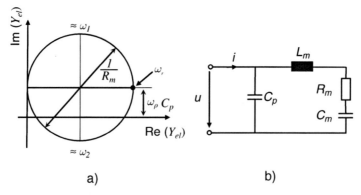

a)   b)

Fig. 4.5 Frequency response locus of the piezoelectric equivalent circuit model in the neighborhood of the resonance with characteristic frequencies ω_1, ω_2, ω_r, mechanical damping $1/R_m$ and offset $\omega_r\,C_p$. (**a**) Frequency response locus for $R = 0$. (**b**) Equivalent circuit model for $R = 0$

and if it is plotted in the complex plane, the response locus has the form shown in Fig. 4.5a. The admittance is mapped on a circle. The circles' offset on the imaginary axis is given by $C_p \, \omega_r$ and thus directly related to the shunted capacitance of the system. The radius r of the circle is:

$$r = \frac{1}{2 \, R_m} \tag{4.14}$$

Circle offset and radius can be easily determined from experimentally determined electrical admittances in the neighborhood of the resonance frequency of the system. With known capacitance C_p and damping R_m, L_m and C_m can be determined using the electrical engineering well-known quality factor for series circuits

$$Q_m = \frac{\omega_r}{\Delta \omega} = \frac{\omega_r}{\omega_2 - \omega_1} = \frac{1}{R_m} \sqrt{\frac{L_m}{C_m}}, \tag{4.15}$$

which is an important dimensionless characteristic parameter for piezoelectric systems. In Eq. 4.15 ω_1 and ω_2 are defined as the frequency values, where the magnitude of the electrical admittance is smaller by approx. $-3\,\mathrm{dB}$ when compared with the maximum at resonance. If the offset of the frequency locus is small, these frequencies can also be found on the locus circle in the imaginary plane at those points where the imaginary part of the electrical admittance has its maximum and minimum value, respectively.

With the resonance frequency

$$\omega_r = \frac{1}{\sqrt{L_m \, C_m}} \tag{4.16}$$

and the quality factor from Eq. 4.15, the parameters R_m, L_m, and C_m can finally be determined by

$$R_m = \frac{1}{2r}; \qquad L_m = \frac{R_m}{\omega_2 - \omega_1}; \qquad C_m = \frac{1}{\omega_r^2 \, L_m}. \tag{4.17}$$

For determining the α, first the radius of the electric admittance locus has to be determined. In the same way, the mechanical admittance locus can be plotted to determine its radius. The coupling factor α can be determined as the ratio of the two radii:

$$Y_{\mathrm{el}}(\mathrm{j}\omega) = \mathrm{j}\omega \left(C_{\mathrm{p}} + \frac{\alpha^2}{c_{\mathrm{m}} + \omega \, (\mathrm{j} \, d_{\mathrm{m}} - \omega \, m_1)} \right) \tag{4.18}$$

$$Y_{\mathrm{mech}}(\mathrm{j}\,\omega) = \frac{v}{u} = \frac{\mathrm{j}\,\omega\,x}{u} = \frac{\mathrm{j}\,\omega\,\alpha}{c_{\mathrm{m}} + \omega \, (\mathrm{j} \, d_{\mathrm{m}} - \omega \, m_1)} \tag{4.19}$$

$$\alpha = \frac{r}{r_{\mathrm{mech}}} \tag{4.20}$$

4.3.5 Case Study

The parameter identification method described in the previous section was applied to a prototype system. The transfer functions of standard cantilever beam bimorphs were measured using a phase-/gain analyzer HP 4192. The current measurement was made using an indirect current probe Tektronics A6312/AM5030 which has the advantage that it is almost free from parasitic feedback to the measured system. For the measurement of the tip velocity, a differential doppler laser vibrometer Polytec OFV-512/PFV-5000 was used.

The measurements were performed on three bimorphs of the type "Sitex module" manufactured by Argillon shown in Fig. 4.1b. Figure 4.6 shows the measured electrical admittance Y_{el} in comparison with that of the model. Here, the parameters of Table 4.2 are used which have been identified by the method mentioned earlier. The dots represent the measured data and the solid lines are the admittances of the piezoelectric equivalent circuit models. A very good agreement between theoretical and experimental values can be observed, which is a necessary condition for the successful validation of the model.

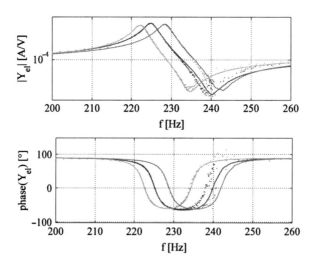

Fig. 4.6 Frequency response of three piezoelectric bending transducer. *Lines* show the identified and *dots* the measured system transfer functions

Table 4.2 Identified "electrical" parameters for three piezoelectric bimorphs

	f_r(Hz)	$R_m(\Omega)$	L_m(H)	C_m(F)	C_p(F)
First Element	222	1.22k	61.02	8.39n	74.93n
Second Element	224	1.08k	50.67	9.88n	71.16n
Third Element	228	1.15k	53.11	9.13n	74.42n

The following Eqs. 4.21ff have been applied for the identification:

$$R_m = \frac{1}{\max[\mathrm{Re}\,(Y_{el})]} \tag{4.21}$$

$$C_p = \frac{\max[\mathrm{Im}\,(Y_{el})] + \min[\mathrm{Im}\,(Y_{el})]}{2\omega_r} \tag{4.22}$$

$$L_m = \frac{R_m}{\omega\,[\min[\mathrm{Im}\,(Y_{el})]] - \omega\,[\max[\mathrm{Im}\,(Y_{el})]]} \tag{4.23}$$

$$\alpha = \frac{1}{R_m\,\max[\mathrm{Re}(Y_{mech})]} \tag{4.24}$$

$$C_m = \frac{1}{L_m\,\omega_r^2} \tag{4.25}$$

A basic modeling assumption is that the system has perfectly linear behavior. In practice, however, the bimorph admittances showed some dependence on the excitation level. Increasing the excitation level resulted in higher damping and a shift of the resonance towards lower frequencies. Nevertheless, for most practical applications, the linear description will be sufficient. If necessary, the model parameters can be identified for different excitation levels, resulting in amplitude-dependent parameter sets of the linear model which can be considered as a best approximation to the nonlinear system behavior in the sense of the harmonic linearization.

4.4 Analytical Determination of the Parameters of the Equivalent Circuit Models

The experimental parameter identification method can only be applied if the system to be considered already exists. Most often, it is necessary to estimate the systems' behavior before a hardware prototype can be built. In this case, analytical or numerical calculations can be applied.

In general, piezoelectric systems with complex structure can be analyzed using the finite element method (Król et al. 2005; Piefort 2001). In this way, the parameters of the piezoelectric equivalent circuit models can be determined. If systems have simple geometries – like bimorph beams – analytical methods can be used to determine the system parameters. Analytical models have the additional advantage that they can be used for optimization. Examples of such analytical models are studied in the following section.

4.4.1 General Procedure for Analytical Bimorph Model

In general, the behavior of piezoelectric bimorph structures is described by the Euler–Bernoulli theory, taking the piezoelectric coupling into account using the material law established in section 4.2. Corresponding analyses have been carried out(e.g., Ballato and Smits 1994, Ballas 2007, Lenk 1977).

The basic idea is to first describe the behavior of a passive beam segment which can be considered as a waveguide, and then superimpose the stresses generated by the piezoelectric material behavior. In order to illustrate the method, a homogeneous beam with height h, width w, density ρ, and Young's modulus Y is considered in the following. The geometric dimensions define the second moment of area I. It can be shown that the transfer functions of an arbitrary segment can be described by the following matrix (Ballas 2007, Lenk 1977):

$$
\begin{pmatrix} v_1 \\ \varXi_1 \\ M_1 \\ F_1 \end{pmatrix} = \frac{1}{2} \begin{pmatrix} \cosh\eta + \cos\eta & \frac{-l_a}{\eta}(\sinh\eta + \sin\eta) & \frac{1}{jz_0 l_a}(\cosh\eta - \cos\eta) & \frac{1}{jz_0\eta}(\sinh\eta - \sin\eta) \\ \frac{-\eta}{l_a}(\sinh\eta - \sin\eta) & \cosh\eta + \cos\eta & \frac{-\eta}{jz_0 l_a^2}(\sinh\eta + \sin\eta) & \frac{-1}{jz_0 l_a}(\cosh\eta - \cos\eta) \\ jz_0 l_a(\cosh\eta - \cos\eta) & \frac{-jz_0 l_a^2}{\eta}(\sinh\eta - \sin\eta) & \cosh\eta + \cos\eta & \frac{l_a}{\eta}(\sinh\eta + \sin\eta) \\ jz_0\eta(\sinh\eta + \sin\eta) & -jz_0 l_a(\cosh\eta - \cos\eta) & \frac{\eta}{l_a}(\sinh\eta - \sin\eta) & \cosh\eta + \cos\eta \end{pmatrix} \begin{pmatrix} v_r \\ \varXi_r \\ M_r \\ F_r \end{pmatrix}
$$

(4.26)

In these equations, the complex amplitudes of the transverse velocities of the left and right segment boundary are denoted by v_1 and v_r, respectively. In the same manner, the angular velocities $\varXi$, moments M, and forces F are described by their complex amplitudes. The length of the element is l_a. Furthermore, it is convenient, to introduce the mass $m = \rho l_a wh$ of the element, its compliance s as well as the non-dimensional variable η. The latter is the ratio between the excitation frequency and the resonance frequency ω_0 of a particular bending mode.

$$
\eta = \kappa l = \sqrt{\omega \sqrt{\rho \, l_a whs}} = \sqrt{\frac{\omega}{\omega_0}}
$$

(4.27)

$$
z_0 = \sqrt{\rho l whs} = \frac{1}{\omega_0 s}
$$

(4.28)

If a voltage is applied to the electrodes of a bimorph element, the resulting stresses can be represented by a differential moment $M_\Delta = M_1 - M_0$ acting at both ends of the segment. Introducing the differential angular velocity $\varXi_\Delta = \varXi_1 - \varXi_0$, the behavior of the piezoelectric bimorph is described by the transfer matrix, cp. Eq. 4.4:

$$
\begin{pmatrix} i \\ M_\Delta \end{pmatrix} = \begin{pmatrix} \frac{4j\omega l w \varepsilon_{33}^T(1-k_{31}^2)}{h} & \frac{whd_{31}}{2s_{11}^E} \\ \frac{whd_{31}}{2s_{11}^E} & -\frac{wh^3(1-0.75k_{31}^2)}{12j\omega l s_{11}^E(1-k_{31}^2)} \end{pmatrix} \begin{pmatrix} U \\ \varXi_\Delta \end{pmatrix}
$$

(4.29)

This equation has been derived for a bimorph with two active layers in parallel configuration (Lenk 1977). In the final step, an active and passive model will be

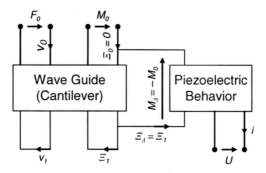

Fig. 4.7 Dynamic piezoelectric cantilever system as black box system according to Lenk (1977)

combined by balancing the state variables M_Δ and Ξ_Δ which enter in both models and have to be identical. Combining the transfer matrix of Eq. 4.29 with that of the waveguide in Eq. 4.26, the coupled electromechanical behavior of the waveguide can be described by the system shown in Fig. 4.7.

4.4.2 Determination of the Parameters of the Piezoelectric Equivalent Circuit Models using the Analytical Model

The parameters of the piezoelectric equivalent circuit can be identified using the analytical model of Section 4.4.1. This method can also be applied to other simple configurations, e.g., bars, beams, or plates. It allows the determination of the parameters of the piezoelectric equivalent circuit model from the geometrical and material properties of the system.

For the case of a cantilever beam with small damping, the results of Lenk (1977) can be used. It must, however, be noted that the model represents the system behavior only in the neighborhood of a single eigenmode. Table 4.3 shows the results for the first bending mode according to Richter et al. (2006).

The system admittances of the equivalent circuit model with the analytical determined parameter set has been compared with the corresponding admittances for the

Table 4.3 Calculation of the lumped parameter set for the first bending mode

Parameter	Analytical expression
M_1	$\frac{1}{4}\rho\, l\, w\, h$
c_m	$\dfrac{b\,h^3(1-0.75\,k_{31}^2)}{3.883\,l^3\,s_{11}^E\,(1-k_{31}^2)}$
d_m	$\dfrac{b\,h^2}{3.941\,Q_m\,l\,\sqrt{\dfrac{s_{11}^E\left(1-k_{31}^2\right)}{\rho\left(1-0.75\,k_{31}^2\right)}}}$
α	$\dfrac{0.668\,b\,h}{l}\sqrt{\dfrac{\varepsilon_{33}^T\,k_{31}^2}{s_{11}^E}}$
C_p	$\dfrac{4\,l\,w\,\varepsilon_{33}^T\left(1-k_{31}^2\right)}{h}$

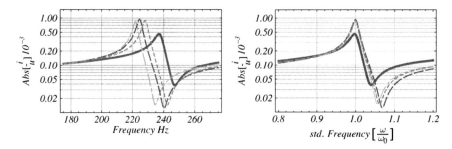

Fig. 4.8 Comparison between models using experimentally identified and analytically determined parameters (*continuous line*)

parameter set that was determined experimentally. As can be seen in Fig. 4.8, a good matching can be observed. Small deviations are due to modeling assumptions, e.g., neglecting the shim layer or the assumption of ideal boundary conditions. Also, nonlinear effects of the material behavior are not considered in the analytical model, but these are present in the experimental results.

4.5 Equivalent Circuit Model for Base Excited Piezoelectric Systems

So far systems with fixed mechanical boundaries have been considered. In practical applications, however, kinematic base excitation is the rule rather than the exception. In the following, a representation of the system dynamics will be developed which takes a kinematically prescribed base motion into account.

Let y be the displacement of the base and x be the modal amplitude of a given eigenmode of the system, then the piezoelectric equivalent circuit model of Fig. 4.9 can be used to describe the dynamics of the system Eq. 4.30. For $y \equiv 0$, it is equivalent to the model studied earlier.

$$m_1 \ddot{x}(t) + d_m(\dot{x}(t) - \dot{y}(t)) + c_m(x(t) - y(t)) = F(t) - \alpha\, U(t)$$
$$R(\alpha\, \dot{x}(t) - \dot{Q}(t)) + \frac{1}{C_p}(\alpha\, x(t) - Q(t)) = U(t) \tag{4.30}$$

The equivalent circuit model described earlier implicitly assumed that the system's excitation acts either via the electrical port or at the bimorph tip. However, most often the excitation of a piezoelectric energy harvesting system is not of that type. Usually energy harvesters will be attached to the vibrating host system via their housing, so that the excitation is by the base motion provided by the host structure. The piezoelectric equivalent circuit models shown in Fig. 4.9a (mechanical) and Fig. 4.9b (electrical) describe the system under dynamic base excitation. This system will be studied in the next sections and the measurements in the latter sections will be used for validation.

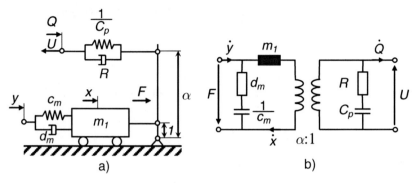

Fig. 4.9 Piezoelectric equivalent circuit model for a base excited system

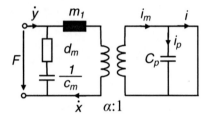

Fig. 4.10 Piezoelectric equivalent circuit model for the base excited system with the short circuited electrodes

Using again the Euler–Bernoulli beam theory for the derivation of the parameters, it can be shown that the parameters of the model are identical to the ones calculated previously. This, fortunately, allows identifying the parameters using the method described there.

Using the equivalent circuit model of Fig. 4.10, which is valid for short circuited electrical port, it becomes obvious that in this case α is the ratio between the electrical current i and the velocity v.

$$\alpha = \frac{i_m}{\dot{x}} = \frac{i}{\dot{x}}\Big|_{i_p \equiv 0} \quad ; \quad i_m = i_p + i \quad (4.31)$$

The velocity $v = \mathrm{d}/\mathrm{d}t\, x$ of the modal coordinate x and the current i can be measured directly providing an alternative method for the identification of the parameter α.

4.6 Overall PEG System Analyses Using Piezoelectric Equivalent Circuit Models

This section shows how the piezoelectric equivalent circuits models can be used to study the effect of the electrical load on the PEGs power capabilities. The application of piezoelectric equivalent circuit models in design optimization will also be addressed.

4.6.1 Piezoelectric Equivalent Circuit Model with Electrical Load

In the following, the mechanical excitation of the PEG is described by the velocity $d/dt\,y$ of the base motion. Dielectric losses are neglected, and the electrical load is modeled by a resistive load R_{Load}. Under these assumptions, Fig. 4.11 gives the overall piezoelectric equivalent circuit model in both mechanical and electrical representations. The equations of motion are:

$$c_m(x(t) - y(t)) + d_m\,(\dot{x}(t) - \dot{y}(t)) + m_1\,\ddot{x}(t) = -\alpha\,F(t)$$

$$\frac{1}{C_p}(\alpha\,x(t) - Q(t)) = F(t) \tag{4.32}$$

$$R_{Load}\,\dot{Q}(t) = F(t)$$

4.6.2 Analysis of the Maximum Power Output

Obviously, one of the most important characteristics of an energy harvesting circuit is the maximum power output. The generated power can be calculated by Eq. 4.33. The voltage amplitude at the output can be expressed by Eq. 4.34, resulting in Eq. 4.35 for the power output.

$$P(j\omega) = U_L(j\,\omega)\,i^*(j\,\omega) \tag{4.33}$$

$$U_L(j\,\omega) = \frac{(c_m + j\,\omega\,d_m)\,\alpha\,R_{Load}\,\dot{y}}{j\,\omega\,R_{Load}\,\alpha^2 + (c_m + j\,\omega\,(d_m + j\,\omega\,m_1))(1 + j\,\omega\,C_p\,R_{Load})} \tag{4.34}$$

$$P(j\,\omega) = U_L(j\,\omega)\left(\frac{U_L(j\,\omega)}{R_{Load}}\right)^*; \qquad \text{using } i = \frac{U}{R} \tag{4.35}$$

Maximum power transfer is achieved if the load resistance matches the impedance of the generator. This can be concluded by introducing a virtual port to the network

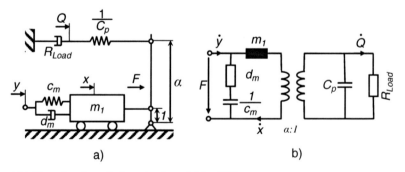

a)   b)

Fig. 4.11 Piezoelectric equivalent circuit model for PEG

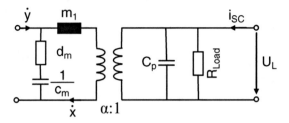

Fig. 4.12 Electric equivalent of loaded piezoelectric element with virtual output for admittance determination

as shown in Fig. 4.12. Using Kirchhoff's law, the current i_{sc} can be evaluated according to Eq. 4.36 and the admittance of the virtual port is given by Eq. 4.37.

$$i_{SC} = \frac{\alpha \, (j \, \omega \, d_m + c_m)}{j \, \omega \, m_1 + d_m + \frac{c_m}{j \, \omega}} \, \dot{y} \tag{4.36}$$

$$Y_L = \frac{i_{SC}}{U_L} = \frac{1}{R_{Load}} + j \, \omega \left(C_p \frac{\alpha^2}{c_m + j \, \omega \, (d_m + j \, \omega \, m_1)} \right) \tag{4.37}$$

Maximum power output occurs if a maximum of current flows over the load and a minimum over the virtual port. As a consequence $Y_L \, (j\omega, \, R_{Load})$ is zero in the optimal case. Solving this equation for ω gives the optimal frequency as a function of the load; and this relation can then be inverted for the determination of the optimal load resistance as a function of frequency. The optimal frequency is plotted as a function of the load resistance in Fig. 4.13a; and the corresponding value for the maximum output power is depicted in Fig. 4.13b. Both diagrams show the results for different levels of mechanical damping.

The influence of the damping is considerable. Weak damping results in two local maxima of the power output and strong damping leads to only one maximum. Looking on the frequencies where the maxima are attained reveals that the two peaks of the power output curve of the weakly damped system are at the resonance and

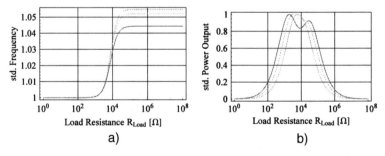

Fig. 4.13 (a) Optimal frequency versus resistive load. (b) Optimal power at the resistor versus its resistance

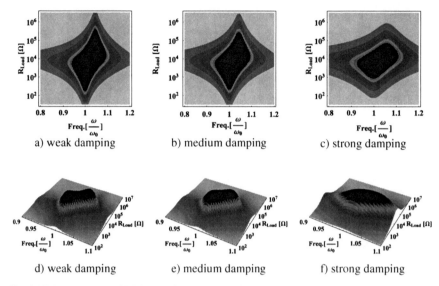

Fig. 4.14 Power output of PEG for different levels of mechanical damping

anti-resonance of the unloaded system. For the strongly damped system, the optimal frequency is between the resonance and anti-resonance frequency of the unloaded system.

Of course, the output power $P(\omega, R_{\text{load}})$, see Eq. 4.35, cannot only be evaluated for the optimal case. It also allows the investigation of the whole working range, as depicted in Fig. 4.14. The system with strong damping can be operated in a wider range but it has higher internal losses, compared with medium and weak damping. The task of the design engineer is to find a good compromise in the trade-off between bandwidth and internal losses.

4.6.3 Experimental Validation of the Piezoelectric Equivalent Circuit Model for Base Excitation

The piezoelectric equivalent circuit models have been applied to calculate output voltage and power using the experimentally identified parameters. These results are compared with the direct measurements of these quantities. Figure 4.15 shows the measured (dots) and estimated (line) voltage and power at three different load resistances for a system with strong damping. Good agreement is observed for small and high load resistance. The system output for optimal load resistance, however, is underestimated.

Figure 4.16 shows the corresponding comparison between theoretical and experimental results for the optimal frequency and for the output power as a function of the load resistance for strongly and weakly damped systems. Note that the output power curve has been normalized to its maximum.

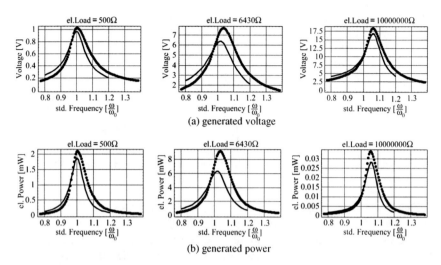

Fig. 4.15 Comparison between measurements (*dots*) and estimated power (*lines*), respectively, voltage for different load resistances as a function of frequency

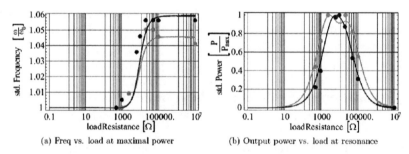

Fig. 4.16 Comparison between experimental (*dots*) and theoretical results (*lines*). (**a**) Optimal frequency and (**b**) power output, as a function of the load resistance

4.6.4 Effect of Geometry

Section 4.6.3 showed a strong dependence of the PEG characteristics on the parameters of the system. The parameters depend on material properties and geometric dimensions of the piezoelectric elements used in the PEG. Therefore, these design parameters can be chosen in such a way that the overall system behavior is of a desired character for a given load resistance and base excitation.

For a bimorph cantilever beam, the parameters of the piezoelectric equivalent circuit model have been determined as a function of the material and geometric properties in chapter 4.4.2. In the following, the effect of the geometry of the piezoelectric element on the power output will be investigated. In this analysis, the following parameters and boundary conditions are assumed to be known a priori: frequency of excitation, velocity of the host system, load impedance.

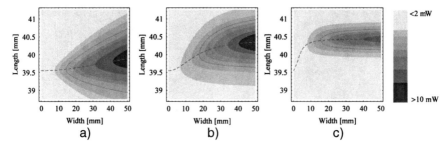

Fig. 4.17 Power output at three resistive loads in a parameter space of length and width

Figure 4.17 shows the output power as a function of the resistive load. The design parameters of the contour plots are width and free length of the bimorph beam transducer. The other geometry and material parameters are constant (thickness of the piezoelectric layer, material parameters, and the properties of the host system). In the diagram, the color of the areas depicts the output power; dark areas show the regions with high-electric power output. The dashed line represents perfect resonance matching of the system.

Figure 4.18 shows the power output as a function of the load resistance and width of the piezoelectric bimorph element. In these calculations, the free length of the bimorph element was held constant.

4.6.5 Modeling of the Coupling Between the PEG and Its Excitation Source, Additional Degrees of Freedom

So far a perfect kinematic base excitation has been assumed. For small or lightweight host structures, this assumption might not be valid and the excitation might be of a force type rather than of the kinematic type. In some instances, it might even be necessary to model the dynamics of the host structure. A model for this situation

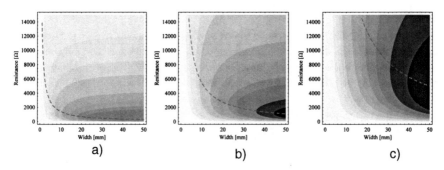

Fig. 4.18 Output power as a function of the load resistance and bimorph width for three transducer of different length

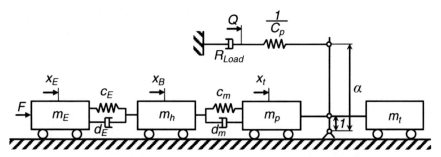

Fig. 4.19 Mechanical representation of a piezoelectric equivalent circuit model of the PEG including excitation source and host structure, modeled by additional degrees of freedom

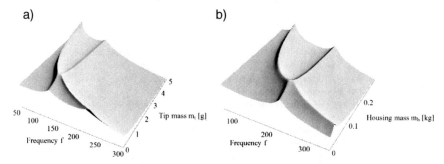

Fig. 4.20 Influence of system's masses to power output. The *left* figure shows the logarithmic power output over the tip mass m_t and the *right* figure depicts the output over the housing mass m_h

is shown in Fig. 4.19 as explained by (Richter and Twiefel 2007). In this system representation, the coordinate x has been renamed to x_t indicating the displacement of the piezoelectric element's tip. According to the abbreviations of the new lumped parameters, the piezoelectric modal mass is indicated by m_p and the equivalent mass of the host structure and excitation source are given by m_h and m_E. A tuning mass m_t has also been introduced.

An optimal working point of the generator can be found when the modal mass of the generator matches the modal mass of the housing in some way. In general, the tip mass as well as the housing mass may be easily changed in the stage of design. Therefore, the power output with altering tip and housing masses is shown in Fig. 4.20a and b. The figures show the logarithmic power output in the frequency domain.

4.7 Summary

The present paper gave a survey on piezoelectric equivalent circuit models for piezoelectric generators (PEG). First the models have been derived and discussed, and then an experimental parameter identification method has been presented. The results of theoretical predictions and actual experimental results showed good agreement. Piezoelectric equivalent circuit models for both electrical and kinematic base excitation have been discussed and applied to the optimization of the power output and other important quantities for PEG.

References

Ballas, R.G. (2007) Piezoelectric Multilayer Beam Bending Actuators: Static and Dynamic Behavior and Aspects of Sensor Integration. Springer, Berlin Heidelberg

Ballato and Smits (1994) Dynamic Admittance Matrix of Piezoelectric Cantilever Bimorphs. In: Journal of Microelectromechanical Systems, Vol. 3, No. 3, pp. 105–112

Cady, W.G. (1922) The piezo-electric resonator. In: Proceedings of the Institute of Radio Engineers, Vol. 10, pp. 83–114

IEEE Std 176–1978 (1978) IEEE Standard on Piezoelectricity. The Institute of Electric and Electronic Engineers

Król, R. Mracek, M. and Redenius, A. (2005) Eine Methodik zur Ableitung einfacher Ersatzmodelle zur automatischen Konfigurierung piezoelektrischer Antriebe mittels der Finite Elementen Methode. In: VDI-Berichte 1892: Mechatronik 2005 Innovative Produktentwicklung, Band 1, S. 137–152.VDI Verlag, Düsseldorf

Lenk, A. (1977) Elektromechanische Systeme. Band 2: Systeme mit verteilten Parametern. VEB Verlag Technik, Berlin

Mason, W.P. (1935) An electromechanical representation of a piezoelectric crystal used as a transducer. In: Proceedings of the Institute of Radio Engineers, Vol. 23, 10, pp. 1252–1263

Piefort, V. (2001) Finite Element Modelling of Piezoelectric Active Structures. PhD-Thesis, Université Libre de Bruxelles.

Richter, B. and Twiefel, J. (2007) On the need of modeling the interdependence between piezoelectric generators and their environmental excitation source. In: Fu-Kuo Chang (ed.) Structural Health Monitoring 2007: Quantification, validation and Implementation. DEStech Publications Inc., Palo Alto, pp. 1749–1756

Richter, B. Twiefel, J. Hemsel, T. and Wallaschek, J. (2006) Model based Design of Piezoelectric Generators utilizing Geometrical and Material Properties. In: 2006 ASME Intl. Eng. Congress and Exposition

Tokin Corp. (1987) Piezo-Electric Bimorph Element. Patent JP63268279A invented by Tamura Mitsuo

Van Dyke, K.S. (1925) The electric network equivalents of a piezoelectric resonator. In: Physical Review. American Institute of Physics. Vol. 25, 895A

Chapter 5
Electromagnetic Energy Harvesting

Stephen P Beeby and Terence O'Donnell

Abstract This chapter focuses on the use of electromagnetic transducers for the harvesting of kinetic (vibration) energy. The chapter introduces the fundamental principals of electromagnetism and describes how the voltage is linked to the product of the flux linkage gradient and the velocity. The flux linkage gradient is largely dependent on the magnets used to produce the field, the arrangement of these magnets, and the area and number of turns for the coil. The characteristics of wire-wound and micro-fabricated coils, and the properties of typical magnetic materials, are reviewed. The scaling of electromagnetic energy harvesters and the design limitations imposed by micro-fabrication processes are discussed in detail. Electromagnetic damping is shown to be proportional to the square of the dimension and analysis shows that the decrease in electromagnetic damping with scale cannot be compensated by increasing the number of turns. For a wire wound coil, the effect of increasing coil turns on EM damping is directly cancelled by an increase in coil resistance. For a planar micro-coil increasing the number of turns results in a greater increase in the coil resistance, resulting in an overall decrease in damping. Increasing coil turns will, however, increase the induced voltage which may be desirable for practical reasons. An analysis is also presented that identifies the optimum conditions that maximise the power in the load. Finally, the chapter concludes with a comprehensive review of electromagnetic harvesters presented to date. This analysis includes a comparison of devices that confirms the theoretical comparison between conventional wound and micro-fabricated coils and the influence of device size on performance.

5.1 Introduction

Electro-magnetism has been used to generate electricity since the early 1930s, not long after Faraday's fundamental breakthrough in electromagnetic induction. The majority of generators used today are based on rotation and are used in numerous applications from the large-scale generation of power to smaller scale applications

S.P. Beeby (✉)
University of Southampton, Highfield, Southampton, SO17 1BJ, UK

S. Priya, D.J. Inman (eds.), *Energy Harvesting Technologies*,
DOI 10.1007/978-0-387-76464-1_5 © Springer Science+Business Media, LLC 2009

in cars to recharge the battery. Electromagnetic generators can also be used to harvest micro- to milli-Watt levels of power using both rotational and linear devices. Provided a generator is correctly designed and not constrained in size, they can be extremely efficient converters of kinetic energy into electrical. Attempts to miniaturise the technique, however, using micro-engineering technology to fabricate a generator, invariably reduce efficiency levels considerably.

This chapter introduces the fundamental principles of electromagnetic induction before exploring the scaling effects that work against successful miniaturisation. Conventional discrete magnets and coils are compared with their micro-machined equivalent and the technical challenges of associated with micro-coils and deposited magnetic materials are highlighted. The chapter concludes with a comprehensive and up to date review and comparison of energy harvesters realised to date. The generators presented demonstrate many of the issues previously discussed.

5.2 Basic Principles

The basic principle on which almost all electromagnetic generators are based is Faraday's law of electromagnetic induction. In 1831, Michael Faraday discovered that when an electric conductor is moved through a magnetic field, a potential difference is induced between the ends of the conductor. The principle of Faraday's law is that the voltage, or electromotive force (emf), induced in a circuit is proportional to the time rate of change of the magnetic flux linkage of that circuit, i.e.

$$V = -\frac{d\phi}{dt} \qquad (5.1)$$

where V is the generated voltage or induced emf and ϕ is the flux linkage. In most generator implementations, the circuit consists of a coil of wire of multiple turns and the magnetic field is created by permanent magnets. In this case, the voltage induced in an N turn coil is given by:

$$V = -\frac{d\Phi}{dt} = -N\frac{d\phi}{dt} \qquad (5.2)$$

where Φ is the total flux linkage of the N turn coil and can be approximated as, $N\phi$, and in this case ϕ can be interpreted as the average flux linkage per turn. In general, the flux linkage for a multiple turn coil should be evaluated as the sum of the linkages for the individual turns, i.e.

$$\Phi = \sum_{i=1}^{N} \int_{A_i} B \cdot dA \qquad (5.3)$$

where B is the magnetic field flux density over the area of the ith turn. In the case where the flux density can be considered uniform over the area of the coil, the integral can be reduced to the product of the coil area, number of turns and the component of flux density normal to the coil area, $\Phi = NBA\sin(\alpha)$, where α is the angle between the coil area and the flux density direction. Consequently, in such a case, the induced voltage is given by:

$$V = -NA\frac{dB}{dt}\sin(\alpha) \tag{5.4}$$

In most linear vibration generators, the motion between the coil and the magnet is in a single direction, e.g., let us assume the x-direction, and the magnetic field, B, is produced by a permanent magnet and has no time variation. For clarity, we restrict the following analysis to this case and the voltage induced in the coil can then be expressed as the product of a flux linkage gradient and the velocity.

$$V = -\frac{d\Phi}{dx}\frac{dx}{dt} = -N\frac{d\phi}{dx}\frac{dx}{dt} \tag{5.5}$$

Power is extracted from the generator by connecting the coil terminals to a load resistance, R_L, and allowing a current to flow in the coil. This current creates its own magnetic field which acts to oppose the field giving rise to it. The interaction between the field caused by the induced current and the field from the magnets gives rise to a force which opposes the motion. It is by acting against this electromagnetic force, F_{em}, that the mechanical energy is transformed into electrical energy. The electromagnetic force is proportional to the current and hence the velocity and is expressed as the product of an electromagnetic damping, D_{em} and the velocity.

$$F_{em} = D_{em}\frac{dx}{dt} \tag{5.6}$$

In order to extract the maximum power in the form of electrical energy, an important goal for the design of a generator is the maximisation of the electromagnetic damping, D_{em}. Therefore, it is important to understand the design parameters which can be used to maximise electromagnetic (EM) damping.

The instantaneous power extracted by the electromagnetic force is given by the product of the force and the velocity shown in Eq. (5.7).

$$P_e = F_{em}(t)dx(t)/dt \tag{5.7}$$

This power is dissipated in the coil and load impedance. Equating the power dissipation in the coil and load to that obtained from the electromagnetic force gives:

$$F_{em}\frac{dx}{dt} = \frac{V^2}{R_L + R_c + j\omega L_c} \tag{5.8}$$

where R_L and R_c are load and coil resistances, respectively, and L_c is the coil inductance. Using Eq. (5.5), the voltage can be expressed as the product of the flux linkage
gradient and the velocity, and substituting Eq. (5.6) for the force, gives the following
expression for the electromagnetic damping:

$$D_{em} = \frac{1}{R_L + R_c + j\omega L_c} \left(\frac{d\Phi}{dx}\right)^2 \tag{5.9}$$

As we can see from Eq. (5.9), this depends on maximising the flux linkage
gradient, and minimising the coil impedance. The flux linkage gradient is largely
dependent on the magnets used to produce the field, the arrangement of these magnets, and the area and number of turns for the coil. The properties of typical magnetic materials are reviewed in the sections below. At the low frequencies generally
encountered in ambient vibrations (typically less than 1 kHz), the coil impedance is
generally dominated by the resistance. The magnitude of the coil resistance depends
on the number of turns and the coil technology. Common technologies for coil fabrication are wire-winding and micro-fabrication. The characteristics of these coil
technologies are also discussed in the sections below.

5.3 Wire-Wound Coil Properties

A fundamental consideration in designing an electromagnetic energy harvester is the
properties of the coil. The number of coil turns and the coil resistance are important
parameters for determining the voltage and useful power developed by a generator.
The number of turns is governed by the geometry of the coil, the diameter of the
wire it is wound from and the density with which the coil wire has been wound.
Insulated circular wire will not fill the coil volume entirely with conductive material
and the percentage of copper within a coil is given by its fill factor. The area of the
wire can be related to the overall cross-sectional area of the coil, A_{coil}, by assuming
a certain copper fill factor, f, i.e. $A_{wire} = f\,A_{coil}/N$. The copper fill factor depends
on tightness of winding, variations in insulation thickness, and winding shape. Most
coils are scramble wound which means the position of the wire is not precisely
controlled and each layer of the coil will not necessarily be completely filled before
the next layer is begun. The fill factor of scramble wound coils will vary but a figure
of 50–60% could be assumed (McLyman 1988). In orthogonal winding, the wire sits
squarely on top of the wire in the layer below. This requires more careful winding
but can give fill factors up to 78% ignoring wire insulation. The highest fill factors
of up to 90% (again ignoring insulation thickness) can be achieved with orthocyclic
winding where the wire sits in the groove between two wires in the layer below.
At first glance, this may seem easy to achieve but in practice problems occur with
underlying wires being pushed apart and the fact that consecutive layers normally
have opposite 'threads', or helical lay, which builds stresses into the coil.

Table 5.1 Fine copper wire gauge number and properties

AWG No.	Copper diameter (μm)	Wire diameter (including insulation) (μm)	Resistance (Ω/m)
58	10	10.6–12.9	0.22
57	11	11.7–14.1	0.18
56	12.5	13.2–16.5	0.14
55	14	14.7–17.8	0.11
54	15.8	16.5–19	0.09
53	17.8	18.5–21.6	0.07
52	20	21.6–25.4	0.056
51	22	24.1–27.8	0.044
50	25	26.7–30.5	0.035
49	28	29.7–33	0.028
48	31.5	32.8–38.1	0.022
47	35.6	36.8–43.2	0.017

The type of wire used in the coil is clearly very important in defining the properties of the coil. Copper wire is available in standard gauges ranging from American Wire Gauge (AWG) number 58 which is 10 μm in diameter to AWG 6/0 which is 14.7 mm in diameter. In miniaturised electromagnetic energy harvesters, fine copper wire is desirable for minimising coil size while maximising the number of turns. The finer the wire, however, the greater its resistance per unit length. Various fine copper wire gauges have been summarised in Table 5.1 to provide an example. Fine copper wire is typically insulated by thin polymer films (e.g. polyurethane, polyester, polyimide) and the specified range of total enamelled wire thickness for each gauge is also given in Table 5.1. Different insulation materials offer different degrees of solvent resistance, operating temperature range and solder-ability. Self-bonding wire is an enamelled wire with an additional adhesive layer that enables wires to be bonded together as the coil is wound. This forms a self-supporting coil and the adhesives are available in different types depending on the method of activation, e.g. heat or solvent.

For a conventionally wound coil with key dimensions shown in Fig. 5.1, Eqs. (5.10), (5.11), (5.12), and (5.13) can be used to predict the number of turns, length of the wire wound within the coil and the coil resistance.

$$V_T = \pi \left(r_o^2 - r_i^2 \right) t \tag{5.10}$$

$$L_w = \frac{4 f V_T}{\pi w_d^2} \tag{5.11}$$

$$R_c = \rho \frac{L_w}{A_w} = \rho \frac{N L_{MT}}{A_w} = \rho \frac{N^2 \pi (r_o + r_i)}{f(r_o - r_i)t} \tag{5.12}$$

$$N = \frac{L_w}{r_i + \frac{(r_o - r_i)}{2}} \tag{5.13}$$

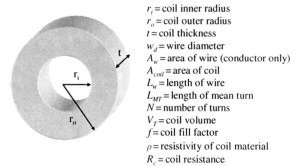

r_i = coil inner radius
r_o = coil outer radius
t = coil thickness
w_d = wire diameter
A_w = area of wire (conductor only)
A_{coil} = area of coil
L_w = length of wire
L_{MT} = length of mean turn
N = number of turns
V_T = coil volume
f = coil fill factor
ρ = resistivity of coil material
R_c = coil resistance

Fig. 5.1 Coil parameters

Typically, the fill factor is an unknown value and has to be calculated by rearranging Eq. (5.11). The length of the wire can be determined by measuring the coil resistance and knowing the diameter of the copper conductor within the enamelled wire or, if the number of turns is known, by applying Eq. (5.13).

The coil inductance can also be expressed as a function of the number of turns and the coil geometry. However, for a typical analysis, the inductance can be neglected as the resistive impedance of the coil is always significantly larger than the inductive impedance at frequencies less than 1 kHz. It can be seen from Eq. (5.12) that, for a constant number of turns, the coil resistance is proportional to the inverse of the coil dimension provided that the copper fill factor does not depend on the scaling.

5.4 Micro-Fabricated Coils

Micro-coils are coils which are fabricated using photolithography techniques to define the coil pattern, most commonly on substrates such as silicon, flex substrates or printed circuit boards (PCBs). Such coils are used for a range of applications such as on-chip inductors, detection coils in sensors, or as a means of producing a magnetic field in actuators. Micro-coils are fabricated by building up layers of planar coils, with each layer typically consisting of a square or a circular spiral coil. The technology which is used to fabricate the spiral coils generally limits the conductor thickness and the minimum spacing achievable between individual coil turns. Therefore, the minimum dimensions of the individual turn in the spiral will depend on the technology used for the fabrication. For example, standard PCB technology may limit the spacing between turns to greater than 150 μm for a thickness of 35 μm, whereas a spacing of 1–2 μm for a typical thickness of 1 μm may be achievable for silicon-based micro-fabrication techniques. As a rough guide for silicon-based coil fabrication, the minimum spacing may be taken to be equal to the conductor thickness, although with more advanced techniques it is feasible to obtain aspect ratios of up to 10 between conductor thickness and minimum spacing.

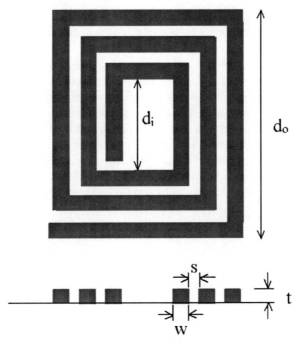

Fig. 5.2 Micro-fabricated square spiral coil

In order to quantify the dependence of coil resistance on dimension for the present analysis, it will be assumed that the micro-coil is a single layer planar square spiral coil. For such a micro-fabricated coil, the resistance is more conveniently expressed in terms of the number of turns, the turn width, w, spacing, s, and thickness, t, and the coil outer, d_o, and inner dimensions, d_i as shown in Fig. 5.2.

If it is assumed that the tracking spacing, track width and thickness are equal then the length of the mean turn, and the wire cross-sectional area would be;

$$L_{MT} = 2(d_o + d_i) \tag{5.14}$$

$$A_{wire} = w^2 = \frac{(d_o - d_i)^2}{(2N - 1)^2} \tag{5.15}$$

Now the coil resistance for a single layer, micro-fabricated coil can be expressed as:

$$R_c = \frac{2\rho_{cu}(d_o + d_i)}{(d_o - d_i)^2}(4N^3 - 4N^2 + N) \tag{5.16}$$

For large N, the N^2 and N terms can be neglected. The dependence of the coil resistance on the cube of the number of turns, arises in this case from the fact that the turn cross-sectional area is dependant on the square of the number of turns, i.e. both track width and thickness decrease with N. Equation (5.15) assumed that the track width, spacing and thickness are equal. It may be possible to achieve a higher aspect ratio, so that track thickness can be several times the width and spacing. This would mean lower coil resistance could be achieved; however, the dependence on the cube of the number of turns would remain. It will be shown later that this dependence of resistance on the cube of the number of turns proves to be the limiting factor for the use of micro-coils in vibration generators. The essential difference between the wire-wound coil and the micro-fabricated coil is that we have assumed that wire-winding is a 3D technology, whereas micro-fabrication is a 2D, planar technology. If there existed a micro-fabrication technology, where the number of coil layers could be arbitrarily high, then the micro-fabricated coil resistance could approach a dependence on the square of the number of turns. However, in practice, micro-fabrication techniques are limited in the number of layers which can be practically achieved.

5.5 Magnetic Materials

The magnetic circuit employed in an electromagnetic generator requires a magnetic field which is generated by the use of permanent (or hard) magnets. Magnetic fields can also be generated by electromagnets but these require a current flow and hence consume power. Therefore, in the case of small-scale low-power devices, the use of electromagnets is not suitable. Permanent magnets are made from ferromagnetic or ferrimagnetic materials that remain magnetic after the application of a magnetisation process.

The atoms of a ferromagnetic material have unpaired electrons and therefore exhibit a net magnetic moment. These atoms are grouped together in large numbers and form magnetic domains within which their magnetic moments are aligned in a particular direction. Taking the magnet as a whole, in the unmagnetised state, the domains are randomly aligned and there is no net magnetic field produced. In magnetised materials, the domains become aligned in the same direction producing a strong magnetic field. Ferrimagnetic materials are subtly different to ferromagnetic in that they contain atoms with opposing magnetic moments. However, the magnitude of these moments is unequal and hence a net magnetic field will exist. Ferrimagnetic materials are of interest since their electrical resistance is typically higher than ferromagnetic materials and therefore eddy current effects are reduced. The net magnetic field associated with a permanent magnet is characterised by north and south poles of equal strength and when placed in close proximity, like pole types repel while opposing poles attract each other.

These materials can be magnetised by the application of a magnetic field of sufficient strength (coercive force). Some materials are classified as 'soft' meaning they can be easily magnetised and also easily demagnetised (e.g. by vibrations). 'Hard' magnetic materials require greater magnetising fields and are also more difficult to demagnetise. Magnets can be magnetised by other permanent magnets with sufficient magnetic field or, in the case of harder materials, electromagnets have to be used. These require high currents to achieve the necessary magnetic fields and are typically pulsed with the current applied for a few milliseconds. During magnetisation, some domains align more easily than others and therefore magnets can exhibit different degrees of magnetisation depending on the coercive force of the magnetising field. Magnets with fully aligned domains are saturated and cannot be magnetised further.

The magnetic field produced by a permanent magnet is typically denoted B. A term of key importance in electrical power generation is magnetic flux (ϕ) which is the product of magnetic field multiplied by the area. It follows that B is also known as the flux density. The magnetic field strength (and magnetising force) is denoted H. The terms B and H are linked by Eq. (5.17) where μ_m is the product of the permeability of free space times the permeability of the material. The units of measure for these factors are given in Table 5.2.

$$B = \mu_m H \qquad (5.17)$$

A useful figure of merit for comparing magnetic materials is the maximum energy product, BH_{MAX}, calculated from a materials magnetic hysteresis loop. At this point in the loop, the volume of material required to deliver a given level of energy into its surroundings is a minimum. Another factor to be considered when comparing magnets is the Curie temperature. This is the maximum operating temperature the material can with withstand before the magnet becomes demagnetised.

Typically, four types of magnet are available: Alnico, ceramic (hard ferrite), samarium cobalt and neodymium iron boron. Each type is subdivided into a range of grades each with its own magnetic properties.

Alnico, developed in the 1940s, is an alloy of varying percentages of aluminium, nickel, cobalt, copper, iron and titanium. It is stable with temperature and can be used in high-temperature applications (up to $\sim 550\,^{\circ}\mathrm{C}$). It is inherently corrosion resistant and has a maximum energy product only surpassed by that of the rare earth magnets. Ceramic or hard ferrite magnets have been commercialised since the 1950s are widely used due to their low cost. They are hard, brittle materials available in a range of compositions, e.g. iron and barium ($BaFe_2O_3$) or strontium oxides ($SrFe_2O_3$). The materials are mixed in powder form and then pressed and sintered to form the magnet geometries.

Table 5.2 Units of measure for magnetic properties

Unit	Symbol	SI	cgs	Conversion factor (cgs to SI)
Flux	ϕ	Weber (Wb)	Maxwell (Mx)	10^8
Flux density	B	Tesla (T)	Gauss (G)	10^{-4}
Magnetic field strength	H	Ampere-turns/m	Oersted (Oe)	$10^3/4\pi$
Permeability	μ	H/m	–	–

Samarium cobalt (SmCo) and neodymium iron boron (NdFeB) are the most recently commercialised magnet materials having been available since the 1970s and 1980s, respectively. They are known as rare earth magnets because they are composed of materials from the lanthanide series within the rare earth group of elements. Both types exhibit much higher magnetic fields than Alnico or ceramic materials. The highest maximum energy product is achieved with NdFeB, but this material suffers from low-working temperatures and poor corrosion resistance. Samarium cobalt magnets exhibit good thermal stability, a maximum working temperature of around 300 °C and are inherently corrosion resistant. Both types are fabricated using similar powder metallurgy processes used with ceramic magnets although NdFeB is also available in bonded form where the powder is held in an epoxy or nylon matrix. Bonded NdFeB magnets can be made in a wide variety of geometries but the magnetic properties are reduced compared to the sintered approach.

For an inertial vibration energy harvesting application, the particular properties of interest relate to the strength of the magnetic field, the flux density and the coercive force. An inertial energy harvester can be relatively easily packaged to minimise risks of corrosion and temperature constraints are also unlikely to be an issue. Therefore, the strongest range of NdFeB magnets are the preferred option in the majority of cases since their use will maximise the magnetic field strength within a given volume. Also they have a high-coercive force and therefore the vibrations of the generator will not depole the magnets. In most generators, it may be beneficial to have the magnet contribute all or part of the inertial mass. Magnetic densities have been included in Table 5.3 and it can be seen that the magnet types have similar densities, apart from ferrites which are $\sim 30\%$ lower.

For small-dimension generators, the deposition and patterning of the magnets using micro-fabrication techniques might be considered. Deposition techniques such as sputtering and electroplating have been used for the fabrication of microscale magnets, with electroplating generally being favoured as the more cost-effective technique for the deposition of thick layers. However, it is the case that the properties (coercive force, Hc, remanence, Br and energy density, BH_{max}) achievable from micro-fabricated magnets are considerably lower than those achievable from bulk-sintered rare earth magnets such as samarium–cobalt or neodymium–iron–boron. For example, the highest coercivity and remanence reported for a 90 μm

Table 5.3 Common magnetic material properties, flux density measured at their pole face with them working at their BH_{max} points (**http://www.magnetsales.co.uk/application_guide/ magnetapplicationguide.html**)

Material	$(BH)_{MAX}$ (kJ/m^3)	Flux density[1] (mT)	Max working temp. (°C)	Curie temp (°C)	Coercive force (Hc)	Density (kg/m^3)
Ceramic	26	100	250	460	High	4980
Alnico	42	130	550	860	Low	7200
SmCo (2:17)	208	350	300	750	High	8400
NdFeB (N38H)	306	450	120	320	High	7470

thick deposited magnet is 160 kA/m and 0.5 T (Ng et al. 2005). This compares to
a values of around 800 kA/m and 1 T for a bulk magnet. Moreover, the properties
tend to degrade with thickness, so that thick (tens of micro metres) micro-fabricated
magnets with good properties are difficult to achieve. Therefore, considerable further work is required on micro-fabricated permanent magnets in order to make them
viable as an alternative to discrete sintered magnets for vibration generators.

5.6 Scaling of Electromagnetic Vibration Generators

For vibration generators, it is of interest to understand how the power generated is
related to the size of the generator, and in particular in the case of the electromagnetic generator, how the power is limited by the interactions of the coils and magnets
as the scale is reduced. In a vibration-powered generator, the available mechanical
energy is associated with the movement of a mass through a certain distance, working against a damping force. Clearly, this will decrease with the dimensions as both
the mass of the moving object and the distance moved is decreased. Generally, the
damping force which controls the movement will consist of parasitic damping and
electromagnetic damping. The electrical energy which can be usefully extracted
from a generator depends on the electromagnetic damping, which as was shown
earlier depends on flux linkage gradient, the number of coil turns, coil impedance
and load impedance. These factors also depend on scale, so that typically as dimension decreases, the magnitude of the magnetic fields decrease and the quality of the
coils decrease and hence the ability to extract electrical energy may be reduced.

In order to investigate how the power achievable from an EM generator scales
with the dimension of the generator, an analysis is presented for an example electromagnetic generator structure shown in Fig. 5.3. This structure consists of a coil
sandwiched between magnets, where the upper and lower magnets consist of two
pairs, oppositely polarised. This opposite polarity creates a flux gradient for the coil
in the direction of movement, which in this case, is in the x-direction. This is a
representation of the type of electromagnetic generator which has been reported in
Glynne-Jones et al. (2004) and Beeby et al. (2007) to have high-power density.

The following assumptions are made in relation to the structure;

- The coil remains fixed while the magnets move in response to the vibration.
  Since the magnets generally have greater mass, m, than the coil, movement of
  the magnets is more beneficial than movement of the coil.
- A cubic volume, i.e. $x = y = z = d$ where d is the dimension of the device is
  assumed.
- Only the magnet and coil dimensions are included in the volume. Any practical
  generator must also include the volume of the housing and the spring. Since the
  housing and the spring can be implemented in many different ways, these are
  neglected for the present analysis. It is assumed that the spring can be implemented so as to allow the required movement of the mass at the frequency of
  interest.

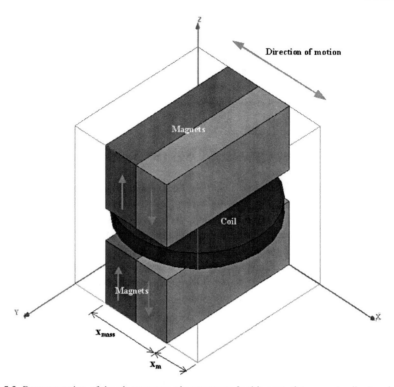

Fig. 5.3 Representation of the electromagnetic generator. In this case, the magnets vibrate relative to the coil in the x-direction. The peak displacement, $x_m = (x - x_{mass})/2$

- The movement of the mass in response to the vibrations is assumed to be sinusoidal and is well represented by the equations of motion for a mass–spring–damper system. The frequency of the input vibrations is assumed to perfectly match the resonant frequency of the device.

For any practical device, the maximum displacement will be constrained by the volume, i.e. the peak displacement, x_m, is given by the difference between the external dimension, d, and the dimension of the mass, x_{mass}. Therefore, a choice can be made to have a thin mass with a large displacement or a wide mass with a small displacement. In fact there exists an optimum ratio of peak displacement, x_m, to mass dimension, x_{mass}, which maximises the mechanical energy. The average power dissipated by the damping force of the mass is:

$$P_D = \frac{1}{T} \int_0^T F_D.U \, dt = \frac{(ma)^2}{2D} \tag{5.18}$$

However at resonance the damping, D controls the peak displacement, according to

$$x_m = \frac{ma}{D\omega} \qquad (5.19)$$

So that the average power can be expressed as:

$$P_D = \frac{m.a.x_m\omega}{2} = \frac{\rho.x_{mass}y.z.(x - x_{mass})\omega}{4} \qquad (5.20)$$

where in the second expression, the peak displacement of the mass is $x_m = (x - x_{mass})/2$, and the mass is expressed as the product of the density of the mass material ρ, and its dimensions, x_{mass}, y and z. Differentiating Eq. (5.20) with respect to x_m and equating to zero gives the optimum mass length for maximum power as:

$$x_{massopt} = \frac{x}{2} \qquad (5.21)$$

From the point of view of the structure in Fig. 5.3, this implies that a single magnet x-dimension should be taken to be one-fourth of the overall dimension. These magnets are assumed to extend for the full y-dimension and the z-dimension of the magnet is taken to be 0.4 times the overall dimension. This leaves the gap between the magnets as one-fifth of the overall dimension. The coil thickness is assumed to occupy half of the gap.

For the scaling analysis, it is assumed that as the dimension is reduced, all of the relative dimensions are retained. Thus, for example, at a dimension of 10 mm the magnets are 4 mm in height, with a 2 mm gap between them. At a dimension of 1 mm the magnets are 0.4 mm in height with a gap of 0.2 mm between them. Now the scaling of the available power is easy to calculate using Eq. (5.20). In order to do this, the total damping is set so as to limit the displacement to be within the volume. Fig. 5.4 plots an example of the scaling of the available power versus the dimension assuming a mass which is made from a sintered rare earth magnetic material, NdFeB, a vibration frequency of 100 Hz and a vibration acceleration level of 1 m/s^2. This graph shows that the power available is proportional to the fourth power of the dimension, arising from the dependence on mass (proportional to cube of dimension) and displacement (proportional to dimension). The corresponding power density is linearly related to dimension.

Fig. 5.4 plots the power dissipated by the total damping force, with the damping set so as to contain the displacement within the volume. This represents a fundamental limit for the achievable power, for the assumptions used. The graph indicates a power density of approximately 2.3 μW/mm^3 (2.3 mW/cm^3) for a device occupying a volume of 1 cm^3. It should be remembered that this analysis does not include the volume of the spring and housing, so that any real generators will necessarily have lower power densities. This graph represents simply the power dissipated by the damping force, but makes no assumption about how that damping force is achieved. To maximise electrical power, ideally we would like all of this damping to be electrical damping, or at least to maximise electrical damping. However, as Eq. (5.9) shows, the magnitude of the electromagnetic damping depends on the

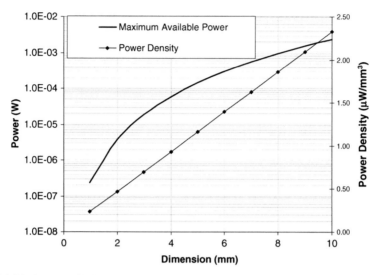

Fig. 5.4 Maximum available and power density versus the dimension for the electromagnetic generator construction in Fig. 5.3, for an acceleration level of $9.81 \, \text{m/s}^2$ and an acceleration frequency of $100 \, \text{Hz}$

magnet parameters and the coil parameters. The magnetic fields may decrease with dimension and the resistance of coils tends to increase. These effects place additional limits on the electrical power which can be achieved and are investigated in the following sections.

5.7 Scaling of Electromagnetic Damping

From the earlier analysis, the electromagnetic damping was derived as:

$$D_{\text{em}} = \frac{1}{R_{\text{c}} + j\omega L + R_{\text{L}}} \left(N \frac{d\varphi}{dx} \right)^2 \tag{5.22}$$

which shows that the damping depends on the coil parameters and the flux linkage which both have a dependence on the dimension.

The flux linkage depends on the gradient of the magnetic field and the area of the coil. Because the dimension of the gap between the magnets scales with the dimension of the magnets, the magnitude of the B field remains constant with scale reduction and hence the gradient of the field would actually increase (Cugat et al., 2003). However, for a vibration generator, the parameter of importance is the gradient of the flux linkage with the coil, which of course depends on the area of the coil. Therefore, even though the flux density gradient is inversely proportional to dimension, the area of the coil will be proportional to the square of the dimension and hence the overall effect is for the flux linkage gradient to be directly proportional to dimension.

The coil parameters such as number of turns, resistance and inductance are not independent. For a fixed coil volume, the coil resistance and inductance depends on the number of turns. In the earlier section, the resistance of the wire-wound coil was defined as:

$$R_c = \frac{\rho N^2 \pi (r_o + r_i)}{f(r_o - r_i)t} \tag{5.23}$$

It can be seen from this expression that resistance of the wire wound coil depends on the square of the number of turns assuming that the fill factor is constant.

For a single layer, micro-fabricated coil, the resistance was expressed as:

$$R_c = \frac{8\rho(d_o + d_i)}{(d_o - d_i)^2}(4N^3 - 4N^2 + N) \tag{5.24}$$

For large N, the N^2 and N terms can be neglected. The dependence of the coil resistance on the cube of the number of turns, arises in this case from the fact that the turn cross-sectional area decreases with the number of turns squared, i.e. both track width and thickness decrease with N.

We can see from both the expressions for resistance that for a constant number of turns, the coil resistance is proportional to the inverse of the scaling factor. However, another interesting point to note is how, using the above expressions, the electromagnetic damping depends on the number of turns. In the case of the wire-wound coil, the resistance is proportional to the square of the number of turns, so that in this case the electromagnetic damping is independent of the number of turns. However, in the case of the micro-coil, the resistance is proportional to the cube of the number of turns, so that the electromagnetic damping is inversely proportional to the number of turns.

Now considering that the flux linkage gradient is directly proportional to the dimension, and the coil resistance is inversely proportional to the dimension (for fixed N), it can be seen that the electromagnetic damping is proportional to the square of the dimension. Moreover, the analysis shows that the decrease in electromagnetic damping with scale cannot be compensated by increasing the number of turns. This is because for a wire wound coil the effect of any increase in the number of turns on EM damping is directly cancelled by an increase in coil resistance. For a planar micro-coil, the situation is worse as for this case, the effect of any increase in the number of turns results in a greater increase in the coil resistance, resulting in an overall decrease in damping. Thus, for a planar micro-coil, the highest EM damping would be achieved for a single turn coil.

It should, however, be noted that the voltage generated always depends directly on the number of turns, so that from a practical point of view a high number of turns may be required in order to obtain an adequate output voltage. For planar micro-coil implementations, this forces a design compromise between achieving high-EM damping and high voltage.

Considering these scaling laws we can now look at how much electrical power might be obtained from our example generator shown earlier in Fig. 5.3 using both a wire-wound coil and a micro-coil. To do this, finite element analysis can be employed to estimate the flux linkage gradient which can be obtained for the coil, using NdFeB magnets. A value of 1.5×10^{-3} Wb/m is obtained for the 10 mm dimension generator. Coil resistance can be estimated for the wire-wound coil using Eq. (5.23), assuming a copper fill factor of 0.5. For the micro-coil, resistance can be estimated using Eq. (5.16). An excitation vibration frequency of 100 Hz with an acceleration level of 1 m/s^2 is again assumed. Note that electromagnetic damping is a maximum when the load resistance is zero, i.e. a short circuit condition. However, from a practical point of view, this is not a useful condition as some load voltage is required. Therefore, we assume that the load and coil resistances are equal. For this illustrative example, Fig. 5.5 shows the maximum available power, the maximum electrical power which can be extracted using a wire-wound coil technology and the maximum electrical power which can be extracted using a planar micro-coil. This graph illustrates the significant difference between the power which can be extracted using the different coil technologies.

In the case of the wire-wound coil technology, for larger dimensions, the entire damping can be provided by electromagnetic damping. However, since the EM damping is proportional to the square of the dimension, at smaller dimensions only a fraction of the damping can be supplied by EM damping so that the electrical power is less than the available power. Note that the number of turns in the coil can

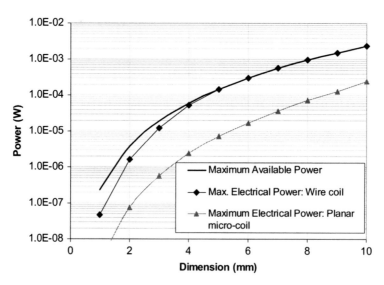

Fig. 5.5 Maximum available power and electrical power which can be extracted using a wire-wound coil and a micro-fabricated coil

be chosen arbitrarily as the damping was shown to be independent of the number of turns.

In the case of the micro-coil, the situation is quite different, and only a fraction of damping can be supplied by the EM damping at all dimensions. This arises directly from the fact that the micro-coil resistance is considerably larger than the wire-wound coil which severely limits the achievable damping. In fact as was shown earlier, the micro-coil resistance depends on the third power of the number of turns in the coil, which would mean damping is maximised for a single turn coil. However, a single turn coil gives rise to an impractically low voltage, therefore to generate the above curve the number of turns are fixed to be 150, which gives a voltage of several hundred milli-volts for the larger dimension devices. The actual value of power extracted may be improved with a different choice of turns but the conclusion regarding the limiting effect of the coil resistance for planar micro-coils remains valid.

5.8 Maximising Power from an EM Generator

In the above analysis, in order to illustrate some theoretical limits, it was assumed that the motion of the mass was displacement limited, i.e. the maximum displacement of the mass was limited by the volume, and that we were free to design the total damping (electrical or mechanical) so as to limit the mass to this displacement. However, in reality, mechanical or parasitic damping, which may arise due to the material or air damping, can be a major limiting factor for the displacement. It is well known that for any vibration generator, irrespective of the working principle, that the maximum electrical power is extracted when the electrical damping is made equal to the parasitic mechanical damping, D_p, which is always present in the system and this maximum electrical power, at resonance, can be expressed as:

$$P_{\max} = \frac{(ma)^2}{8D_p} \qquad (5.25)$$

In such a case where the generator is limited by parasitic damping, the optimisation strategy to extract maximum electrical power is to try to make the EM damping equal to the parasitic damping, by choosing an optimum load resistance (assuming that the design already maximises flux linkage gradient). Equating the expression for EM damping to parasitic damping and rearranging to obtain the load resistance, gives the optimum load resistance which maximises generated electrical power as:

$$R_l = \frac{N^2 \left(\frac{\mathrm{d}\phi}{\mathrm{d}x}\right)^2}{D_p} - R_c \qquad (5.26)$$

However, in a practical situation, we want to maximise load power as opposed to generated power. The optimum load resistance which maximises load power can be shown to be given by:

$$R_l = R_c + \frac{N^2 \left(\frac{d\phi}{dx} \right)^2}{D_p} \tag{5.27}$$

Note that in the case where parasitic damping is much greater than electromagnetic damping (parasitic damping much greater than flux linkage gradient term), the optimum load resistance becomes the coil resistance.

5.9 Review of Existing Devices

This section reviews electromagnetic energy harvesting devices reported to date. The devices have been grouped according to size with a further section describing commercial devices. Many of the principles discussed in this chapter are illustrated in these examples; in particular, the issues relating to device size and fabrication constraints are clearly demonstrated. The review also highlights some of the linear magnetic circuits that may be employed in energy harvesting applications. For other reviews, the articles by Beeby et al. (2006) and Arnold (2007) are recommended.

5.10 Microscale Implementations

The earliest microscale device was reported by researchers from the University of Sheffield, UK (Williams et al. 1996, 2001, Shearwood and Yates 1997). Figure 5.7 shows the electromagnetic approach. It consists of two parts: an upper gallium arsenide (GaAs) wafer containing the spring mass arrangement and a lower wafer that incorporates the planar integrated coil. The seismic mass consists of a vertically polarised 1 mm × 1 mm × 0.3 mm samarium–cobalt magnet of mass 2.4×10^{-3} kg which is attached to a 7 μm thick cured polyimide circular membrane that forms the spring element. The planar coil is formed on the bottom wafer from a 2.5 μm thick gold (Au) thin film layer is patterned using a lift off process which yielded 13 turns of 20 μm line width and 5 μm spacing. The two wafers were bonded together using silver epoxy to form the assembled generator. The overall size of the electromagnetic transducer described by is around 5 mm × 5 mm × 1 mm and a schematic is shown in Fig. 5.6.

When the generator is subject to external vibrations, the mass moves vertically out of phase with the generator housing producing a net movement between the magnet and the coil. Early design work presented the familiar equation predicting power output from such a device (Williams and Yates, 1996). This analysis predicted that $+/-50$ μm inertial mass displacement would produce 1 μW at 70 Hz and 100 μW at 330 Hz. In these cases, the driving amplitude, Y, was fixed at 30 μm and, since acceleration $A = \omega^2 Y$, the corresponding acceleration levels are 5.8 and 129 m/s^2 respectively. The device was tested and generated 0.3 μW at excitation

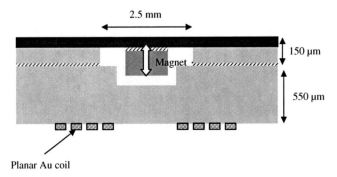

Fig. 5.6 Cross section of the electromagnetic generator proposed by Williams et al. (2001)

frequency of 4.4 kHz with external vibrations, Y, equalling 0.5 μm equating to an acceleration of 382 m/s^2. The measured electrical power output was considerably lower than the predicted value and this was thought to be due to the non-linear effects arising from the membrane spring. This exhibited the hard-spring effect where the resonant frequency increases with increasing excitation amplitude. More significantly, however, is the low electromagnetic damping which would reduce the practical power output achieved. The damping could be improved by bringing the coil closer to the magnets. However, the low number of coil turns would always limit the voltage achievable and increasing the number of turns would only result in higher coil resistance and decreased EM damping. This highlights the key fabrication challenge facing microscale electromagnetic devices. Another consideration highlighted by this design is the high-resonant frequency achieved in practice. This arises from the small device size and is typical for micro-machined generators.

Similar designs to the Sheffield generator have been demonstrated by other researchers. Huang et al. (2003) have presented a two wafer device that uses an electroplated nickel–iron suspension spring element upon which the magnet is mounted. The second wafer hosts the integrated coils which were fabricated by electroplating copper into a 100 μm thick photoresist mould. The coil has an internal resistance of 2 Ω although the number of turns is not stated. The device is targeted at generating power from human motion and is reported to generate 0.16 μW from a 'finger tap'. The device has a resonant frequency of 100 Hz and when operated at resonance with an excitation amplitude of 50 μm (19.7 m/s^2) it produces 1.4 μW.

Researchers from the University of Barcelona have described a similar structure in the paper by Pérez-Rodríguez et al. (2005). They have bonded a neodymium iron boron (NdFeB) magnet to a polyimide membrane to form the spring–mass system with the polyimide membranes being attached to a PCB square frame around which the coil has been formed. Initial results were from a planar coil made from a 1.5 μm thick aluminium layer. The power output is reported to be 1.44 μW for an excitation displacement of 10 μm and a resonant frequency of 400 Hz, which equates to 63 m/s^2. It was observed that the parasitic damping in this device was far greater than the electromagnetic damping factor and therefore maximum power output was

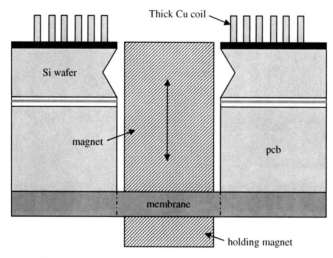

Fig. 5.7 Improved EM generator with electroplated copper coil (after Serre et al. 2008)

not achieved. This is due to the poor method of assembly introducing large damping effects and the poor electromagnetic coupling factor due to the geometry of the device. A later paper by the group reported on some improvements and the revised design is shown in Fig. 5.7 (Serre et al. 2008). The device has a top 1 mm thick silicon (Si) (100) wafer which contains the 15 μm thick, 20 μm wide electroplated copper tracks which form a coil with 52 turns and a resistance of 100 Ω. The spring–mass element is formed by bonding a Kapton membrane between two PCB frames with the NeFeB magnet held at the centre of the membrane by a second smaller holding magnet. The improved version of the device generates about 55 μW at 380 Hz from a 5 μm excitation amplitude which equates to 29 m/s². Unfortunately, the device still demonstrates stress stiffening and therefore hysteresis. The use of polymer membranes, such as Kapton, as the spring element in these devices also suffers from the potential for creep and fatigue failure.

A similar electromagnetic circuit has more recently been described by Wang et al. (2007). They have developed a micro-machined electromagnetic generator that combines an electroplated copper planar spring with an integrated coil. The electroplated spring is formed on a silicon substrate with the silicon being etched away in KOH to leave the freestanding copper structure. A NeFeB permanent magnet is manually glued to the centre of the spring and the overall structure has resonant frequencies of around 55, 121 and 122 Hz. A two layer coil was fabricated on a glass wafer using a combination of spin-coated resists, copper electroplating and polyimide coating to insulate the two layers. The coil and magnet spring layers were simply glued together to assemble the device. Maximum output at modes 2 and 3 where around 60 mV generated from 14.7 m/s².

A team of researchers from The Chinese University of Hong Kong have developed a magnet spring arrangement which comprises of a NdFeB magnet mounted on

laser-micro-machined spiral copper spring (Li et al. 2000a, b). Meandering and spiral spring designs were simulated and the spiral design was found to have twice the
deflection. The springs were laser machined from 110 μm thick copper had diam-
eters ranging from 4 to 10 mm with width/gap dimensions ranging from 40/40 μm
to 100/100 μm. The spring–mass structure has a resonant frequency of 64 Hz. The
wire-wound coil is fixed in position on the housing of the structure. The device has
an overall volume of around 1 cm^3 and produces 2 V at its resonant frequency. This
is a power output of 10 μW for an excitation amplitude of 100 μm (16.2 m/s^2). A
later publication by Ching et al. (2002) describes a similar generator fabricated on
a printed circuit board. An improved spiral spring resulted in a peak-to-peak output
voltage of up to 4.4 V and a maximum power of 830 μW when driven by a 200 μm
displacement in its third mode with a frequency of 110 Hz (equates to 95.5 m/s^2).

Mizuno and Chetwynd (2003) proposed an array of micro-machined cantilever-
based electromagnetic generators to boost total power output. Each cantilever
element comprises an inertial mass, an integrated coil and is aligned adjacent to
a common fixed external magnet. The dimensions of the proposed micro-cantilever
beam were 500 μm × 100 μm × 20 μm. The theoretical analysis predicted a power
output of only 6 nW and a voltage output of 1.4 mV for a typical single cantilever
element at a resonant frequency of 58 kHz. For evaluation, the authors fabricated a
larger scale version of their proposed device. The size of the cantilever beam was
increased to 25 mm × 10 mm × 1 mm which was placed next to a fixed NdFeB
magnet whose size was 30 mm × 10 mm × 6 mm. The cantilever contained four
coil turns fabricated from 50 μm thick gold with a track width of 1 mm. The resonant frequency was 700 Hz. For an input vibration of 0.64 μm (corresponding to
12.4 m/s^2 at 700 Hz), the output power was found to be 320 μV and 0.4 nW across
a 128Ω resistive load (which matched the four turn coil resistance). Suggested
methods to improve the power output include implementing the array of cantilevers
as described initially and reducing the coil magnet gap. The author's conclusion
is pessimistic regarding the opportunities for microscale electromagnetic devices,
especially considering the low-output voltage. The device presented does indeed
lead to this conclusion due, as with the Sheffield generator described above, to the
very low-electromagnetic damping achieved with their design.

Another two part micro-machined silicon-based generator has been described by
Kulah and Najafi (2004). The device consists of two micro machined resonant systems that are bought together in order to achieve mechanical up-frequency conversion. This approach is designed to overcome the high inherent frequency exhibited
by the majority of microscale devices which is not compatible with the majority
of ambient vibration frequencies which are often less than 100 Hz. The generator
is fabricated as two separate chips. The upper chip consists of an NdFeB magnet
bonded to a low stiffness parylene diaphragm giving low resonant frequencies in
the range of 1–100 Hz. The NdFeB magnet is used to excite a lower structure into
resonance through magnetic attraction. The lower chip consists of an array of resonating cantilevers again fabricated from parylene. These cantilevers also include
metal tracks which form the coil. These metal tracks are formed using sputtered
nickel the ferromagnetic properties of which also enable the magnetic excitation.

The microscale designed cantilevers have a resonant frequency of 11.4 kHz. The analysis of the microscale design predicts a theoretical maximum power of 2.5 μW, however, a larger millimetre scale mock up only delivered 4 nW in practice (the level of input mechanical excitation was not quoted). This mock up validated the frequency conversion principle and little attention was paid to the design of the electromagnetic energy harvesting circuit.

One of the main limitations of microscale electromagnetic generators is the limited number of coil turns possible with integrated circuit technology. Scherrer et al. (2005) have evaluated the potential of using low-temperature co-fired ceramics (LTCC) to fabricate a multi-layer screen printed coil. An LTCC device is made up from layers of flexible, un-fired ceramic tape. Conductive tracks can be simply fabricated on the substrate using conventional screen printing. The unfired layers can also be easily machined to form buried channels or vias for interlayer connection. The layers are then stacked, compressed and to form a laminated structure which is finally fired in a furnace. The result is a rigid, strong multilayer device which, in this case, comprises a coil made up of 96 tape layers and has a total of 576 turns. This stack is located between two beryllium copper springs which enable it to move vertically in response to an input excitation with a resonant frequency of 35 Hz. The oscillating stack cuts the flux lines of externally mounted magnets and the theoretical maximum output power was predicted to be 7 mW. This work is of interest as the LTCC technology employed has the potential to fabricate many coil layers and hence to overcome the limitations of micro-fabrication techniques, but still maintaining the advantages of batch fabrication.

In another attempt to overcome the limited number of turns possible with integrated coils, Beeby et al. (2005) attempted to integrate a traditionally wound coil within a micro-machined silicon-resonant structure. The silicon structure consisted of a cantilever beam supporting a 'paddle' in which a circular recess had been etched to accommodate the coil. This structure was deep-reactive ion etched through a standard 525 μm thick silicon wafer and a 600 turn coil wound with 25 μm thick enamelled copper wire was placed in the recess. The silicon coil layer is sandwiched between two perspex chips which house four NeFeB magnets, as shown in Fig. 5.8. The coil and paddle structure is designed to vibrate laterally in the plane of the paddle. The device has a resonant frequency of 9.5 kHz and has been shown to generate 21 nW of electrical power from and acceleration level of 1.92 m/s^2. This low-power output is due to the high-damping levels that arise from the frictional losses from the loose wires that come out from the coil to the edge of device. A preferred arrangement would be to have the magnets on the moving component and the coil fixed. Alternatively, integrated coils could be incorporated on the paddle. This approach was simulated along with various electroplated magnet configurations by Kulkarni et al. (2006). Theoretically, a 250 turn coil vibrating at 7.4 kHz, moving in the magnetic field with an amplitude of 240 μm could produce 950 mV and 85 μW. However, this analysis ignores parasitic damping and assumes high-quality electroplated magnetic films with a flux density of 1.6 T.

A device described by Sari et al. (2007) contains an array of parylene cantilevers arranged around a central square permanent magnet 8 mm × 8 mm × 8 mm in size.

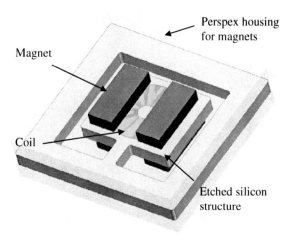

Fig. 5.8 A silicon electromagnetic generator with discrete coil (after Beeby et al. 2005)

Each cantilever has a 10 turn coil sandwiched between three parylene layers giving an overall cantilever thickness of about 15 μm. The cantilevers are released by deep-reactive ion etching from the back which leaves an opening for the magnet. Forty cantilevers are fabricated simultaneously in this manner and the key to this design is that each one has a fractionally different length. This produces a wideband generator that was shown to deliver 0.5 μW power over a frequency band from 3.3 to 3.6 kHz. The objective of realising a wideband response was achieved but the fact each cantilever generator resonates independently means the voltage output is just 20 mV. Also, this approach results in a large overall device size of 1400 mm³.

5.11 Macro-Scale Implementations

This section of the review deals with devices that range in size from 150 mm³ to above 30 cm³. While some of these devices are indeed similar in size to the devices presented in Section 5.10, they are all fabricated using discrete components rather than using micro-machining or MEMS processes.

Amirtharajah and Chandrakasan (1998) from the Massachusetts Institute of Technology describe a self-powered DSP system powered by an electromagnetic generator. The generator consists of a cylindrical housing in which a cylindrical mass is attached to a spring and fixed to one end. A permanent magnet is attached to the other end of the housing and the mass, which is free to oscillate vertically within the housing, has a coil attached to it. As the mass moves the coil cuts the magnetic flux and a voltage is generated. The mass was 0.5 g which, for a spring constant of 174 N/m, gives a resonant frequency of 94 Hz. The peak output voltage was measured at 180 mV which was too low to be rectified by a diode. The authors simulated the generator output in a human-powered application and predicted that an

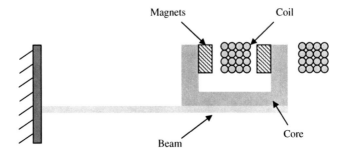

Fig. 5.9 The electromagnetic generator described by El-Hami et al. (2000)

average of 400 μW could be generated from 2 cm movement at 2 Hz, which equates to 3.2 m/s^2.

El-Hami et al. at the University of Southampton, UK (2000, 2001) presented the simulation, modelling, fabrication and characterisation of a cantilever spring-based electromagnetic generator. The cantilever beam is clamped at one end and has a pair of NdFeB magnets attached to a c-shaped core located at the free end. The coil is made up of 27 turns of 0.2 mm diameter enamelled copper wire and is fixed in position between the poles of the magnets. The device volume is 240 mm^3 and is shown in Fig. 5.9. The device was found to produce 0.53 mW at a vibration frequency of 322 Hz and an excitation amplitude of 25 μm (102 m/s^2).

The team at Southampton further developed the device concentrating on improving the magnetic circuit and therefore the power generated. The paper by Glynne-Jones et al. (2004) describes a cantilever-based prototype with a four magnet and fixed coil arrangement similar to that shown in Fig. 5.3. The overall generator

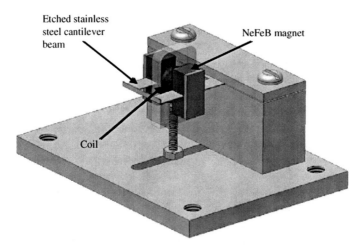

Fig. 5.10 Schematic of the four magnet generator arrangement demonstrated by Glynne-Jones (2004)

volume is 3.15 cm^3 and is shown in Fig. 5.10. At its resonant frequency of 106 Hz and an acceleration level of 2.6 m/s^2, the improved generator produced an output voltage of 1 V, compared with $\sim$ 150 mV for the first arrangement. The generator was mounted on the engine block of a car and its effectiveness naturally depended on the frequency of vibrations which in turn depended on the engine speed. The maximum peak power produced was 4 mW, while the average power was found to be 157 μW over a journey of 1.24 km.

This cantilever-based four magnet arrangement has more recently been realised in miniaturised form during an EU funded Framework 6 project 'Vibration Energy Scavenging' (VIBES). The miniature device, described by Beeby et al. (2007), utilises discrete magnets, conventionally wound coils and machined components. The use of wire erosion and a CNC micro-mill to fabricate these parts has, however, enabled components with sub-millimetre features to be realised. The use of bulk magnets maintains high-flux densities and a coil wound from enamelled copper wire as thin as 12 μm in diameter has resulted in useful levels of energy and power from a device volume of 150 mm^3. The generator produced 46 μW and 428 mV at its resonant frequency of 52 Hz and an acceleration level of 0.59 m/s^2 (excitation amplitude 5.5 μm). Further improvements enabling a degree of manual tunability, increased magnet size and rearrangement of the centre of gravity increased the power output to 58 μW from the same excitation levels (Torah et al. 2007). This device is shown in Fig. 5.11.

A linear electromagnetic generator designed to be driven by human motion has been described by von Buren and Troster (2007). The design consists of a tubular translator which contains a number of cylindrical magnets separated by spacers. The translator moves vertically within a series of stator coils and the spacing and dimensions of the components was optimised using finite element analysis. The vertical motion of the translator is controlled by a parallel spring stage consisting of two parallel beams. The stator and translator designs were realised with six magnets, five coils and a volume of 0.5 cm^3, while the total device volume is 30.4 cm^3. The fabricated prototype produced an average power output of 35 μW when located just below a subject's knee.

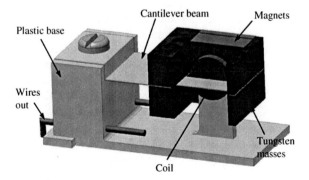

Fig. 5.11 Miniaturised four magnet cantilever generator (cut away)

A similar linear arrangement has been demonstrated by Yuen et al. (2007). The device uses and electroplated copper spring which is released from a substrate on to which an NeFeB magnet is attached, mass 192 mg. The spring mass assembly is located within a conventionally wound coil. Two of these generators have been packaged together with a voltage conducting circuit within a AA battery-sized housing. Two generators in this arrangement were reported to deliver 120 μW power at 70.5 Hz. The acceleration level was measured at 4.63 m/s^2, but this is not consistent with the stated excitation amplitude which was also observed at 250 μm and equates to 49 m/s^2.

An electromagnetic vibration powered generator has been developed by Hadas et al. (2007) as part of a European framework six projects 'WISE – Integrated Wireless Sensing'. This project is concerned with wireless aircraft-monitoring systems and as such the vibration powered generator is designed to operate at relatively low frequencies in the region of 30–40 Hz. The device presented is 45 cm^3 in volume, resonates at 34.5 Hz and delivers around 3.5 mW from 3.1 m/s^2 acceleration levels. The magnetic circuit is not described in this reference but the device is clearly well designed producing good power levels from low-vibration levels. The device is large in size and would have to be miniaturised for the intended application.

5.12 Commercial Devices

The harvesting of kinetic energy for the powering of watches is well known and relies on the conversion of human motion into rotary motion at the generator. A Dutch company Kinetron (www.kinetron.nl) specialise in precision electromechanical products and have also developed a range of rotational electromagnetic generators capable of producing output powers in the range 10–140 mW depending on the rotational speed available. These have been developed for a range of applications including self-powered pedal lights for bicycles and water turbines. Perpetuum Ltd. (www.perpetuum.co.uk) is a UK-based spin out from the University of Southampton. They have developed a series of vibration-powered electromagnetic generators for different applications. The PMG17 generator, designed for industrial condition monitoring applications, is available in 100 or 120 Hz versions. This generator can deliver 1 mW of power, after power conditioning, from acceleration levels of just 0.25 m/s^2. As vibration levels increase the electrical damping levels also rises such that at 2.5 m/s^2 the generator delivers 2 mW over a 20 Hz bandwidth with a peak output of over 10 mW. The device is 135 cm^3 in volume and is certified for use in hazardous environments. A competing product has been developed by Ferro Solution in the United States. The VEH-360 generator has a volume of 250 cm^3 including the evaluation circuit and delivers an unconditioned peak power of 0.8 or 0.43 mW after power conditioning at 0.25 m/s^2 and 60 Hz frequency (http://www.ferrosi.com/files/VEH360_datasheet.pdf). The final commercially available device has been developed by Lumedyne Technologies who claim the best known power density value for a vibration energy harvester of 40 mW/cc per g acceleration (http://www.lumedynetechnologies.com/Energy%20Harvester.html).

This claim, however, is not well qualified in that little information is supplied in the data sheet regarding power output over a range of frequencies and acceleration levels. A more recent whitepaper does contain a graph of power output versus frequency where the power output is about 1 mW for 1 m/s^2 acceleration at a resonant frequency of around 53 Hz (Waters et al. 2008). This absolute value is low in comparison to Perpetuum's device and the device also exhibits a narrow bandwidth which is not desirable.

Table 5.4 contains a summary of the performance of electromagnetic generators reported to date. Comparing different devices is not straightforward since the data presented in published works vary considerably and selecting a suitable factor is problematic. We have used a figure for normalised power (NP) and normalised power density (NPD) to apply a simple comparison between devices. Normalised power is the stated power output of the device normalised to 1 m/s^2 acceleration level since power output varies with acceleration squared. Normalised power density is simply the normalised power divided by the volume. These factors are not perfect since they ignore frequency (which is governed by the application) and, more importantly, it does not reflect the bandwidth of the generator. In practice, the bandwidth of a device is as an important consideration as peak power output in determining the suitability of a generator for a given application. The following analysis does not, therefore, attempt to judge which is the best generator, rather it is useful in identifying trends and highlighting broader comparisons such as integrated versus discrete coil.

Normalised power versus device size has been plotted in Fig. 5.12 for both integrated and discrete coil devices. This figure clearly shows a trend of reducing power output with reducing device size for a given acceleration level. This is largely in

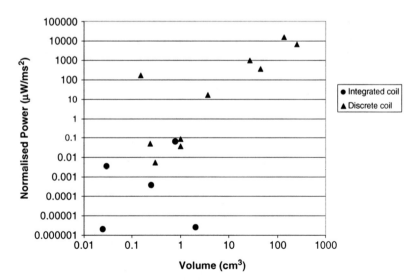

Fig. 5.12 Normalised power versus device volume

Table 5.4 Summary table comparing electromagnetic generators demonstrated to date

Reference	Year	P (µW)	F (Hz)	A (ms^{-2})	Volume (cm^3)	NP	NPD
Shearwood 1997 Sheffield University (UK)	1997	0.3	4400	382	0.025	2×10^{-6}	8×10^{-5}
Amarithajah 1998 MIT (US)	1998	400[1]	94	–	–	–	–
Li Hong Kong University (China)	2000	10	64	16.16	1	0.04	0.04
El-hami University of Southampton (UK)	2001	530	322	102	0.24	0.05	0.21
Ching Hong Kong University (China)	2002	830	110	95.5	1	0.09	0.09
Mizuno Warwick University (UK)	2003	0.0004	700	12.4	2.1*	3×10^{-6}	1×10^{-6}
Huang Tsing Hua University (Taiwan)	2003	1.4	100	19.7	0.03*	0.004	0.12
Glynne-Jones (2001) University of Southampton (UK)	2004	2800	106	13	3.66	16.6	4.5
Kulah Michigan University (US)	2004	0.004	25	–	2	–	–
Pérez-Rodríguez Univerity of Barcelona (Spain)	2005	1.44	400	63	0.25	4×10^{-4}	1×10^{-3}
Beeby University of Southampton (UK)	2005	0.021	9500	1.92	0.3	5×10^{-3}	0.02
Scherrer Boise State University (US)	2005	7000**	35	–	9	–	–

Table 5.4 (Continued)

Reference	Year	P (μW)	F (Hz)	A (ms^{-2})	Volume (cm^3)	NP	NPD
Beeby University of Southampton (UK)	2007	58	52	0.59	0.15	166	1110
Wang Shanghai Jiao Tong University (China)	2007	-	121	14.7	0.1*	-	-
Hadas Brno University of Technology (Czech Republic)	2007	3500	34.5	3.1	45	364	8.1
Sari Middle East Technical University (Turkey)	2007	0.5	3300–3600	-	1.4	-	-
Serrer University of Barcelona (Spain)	2008	55	380	29	0.8*	0.07	0.08
Perpetuum PMG-17 UK company	2008	1000	100	0.25	135	16000	118
Ferro Solutions VEH-360 US Company	2008	430	60	0.25	250	6880	28
Lumedyne Technologies US Company	2008	1000	53	1	27	1000	37

*Estimated value from literature, **simulated value

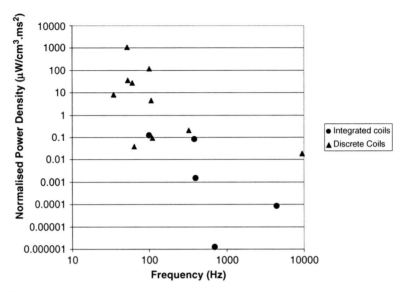

Fig. 5.13 Normalised power density versus device frequency

agreement with the scaling analysis presented previously. As a point of comparison, Fig. 5.5 predicted a normalised power of approximately 2000 μW for a volume of 1 cm³, which compares with the actual achieved powers of approximately 1 μW. This comparison cannot be taken to be very accurate, as many of the fabricated generators are very different in construction, the volume includes the spring and housing and they operate at various frequencies. Nevertheless, the three orders of magnitude difference might suggest that there is still significant room for optimisation in many of the devices fabricated to date. The integrated coil devices naturally tend to be smaller in size and also deliver lower power levels (and voltage levels). Normalised power density has been plotted against device frequency in Fig. 5.13. This shows that lower frequency device tend to have higher power densities despite their tendency for larger device size. Once again, discrete coils offer higher performance levels than their integrated counterpart.

5.13 Conclusions

Electromagnetic power generation is an established technology and the use of this transduction mechanism in small-scale energy harvesting applications is well researched. It is clear from the fundamental principles, and from the analysis of generators tested to date, that electromagnetic devices do not favourably scale down in size. Future applications for microscale vibration energy harvesters will be best served by piezoelectric and electrostatic transduction mechanisms that can more easily be realised using MEMS technology. However, where size is not a constraint and conventional discrete coils and magnets can be employed, electromagnetic generators provide the highest levels of performance achievable to date. Electromagnetic energy conversion relies only on relative velocity and change in magnetic flux to generate electricity and therefore an electromagnetic device will not be limited in amplitude by the fatigue strength of, for example, piezoelectric materials. In an electromagnetic generator, the design of the spring element can be chosen purely on the basis of the best spring material and will not be compromised by the inferior spring properties of a piezoelectric material. The high level of electromagnetic damping achievable also means these devices demonstrate the broadest frequency bandwidth over which energy can be harvested compared with other transduction mechanisms. While this is an important consideration, challenges remain to maximise the conversion efficiency of small-scale generators and to realise adaptive generators that can track frequency changes in the environment. Magnetics may find other uses in these generators as a means to adjust the stiffness of the spring element and achieve a degree of frequency adjustment in the device.

References

Amirtharajah R, Chandrakasan AP (1998) Self-powered signal processing using vibration-based power generation, IEEE J. Solid-State Circuits 33: 687–695

Arnold DP (2007) Review of microscale magnetic power generation. IEEE Trans. Magnetics 43(11): 3940–3951

Beeby SP, Torah RN, Tudor MJ, Glynne-Jones P, O'Donnell T, Saha CR, Roy S (2007) A micro electromagnetic generator for vibration energy harvesting. J. Micromech. Microeng. 17: 1257–1265

Beeby SP, Tudor MJ, Koukharenko E, White NM, O'Donnell T, Saha C, Kulkarni S, Roy S (2005) Micromachined silicon generator for harvesting power from vibrations. Proceedings of Transducers 2005, Seoul, Korea: 780–783.

Beeby SP, Tudor MJ, White NM (2006) Energy harvesting vibration sources for microsystems applications. Mease. Sci. Technol. 17: R175–R195.

Ching NNH, Wong HY, Li WJ, Leong PHW, Wen Z (2002) A laser-micromachined vibrational to electrical power transducer for wireless sensing systems. Sens Actuators A 97–98: 685–690

Cugat O, Delamare J, Reyne G (2003) Magnetic Micro-Actuators and Systems (MAGMAS), IEEE Trans. Magnetics 39(5).

El-Hami M, Glynne-Jones P, White NM, Hill M, Beeby S, James E, Brown AD, Ross JN (2000) A new approach towards the design of a vibration-based microelectromechanical generator. Proc. 14th European Conference on Solid-State Transducers, Copenhagen, August 27–30: 483–486

El-Hami M, Glynne-Jones P, White NM, Hill M, Beeby S, James E, Brown AD, Ross JN (2001) Design and fabrication of a new vibration-based electromechanical power generator. Sens. Actuators A 92: 335–342

Glynne-Jones P (2001) Vibration powered generators for self-powered microsystems, PhD Thesis, University of Southampton

Glynne-Jones P, Tudor MJ, Beeby SP, White NM (2004) An electromagnetic, vibration-powered generator for intelligent sensor systems. Sens. Actuators A110(1–3): 344–349

Hadas Z, Kluge M, Singule V, Ondrusek C (2007) Electromagnetic Vibration Power Generator. IEEE Int Symp Diagnostics for Electric Machines, Power Electronics and Drives: 451–455

http://www.ferrosi.com/files/VEH360_datasheet.pdf

http://www.lumedynetechnologies.com/Energy%20Harvester.html

http://www.magnetsales.co.uk/application_guide/magnetapplicationguide.html

Huang WS, Tzeng KE, Cheng MC, Huang RS (2003) Design and fabrication of a vibrational micro-generator for wearable MEMS. Proceedings of Eurosensors XVII, Guimaraes, Portugal 695–697

Kulah H, Najafi K (2004) An electromagnetic micro power generator for low-frequency environmental vibrations. Micro Electro Mechanical Systems: 17th IEEE Conference on MEMS, Maastricht: 237–240

Kulkarni S, Roy S, O'Donnell T, Beeby S, Tudor J (2006) Vibration based electromagnetic micropower generator on silicon. J. Appl. Phys. 99: 08P511

Li WJ, Ho TCH, Chan GMH, Leong PHW, Wong HY (2000a) Infrared signal transmission by a laser-micromachined vibration induced power generator. Proc. 43rd Midwest Symp. on Circuits and Systems, Aug 8–11: 236–239

Li WJ, Wen Z, Wong PK, Chan GMH, Leong PHW (2000b) A micromachined vibration-induced power generator for low power sensors of robotic systems. Proc of the World Automation Congress: 8th International Symposium on Robotics with Applications, Hawaii, June 11–14

McLyman W.T, (1988) Transformer and Inductor Design Handbook, Second Edition, Marcel Dekker Inc.New York

Mizuno M, Chetwynd D (2003) Investigation of a resonance microgenerator. J. Micromech. Microeng. 13: 209–216

Ng WB, Takado A, and Okada K (2005) Electrodeposited CoNiReWP thick array of high vertical magnetic anisotropy, IEEE Trans. Magn. 41(10): 3886–3888

Pérez-Rodríguez A, Serre C, Fondevilla N, Cereceda C, Morante JR, Esteve J, Montserrat J (2005) Design of electromagnetic inertial generators for energy scavenging applications Proceeedings of Eurosensors XIX, Barcelona, Spain, paper MC5

Sari I, Balkan T, Kulah H (2007) A wideband electromagnetic micro power generator for wireless microsystems. Transducers '07 & Eurosensors XXI, Digest of Technical papers, Vol. 1: 275–278

Scherrer S, Plumlee DG, Moll AJ (2005) Energy scavenging device in LTCC materials. IEEE workshop on Microelectronics and Electron Devices, WMED '05: 77–78

Serre C, Perez-Rodriguez A, Fondevilla N, Martincic E, Martinez S, Morante JR, Montserrat J, Esteve J (2008) Design and implementation of mechanical resonators for optimised inertial electromagnetic generators. Microsys. Technol. 14: 653–658

Shearwood C, Yates RB (1997) Development of an electromagnetic micro-generator, Electron. Lett. 33: 1883–1884

Torah RN, Glynne-Jones P, Tudor MJ, Beeby SP (2007) Energy aware wireless microsystem powered by vibration energy harvesting. Proc. 7th Int. Workshop on Micro and Nanotechnology for Power Generation and Energy Conversion Applications (PowerMEMS 2007), November 28–29th, Freiburg Germany: 323–326

von Buren T, Troster G (2007) Design and optimisation of a linear vibration-driven electromagnetic micro-power generator. Sens. Actuators A 135: 765–75

Wang P-H, Dai X-H, Fang D-M, Zhao X-L (2007) Design, fabrication and performance of a new vibration based electromagnetic micro power generator. Microelectronics J 38: 1175–1180

Waters RL, Chisum B, Jazo H, Fralick M (2008) Development of an electro-magnetic transducer for energy harvesting of kinetic energy and its applicability to a MEMS-scale device. Proc nanoPower Forum 2008

Williams CB, Shearwood C, Harradine MA, Mellor PH, Birch TS, Yates RB (2001) Development of an electromagnetic micro-generator. IEE Proc.-Circuits Devices Syst. 148(6): 337–342

Williams CW, Woods RC, Yates RB (1996) Feasibility of a vibration powered micro-electric generator. IEE Colloquium on Compact Power Sources: 7/1–7/3

Williams CW, Yates RB (1996) Analysis of a micro-electric generator for microsystems. Sens Actuators A 52: 8–11
www.kinetron.nl
www.perpetuum.co.uk
Yuen SCL, Lee JMH, Li WJ, Leong PHW (2007) An AA-sized vibration-based microgenerator for wireless systems. IEEE Pervasive Computing 6: 64–72

Part II
Energy Harvesting Circuits and Architectures

Chapter 6
On the Optimal Energy Harvesting from a Vibration Source Using a Piezoelectric Stack

Jamil M. Renno, Mohammed F. Daqaq and Daniel J. Inman

Abstract The modeling of an energy harvesting device consisting of a piezoelectric based stack is presented. In addition, the optimization of the power acquired from the energy harvester is considered. The harvesting device is a piezoceramic element in a stack configuration, which scavenges mechanical energy emanating from a 1D-sinusoidal-base excitation. The device is connected to a harvesting circuit, which employs an inductor and a resistive load. This circuit represents a generalization of the purely resistive circuit, which has received considerable attention in the literature. The optimization problem is formulated as a nonlinear programming problem, wherein the Karush-Kuhn-Tucker (KKT) conditions are stated and the various resulting cases are treated. One of these cases is that of a purely resistive circuit. For resistive circuits, researchers usually neglect the effect of mechanical damping in their optimization procedures. However, in this chapter, we specifically explore the role of damping and electromechanical coupling on the optimization of circuit parameters. We show that mechanical damping has a qualitative effect on the optimal circuit parameters. Further, we observe that beyond an optimal coupling coefficient, the harvested power decreases. This result challenges previously published results suggesting that larger coupling coefficients culminate in more efficient energy harvesters. As for the harvesting circuit, the addition of the inductor provides substantial improvement to the performance of the energy harvesting device. More specifically, at the optimal circuit parameters, optimal power values obtained through a purely resistive circuit at optimal excitation frequencies can be obtained at any excitation frequency. Moreover, simulations reveal that the optimal harvested power is independent of the coupling coefficient (within realistic values of the coupling coefficient); a result that supports our previous findings for a purely resistive circuit.

Keywords Piezoelectricity · Energy harvesting · Optimization · Damping

---

D.J. Inman (✉)

Center for Intelligent Material Systems and Structures, 310 Durham Hall, Blacksburg, VA 24061-0261

email: dinman@vt.edu.

S. Priya, D.J. Inman (eds.), *Energy Harvesting Technologies,*   165
DOI 10.1007/978-0-387-76464-1_6 © Springer Science+Business Media, LLC 2009

Nomenclature

A area, m^2
C capacitance of piezoelectric layer, F
I current magnitude, A
M proof mass, kg
M_p mass of piezoelectric layer, kg
P magnitude of harvested power, W
R_l electrical resistance, Ω
V voltage magnitude, V
X displacement magnitude, m
c mechanical modal damping coefficient, N s/m
c_{33} elasticity coefficient of the piezoelectric layer in the $\{33\}$ direction, N/m^2
e_{33} piezoelectric constant in the $\{33\}$ direction, C/m^2
k effective stiffness of the harvester, N/m
L inductance, H
m approximate effective mass of the harvester, kg
t thickness, m
x piezoelectric layer displacement, m
x_B base displacement, m
$\bar{\varepsilon}$ mechanical strain, dimensionless
ϵ_{33} permittivity of the piezoelectric layer in the $\{33\}$ direction, F/m
θ electromechanical coupling coefficient, N/V
Subscript
e   electrode layer property
p   piezoelectric layer property
Superscript
E   property measured at zero electric field
S   property measured at zero strain
T   property measured at zero stress

6.1 Introduction

This chapter focuses on some modeling and circuit issues associated with the design of a piezoelectric stack with polling voltage acting in the same direction as the applied strain. Sodano et al. (2004) provided a review of piezoceramic-based harvesting research up until the end of 2003. Sodano et al. (2005a,b) investigated the effects of various electrode and packaging configurations for layered and cantilevered piezoceramic harvesters (generated field perpendicular to the applied strain). Beeby et al. (2006) provided a review of piezoelectric-based harvesting research up to about 2005 and included stack-type configurations. Many researchers have considered the design and performance optimization of energy harvesters. Researchers such as Roundy (2005), duToit (2005) investigated possible ways to maximize the harvested power by analyzing the effect of the design parameters such as the load resistance and the electromechanical coupling on the extracted power.

One of the most efficient techniques to maximize the harvested energy is to tune the impedance of the harvesting circuit to that of the mechanical system. Along that line, Wu et al. (2006) developed a tunable resonant frequency power-harvesting device in the form of a cantilever beam. The device utilized a variable capacitive load to shift the gain curve of the cantilever beam such that it matches the frequency of the external vibration in real time. Twiefel et al. (2006) also presented a model that can be tuned to the external excitation frequency. Further, Johnson et al. (2006) presented the design of a unimorph piezoelectric cantilever beam tuned to harvest optimal energy from a specific machine application. Grisso and Inman (2006) proposed an integrated sensor "patch" that can harvest energy from ambient vibration and temperature gradients. This patch would be able to take measurements, compute the structural health of the host, and broadcast the state of the system when necessary.

Rastegar et al. (2006) presented a class of efficient energy harvesting devices that are mounted on platforms vibrating at low frequencies. Priya et al. (2006) reported on the advances of powering stationary and mobile untethered sensors using a fusion of energy harvesting approaches. Lefeuvre et al. (2006) compared four vibration-powered generators with the purpose of powering standalone systems. They proposed an approach based on processing the voltage delivered by the piezoelectric material that would enhance the electromechanical conversion. von Büren et al. (2006) presented a comparison of the simulated performance of optimized configurations of different harvesting architectures using measured acceleration data from walking motion of human subjects. Badel et al. (2006) considered the addition of an electrical switching device connected in parallel with the piezoelectric element. The switch is triggered on the maxima or minima of the displacement realizing a voltage inversion through an inductor, hence allowing an artificial increase of the piezoelectric element's output voltage. Anderson and Sexton (2006) presented a model for piezoelectric energy harvesting with a cantilever beam configuration. The model incorporated an expression for variable geometry, tip mass, and material constants. For further literature review, we refer the reader to Anton and Sodano (2007) who presented a comprehensive review of research published in the area of power harvesting between 2003 and 2006.

When optimizing energy harvesting devices, most of the previous research efforts were simplified by assuming that the maximum power occurs at resonance (Stephen 2006). duToit (2005) showed that this assumption hides essential features of the coupled electromechanical response of the device. This coupling results in another optimal frequency at the anti-resonance. To obtain the complete features of the coupled response, duToit et al. (2005) used a linearly coupled model to derive an expression for the extracted power. In the resulting expression, they set the mechanical damping to zero and analyzed the energy harvesting circuit at short- and open-circuit conditions to obtain both of the optimal frequencies.

Here, we consider a stack-harvesting device similar to that analyzed by duToit et al. (2005). However, we propose a harvesting circuit wherein an inductor is employed with the load resistor (in parallel and in series). Lesieutre et al. (2004) employed an inductor in the harvester circuit. However, their work focused on the analysis of the damping resulting from energy harvesting.

By utilizing the Karush-Kuhn-Tucker (KKT) method (Winston 2003), we investigate the problem of optimizing the circuit parameters to realize maximum power at any external excitation frequency. Unlike previous research effort, we account for the mechanical damping and obtain exact analytical expressions for the optimal frequency ratios as a function of the damping and electromechanical coupling. Further, we show that neglecting the mechanical damping may result in qualitatively and quantitatively erroneous predictions especially for small quality factors and/or small electromechanical coupling coefficients. It is also shown that the proposed circuit can provide a superior performance to a purely resistive circuit. More specifically, at the optimal electrical elements, the proposed circuit can harvest maximum power at any excitation frequency. However, the performance of this arrangement cannot be always optimized. This stems from the fact that, the KKT conditions are violated at low damping ratios (below a *bifurcation damping value*), and hence the purely resistive circuit is the only way to optimize the power. At high-damping ratios (beyond a *bifurcation damping value*) however, this singularity ceases to exit, and the device delivers more power than the maximum power achieved by the purely resistive circuit at any excitation frequency.

In summary, our contribution lies in optimizing the energy harvested from a piezoelectric-energy-based harvesting device, while including inductance in the circuit and damping in the structure. Using this optimal circuit results in broad band optimal harvesting, offering an improvement over previous results which provide optimal harvesting at discrete frequency values.

Along the way, we present a thorough analysis of traditional purely resistive circuits. To this end, the role of damping and electromechanical coupling are explored, and it is shown that there exists an optimal coupling coefficient beyond which the power cannot increase. In this spirit, this work fills a gap in the literature addressing the optimization of 1D energy harvesters. This work also questions the suggestion that a higher coupling coefficient results in higher power values.

The remainder of the chapter is organized as follows. Section 6.2 presents the analytical model of the system under consideration. We first treat the parallel configuration of the inductor and resistor. The optimization problem is stated in Section 6.3. Section 6.4.1 studies the harvesting device equipped with a purely resistive circuit. The section presents exact analytical solutions for the optimal frequencies, and presents an analysis of the role of damping and coupling coefficient. In Section 6.4.2, we consider the new circuit (parallel configuration). In Section 6.5, we treat the series configuration of the harvesting circuit. Finally, conclusions for this work are drawn in Section 6.6.

6.2 One-dimensional Electromechanical Analytic Model

A stack-type piezoelectric-harvesting device is considered here. Figure 6.1 shows a schematic of a harvesting device with a generalized circuit that employs inductors in series and in parallel. A purely resistive circuit is just a special case of Fig. 6.1, and

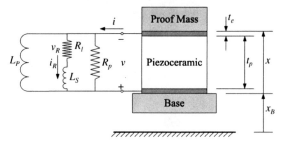

Fig. 6.1 Schematic of piezoelectric energy harvesters with a general harvesting circuit

was studied by duToit (2005) and Daqaq et al. (2007). Most circuits considered in
the literature did not involve inductors. In this section, we will derive the equations
governing the behavior of the harvesting device of Fig. 6.1.

The 3D, linear-elastic-small signal constitutive law of piezoelectricity is
given as,

$$\begin{Bmatrix} T \\ D \end{Bmatrix} = \begin{bmatrix} c^E & -e^T \\ e & \epsilon^S \end{bmatrix} \begin{Bmatrix} S \\ E \end{Bmatrix} \tag{6.1}$$

where

$$T = [\, T_{11} \ T_{22} \ T_{33} \ T_{23} \ T_{13} \ T_{12} \,]^T \ , \quad S = [\, S_{11} \ S_{22} \ S_{33} \ S_{23} \ S_{13} \ S_{12} \,]^T \ ,$$
$$D = [\, D_1 \ D_2 \ D_3 \,]^T \ , \quad E = [\, E_1 \ E_2 \ E_3 \,]^T \ ,$$
$$c^E = [c^E_{ij}] \quad \text{with} \quad i, j = 1, \cdots, 6 \ ,$$
$$\epsilon^S = [\epsilon_{ij}] \quad \text{with} \quad i, j = 1, \cdots, 3 \ ,$$
$$e = [e_{ij}] \quad \text{with} \quad i = 1, \cdots, 3 \text{ and } j = 1, \cdots, 6 \ . \tag{6.2}$$

Since the stress and strain are symmetric tensors, their subscripts are usually
relabeled per Voigt's notation as,

$$ii \rightarrow i \quad \text{and} \quad 23 \rightarrow 4 \,, 13 \rightarrow 5 \,, 12 \rightarrow 6 \ . \tag{6.3}$$

In (6.1), T, D, S and E present the stress, electric displacement, strain, and
electric field, respectively. Moreover, ϵ^S is the permitivity of the piezoelectric layer,
measured at zero strain, and e is the piezoelectric material coupling constant relating
the charge density and the strain. The stiffness of the piezoelectric material, measured at zero electric field, is c^E. The device at hand is 1D, and assuming that the
piezoelectric material is poled in the direction of the displacement, x, the constitutive equations reduce to

$$T_3 = c^E_{33} S_3 - e_{33} E_3 \ , \tag{6.4}$$
$$D_3 = e_{33} S_3 + \epsilon^S_{33} E_3 \ . \tag{6.5}$$

The strain, stress, electric field, and electric displacement are related to the device parameters through,

$$S_3 = \frac{x}{t_p} \quad \text{and} \quad T_3 = -\frac{m(\ddot{x} + \ddot{x}_B)}{A} \quad \text{where} \quad m = M + \frac{1}{3}M_p,$$

$$E_3 = -\frac{v}{t_p} \quad \text{and} \quad D_3 = \frac{q}{A}. \tag{6.6}$$

Here, the approximate total mass of the system, m, is taken to be analogous to the effective mass used in a single-degree-of-freedom model of a cantilevered rod with a tip mass in longitudinal vibrations, duToit (2005). This approximation has no effect on the analysis presented here, as the system at hand is indeed governed by a single mechanical degree of freedom. A continuum model would certainly present a better approximation of the system dynamics (Erturk and Inman 2008a,b). We neglect the electrode thickness and set t_e equal to zero.

Substituting the expressions for T_3, S_3, and E_3 of (6.6) in (6.4) yields,

$$-\frac{m}{A}(\ddot{x} + \ddot{x}_B) = c_{33}^E \frac{x}{t_p} + e_{33}\frac{v}{t_p}. \tag{6.7}$$

Equation (6.7) is multiplied by A and the terms are rearranged to yield,

$$m\ddot{x} + kx - \theta v = -m\ddot{x}_B \quad \text{where} \quad k = \frac{c_{33}^E A}{t_p}, \theta = -\frac{e_{33}A}{t_p}. \tag{6.8}$$

Now, substituting the expressions for D_3, S_3, and E_3 of (6.6) into (6.5) and multiplying by A yields,

$$q = -\theta x - C_p v \quad \text{where} \quad i = \frac{dq}{dt}, C_p = \frac{\epsilon_{33}^S A}{t_p}. \tag{6.9}$$

Turning our attention to the circuit of Fig. 6.1, and recalling the principles of electric circuits and Kirchhoff's current law, we can state that

$$v = R_l i_R + L_s \dot{i}_R,$$

$$i = i_R + \frac{v}{R_p} + \frac{1}{L_P}\int_0^t v(\tau)d\tau. \tag{6.10}$$

To substitute (6.10) in (6.9), (6.9) should be differentiated with respect to time. This substitution will allow (6.9) to be formulated in terms of the current passing through the load resistance, R_l, and its time derivatives only,

$$i_R + \frac{(R_1 i_R + L_S \dot{i}_R)}{R_p} + \frac{1}{L_p} \int_0^t \left(R_1 i_R(\tau) + L_S \dot{i}_R(\tau) \right) d\tau$$

$$= -\theta \dot{x} - C_p(R_1 \dot{i}_R + L_S \ddot{i}_R) , \tag{6.11}$$

which can be differentiated with respect to time one more time, and reorganized. Moreover, proportional damping could be added to (6.8), and the coupled governing equations are,

$$m\ddot{x} + c\dot{x} + kx - \theta(R_1 i_R + L_S \dot{i}_R) = -m\ddot{x}_B ,$$

$$C_p L_S \, \dddot{i}_R + \left(C_p R_1 + \frac{L_S}{R_p} \right) \ddot{i}_R + \left(\frac{L_S}{L_P} + \frac{R_1}{R_p} + 1 \right) \dot{i}_R + \frac{R_1}{L_P} i_R + \theta \ddot{x} = 0. \tag{6.12}$$

In terms of the voltage across the load resistance, v_R, (6.12) is expressed as

$$m\ddot{x} + c\dot{x} + kx - \theta\left(v_R + \frac{L_S}{R_1} \dot{v}_R \right) = -m\ddot{x}_B$$

$$\theta \ddot{x} + \frac{C_p L_S}{R_1} \dddot{v}_R + \left(C_p + \frac{L_S}{R_p R_1} \right) \ddot{v}_R + \left(\frac{L_S}{L_P R_1} + \frac{1}{R_{eq}} \right) \dot{v}_R + \frac{1}{L_P} v_R = 0. \tag{6.13}$$

In the absence of both inductors, L_S and L_P, (6.13) identically reduces to the model derived previously by duToit (2005). Table 6.1 presents numerical values of the parameters used in the simulations presented in this chapter.

In order to obtain a closed-form analytical solution of the system's displacement and the magnitudes (half peak-to-peak value) of the voltage developed and the power extracted on can proceed as follows. First, we obtain the Laplace transform of the displacement and voltage. These are obtained as,

$$X(s) = \frac{N_x(s)}{D(s)} \quad \text{and} \quad V_R(s) = \frac{N_v(s)}{D(s)} , \tag{6.14}$$

Table 6.1 Data used to simulate the energy harvester

Piezoceramic harvester's properties	
Proof mass, M (kg)	0.01
Thickness, t_p (m)	0.01
Cross-sectional area, A_p (m^2)	0.0001
Mass, M_p (kg)	0.00075
Permittivity, ϵ_{33} (F/m)	1.137×10^{-8}
Coupling coefficient, k_{33} (dimensionless)	0.75
Leakage resistance, R_p (Ω)	5×10^9
Base acceleration magnitude	$1g$ ($g = 9.81$ m/s^2)

where

$$N_x(s) = -mX_B(s)s^2(\varphi(s) + \psi(s)) \,,$$
$$N_v(s) = L_P m R_1 R_p \theta X_B(s)s^4 \,,$$
$$D(s) = L_P R_p s^2 \theta^2 (R_1 + L_S s) + m(\varphi(s) + \psi(s))(s^2 + 2\zeta\omega_n s + \omega_n^2) \,,$$
$$\varphi(s) = R_l(R_p + L_P s + C_p L_P R_p s^2) \,,$$
$$\psi(s) = s(L_S R_p + L_P(R_p + L_S s + C_p L_S R_p s^2)) \,.$$

Now, assuming a harmonic base excitation of amplitude $\bar{X}_B$ and frequency ω, the displacement and voltage amplitudes can be obtained by finding the magnitude of the complex quantities $X(j\omega)$ and $V(j\omega)$. Moreover, these amplitudes can be normalized with respect to the base acceleration amplitude, $\bar{X}_B\omega^2$. Our choice of normalization is motivated by practical intuition, as the measurement of acceleration is widely used in experimental settings (e.g., using accelerometers).

6.3 Power Optimization

In this section, we address the problem of maximizing the flow of energy from a vibrating structure to a given electric load. In specific, we consider optimizing the load resistance, R_1, and either one of the inductors. Towards that end, we cast the optimization problem as a nonlinear programming problem with the Karush-Kuhn-Tucker (KKT) conditions. The KKT technique is a generalized form of the long celebrated method of Lagrange multipliers (see, for instance, Winston 2003). The objective function to be optimized is a nonlinear function of two variables, $\mathcal{P}(\alpha, \beta)$ and the optimal values for α and β are constrained to be nonnegative. Hence, the nonlinear optimization problem and the associated constraints can be stated as follows,

$$\text{Find the } \min_{\alpha,\beta}(-\mathcal{P}) \text{ subject to } g_i(\alpha, \beta) \le 0, \qquad i = 1, 2 \,. \tag{6.15}$$

The negative sign in (6.15) indicates that we are seeking a maximum of $\mathcal{P}(\alpha, \beta)$. The constrain functions $g_i(\alpha, \beta)$ declare the nonnegativity of α and β, hence they are given by

$$g_1(\alpha, \beta) = -\alpha \quad \text{and} \quad g_2(\alpha, \beta) = -\beta \,. \tag{6.16}$$

The necessary KKT conditions are stated as follows: if the pair $(\alpha_{opt}, \beta_{opt})$ is a local optimum, then there exists constants $\mu_i \ge 0$ $(i = 1, 2)$, such that

$$-\nabla\mathcal{P}(\alpha_{opt}, \beta_{opt}) + \boldsymbol{\mu}^T.\nabla\boldsymbol{g}(\alpha_{opt}, \beta_{opt}) = 0 \text{ and } \mu_i g_i(\alpha_{opt}, \beta_{opt}) = 0, \quad i = 1, 2 \,, \tag{6.17}$$

where ∇P is the gradient operator producing a column vector when operating on a scalar function ($P(\alpha, \beta)$), and a square matrix when operating on a column vector ($g(\alpha, \beta)$). Expanding the above expression yields

$$\frac{\partial P}{\partial \alpha}\bigg|_{(\alpha_{opt}, \beta_{opt})} + \mu_1 = 0 \quad \text{and} \quad \mu_1 \alpha_{opt} = 0, \tag{6.18a}$$

and

$$\frac{\partial P}{\partial \beta}\bigg|_{(\alpha_{opt}, \beta_{opt})} + \mu_2 = 0 \quad \text{and} \quad \mu_2 \beta_{opt} = 0. \tag{6.18b}$$

Moreover, the KKT necessary conditions state that if the sufficient conditions, stated in (6.18), are satisfied, and the objective function $P(\alpha, \beta)$ and the constrain functions $g_i(\alpha, \beta)$ are convex, then the pair ($\alpha_{opt}, \beta_{opt}$) is a global optimum. The functions $g_i(\alpha, \beta)$ are convex. Moreover, one can obtain the Hessian of $P(\alpha, \beta)$ and show that it is positive definite. The Hessian of P is not presentable, and hence will not be stated here. However, the global optimality is guaranteed through the necessary and sufficient conditions of the KKT conditions.

6.4 Optimality of the Parallel-RL Circuit

In the absence of the series inductance L_S, the objective function to be optimized, P, in this case presenting the magnitude of the harvested power, reduces to

$$P = \frac{(\omega^2 \bar{X}_b)^2}{\omega_n^3} \frac{k\alpha\beta^2\kappa^2\Omega^4}{B}. \tag{6.19}$$

Furthermore, the normalized steady-state amplitude of the stack displacement and voltage across the resistive load are given by

$$\left|\frac{X}{\omega^2 \bar{X}_b}\right| = \frac{1}{\omega_n^2} \frac{\sqrt{\beta^2\Omega^2 + \alpha^2(\beta\Omega^2 - 1)^2}}{\sqrt{B}}, \tag{6.20}$$

$$\left|\frac{V_R}{\omega^2 \bar{X}_b}\right| = \frac{1}{|\theta|} \frac{m\alpha\beta\kappa^2\Omega^2}{\sqrt{B}}, \tag{6.21}$$

where

$$B = ((\beta + 2\alpha\zeta)\Omega - \beta(1 + 2\alpha\zeta)\Omega^3)^2$$
$$+ (\alpha - (2\beta\zeta + \alpha(1 + \beta + \beta\kappa^2))\Omega^2 + \alpha\beta\Omega^4)^2. \tag{6.22}$$

Here, ω_n is the natural frequency of the mechanical system, α and β are dimensionless time constants, κ is an alternative electromechanical coupling coefficient,

Ω is the frequency ratio, and ζ is the modal damping ratio of the mechanical system. These quantities are given below,

$$\omega_n = \sqrt{\frac{k}{m}} \qquad \alpha = R_{eq}\omega_n C_p \qquad \beta = \omega_n^2 L_p C_p ,$$

$$\kappa^2 = \frac{\theta^2}{kC_p} \qquad \Omega = \frac{\omega}{\omega_n} \qquad \zeta = \frac{c}{2m\omega_n} . \tag{6.23}$$

The equivalent resistance R_{eq} represents the parallel resistance of the load and leakage resistances, R_l and R_p, respectively. The leakage resistance, Table 6.1, is usually much higher than the load resistance (Sood 2003). Consequently, in the remainder of this work, we assume $R_{eq} \approx R_l$, and we only refer to the load resistance R_l. A note regarding the quantity κ is also in order. Throughout this work, κ will be loosely referred to as the *coupling coefficient*. This quantity is not to be confused with the more traditional coupling coefficient referred to in the literature, k_{33}, given by

$$k_{33}^2 = \frac{d_{33}^2}{s_{33}^E \epsilon_{33}^T} . \tag{6.24}$$

Both quantities, k_{33} and κ, are positive. However, k_{33} cannot exceed 1, whereas κ is not bounded by $1(0 < k_{33} < 1, 0 < \kappa)$. Both coefficients are related through duToit (2005),

$$\kappa^2 = \frac{k_{33}^2}{1 - k_{33}^2} = \frac{e_{33}^2}{c_{33}^E \epsilon_{33}^S} . $$

Lesieutre and Davis (1997) provide a useful discussion on the coupling coefficient of a piezoelectric device and that of its active material.

Next, we turn our attention to the optimization problem. The resulting system of equalities and inequalities obtained through the KKT conditions requires thorough analysis to obtain the possible optimal solutions for the couple $(\alpha_{opt}, \beta_{opt})$. A preliminary investigation reveals the existence of the following cases:

(a) CASE I ($\mu_1 = 0$ and $\mu_2 \neq 0$): In this case, $\beta_{opt} = 0$, which yields a purely resistive circuit. As such, the optimization problem reduces to solve the following equation:

$$\left. \frac{\partial \mathcal{P}}{\partial \alpha} \right|_{\alpha_{opt}} = 0 . \tag{6.25}$$

(b) CASE II ($\mu_1 = 0$ and $\mu_2 = 0$): Here, $\alpha_{opt} \neq 0$ and $\beta_{opt} \neq 0$ which represent a parallel RL circuit. In this case, the optimization problem reduces to solving the following nonlinear equations,

$$\frac{\partial P}{\partial \alpha}\bigg|_{(\alpha_{opt},\beta_{opt})} = 0 \quad \text{and} \quad \frac{\partial P}{\partial \beta}\bigg|_{(\alpha_{opt},\beta_{opt})} = 0 . \qquad (6.26)$$

It will be shown that this case is a separable optimization problem that would yield an interesting expression for the optimal magnitude of the harvested power.

(c) CASE III ($\mu_1 \neq 0$ and $\mu_2 \neq 0$): This yields $\alpha_{opt} = \beta_{opt} = 0$ as per (6.18). This result violates the purpose of the problem, as no power can be extracted in the absence of an electric circuit.

(d) CASE IV ($\mu_1 \neq 0$ and $\mu_2 = 0$): Accordingly, α_{opt} must be 0, and the problem reduces to

$$\frac{\partial P}{\partial \alpha}\bigg|_{(0,\beta_{opt})} = 0 \quad \text{and} \quad \frac{\partial P}{\partial \beta}\bigg|_{(0,\beta_{opt})} = 0 .$$

An optimal inductor value β_{opt} can be obtained, however, since we are seeking to optimize the flow of energy to the load which is the element that extracts the power in this case, this solution is not meaningful.

By virtue of the preceding discussion only, CASE I and CASE II should be further analyzed as they are the only physically meaningful among the four possible cases. In the following, CASE I and CASE II will be treated in detail.

6.4.1 The Purely Resistive Circuit

CASE I suggests that one can utilize a purely resistive circuit to optimize the flow of energy. In this case, the solution of (6.25) yields

$$\alpha_{opt}^2 = \frac{1}{\Omega^2} \frac{(1 - \Omega^2)^2 + (2\zeta\Omega)^2}{([1 + \kappa^2] - \Omega^2)^2 + (2\zeta\Omega)^2} . \qquad (6.27)$$

The preceding expression suggests that for a given frequency ratio, Ω, one can maximize the flow of energy to the load by choosing the optimal resistance according to (6.27) and (6.23). The question that becomes legitimate here is what frequency ratio would yield optimal power. The answer can be obtained by differentiating equation (6.19) with respect to Ω and solving the resulting equation together with equation (6.27) for the optimal frequency ratios, Ω_{opt}, and the optimal load resistance, α_{opt}. Having said that, it turns out that the solution cannot be easily obtained as the resulting optimization problem yields a polynomial of the 12th degree in Ω_{opt}. One way to facilitate the solution of the resulting equation is to assume a small damping ratio, ζ. In the following, we present a detailed analysis based on the small damping assumption.

6.4.1.1 Small Damping Assumption

To facilitate the optimization problem, duToit (2005) among others (Stephen 2006), (Wu et al., 2006) set the damping ratio, ζ, to zero then analyzed the harvesting circuit at short- and open-circuit conditions by letting the dimensionless time constant α approach 0 and ∞, respectively, to obtain two optimal frequency ratios for which the amplitude of the harvested power is maximum. These two frequency ratios are denoted as the resonance frequency, Ω_r, and the anti-resonance frequency, Ω_{ar}, given by

$$\Omega_r = 1 \quad \text{and} \quad \Omega_{ar} = \sqrt{1 + \kappa^2} \, . \tag{6.28}$$

At zero damping ratio, the preceding expressions accurately describe the optimal frequency ratios for which the power harvested is indeed maximum. In fact, it has been shown previously that for very small damping ratio, ζ, one can utilize the ratio Ω_r / Ω_{ar} to experimentally obtain the coupling coefficient κ of a given piezoceramic. At these frequencies, the optimal load time constant reduces to

$$\alpha_{opt,r}^2 = \frac{4\zeta^2}{4\zeta^2 + \kappa^4}, \tag{6.29}$$

$$\alpha_{opt,ar}^2 = \frac{(1 + \kappa^2)^2 - 2(1 + \kappa^2) + 4\zeta^2(1 + \kappa^2) + 1}{4\zeta^2(1 + \kappa^2)^2} \, . \tag{6.30}$$

Now, assuming that ζ is at least one order of magnitude smaller than κ^2, i.e., $\zeta / \kappa^2 << 1$; equation (6.29) and (6.30) can be further simplified as

$$\alpha_{opt,r}^2 = \frac{2\zeta}{\kappa^2} \, , \tag{6.31}$$

$$\alpha_{opt,ar}^2 = \frac{\kappa^2}{2\zeta(1 + \kappa^2)} \, . \tag{6.32}$$

Substituting (6.28), (6.31), and (6.32) into (6.19) yield the following expression for the optimal power

$$\mathcal{P}_{opt}\big|_{\Omega = \Omega_r, \Omega_{ar}} = \frac{\omega^2 \bar{X}_b k}{8\zeta \omega_n^3} \, . \tag{6.33}$$

It is worth noting that the preceding expression is not a function of the coupling coefficient κ. This implies that for very small damping and $\zeta << \kappa^2$, the maximum power obtained by optimizing the load resistance at the resonance and anti-resonance frequencies does not depend on the coupling coefficient. In other words, the process of optimizing the load while tuning the excitation frequency to the resonance and/or anti-resonance frequency of the system completely eliminates the effect of electromechanical coupling on the output power. This result is very interesting as it suggests that if one can design a circuit which can tune itself to the resonance and anti-resonance frequency while optimizing its load in real time, then the output power would be the same, (6.33), regardless of whether the material has high or low electromechanical coupling, provided that $\zeta \ll \kappa^2$. However, one has to keep in mind that the optimal value for the resistive load depends on κ. For instance, at resonance, the optimal load is inversely proportional to κ^2. It is also worth noting that, for very small damping and $\zeta \ll \kappa^2$, the optimal power obtained at the resonance and anti-resonance frequencies is equal.

6.4.1.2 Effect of Damping

While the preceding analysis accurately describes the optimal harvesting problem for very small damping, it is imperative to study and understand the effect of damping on the optimal parameters. More specifically, it is well understood that, for many mechanical systems, the assumption of small damping is not always valid, thereby such assumption could yield erroneous predictions of the optimal values and might hide many of the essential features of the actual system. For instance, (6.28) suggests that the optimal frequency ratios remain constant as the damping ratio is varied. In addition, the goal of harvesting is to subtract energy from the structure, which in turn serves to increase the effective damping. In the following, it will be shown that this suggestion is inaccurate. As such, and to gain a better understanding of the effect of damping on the optimal parameters, we plot the numerical solutions of the optimal frequency ratios versus the damping ratio, ζ, at a given coupling coefficient κ, as displayed in Fig. 6.2a.

At $\zeta = 0$, the system has three real positive extrema: two maxima and a minimum. The maxima are $\Omega_r = 1$ and $\Omega_{ar} = \sqrt{1 + \kappa^2}$ as predicted by duToit (2005). The third root is a minimum, given as

$$\Omega_{min} = \frac{1}{\sqrt{6}}\sqrt{2 + \kappa^2 + \sqrt{16 + 16\kappa^2 + \kappa^4}}. \tag{6.34}$$

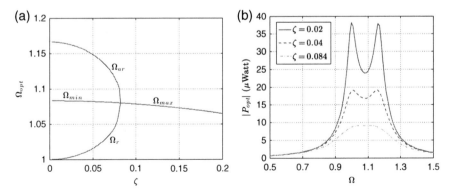

Fig. 6.2 Optimal frequency ratios and harvested power at $\kappa = 0.6$ and corresponding α_{opt}

Examining Fig. 6.2a, the following can be inferred. As the damping ratio is increased, the value of Ω_r increases whereas the value of Ω_{ar} decreases. On the other hand, the value of the minimum, Ω_{min}, decreases slowly relative to changing values of ζ. The variation of these three roots strongly resembles a subcritical pitchfork bifurcation with the damping ratio being the bifurcation parameter. More specifically, for damping ratios that are less than a particular damping ratio (*bifurcation damping ratio ζ_b*), the optimal power resembles a double-peak single-well potential function with the double peaks representing the two maxima and the well representing the minimum, Fig. 6.2b. At ζ_b, the power expression becomes of the single-peak potential-type and the values of the three extrema coincide. As such, one extremum, a maximum exists only. This solution is an extension of the minimum solution prior to crossing ζ_b. Further increase in ζ beyond ζ_b results in a decrease in the optimal frequency ratio. For the case considered here, the bifurcation damping ratio was found to be $\zeta_b = 0.084$.

The associated magnitudes of the voltage and displacement are displayed in Fig. 6.3a and 6.3b, respectively. Figure 6.4a displays the magnitude of the current passing through R_1. It is worth noting that, at the resonance frequency ratio, Ω_r, the current magnitude is at a maximum, whereas the voltage magnitude is at a minimum. Hence, Ω_r is a suitable operating point for applications requiring high current, such as charging a storage device, (duToit, 2005). This observation is reversed at the anti-resonance frequency ratio, Ω_{ar}. At Ω_{ar}, the voltage is at a maximum, whereas the current is at a minimum. This operating condition is suitable for applications requiring high voltage, such as wireless sensors (duToit, 2005) which use mainly diodes and transistors. Furthermore, the optimal value of the displacement behaves similar to the voltage. It assumes a small value at the resonance frequency, Fig. 6.4b. Recall that the time constant α is proportional to the load resistance, R_1. However, the power is inversely proportional to the load resistance. Hence, to maintain the same harvested power, the voltage and the resistance have to behave similarly. The magnitude of the voltage and R_1 are both small at Ω_r and both attain maxima at Ω_{ar}.

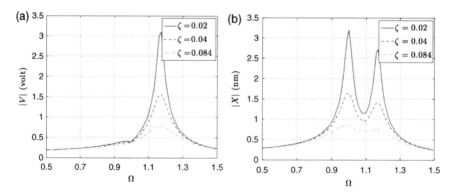

Fig. 6.3 Voltage and displacement magnitude at different damping ratios with $\kappa = 0.6$

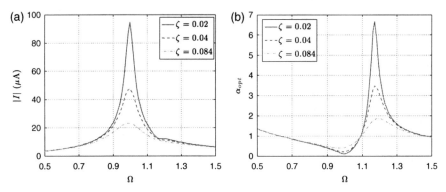

Fig. 6.4 Current magnitude and optimal time constant at different damping ratios with $\kappa = 0.6$

Exact solution for optimal frequencies: Using the numerical bifurcation diagram of Fig. 6.2a, exact analytical expressions for the resonance, Ω_r, and antiresonance, Ω_{ar}, frequency ratios can be obtained. The process of determining these roots involved numerous trail-and-error approaches. These roots are,

$$\Omega_r = \frac{\sqrt{2 - 4\zeta^2 + \kappa^2 - \sqrt{16\zeta^4 - 16\zeta^2 - 8\zeta^2\kappa^2 + \kappa^4}}}{\sqrt{2}},$$

$$\Omega_{ar} = \frac{\sqrt{2 - 4\zeta^2 + \kappa^2 + \sqrt{16\zeta^4 - 16\zeta^2 - 8\zeta^2\kappa^2 + \kappa^4}}}{\sqrt{2}}. \tag{6.35}$$

Now, one has to find the exact solution for Ω_{max} which represents the middle branch of the bifurcation diagram displayed in Fig. 6.2a. Equation (6.35) yields complex quantities when $\zeta > \zeta_b$. This can be exploited to solve for Ω_{max}, since the two roots obtained so far, Ω_r and Ω_{ar} constitute four roots as the conjugate of a complex root is a root. Moreover, the opposite of these roots and their conjugates are also roots of the ensuing equation. Hence, by eliminating the eight known roots, one can solve a polynomial of the 4th degree instead of solving a polynomial of the 12th degree. The root presenting the middle branch turns out to be,

$$\Omega_{min} = \frac{\sqrt{6}\sqrt{2 - 4\zeta^2 + \kappa^2 + \sqrt{16 + 16\zeta^4 + 16\kappa^2 + \kappa^4 - 8\zeta^2(2 + \kappa^2)}}}{6}. \tag{6.36}$$

As the three branches of the bifurcation diagram are obtained analytically, further analysis can be conducted to expand our understanding of optimal energy harvesters.

Criticality treatment: The expressions of Ω_r and Ω_{ar} are interesting and provoke some investigation. For instance, note that these quantities are complex when,

$$f(\zeta, \kappa) = 16\zeta^4 - 16\zeta^2 - 8\zeta^2\kappa^2 + \kappa^4 < 0. \tag{6.37}$$

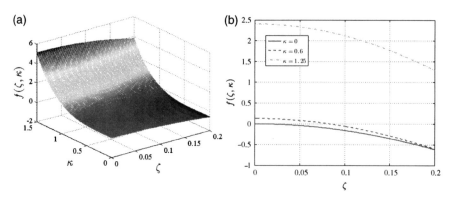

Fig. 6.5 (a) 3D display of the function $f(\zeta, \kappa)$. (b) $f(\zeta, \kappa)$ versus ζ for constant values of κ

Figure 6.5a displays a surface representing the variation of this function with varying values of ζ and κ. As ζ is increased for constant values of κ, f varies from being large and positive (*three optimal frequency ratios*) to being small and negative (*only one optimal frequency ratio*). At $f(\zeta, \kappa)$ equal to zero, equation (6.37) can be solved to yield an exact value for the bifurcation damping discussed previously

$$\zeta_b = \frac{1}{2}(-1 + \sqrt{1 + \kappa^2}). \tag{6.38}$$

Equation (6.38) reveals that the bifurcation damping ratio is independent of the harvesting circuit parameters since α does not appear in the equation. Further, it is evident that ζ_b is only a function of the coupling coefficient κ, and hence depends only on the properties of the piezoceramic used. As κ approaches zero, ζ_b approaches zero, and hence the transition bifurcation takes place at lower values of ζ. On the other hand, as the coupling coefficient increases, the bifurcation damping increases. In other words, the two branches of Ω_r and Ω_{ar} coexist for larger values of damping, Fig. 6.6a. This is favorable as the existence of these two branches provides an operating point, Ω_r for low-voltage/high-current, and another operating point, Ω_{ar}, for high-voltage/low-current as discussed earlier.

Neglecting the effect of mechanical damping when optimizing the frequency ratio does not only affect the optimal frequency ratios; it affects also the optimal time constant α_{opt} and hence the value of the optimal load resistance. Figure 6.6b illustrates variation of the optimal time constant α_{opt} with the damping ratio ζ at $\kappa = 0.6$. As predicted earlier by duToit (2005), for $\zeta = 0$, short-circuit condition ($R_l = 0$) maximizes the power at resonance, whereas open-circuit condition ($R_l \rightarrow \infty$) maximizes the power at anti-resonance. However, as ζ is increased, the optimal time constant obtained via equations (6.31) and (6.32) deviates from the actual values. More specifically, for values of ζ beyond ζ_b, the negligible damping assumption predicts that two values of R_l maximize the power. However, for damping ratios beyond ζ_b, there is only one value for R_l that maximizes the power. This value can be obtained analytically by substituting the solution of the optimal

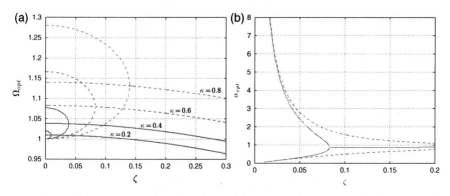

Fig. 6.6 (a) Bifurcation diagrams of the optimal frequency ratios versus the damping ratio ζ for different values of κ. (b) Variation of the optimal time constant α_{opt} with the damping ratio ζ at $\kappa = 0.6$. *Dashed lines* represent results obtained via the small damping assumption

frequency ratio for $\zeta > \zeta_b$, (6.36), in the expression determined for the optimal time constant α_{opt}, (6.27). The optimal resistance values can then be obtained.

Similar to the bifurcation damping ratio, one can also define a bifurcation coupling coefficient κ_b. This parameter is given by solving (6.37) for κ and obtaining

$$\kappa_b = 2\sqrt{\zeta + \zeta^2} . \tag{6.39}$$

For constant ζ values, $f(\zeta, \kappa)$ is negative for small values of the coupling coefficient and becomes positive when the coupling coefficient crosses the bifurcation coupling coefficient value, κ_b. This behavior can be concluded from Fig. 6.7a and indicates the existence of another bifurcation diagram that is "*opposite*" to the one obtained previously. It turns out that, for small values of κ, one extremum, a maximum, only exists. Beyond the value, this maximum becomes a minimum, and two roots branch out. These two branches represent maxima. Fig. 6.7b illustrates bifurcation diagrams that are obtained by varying κ and solving (6.35) and (6.36) for

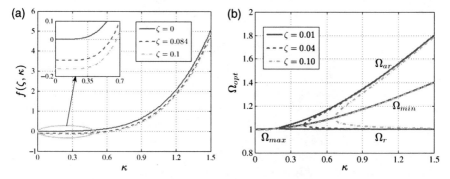

Fig. 6.7 (a) $f(\zeta, \kappa)$ versus ζ for constant values of κ. (b) Bifurcation diagrams of the optimal frequency ratios versus the coupling coefficient κ for different values of ζ

different values of ζ. It is noted that for all the values of damping ratio used, all the branches of the bifurcation diagram coincide for high values of the coupling coefficient. Moreover, the middle branches (Ω_{max} before κ_b, Ω_{min} after κ_b) differ slightly before κ_b, and practically coincide after crossing κ_b. Additionally, the bifurcation coupling coefficient increases as the damping increases, which is also evident from (6.39).

6.4.1.3 Effect of Electromechanical Coupling

In order to gain a better understanding of the effect of the electromechanical coupling coefficient κ on the power harvested from a purely resistive circuit, we use (6.12) to relate the power harvested directly to the displacement of the system. This is obtained as,

$$|\mathcal{P}| = \frac{k\alpha\kappa^2\Omega^2\omega_n}{1+\alpha^2\Omega^2}|X|^2 \tag{6.40}$$

Equation (6.40) clearly states that for a constant Ω, α, and $|X|$, the harvested power increases as κ increases. However, although Ω and α can be maintained constant, it is almost impossible to maintain the displacement magnitude $|X|$ constant. More specifically, equation (6.20) indicates that, for constant input acceleration amplitude, as κ increases, the displacement magnitude $|X|$ decreases. Therefore, for constant input acceleration amplitude, an increase in κ is not necessarily accompanied by an increase in the harvested power.

Furthermore, according to (6.20), as κ approaches zero, $|X|$ is maximized, therefore all the energy supplied by the environment is transferred to the structure. However, this energy cannot be harvested as per (6.19).

The above discussion suggests that a larger κ does not always imply a more efficient energy harvester. The coupling coefficient κ also acts as a damping term that minimizes the flow of energy from the environment to the harvesting device. For a given frequency ratio and time constant, Ω and α, the optimal coupling coefficient can be obtained by setting the derivative of (6.19) with respect to κ equal to zero, and then solving for κ to obtain

$$\kappa_{opt}^2 = \frac{1}{\alpha\Omega}\sqrt{(1+\alpha^2\Omega^2)(1+(4\zeta^2-2)\Omega^2+\Omega^4)}. \tag{6.41}$$

Note that (6.41) has a solution for every possible physical value of α, Ω, and ζ. This stems from the fact that the quantity under the square root is always positive, unless $\zeta = 0$ and $\Omega = 1$ (note that $\Omega = \Omega_r = 1$ at $\zeta = 0$, (6.35)). But, in this case, $\kappa_{opt} = 0$ and in turn, the harvested power approaches infinity according to (6.19). This is physically unrealizable and emanates from the fact that the displacement amplitude $|X|$ unrealistically approaches infinity at resonance for zero damping ratio and zero coupling coefficient. Figure 6.8a shows the variation of the harvested power with the coupling coefficient for different values of load resistance at the resonance

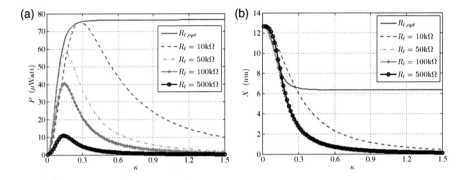

Fig. 6.8 Variation of the power and displacement with the coupling coefficient for different values of load resistance, R_1, at $\Omega = \Omega_r$ and $\zeta = 0.01$

frequency and with $\zeta = 0.01$. For a given R_1, as the coupling coefficient increases, the harvested power increases until it reaches a maximum value at κ_{opt}. Further increase in κ results in a decrease in the harvested power which approaches zero as the coupling coefficient increases.

At the optimal load resistance, $R_{1,opt}$, the harvested power reaches a saturation limit at the optimal coupling coefficient, κ_{opt}. Moreover, the displacement reaches a limit value as well, Fig. 6.8b. These results might be surprising, yet they are supported by the equations governing the behavior of the system. Replace the expression of the optimal time constant, (6.27), in the expression of the power and displacement magnitude, (6.19) and (6.20). Then, compute the limit of these expressions as κ approaches infinity,

$$|P_{\kappa\to\infty}| = \frac{k\Omega}{2(2\zeta\Omega + \sqrt{1+(4\zeta^2-2)\Omega^2+\Omega^4})\omega_n^3} \, ,$$

$$|X_{\kappa\to\infty}| = \frac{1}{\omega_n^2\sqrt{2+2\Omega((4\zeta^2-2)\Omega+\Omega^3+2\zeta\sqrt{1+(4\zeta^2-2)\Omega^2+\Omega^4}}} \, . \tag{6.42}$$

The expressions in (6.42) can be used to verify the saturation limits displayed in Fig. 6.8a and 6.8b.

6.4.2 The Parallel-RL Circuit

The second case resulting from the power optimization problem suggests utilizing an inductor in parallel with the resistive load. In this case, we solve (6.26) which yields the following optimal values for α and β:

$$\alpha_{opt} = \frac{\Omega^4 + (4\zeta^2 - 2)\Omega^2 + 1}{2\zeta\kappa^2\Omega^2} \, , \tag{6.43a}$$

and

$$\beta_{\text{opt}}^{(1)} = 0 \geq 0 \quad \text{and} \quad \beta_{\text{opt}}^{(2)} = \frac{\Omega^4 + (4\zeta^2 - 2)\Omega^2 + 1}{\Omega^2(\Omega^4 - (2 + \kappa^2 - 4\zeta^2)\Omega^2 + \kappa^2 + 1)} \geq 0 .$$

(6.43b)

Note that α_{opt} and β_{opt} are independent of each other. In other words, the optimization problem at hand is of an additive (separable) nature. This implies that the power can be independently optimized with respect to each electric element. Now, turning the attention to the solutions obtained for β_{opt}, one realizes that there are two solutions. The first solution, $\beta_{\text{opt}}^{(1)}$, implies that no inductor is added to the circuit. Section 6.4.1 treated this case in details. The second solution, $\beta_{\text{opt}}^{(2)}$, reveals interesting results and requires further attention. For instance, while the first solution obtained, $\beta_{\text{opt}}^{(1)}$, always satisfies the constraint $\beta_{\text{opt}} \geq 0$, the second solution $\beta_{\text{opt}}^{(2)}$ is less than zero and hence violates the KKT conditions when

$$\Omega_1 < \Omega < \Omega_2 \quad \text{and} \quad \zeta \leq \zeta_b ,$$

(6.44)

where

$$\Omega_1 = \frac{\sqrt{2 - 4\zeta^2 + \kappa^2 - \sqrt{16\zeta^4 - 16\zeta^2 - 8\zeta^2\kappa^2 + \kappa^4}}}{\sqrt{2}} ,$$

$$\Omega_2 = \frac{\sqrt{2 - 4\zeta^2 + \kappa^2 + \sqrt{16\zeta^4 - 16\zeta^2 - 8\zeta^2\kappa^2 + \kappa^4}}}{\sqrt{2}} .$$

(6.45)

It turns out that the values of Ω_1 and Ω_2 are nothing but the resonance and anti-resonance frequency ratios obtained using the purely resistive circuit as demonstrated in equation (6.35). Hence, $\beta_{\text{opt}}^{(1)}$ optimizes the system, always satisfies the KKT conditions, and was treated in Section 6.4.1. The other optimal solution, $\beta_{\text{opt}}^{(2)}$ violates the KKT conditions when $\Omega_1 \leq \Omega \leq \Omega_2$ and $\zeta \leq \zeta_b$, and in such circumstance, the first optimal value $\beta_{\text{opt}}^{(1)}$ is used to optimize the power.

Through the rest of this work, β_{opt} will refer solely to $\beta_{\text{opt}}^{(2)}$ unless specified otherwise. Based on these findings, an optimal power expression can be obtained by substituting α_{opt} and β_{opt} in (6.19), yielding

$$P_{\text{opt}} = \frac{k}{8\zeta\omega_n^3} .$$

(6.46)

Equation (6.46) is extremely important and quite interesting as it states the following:

(a) The optimal power is independent of the coupling coefficient κ. This idea was discussed in Section 6.4.1.3, where it was shown that increasing the coupling coefficient κ does not necessarily cause an increase in the harvested power.

Moreover, it was shown in Daqaq et al. (2007) that, at the optimal resistance of the purely resistive energy harvesting circuit, the power saturates and does not increase as κ increases. The result obtained here is along these lines. However, (6.46) suggests that the harvested power would remain constant even when the coupling coefficient κ becomes arbitrarily small or arbitrarily large and regardless of the value of ζ. To investigate this, we examine the expressions of the optimal voltage and current,

$$\left| \frac{V_{opt}}{\omega^2 \bar{X}_b} \right| = \frac{k\sqrt{C}(1 + \kappa^2 - (2 + \kappa^2 - 4\zeta^2)\Omega^2 + \Omega^4)}{4C_p\zeta\kappa(1 + (4\zeta^2 - 2)\Omega^2 + \Omega^4)\omega_n^2} , \qquad (6.47a)$$

$$\left| \frac{I_{opt}}{\omega^2 \bar{X}_b} \right| = \frac{C_p\kappa(1 + (4\zeta^2 - 2)\Omega^2 + \Omega^4)}{2\sqrt{C}(1 + \kappa^2 - (2 + \kappa^2 - 4\zeta^2)\Omega^2 + \Omega^4)\omega_n} , \qquad (6.47b)$$

where

$$C = \frac{C_p(1 + (4\zeta^2 - 2)\Omega^2 + \Omega^4)^3}{k\Omega^2(1 + \kappa^2 - (\kappa^2 - 4\zeta^2 + 2)\Omega^2 + \Omega^4)^2} . \qquad (6.47c)$$

Examining (6.47) shows that,

$$\lim_{\kappa \to 0} |V_{opt}| = \infty \quad \text{and} \quad \lim_{\kappa \to 0} |I_{opt}| = 0 , \qquad (6.48a)$$

and

$$\lim_{\kappa \to \infty} |V_{opt}| = 0 \quad \text{and} \quad \lim_{\kappa \to \infty} |I_{opt}| = \infty . \qquad (6.48b)$$

Equation (6.48) remove any potential confusion, and elaborate that (6.46) is only valid for realistic values of κ. For very small (or very large) values of κ, the voltage and current values are not useful.

(b) The optimal power is independent of the frequency ratio. Previous works usually concluded that maximum power can be achieved at the resonance frequency (Stephen 2006). This concept was advanced with the work of duToit (2005) where it was shown that two frequency ratios, resonance and anti-resonance, maximize the value of the harvested power. Further, it is worth noting that the optimal power expression in (6.46) is the same approximate expression obtained for the purely resistive circuit at the resonance and anti-resonance frequency ratios for small values of ζ and under the assumption $(\zeta/\kappa^2) \ll 1$, (6.33). However, while the expression derived earlier is obtained at the resonance and the anti-resonance frequency and is based on small values for (ζ/κ^2), (6.46) is valid for all frequency ratios and any value of (ζ/κ^2) provided that $\beta_{opt} > 0$. This implies that, adding and inductor to the circuit provides a new degree-of-freedom that can be optimzied and hence eliminates the effect of the frequency ratio from the optimal power, thereby tuning the inductor and resistor of the energy harvesting circuit to their optimal values which depend on the frequency ratio, it is possible to harvest optimal power everywhere in the frequency domain making the proposed circuit superior to the purely resistive circuit.

(c) Another interesting notion is realized by deriving the expression for the optimal strain rate within the piezoelectric element. This can be obtained by calculating the magnitude of $\dot{\epsilon} = \dot{x}/t_p$, and substituting the optimal circuit elements values α_{opt} and β_{opt} which simplifies to

$$\left| \frac{\dot{\epsilon}}{\omega^2 \bar{X}_b} \right| = \frac{1}{4 t_p \zeta \omega_n} .$$

The above expression implies that, for constant base acceleration (force), to achieve optimal power everywhere in the frequency domain, the harvesting circuit has to maintain a constant optimal strain-rate (velocity). Therefore, it can be concluded that the power optimization problem can be reduced to a strain optimization problem because simply the power is the product of force and velocity. Therefore, a circuit that could tune itself in real time to maximize the power by changing α_{opt} and β_{opt} according to (6.43a) and (6.43b) would in essence change the effective stiffness and damping of the system such that the piezoceramic strain rate is constant.

6.4.2.1 Analysis and Practicality of Optimization Results

Figure 6.9a displays the optimal power at a coupling coefficient, $\kappa = 0.6$, and a damping ratio $\zeta = 0.04 < \zeta_b = 0.085$. The presence of the inductor helps transform the power figure of Fig. 6.8a into a broadband–constant curve as shown in Fig. 6.9a. Figure 6.10a and 6.10b display the voltage and current, respectively. It is worth noting that, the power, voltage, and current coincide with those obtained for CASE I when $\Omega_r < \Omega < \Omega_{ar}$ ($\beta_{opt} = 0$). To further examine the practicality of the optimal results, one has to check the values obtained for the optimal resistance

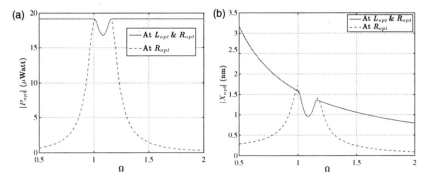

Fig. 6.9 (a) Optimal harvested power at $\kappa = 0.6$ and $\zeta = 0.04 < \zeta_b$ (inductor in parallel). (b) Optimal displacement at $\kappa = 0.6$ and $\zeta = 0.04 < \zeta_b$ (inductor in parallel)

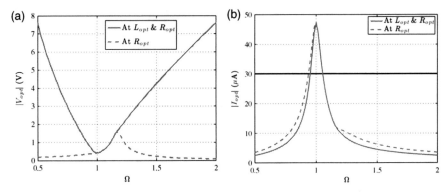

Fig. 6.10 (a) Optimal voltage at $\kappa = 0.6$ and $\zeta = 0.04 < \zeta_b$ (inductor in parallel). (b) Optimal current at $\kappa = 0.6$ and $\zeta = 0.04 < \zeta_b$ (inductor in parallel)

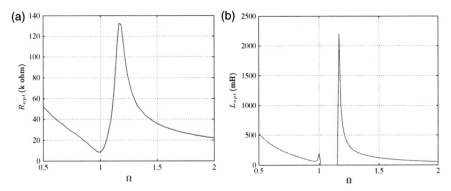

Fig. 6.11 (a) Optimal resistance at $\kappa = 0.6$ and $\zeta = 0.04 < \zeta_b$ (inductor in parallel). (b) Optimal inductance at $\kappa = 0.6$ and $\zeta = 0.04 < \zeta_b$ (inductor in parallel)

and inductance. As illustrated in Fig. 6.11a and 6.11b, these values are for the most part practical quantities that can be found off the shelf or even manufactured, if needed. For instance, at $\Omega = 2$, the optimal values for the load resistance and inductance are, respectively, 22 k and 55 mH which are achievable values. It is worth noting that, near the anti-resonance and for very small excitation frequencies, the optimal inductance can be rather large, however, it can be achieved by connecting a number of inductors in series.

The behavior of the optimal time constants α_{opt} and β_{opt} is depicted in Fig. 6.12. As the damping ratio ζ increases, we notice that the optimal time constant α decreases throughout the frequency ratio span, Fig. 6.12a. On the other hand, the outcome of the optimization problem changes as ζ increases and crosses ζ_b. We notice that a solution for β_{opt} exists throughout the frequency ratio range. However, the values of β_{opt} are very close when $\Omega < \Omega_r$ and $\Omega > \Omega_{ar}$.

Figure 6.13 shows the optimal power with the same coupling coefficient used previously ($\kappa = 0.6$) and a damping ratio $\zeta = 0.2 > \zeta_b$. A constant optimal power

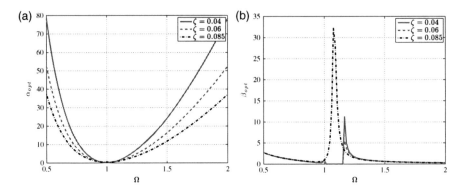

Fig. 6.12 Optimal time constants α_{opt} and β_{opt} at $\kappa = 0.6$ and $\zeta = 0.04, \ 0.06,$ and 0.085

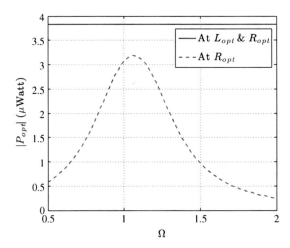

Fig. 6.13 Optimal power at $\kappa = 0.6$ and $\zeta = 0.2 > \zeta_b$ (inductor in parallel)

can still be harvested if optimal circuit elements are used, as per (6.46). Moreover, the power harvested via the proposed circuit is higher than the maximum power that can be obtained via the circuit utilized by duToit et al. (2005). It is also worth noting that the expression of (6.46) cannot describe the optimal power harvested using the circuit of duToit (2005) as the ratio (ζ/κ^2) is not small any more.

6.5 The Series-RL Circuit

For the purpose of completion and comparison, we consider in this section, the optimization problem when the inductor is connected in series with the resistive load (Fig. 6.1 in the absence of L_P). In this case, the governing equations are

$$m\ddot{x} + c\dot{x} + kx - \theta(\dot{q}R_{eq} + \ddot{q}L_S) = -m\ddot{x}_b$$

$$C_p L_S \ddot{q} + C_p R_{eq} \dot{q} + q\theta x = 0 \quad \text{where} \quad i = \frac{dq}{dt} \tag{6.49}$$

For sinusoidal base excitation, the magnitude of the response is given as

$$\left|\frac{X}{\omega^2 X_b}\right| = \frac{\sqrt{\alpha^2 \Omega^2 + (\beta\Omega^2 - 1)^2}}{\omega_n^2 \sqrt{C}} \tag{6.50}$$

$$\left|\frac{I}{\omega^2 X_b}\right| = \frac{\kappa \sqrt{C_p k \Omega^2}}{\omega_n \sqrt{C}} \tag{6.51}$$

$$\left|\frac{P}{\omega^2 X_b}\right| = \frac{k\alpha\kappa^2 \Omega^2}{\omega_n^3 C} \tag{6.52}$$

where

$$C = \Omega^2(\alpha + 2\zeta + \alpha\kappa^2 - (\alpha + 2\beta\zeta)\Omega^2)^2 + (1 - (1 + \beta + 2\alpha\zeta + \beta\kappa^2)\Omega^2 + \beta\Omega^4)^2 \tag{6.53}$$

Again, we cast the same optimization problem with the KKT conditions of (6.15) through (6.18) with $\mathcal{P}$ given by

$$\mathcal{P} = \frac{k\alpha\kappa^2 \Omega^2}{\Omega^2(\alpha + 2\zeta + \alpha\kappa^2 - (\alpha + 2\beta\zeta)\Omega^2)^2 + (1 - (1 + \beta + 2\alpha\zeta + \beta\kappa^2)\Omega^2 + \beta\Omega^4)^2}$$

The analysis gives rise to four cases (similar to the analysis in (6.4.1) and (6.4.2)). The results are presented in Section 6.5.1.

6.5.1 Optimality Results for Series RL-Circuit

The presence of the inductor in series with the resistor leads to the same optimal power achieved in Section 6.4.2. The optimal power expression obtained here is identical to that of (6.46). As such, the results illustrated in Fig. 6.14a are identical to those shown in Fig. 6.9a. Similarly, as demonstrated in Fig. 6.9b and Fig. 6.11b, the displacement behavior remains unchanged. Having said that, one would expect that the voltage and current results will change due to the fact that the parallel configuration acts as a current divider while the series configuration represents a voltage divider. As a result, comparing Fig. 6.10a and Fig. 6.15a, it is evident that, in general, the optimal voltage is higher for the case of an inductor in parallel. On the other hand, this notion is reversed for the magnitude of the current as observed by comparing Fig. 6.10b and Fig. 6.15b. These observations suggest that a series connection is more favorable when charging a storage device, which requires high current. However, the parallel configuration is favored when supporting applications that require high voltage such as wireless sensors.

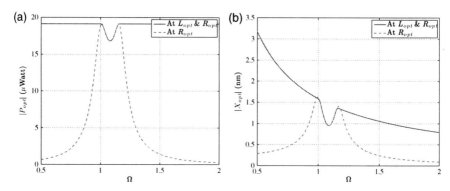

Fig. 6.14 (a) Optimal harvested power at $\kappa = 0.6$ and $\zeta = 0.04 < \zeta_b$ (inductor in series).
(b) Optimal displacement at $\kappa = 0.6$ and $\zeta = 0.04 < \zeta_b$ (inductor in series)

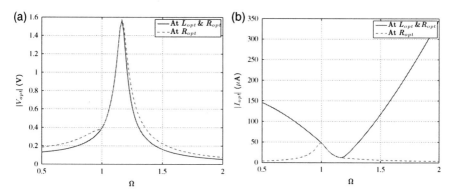

Fig. 6.15 (a) Optimal voltage at $\kappa = 0.6$ and $\zeta = 0.04 < \zeta_b$ (inductor in series). (b) Optimal
current at $\kappa = 0.6$ and $\zeta = 0.04 < \zeta_b$ (inductor in series)

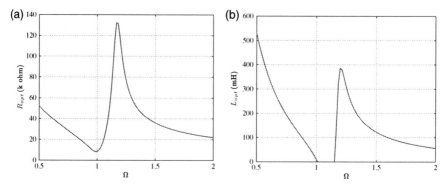

Fig. 6.16 (a) Optimal resistance at $\kappa = 0.6$ and $\zeta = 0.04 < \zeta_b$ (inductor in series). (b) Optimal
inductance at $\kappa = 0.6$ and $\zeta = 0.04 < \zeta_b$ (inductor in series)

The optimal resistance and inductance values are shown in Fig. 6.16. When the damping ratio is lower than ζ_b, we see that the optimization procedure fails when $\Omega_r < \Omega < \Omega_{ar}$, Fig. 6.16b. Moreover, comparing Fig. 6.16b and Fig. 6.11b, we notice that the values of the optimal inductance are generally lower for the series connection case. On the other hand, the behavior of the optimal resistance values is the same for the parallel and inductor series, Fig. 6.16a and Fig. 6.11a.

Figure 6.17 shows the behavior of the power and optimal inductance when $\zeta = \zeta_b$. In this case, the optimization problem yields values for the complete frequency range, Fig. 6.17b. Moreover, and once again, the presence of the inductor proves to provide higher power than the maximum power achieved by the purely resistive circuit, Fig. 6.18 and Fig. 6.13.

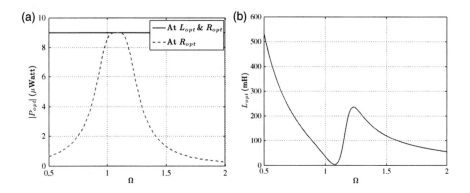

Fig. 6.17 (a) Optimal harvested power at $\kappa = 0.6$ and $\zeta = 0.085 = \zeta_b$ (inductor in series). (b) Optimal inductance at $\kappa = 0.6$ and $\zeta = 0.085 = \zeta_b$ (inductor in series)

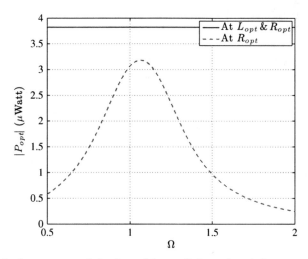

Fig. 6.18 Optimal power at $\kappa = 0.6$ and $\zeta = 0.2 > \zeta_b$ (inductor in series)

6.6 Conclusions

This chapter treats the optimization problem of a vibration-based energy harvester that utilizes a circuit design consisting of a resistor and an inductor. Utilizing the Karush–Kuhn–Tucker (KKT) technique, the optimization problem is cast as a non-linear programing problem with two constraints that guarantee the nonnegativity of the optimal values for the circuit parameters. When treating the parallel and series configurations, the optimization problem gives rise to two physically meaningful cases. The first of which suggests utilizing a purely resistive circuit. While this case has received considerable attention in the literature, most of the previous research efforts neglected the effect of mechanical damping on the optimal parameters. As such, in this work, we account for mechanical damping and demonstrate its imperative role. More specifically, it is shown that for damping ratios that are below a bifurcation damping ratio, the power has two maxima (at the *resonance and anti-resonance frequencies*) and one minimum. On the other hand, beyond the bifurcation damping ratio, the power exhibits only one maximum. In addition, we explore the effect of electromechanical coupling on the optimal power and show that materials with higher electromechanical coupling coefficients do not necessarily yield higher output power.

The second case resulting from the optimization problem suggests that by employing an optimal inductor in the circuit, one can substantially enhance the harvested power. Specifically, we demonstrate that for damping ratios that are less than the bifurcation damping ratio, one can acquire the maximum power obtained at the resonance and antiresonance frequency for the purely resistive circuit everywhere in the frequency domain except for excitation frequencies between the resonance and anti-resonance. On the other hand, when the damping ratio is higher than the bifurcation damping ratio, the harvested power through the proposed circuit can be much higher than that obtained via a purely resistive circuit. The results obtained in this chapter have critical implications as they suggest the following. First, adding an inductor to the circuit allows for tuning the energy harvesting device to scavenge the optimal power for a broad range of excitation frequencies. This implies that it is not necessary to tune the natural frequency of the mechanical element to the resonance and anti-resonance frequency to obtain optimal power. Second, in order to maintain optimal power for any excitation frequency it is essential to maintain an optimal strain rate. As such, the power optimization problem is equivalent to optimizing the strain rate of the mechanical element and is not related to the magnitude of the strain itself. In summary, we believe the work presented herein provides a solid bases for more involved optimization studies of other energy harvesting configurations such as unimorphs and bimorphs.

References

Anderson T, Sexton D (2006) A vibration energy harvesting sensor platform for increased indus-trial efficiency. In: Tomizuka M (ed) Smart Structures and Materials: Sensors and Smart Structures Technologies for Civil, Mechanical and Aerospace Systems, San Diego, CA, Proceedings of SPIE, vol 6174, pp 621–629

Anton S, Sodano H (2007) A review of power harvesting using piezoelectric materials (2003–2006). Smart Materials and Structures 16:1–21

Badel A, Guyomar D, Lefeuvre E, Richard C (2006) Piezoelectric energy harvesting using a synchronized switch technique. Journal of Intelligent Material Systems and Structures 17: 831–839

Beeby S, Tudor M, White N (2006) Energy harvesting vibration sources for microsystems applications. Measurement Science and Technology 17:R175–R195

von Büren T, Mitcheson P, Green T, Yeatman E, Holmes A, Tröster G (2006) Optimization of inertial micropower generators for human walking motion. IEEE Sensors Journal 6:28–38

Daqaq M, Renno J, Farmer J, Inman D (2007) Effects of system parameters and damping on an optimal vibration-based energy harvester. In: Proceedings of 48th AIAA/ASME/ ASCE/AHS/ASC Structures, Structural Dynamics, and Material Conference, Waikiki, HI, vol 8, pp 7901–7911

duToit N (2005) Modeling and design of a mems piezoelectric vibration energy harvester. Master's thesis, Massachusets Institute of Technology, Cambridge, MA

duToit N, Wardle B, Kim SG (2005) Design considerations for mems-scale piezoelectric mechanical vibration energy harvesters. Integrated Ferroelectrics 71:121–160

Erturk A, Inman D (2008a) A distributed parameter electromechanical model for cantilevered piezoelectric energy harvesters. Journal of Vibration and Acoustics, Transaction of ASME (to appear April 2008)

Erturk A, Inman D (2008b) On mechanical modeling of cantilevered piezoelectric vibration energy harvesters. Journal of Intelligent Material Systems and Structures (to appear) DOI 10.1177/1045389X07085639

Grisso B, Inman D (2006) Towards autonomous sensing. In: Tomizuka M (ed) Smart Structures and Materials: Sensors and Smart Structures Technologies for Civil, Mechanical and Aerospace Systems, San Diego, CA, Proceedings of SPIE, vol 6174, pp 248–254

Johnson T, Charnegie D, Clark W, Buric M, Kusic G (2006) Energy harvesting from mechanical vibrations using piezoelectric cantilever beams. In: Clark W, Ahmadian M, Lumsdaine A (eds) Smart Structures and Materials: Damping and Isolation, San Diego, CA, Proceedings of SPIE, vol 6169, pp 81–92

Lefeuvre E, Badel A, Richard C, Petit L, Guyomar D (2006) A comparison between several vibration-powered piezoelectric generators for standalone systems. Sensors and Actuators 126:405–416

Lesieutre G, Davis C (1997) Can a coupling coefficient of a piezoelectric device be higher than those of its active material? Journal of Intelligent Material Systems and Structures 8: 859–867

Lesieutre G, Ottman G, Hofmann H (2004) Damping as a result of piezoelectric energy harvesting. Journal of Sound and Vibration 269:991–1001

Priya S, Popa D, Lewis F (2006) Energy efficient mobile wireless sensor networks. In: Proceedings of ASME International Mechanical Engineering Congress & Exposition, Chicago, IL, IMECE 2006-14078

Rastegar J, Pereira C, Nguyen HL (2006) Piezoelectric-based power sources for harvesting energy from platforms with low frequency vibration. In: White E (ed) Smart Structures and Materials: Damping and Isolation, San Diego, CA, Proceedings of SPIE, vol 6171, pp 1–7

Roundy S (2005) On the effectiveness of vibration-based energy harvesting. Journal of Intelligent Material Systems and Structures 16:808–425

Sodano H, Inman D, Park G (2004) A review of power harvesting from vibration using piezoelectric materials. The Shock and Vibration Digest 36:197–205

Sodano H, Inman D, Park G (2005a) Comparison of piezoelectric energy harvesting devices for recharging batteries. Journal of Intelligent Material Systems and Structures 16:799–807

Sodano H, Inman D, Park G (2005b) Generation and storage of electricity from power harvesting devices. Journal of Intelligent Material Systems and Structures 16:67–75

Sood R (2003) Piezoelectric micro power generator (PMPG): A MEMS-based energy scavenger. Master's thesis, Massachusetts Institute of Technology, Cambridge, MA

Stephen N (2006) On energy harvesting from ambient vibration. Journal of Sound and Vibration 293(1):409–425

Twiefel J, Richter B, Hemsel T, Wallaschek (2006) Model based design of piezoelectric energy harvesting systems. In: Clark W, Ahmadian M, Lumsdaine A (eds) Smart Structures and Materials: Damping and Isolation, San Diego, CA, Proceedings of SPIE, vol 6169, p 616909

Winston WL (2003) Operations Research: Applications and Algorithms, 4th edn. Duxbury Press, Pacific Grove, CA

Wu WJ, Chen YY, Lee BS, He JJ, Pen YT (2006) Tunable resonant frequency power harvesting devices. In: Clark W, Ahmadian M, Lumsdaine A (eds) Smart Structures and Materials: Damping and Isolation, San Diego, CA, Proceedings of SPIE, vol 6169, p 61690A

Chapter 7
Energy Harvesting Wireless Sensors

S.W. Arms, C.P. Townsend, D.L. Churchill, M.J. Hamel, M. Augustin, D. Yeary, and N. Phan

Abstract Breaking down the barriers of traditional sensors, MicroStrain's energy harvesting wireless sensors eliminate long cable runs as well as battery maintenance. Combining processors with sensors, the wireless nodes can record and transmit data, use energy in an intelligent manner, and automatically change their operating modes as the application may demand. Harvesting energy from ambient motion, strain, or light, they use background recharging to maintain an energy reserve. Recent applications include piezoelectric powered damage tracking nodes for helicopters as well as solar powered strain and seismic sensor networks for bridges.

Keywords Energy Harvesting · Wireless Sensors · Piezoelectric · Mote · Node

7.1 Introduction

Recent developments in combining sensors, microprocessors, and radio frequency (RF) communications holds the potential to revolutionize the way we monitor and maintain critical systems (The Economist 2007). In the future, literally billions of wireless sensors may become deeply embedded within machines, structures, and the environment. Sensed information will be automatically collected, compressed, and forwarded for condition-based maintenance.

But who will change the billions of dead batteries?

We believe the answer is to harvest and store energy from the environment – using strain, vibration, light, and motion to generate the energy for sensing and communications. Combined with strict power management, smart wireless sensing networks can operate indefinitely, without the need for battery maintenance.

S.W. Arms (✉)

MicroStrain, Inc., 459 Hurricane Lane, Williston, Vermont 05495, USA

e-mail: swarms@microstrain.com

S. Priya, D.J. Inman (eds.), *Energy Harvesting Technologies*,

DOI 10.1007/978-0-387-76464-1_7 © Springer Science+Business Media, LLC 2009

7.2 Background

MicroStrain has been active in adaptive energy harvesting electronics for wireless sensor networks. Our electronics feature smart comparators – switches that consume only nanoampere levels of current – to control when to permit a wireless sensing node to operate. This insures that the energy "checkbook" is balanced. In other words, the system waits till there is a sufficient energy to perform a programed task. Only when the stored energy is high enough, the nanoamp comparator switch will allow the wireless sensor to draw current (Hamel et al., 2003). This is critical for applications where the ambient energy levels may be low or intermittent. Without this switch, the system may never successfully start up, because stored energy levels may always remain insufficient for the task at hand (Arms et al. 2005).

MicroStrain's miniaturized energy harvesting sensing nodes feature a precision time keeper, non-volatile memory for on board data logging, and frequency agile IEEE 802.15.4 transceiver (Fig. 7.1). Sampling rates, sample durations, sensor offsets, sensor gains, and on board shunt calibration are all wirelessly programmable.

7.3 Tracking Helicopter Component Loads with Energy Harvesting Wireless Sensors

Traditionally, helicopter component loads were monitored on only one or two strain gauge instrumented flight test aircraft, using slip rings. These techniques are too costly and difficult to maintain for application to an entire fleet of vehicles. The fatigue life of critical components, such as the pitch links, pitch horns, swash plate, yoke, and rotor were conservatively estimated based on the aircraft's flight hours,

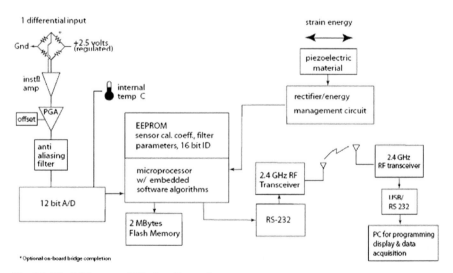

Fig. 7.1 Block Diagram of Wireless Transceiver

rather than actual usage (Moon & Simmerman 2005). Energy harvesting, combined with advanced, micropower wireless sensing electronics, will enable the realization of direct tracking of the operational loads on these critical structures, resulting in improved condition-based maintenance, and enhanced safety (Arms et al. 2006, Maley et al. 2007).

The design of an optimized energy harvesting system for sensing requires a comprehensive knowledge of the strain and/or vibration energies available in the operational environment. Furthermore, the sensing requirements must be well defined, including the sensor types and number of channels, their burst sample rates, duty cycles, and wireless data communications rates and range.

A block diagram of the electronics used to sense, record, transmit, and receive wireless data from an array of strain gauges is shown in Fig. 7.1. Circuits for converting piezoelectric material strain into useable electrical energy have been previously described (Churchill et al. 2003, Hamel et al. 2003, Arms et al. 2005). Key elements for piezoelectric energy conversion include impedance-matching circuits and an energy management circuit, which monitor the voltage levels on the energy storage elements. The energy storage monitor was comprised of a programmable comparator, or "smart switch", which itself drew only nanoampere current levels.

The programmable comparator's mission was to prevent the system from drawing energy from the storage elements, unless sufficient energy was stored to perform the pre-programmed sensing/processing/logging/communication task(s). This technique allowed the energy storage elements, which may include a rechargeable battery and/or a super-capacitor, to store up enough energy to perform for the specific task, *before allowing the task to actually be performed*. This approach enabled adaptive sensing, which will operate in applications where the ambient energy levels are low, variable, or intermittent.

Time-stamped sensor data may be logged to non-volatile memory at programmable sample rates or they may be logged and transmitted over a wireless link. Two megabytes of non-volatile, electrically erasable, programmable memory (EEPROM) were included to enable on board data logging. The bi-directional wireless communications link was comprised of a programmable, frequency agile IEEE 802.15.4 standard spread spectrum radio transceiver, operating in the license-free 2.4 GHz band. The base station was comprised of a notebook personal computer operating Windows XP and connected over the Universal Serial Bus (USB) port to a standard base station from MicroStrain, Inc. (Williston, Vermont).

The energy harvesting wireless pitch link was installed on a Bell Model 412 experimental rotorcraft and successfully flight tested at the Bell XworX facilities in Fort Worth, Texas during February 2007. The pitch link is a critical component on rotorcraft, as it is subject to large compressive and tensile loads (to thousands of pounds), and these loads increase dramatically during severe usage conditions. For example, it has been reported that the Sikorsky H-60 pitch link loads are six to eight times higher during gunnery turns and pull-ups when compared with the straight and level flight (Moon et al. 1996). Therefore, the pitch link provides a strong measure of usage severity, and is therefore a good candidate for wireless structural health monitoring (SHM).

Piezoelectric composite fiber material (PZT, type P2, Smart Material Corp., Sarasota, Florida) was epoxy-bonded to the shaft of the pitch link to support strain energy harvesting. Each patch measured approximately 3 in. by 1 in. (75 mm by 25 mm), and was ∼ .010 in. (250 μm) thin. This flexible composite is composed of many PZT fibers, and is highly damage tolerant and fatigue resistant. The cyclic loads on the Bell M412 pitch link are dominated by the cyclic revolution rate of the rotor, which is ∼ 5 Hz. The PZT supplier has reported ongoing fatigue test results which indicated that these patches can tolerate over 10 billion cycles at ±500 microstrain. This translates into ∼ 27 years operational life at 10 cycles/sec. The strain levels of the Bell M412 pitch link were reported to be relatively low during straight and level flight (±35 microstrain), and the component itself is relatively small in size. These limitations required that multiple PZT elements be bonded to the M412 pitch link in order to generate enough energy to support continuous loads monitoring.

The pitch link system featured wirelessly software programmable strain gauge offsets, amplifier gain, and shunt calibration as well as pulsed and regulated bridge excitation. The electronics module (Fig. 7.2) weighed 8.2 g, the PZT weight was 4.3 g/patch, with a total of 12 patches to weigh 52 g in total.

Strain gauges bonded to the pitch link were monitored by both wireless and conventional methods, both in the lab and during flight tests, as a means of checking the wireless results. The strain gauges were arranged on the pitch link to amplify tension and compression, and to cancel bending and thermal effects. The accuracy of MicroStrains' wireless strain measurement node was measured against a hard-wired reference system (HBM model MGC plus, Darmstadt, Germany) using a Baldwin precision calibrator as a strain reference. The accuracy of the wireless strain sensing node was measured at ±2.5 microstrain, with offset temperature coefficient of −.007%/°C, and span temperature coefficient of 004%/°C.

The pitch link module included a precision nanopower real-time clock (RTC) with ±10 PPM time reference accuracy. The RTC time was synchronized at the beginning of the flight test to the time provided by the hard-wired instrumentation,

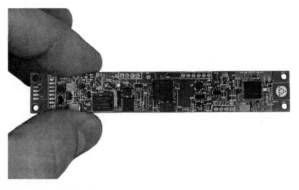

Fig. 7.2 ESG-Link Wireless Module

which used the global positioning system (GPS) for a time reference. The worst case accumulated time error for the RTC over 1.5 h flight test may be estimated to total 54 ms without any compensation. Compensation for RTC clock drift was provided by an end of flight RTC synchronization to GPS time and then assuming a linear error accumulation.

Environmental packaging for the pitch link was based on polyurethane sealant (Hi-Tak, Av-DEC, Aviation Devices and Electronic Components, LLC, Fort Worth, Texas) combined with cold shrink tubing (3M, Minneapolis, Minnesota) which weighs 7 g/in. × 8 in., with a total of 56 g. However, Bell Helicopter's flight test engineers chose to use a method based on clamps with a counterbalance rather than the cold shrink method in order to facilitate servicing if required. A photograph of the completed Bell M412 pitch link assembly, including the microelectronics module, bonded PZT material, and clamping system have been previously described (Arms et al. 2007).

Electromagnetic interference (EMI) testing was performed prior to flight testing. Flight test data were collected over a 1 h period. Data were sampled, logged, and transmitted wirelessly to a base station mounted on the ceiling of the cockpit, behind the pilot. A photograph of the installed energy harvesting pitch link, installed on the rotating structures of the Bell M412 helicopter, is shown in Fig. 7.3.

In addition to the flight testing on helicopters, MicroStrain, Inc. has also demonstrated vibration energy harvesting wireless sensors for use aboard Navy ships. The application identified by NAVSEA program managers was sensing of heat stress within various compartments of aircraft carriers. Traditionally, heat stress sensing has been performed using a suite of sensor, hand carried to various compartments

Fig. 7.3 Pitch-Link Installed on Bell 412 Helicopter

of the ship, and readings noted by hand. These sensors included a blackbody temperature sensor (to measure heating by radiation), a dry, and a wet bulb temperature sensor. Existing commercial wet bulb sensors require a reservoir of distilled water, which evaporates over a wick (comprised of a tubular knitted "sock") which completely covers the wet bulb temperature sensor (Model HS-32, Metrosonics, Oconomowoc, WI). Moisture climbs up the wick due to capillary action, and as this moisture evaporates, this sensor provides a measure of evaporative cooling. Therefore, a major drawback to existing commercial heat stress nodes is the requirement that the user maintain the reservoir with distilled water. A second limitation is that no sensor network exists to automate the data collection process.

To facilitate shipboard automated monitoring, MicroStrain engineers developed new wireless heat stress nodes (HSNs) which eliminated the wet bulb sensor (and its reservoir and wick) in favor of a micromachined relative humidity (RH) sensor (Sensirion, Model SHT15, Westlake Village, CA). The tiny RH sensor was protected from direct contact with liquid water but permitted contact with water vapor using a filter cap (Model SF1, Sensirion). In addition to the miniature RH sensor, a dry bulb temperature sensor and blackbody temperature sensor was provided. These sensor data were combined and processed using on board algorithms to compute each ship compartments' wet bulb globe temperature (WBGT) and heat stress indexes.

Navy stay time tables for various work tasks within the ship were wirelessly uploaded to each heat stress node to allow automated heat stress recording and reporting from various instrumented compartments. MicroStrain's prototype HSN nodes were directly compared with Metrosonics' model HS-32 measurements of Wet Bulb, Dry Bulb, and Black Body Globe temperatures. The wireless HSNs were found to be accurate to within $\pm 1.1\,^\circ$C, which met the US Navy's requisite error band for all measured and calculated temperatures. A successful demonstration of the heat stress nodes' wireless connectivity was also provided on the USS George Washington aircraft carrier (CVN-73) at VASCIC in collaboration with Northrop Grumman at Newport News, VA.

Photographs of the vibration energy harvesting heat stress nodes are provided in Fig. 7.4a & b. The vibration harvester was tuned to the predominant vibration frequency of the machine to which it was designed to be affixed, which was an air compressor. This represented a very challenging application, as the available energy for harvesting was approximately equivalent to a 52 Hz sinusoid at vibration amplitudes of only 30–40 mg.

Results showed that a balance of the energy checkbook was achieved through a combination of strategies, including the development of embedded, energy aware operational modes. Micropower sleep modes were used as much as possible, including using sleeping between samples. The average pitch link microelectronics power consumption was 34 μA at 3 VDC excitation (102 μW), w/4500 Ω strain gauge bridge logged at 32 samples/sec. This leads to a power consumption "figure of merit" of $\sim 3.2\,\mu$W/sample/sec, where sample/sec is the programmable strain gauge bridge sampling rate. Time-stamped pitch link loads were recorded/

Fig. 7.4 a & b Wireless Heat Stress Node

transmitted at programmable sample rates. Power consumption results for four operational software modes are described below.

Mode 1: Real Time Transmission: Highest power mode – device was programmed to both transmit and log data to non-volatile (flash) memory. Device logs data at a specified rate, and once 100 samples are acquired, the system transmits data. Power consumption at 32 samples/sec was $\sim 250\,\mu\mathrm{W}$.

Mode 2: Real Time Transmission with Energy Aware Option: Device logs data at specified rate, once 100 samples are acquired, the system checks to see if enough energy has been stored to permit transmission of data over the RF link. If enough power is available, data are transmitted, if not, data are logged to memory and are not transmitted. Power consumption at 32 samples/sec was $\sim 250\,\mu\mathrm{W}$ provided enough power was available. Otherwise consumption varies with available energy.

Mode 3: Real Time Data Logging: Lowest power mode – data were logged to memory for download at the end of the flight test. Power consumption at 32 samples/sec was $\sim 100\,\mu\mathrm{W}$.

Mode 4: Data Transmission When Storage Capacitor Reaches Threshold: This mode simply charges an energy storage capacitor to a certain level; once this level was reached, a nanopower comparator switch turned the circuit on, and then a predetermined amount of data were transmitted. This differs from Mode 2 in that no data were logged in the background and the system cannot consume any power until sufficient energy has been stored. Average power consumption varies with available energy, timekeeper draws $9\,\mu\mathrm{W}$.

The average power generated by the pitch link strain harvester representative of straight and level flight, was measured at $280\,\mu\mathrm{W}$. Note, however, that the actual energy generation is higher, due to the presence of higher frequency components on the actual pitch link when in flight. Furthermore, due to time constraints, the M412 flight tests were performed with the aircraft not fully loaded (i.e., underweighted). A fully loaded vehicle would generate higher pitch link strains, and therefore, it would produce more energy within the strain-harvesting elements.

When power generation tests were replicated using the 500 Hz data collected from the pitch link's hard-wired slip ring system, the strain harvester produced $\sim$40% more energy. For the maximally loaded pitch link (dive test), the power output was $34\,\mu\mathrm{W}/\mathrm{patch} \times 12$ patches $= 408\,\mu\mathrm{W}$, which supports continuous logging of data at 128 Hz. For conditions of low load (straight and level flight), the power output was $25\,\mu\mathrm{W}/\mathrm{patch} \times 12$ patches $= 300\,\mu\mathrm{W}$, which supports logging of data at 90 Hz.

EMI test results. The wireless pitch link transmitter exhibited no interference with flight test instrumentation, radios, or control systems, including the auto pilot. The high-power telemetry system that communicates data from the flight instrumentation aboard the helicopter to a ground test station with 50 mile range had no effect on the wireless pitch link's data transmission.

Flight test results. The wireless pitch link was programmed to record the time-stamped pitch link loads during flight, and to periodically transmit data to a radio transceiver in the cabin (Fig. 7.6). Pitch link load data were collected for approximately 1 hr of flight for a variety of typical flight maneuvers. Data were communicated from the rotating pitch link to the aircraft cabin with less than 1% digital packet error rate. In addition, all flight data were logged internally on the sensor and downloaded for post-flight data verification. Wireless data were collected at 64 Hz, while hard-wired data were collected (through slip rings) at 500 Hz. Wireless data tracked hard-wired data, but higher sampling rates would improve wireless fidelity (Fig. 7.5). A lab test was performed using an electro-dynamic shaker which replicated the pitch link's complex waveform. Wireless strain data were sampled at 512 Hz and hard-wired data at 4 kHz. At these higher sampling rates, the wireless data were essentially indistinguishable from hard-wired data.

Shipboard test results. Each heat stress node update required that the node sample all three sensors (RH, dry bulb, and black body), compute a heat stress result, and then transmit the data. Power consumption for the wireless heat stress nodes, using strict power management, was under 200 μW with an update rate of once every 2 min. The vibration harvester was tuned to the fundamental rotation frequency of the air compressor's electric motor (52 Hz), and produced 3–4 mW at 150 mg. Power generation for the vibration energy harvester was dependent on the frequency and level of ambient vibration. For extremely low levels of vibration, such as the shipboard air compressor, operating at only 30 mg, the nanoamp comparator switch was essential for reliable heat stress node operation, because the sensing elements must not be allowed to draw power until sufficient energy is stored. Using software *mode 4*, and operating at 30 mg, HSNs would update every 5 min, which was satisfactory for the shipboard heat stress-monitoring applications. Successful wireless tests were performed aboard the George Washington aircraft carrier (CVN-73) using a

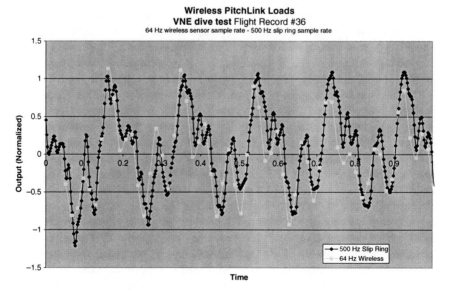

Fig. 7.5 Comparison of Data From Wireless Energy Harvesting Node Versus Hardwired Data

battery-powered version of the heat stress nodes, and supported by a multi-hop "Zig-bee" networking scheme. A single AA Li–ion battery (Tadiran Corp, Port Washing-ton, NY) has 2400 mA h capacity. Within a battery-powered HSN updating every 2 min, and consuming only 67 μA, the HSN would operate for 4 years.

In conclusion, energy harvesting using piezoelectric materials to power energy-aware wireless sensor nodes was demonstrated. Self-powered wireless pitch link load sensors and shipboard heat stress nodes were developed and tested aboard a helicopter, and an aircraft carrier. Under the conditions of low-ambient energy levels for harvesting, the amount of energy consumed was less than the amount of energy harvested. This enables on board wireless sensors to operate perpetually without battery maintenance. Our current and planned research includes the development and testing of highly efficient energy generators to support those applications that demand higher sampling rates.

7.4 Monitoring Large Bridge Spans with Solar-Powered Wireless Sensors

These techniques can also be applied to other applications, such as monitoring large civil structures. Recently, sudden structural failures of large bridge spans, such as the Interstate 35 W Bridge in Minneapolis, and the Chan Tho Bridge in Vietnam have resulted in the tragic loss of lives. Three years ago, the Federal Highway Administra-tion reported that ∼ 20% of the US interstate bridges (nearly 12,000 bridges) were rated as deficient. Developing and deploying cost-efficient methods for monitoring bridges – and for determining which bridges require immediate attention – should be an important priority for the United States and the world.

Wireless sensor networks have the potential to enable cost-efficient, scaleable-monitoring systems that could be tailored for each particular bridge's requirements. Eliminating long runs of wiring from each sensor location greatly simplifies system installation and allows a large array of sensor nodes to be rapidly deployed.

We have recently supported two major wireless installations that are actively monitoring the structural strains and seismic activity of major spans. Leveraging energy harvesting technology supported by the US Navy, these wireless sensor net-works are powered by the sun, and therefore do not require battery maintenance.

MicroStrain has previously described battery-powered wireless strain sensors for structural health monitoring (Townsend et al. 2002, Galbreath et al. 2003, Arms et al. 2004). One example is the Ben Franklin Bridge, which spans the Delaware River from Philadelphia, PA to Camden, NJ. The monitoring system was accessed remotely over commercial cellular telephone networks, and sensor data were pro-vided to the customer via secure access to a web-based server. The wireless nodes measured structural strains in the cantilever beams as passenger trains traversed the span. Measurements taken over several months' time, under contract from the Delaware River Port Authority (DRPA), were used to document the bridge's cyclic structural strains.

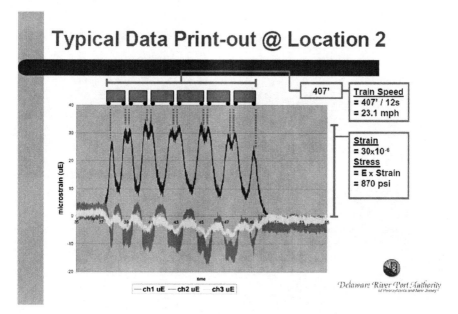

Fig. 7.6 Typical strain data collected from location #2 from the Ben Franklin Bridge with train crossing (slide courtesy DRPA, reference 8). The strain pattern reflects the dynamic loads resulting from the train's axles into the bridge's instrumented cantilever beam steel-supporting structure

At the locations tested, the measured strains and calculated stresses were far below the endurance limit (Fig. 7.6). Repeated cyclic stress above the endurance limit results in the gradual reduction of strength, or fatigue. Therefore, bridges are designed to operate at stress levels below this limit. From the information automatically collected by the wireless strain sensors, DRPA engineers concluded that cyclic stress fatigue due to train crossings was not a problem (Rong & Cuffari 2004)..

MicroStrain's first solar-powered bridge installation was recently made in Corinth, Greece. This system uses arrays of wireless tri-axial accelerometer nodes to monitor the span's background vibration levels at all times. Each node and solar panel is packaged within watertight enclosures for outdoor use. In the event that seismic activity is detected at any one of the nodes, the entire wireless network of nodes is alerted, and data are collected simultaneously from the entire network. Photographs of this bridge and the wireless G-LINK® nodes as installed in Corinth are shown in Fig. 7.7.

The second solar-powered installation is on the Goldstar bridge in New London, Connecticut, in collaboration with John DeWolf, Ph.D. of the University of Connecticut. This system monitors not only vibration, but also the strains and temperatures from key structural elements of the span (Fig. 7.8). Intended for long-term monitoring, these new installations overcome the limitations of older types, which required that the wireless node's batteries be replaced or recharged periodically. Maintenance of batteries is simply not practical on bridges, where sensor nodes must

(a)   (b)

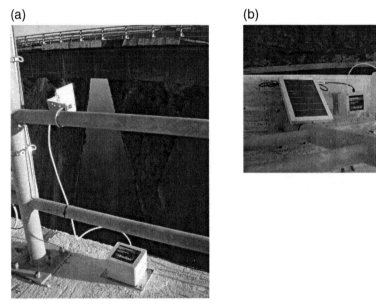

Fig. 7.7 Solar-powered wireless G-Link ® seismic sensors on Corinth Bridge, Greece

Fig. 7.8 Goldstar Bridge over the Thames River in New London, Connecticut. Multiple solar-powered nodes monitor strains and vibrations at key locations on this structure

be placed on, under, and within the structure in locations which may be extremely difficult to access.

The Connecticut Goldstar bridge program is a long-term project developed to learn how bridge-monitoring systems can be used for evaluation of in-service behavior, for long-term structural health monitoring of each bridge, and for assisting the Connecticut Department of Transportation to manage the State's bridge infrastructure (DeWolf 2006).

7.5 About MicroStrain Inc.

MicroStrain is a privately held corporation based in Williston, Vermont. MicroStrain produces smart, wireless, microminiature displacement, orientation, and strain sensors. Applications include advanced automotive controls, health monitoring, inspection of machines and civil structures, smart medical devices, and navigation/control systems for unmanned vehicles. For further information, please visit MicroStrain's website at www.microstrain.com or call 802-862-6629.

References

Arms et al., "Power Management for Energy Harvesting Wireless Sensors", Proceedings SPIE Int'l Symposium on Smart Structures and Smart Materials, San Diego, CA, March 9th, 2005

Arms et al., "Energy Harvesting Wireless Sensors for Helicopter Damage Tracking", AHS 62, Phoenix, AZ, May 11th, 2006

Arms et al., "Tracking Pitch Link Dynamic Loads with Energy Harvesting Wireless Sensors", AHS 63, Virginia Beach, VA, May 2007

Arms S.W. et al., "Vibration Energy Harvesting for Wireless Health Monitoring Sensors", Proceedings IWSHM, Stanford, CA, Sep 2005

Arms S.W., Galbreath J.H., Newhard A.T., Townsend C.P., "Remotely Reprogrammable Sensors for Structural Health Monitoring", Proceedings NDE/NDT for Highways and Bridges, Structural Materials Technology VI, pp. 331–338, NDE/NDT for Highways & Bridges, Buffalo, NY, 16 Sep 2004

Churchill D.L. et al., "Strain Energy Harvesting for Wireless Sensor Networks", Proceedings SPIE Smart Structures and Smart Materials, vol. 5005, pp. 319–327, 2003

DeWolf, J.T., "The Long Term Structural Health Monitoring of Bridges in the State of Connecticut", Third European Workshop on Structural Health Monitoring, Granada, Spain, July 5–7, 2006 (abstract hyperlink: http://atlas-conferences.com/c/a/r/b/96.htm)

Galbreath J.H., Townsend C.P., Mundell S.W., Arms S.W., "Civil Structure Strain Monitoring with Power-Efficient High-Speed Wireless Sensor Networks", International Workshop for Structural Health Monitoring, by invitation, Stanford, CA, September 2003

Hamel et al., "Energy Harvesting for Wireless Sensor Operation and Data Transmission", US Utility Patent, number 7081693, filed March 2003

Hamel M.J. et al., "Energy Harvesting for Wireless Sensor Operation and Data Transmission", US Patent Appl. Publ. *US 2004/0078662A1*, filed March 2003

Maley et al., "US Navy Roadmap to Structural Health and Usage Monitoring", AHS 63, Virginia Beach, VA, May 1–3, 2007

Moon S., Menon D., Barndt G., Fatigue Life Reliability Based on Measured Usage, *AHS 52*, Washington, DC, June 4–6, 1996

Moon S., & Simmerman C., "The Art of Helicopter Usage Spectrum Dev.", Am. Helicopter Soc. (AHS) 61st Annual Forum, Grapevine, TX June 1–3, 2005

Rong A.Y., & Cuffari M.A., "Structural Health Monitoring of a Steel Bridge Using Wireless Strain Gauges" Structural Materials Technology VI, pp. 327–330, NDE/NDT for Highways & Bridges, Buffalo, NY, 16 Sep 2004

The Economist, "The Coming Wireless Revolution: When Everything Connects", April 28th-May 4th 2007

Townsend C.P., Hamel, M.J., Arms, S.W., "Scaleable Wireless Web Enabled Sensor Networks", Proceedings SPIE's 9th Int'l Symposium on Smart Structures & Materials and 7th Int'l Symposium on Nondestructive Evaluation and Health Monitoring of Aging Infrastructure, San Diego, CA, paper presented 17–21 March, 2002

Chapter 8
Energy Harvesting using Non-linear Techniques

Daniel Guyomar, Claude Richard, Adrien Badel, Elie Lefeuvre
and Mickaël Lallart

Abstract Recent progresses in both microelectronic and energy conversion fields
have made the conception of truly self-powered, wireless systems no longer chimeri-
cal. Combined with the increasing demands from industries for left-behind sensors
and sensor networks, such advances therefore led to an imminent technological
breakthrough in terms of autonomous devices. Whereas some of such systems are
commercially available, optimization of microgenerators that harvest their energy
from their near environment is still an issue for giving a positive energy balance to
electronic circuits that feature complex functions, or for minimizing the amount of
needed active material. Many sources are available for energy harvesting (thermal,
solar, and so on), but vibrations are one of the most commonly available sources
and present a significant energy amount. For such a source, piezoelectric elements
are very good agents for energy conversion, as they present relatively high coupling
coefficient as well as high power densities.

Several ways for optimization can be explored, but the two main issues concern
the increase of the converted and extracted energies, and the independency of the
harvested power from the load connected to the harvester.

Particularly, applying an original nonlinear treatment has been shown to be an
efficient way for artificially increasing the conversion potential of piezoelectric element applied to the vibration damping problem. It is therefore possible to extend
such principles to energy harvesting, allowing a significant increase in terms of
extracted and harvested energy, and/or allowing a decoupling of the extraction and
storage stage.

The purposes of the following developments consist in demonstrating the ability
of such microgenerators to convert ambient vibrations into electrical energy in an
efficient manner. As well, when designing an energy harvester for industrial application, one has to keep in mind that the microgenerator also must be self-powered
itself, and needs to present a positive energy balance. Therefore, in addition to the
theoretical developments and experimental validations, some technological considerations will be presented, and solutions to perform the proposed processing using a

D. Guyomar (✉)
LGEF, INSA Lyon, 8 rue de la Physique, F-69621, France
e-mail: daniel.guyomar@insa-lyon.fr

S. Priya, D.J. Inman (eds.), *Energy Harvesting Technologies*,
DOI 10.1007/978-0-387-76464-1_8 © Springer Science+Business Media, LLC 2009

negligible part of the available energy will be proposed. Moreover, the behavior of the exposed technique under realistic vibrations will be investigated.

8.1 Introduction

The last two decades of the 20th century have been particularly prolific in the semiconductor industry. The developments have been incredibly strong toward an increase in the computing and speed capabilities as well as integration and decrease of supply voltage and power. The development of GSM technology, which increased from 1 million user market in 1993 to nearly 600 millions by 2001, and marketing trends for portable or nomad equipments, specifically battery-powered, had also pushed forward even more the development of low-power electronics. Under this market pressure, digital electronics as well as analog circuits or RF chips underwent similar progresses. As a matter of example, the MC6805 microcontroller from Motorola, universally used in the late 1980s, needed approximately 25 mW of power supply under 5 V with a computing capability of about one MIPS. Recent microcontroller such as EM6680 from EM Microelectronics features a 300 μW power need for a 3 V supply voltage for similar computing performances. This constant decrease in power requirements for electronic circuitry opened-up a very exciting research field consisting in the developments of techniques allowing the gathering of energy in the immediate surrounding of the electronic device. Exploiting either ambient light or vibration or heat enables the design of completely self-powered, standalone functions that can even be part of a network using low-power RF links. Among the many suitable technologies, the use of piezoelectric systems allowing the conversion of structural vibrations or motions into usable energy is of major interest. The idea of using piezoelectrics as an electric generator is not new, gas igniters using user pressure have been one of the most common portable energy generators for years. In the same way, many military applications relied on piezoelectrics to generate the energy needed for various ammunitions. Inversely, using piezoelectrics for supplying complex functions is a very recent field and the perspectives of these new applications are very demanding as a matter of optimization of these generators.

A lot of effort has been invested in two fields. On one hand, the design of suitable electromechanical devices, comprising a mechanical system allowing the conversion of ambient motions into usable strain applied on the piezoelement. On the other hand, the architecture of the electrical network, connected on the piezoelement, allowing the management of the converted energy. This paper will focus mainly on this later field. Most of the developed piezoelectric energy management devices were based on the same scheme. It is composed of a rectifier bridge which allows charging a capacitance for filtering purpose and then a DC–DC converter provides the energy to the load, adjusting both the voltage and the global electrical impedance in order to optimize the power flow. Very few works in fact were devoted to optimize the energy management device in order to increase the power extracted from the mechanical structure.

In the development proposed here, a nonlinear technique is proposed. It consists in adding in parallel with the piezoelement, a switching device leading to a distorting of the generated piezovoltage in order to shape it as a nearly square signal in phase with the derivative of the strain. This technique which was primarily applied for mechanical resonance suppression was extended to the field of energy harvesting. It resulted in an extraordinary increase in the extracted energy, especially for non-resonant structures.

The first chapter is an introduction to the nonlinear technique starting from the point of view of vibration damping. Structural models used along this paper and basic technological considerations are detailed. Then, the nonlinear energy-harvesting technique is developed in the case of steady-state harmonic vibrations. Standard and nonlinear techniques are compared at and out of the mechanical resonances, taking into account the damping effect due to the modified energy balance. The comparison is extended in the case of pulsed operation corresponding to many implementation cases. Different architectures relying on nonlinear voltage processing for the optimization of the extracted energy are described and discussed. Finally, the discussion is completed with the case of broadband signals for which the switching control relies on a statistic analysis of the voltage or mechanical strain signal.

8.2 Introduction to Nonlinear Techniques and their Application to Vibration Control

8.2.1 Principles

As outlined in the previous paragraph, the converted energy of a piezoelectric element can either be stored or directly dissipated. In a mechanical point of view, both of these processes are equivalent to mechanical losses. However, the control strategies are different whether the final aim is to control vibrations or use the converted energy in order to supply an electronic circuit.

In this section, a brief overview of the nonlinear vibration control techniques, in order to have a comprehensive view of the underlying principles of the electromechanical conversion, is proposed. As it will be seen, the vibration control principles are very close to those of energy harvesting. As well, this section proposes a simple, but realistic modeling of an electromechanical structure excited around one of its resonance frequencies. This model is sufficient for the following analyses as the large majority of the electromechanical energy is within the first resonance frequency bands.

8.2.1.1 Modeling

It is proposed here a simple lumped model for an instrumented structure as shown in Fig. 8.1. This model is derived from the piezoelectric constitutive equations recalled

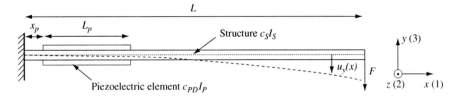

Fig. 8.1 Vibrating structure with piezoelectric elements bonded on the surface

in Eq. (8.1), with S, T, E, and D the strain tensor, stress tensor, electric field, and induction field, respectively, and s^E, d, and ε^T the mechanical compliance tensor of the short-circuited piezoelectric element, the piezoelectric constant, and the dielectric constant of the piezoelectric material at constant stress, respectively. Such a model permits an analytical analysis for vibration damping and energy harvesting techniques, while being simple but realistic. A detailed analysis of this model can be found in Badel et al. (2007).

$$\begin{cases} T = c^E S - e^t E \\ D = \varepsilon^S E + eS \end{cases} \tag{8.1}$$

It can be shown (Badel et al. 2007) that the equivalent lumped model for a monomodal structure is given by a simple spring–mass–damper system with an electromechanical coupling described by Eq. (8.2) and represented in Fig. 8.2. u, F, V, and I are the flexural displacement, the applied force, the piezoelectric voltage, and the current flowing out the piezoelectric element, respectively. Definitions of the model parameters are given in Table 8.1. The system presents one mechanical degree of freedom (u) and one electrical degree of freedom (V).

The electromechanical coupling coefficient k can also be expressed using the model parameters, yielding Eq. (8.3).

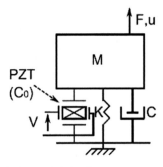

Fig. 8.2 Lumped model for an instrumented structure

Table 8.1 Model parameter

Parameter	Denomination
M	Dynamic mass
C	Structural damping coefficient
K_E	Short-circuit stiffness
α	Force factor
C_0	Clamped capacitance of the piezoelectric element

$$\begin{cases} M\ddot{u} + C\dot{u} + K_E u = F - \alpha V \\ I = \alpha \dot{u} - C_0 \dot{V} \end{cases} \tag{8.2}$$

$$k^2 = \frac{\alpha^2}{C_0 K_E + \alpha^2} \tag{8.3}$$

It is important to note that this model represents a simple, but realistic behavior of the electromechanical structure near a resonance frequency. This is sufficient for the following developments as:

– In the case of vibration control, only the forces that drive the system at one of its resonance frequencies lead to significant displacement.
– In the case of energy harvesting, the large majority of the energy that can be harvested is concentrated within the resonance frequency bands.

8.2.1.2 Vibration Damping Principles

This section exposes the general principles of the so-called "Synchronized Switch Damping" (SSD) method for vibration control, depicted in Fig. 8.3. According to the switching network, several techniques can be achieved: SSDS (short-circuit: Richard et al. 1998), SSDI (inductor: Richard et al. 2000), SSDV (voltage source: Lefeuvre et al. 2006a), adaptive SSDV (voltage source tuned on the displacement: Badel et al. 2006a).

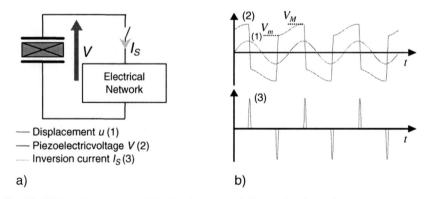

— Displacement u (1)
— Piezoelectricvoltage V (2)
— Inversion current I_S (3)

a)   b)

Fig. 8.3 SSD method: (**a**) general block schematic and (**b**) associated waveforms

This section proposes a common theoretical development for all of these methods. The main difference between all of these method is that SSDV and adaptive SSDV methods require an external energy supply (semi-active techniques), but the SSDI and SSDS are not (semi-passive techniques). The latter can beside be made self-powered.

Starting from the constitutive equations (Eq. (8.2)) of the electromechanical structure, the uncontrolled system with the piezoelement in open-circuit condition ($I = 0$) and considering null initial conditions leads to the simplified expression (Eq. (8.4)), which can be expressed in the Fourier domain as Eq. (8.5). K_D is given as the open-circuit stiffness. Considering a weakly damped system allows to consider the resonance frequency very close to the natural frequency, leading to the expression of the angular resonance frequency ω_0 given by Eq. (8.6). Therefore, in the case of the uncontrolled system, excited by a driving force of amplitude F_M and considering that the speed and the force are in phase at the resonance[1], the expression of the uncontrolled displacement magnitude u_M yields Eq.(8.7).

$$M\ddot{u} + C\dot{u} + \left(K_E + \frac{\alpha^2}{C_0} \right) u = F$$

$$\Leftrightarrow M\ddot{u} + C\dot{u} + K_D u = F \tag{8.4}$$

$$\frac{U(\omega)}{F(\omega)} = \frac{1}{-M\omega^2 + jC\omega + K_D} \tag{8.5}$$

$$\omega_0 = \sqrt{\frac{K_D}{M}} \tag{8.6}$$

$$(u_M)_{\text{uncont}} = \frac{1}{C\omega_0} F_M \tag{8.7}$$

The general circuit for the SSD techniques is presented in Fig. 8.4. According to the chosen configuration, L and/or V_S are null or not. The principles of the SSD are to close for a very short time period, the switch S_{W1} or S_{W2} in each time an extremum of electrostatic energy is reached. As the switching time period is very brief, the piezoelement is almost always in open-circuit condition. As well as the charges available on the piezoelectric element are matched with the voltage and displacement, the extrema actually occur when the displacement or the voltage

---

[1] This is a good approximation for weakly damped systems

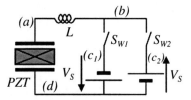

Fig. 8.4 General SSD circuit

reaches an extremum. Typically, S_{W1} is closed when the voltage or the displacement reaches a maximum and S_{W2} is closed when the voltage or the displacement
reaches a minimum. When S_{W1} is closed, there is therefore a current flow through
(a) → (b) → (c$_1$) → (d), and through (d) → (c$_2$) → (b) → (a) when S_{W2} is
closed. Obviously, when no voltage source is used, only one switch is needed for
both minima and maxima.

The effect of the switch is a quick inversion of the piezoelectric voltage, as shown
in Fig. 8.3(b). The time duration of the inversion is typically given by the half-
pseudo-period (Eq. (8.8)) of the resonant electric circuit shaped by the piezoelement
(with a blocking capacitor C_0) and the inductor L. After this time period, the switch
is opened and the piezoelement let in open-circuit condition until its voltage reaches
a new extremum.

This inversion is, however, not perfect due to internal losses in the switching circuit, and is characterized by the inversion ratio γ, that can be expressed as Eq. (8.9)
considering the electrical quality factor Q_i of the inversion circuit.

$$t_i = \pi \sqrt{LC_0} \tag{8.8}$$

$$\gamma = e^{-(\pi/2Q_i)} \tag{8.9}$$

In the steady-state case, the controlled voltage can be seen as the sum of a voltage
proportional to the displacement and a piecewise function h with amplitude H, as
defined in Eq. (8.10). In other terms, the voltage proportional to the displacement
corresponds to the open-circuit condition ($I = 0$), while the piecewise function h
results from the change in initial conditions induced by the switch event. h can also
be expressed considering the speed in Eq. (8.11) as follows:

$$V = \frac{\alpha}{C_0}(u + h) \tag{8.10}$$

$$h = H \times \text{sign}(\dot{u}) \tag{8.11}$$

Considering the absolute value of the voltage before and after the inversion, noted
V_M and V_m, respectively, yields Eq. (8.12). This leads to the expression of h defined
as Eq. (8.13).

$$\begin{cases} (V_m - V_S) = \gamma(V_M + V_S) \\ V_M - V_m = 2\frac{\alpha}{C_0}u_M \\ \frac{1}{2}(V_M + V_m) = \frac{\alpha}{C_0}H \end{cases} \qquad (8.12)$$

$$h = \frac{1+\gamma}{1-\gamma}\left(u_M + \frac{C_0}{\alpha}V_S\right)\text{sign}(\dot{u}) \qquad (8.13)$$

Therefore, considering that the displacement remains purely sinusoidal allows approximating the piecewise function by its first harmonic in the Fourier space as Eq. (8.14). Thus, expressing the motion equation of the structure given by Eq. (8.2) yields Eq. (8.15) in the Fourier space, or equivalently Eq. (8.16). Consequently, it is possible to express the controlled displacement magnitude u_M at the resonance frequency as Eq. (8.17).

$$h \approx j\frac{4}{\pi}\frac{1+\gamma}{1-\gamma}\left(U(\omega) + \frac{C_0}{\alpha}V_S\right) \qquad (8.14)$$

$$\left(-M\omega^2 + jC\omega + K_E\right)U(\omega)$$

$$= F(\omega) - \frac{\alpha^2}{C_0}\left(1 + j\frac{4}{\pi}\frac{1+\gamma}{1-\gamma}\right)U(\omega) - j\alpha\frac{4}{\pi}\frac{1+\gamma}{1-\gamma}V_S \qquad (8.15)$$

$$\left[-M\omega^2 + j\left(C\omega + \frac{\alpha^2}{C_0}\frac{4}{\pi}\frac{1+\gamma}{1-\gamma}\right) + K_D\right]U(\omega)$$

$$= F(\omega) - j\alpha\frac{4}{\pi}\frac{1+\gamma}{1-\gamma}V_S \qquad (8.16)$$

$$(u_M)_{\text{cont}} = \frac{1}{C\omega_0 + \frac{4}{\pi}\frac{\alpha^2}{C_0}\frac{1+\gamma}{1-\gamma}}\left(F_M - \alpha\frac{4}{\pi}\frac{1+\gamma}{1-\gamma}V_S\right) \qquad (8.17)$$

The attenuation is obtained from Eqs. (8.7) and (8.17) and yields Eq. (8.18). This attenuation can also be expressed considering the overall coupling coefficient of the electromechanical structure k and its mechanical quality factor Q_M, leading to Eq. (8.19).

$$A = \frac{(u_M)_{\text{cont}}}{(u_M)_{\text{uncont}}} = \frac{1}{1 + \frac{1}{C\omega_0}\frac{4}{\pi}\frac{\alpha^2}{C_0}\frac{1+\gamma}{1-\gamma}}\left(1 - \alpha\frac{4}{\pi}\frac{1+\gamma}{1-\gamma}\frac{V_S}{F_M}\right) \qquad (8.18)$$

$$A = \frac{(u_M)_{\text{cont}}}{(u_M)_{\text{uncont}}} = \frac{1}{1 + \frac{4}{\pi}\frac{1+\gamma}{1-\gamma}k^2 Q_M}\left(1 - \alpha\frac{4}{\pi}\frac{1+\gamma}{1-\gamma}\frac{V_S}{F_M}\right) \qquad (8.19)$$

Table 8.2 gives the parameters according to the considered technique and the corresponding attenuation at the resonance. β is given as the control gain of the

Table 8.2 Synchronized Switch Damping: parameters of derived methods and related attenuations

Method	Inversion coefficient	Voltage source	Attenuation
SSDS	0	0	$\dfrac{1}{1+\dfrac{4}{\pi}k^2 Q_M}$
SSDI	γ	0	$\dfrac{1}{1+\dfrac{4}{\pi}\dfrac{1+\gamma}{1-\gamma}k^2 Q_M}$
SSDV	γ	V_S	$\dfrac{1-\alpha\dfrac{4}{\pi}\dfrac{1+\gamma}{1-\gamma}\dfrac{V_S}{F_M}}{1+\dfrac{4}{\pi}\dfrac{1+\gamma}{1-\gamma}k^2 Q_M}$
Adaptive SSDV	$\dfrac{(1+\beta)(1+\gamma)+\gamma-1}{(1+\beta)(1+\gamma)-\gamma+1}$	0	$\dfrac{1}{1+(1+\beta)\dfrac{4}{\pi}\dfrac{1+\gamma}{1-\gamma}k^2 Q_M}$

adaptive SSDV defined as Eq. (8.20). Particularly, it can be noted that the SSDV technique admits a minimum force magnitude given by Eq. (8.21), and therefore can leads to instability problems, as exposed in Badel et al. (2006a) and Lallart et al. (2008).

$$V_S = \beta\frac{\alpha}{C_0}u_M \tag{8.20}$$

$$(F_M)_{min} = \alpha\frac{4}{\pi}\frac{1+\gamma}{1-\gamma}V_S \tag{8.21}$$

8.2.1.3 Physical Interpretation

The effect of the nonlinear treatment applied to the voltage by the switching process thus creates a force that is opposed to the speed and the driving force at the resonance frequency (Fig. 8.5). Compared with the uncontrolled system where the piezoelectric elements are left in open-circuit condition, this opposed force is applied by the piecewise function h. In other terms, that means that the switch effect can be seen as a dry friction. The losses induced by this kind of friction are thus independent of the frequency. However, the difference with pure dry friction is that the amplitude of the force created by the switch is proportional to the displacement magnitude.

The magnitude of the generated force depends not only on the electromechanical figure of merit $k^2 Q_M$, but also on the type of the used technique. Typically, the inductor used in SSDI aims at artificially increasing the opposite force when compared with the SSDS technique, and the voltage source in SSDV consists of enhancing the inversion. The adaptive SSDV technique aims at stabilizing the SSDV.

In an energy point of view, the energy balance is obtained by multiplying the motion equation (Eq. (8.2)) by the velocity $\dot{u}$, and integrating over the time, yielding

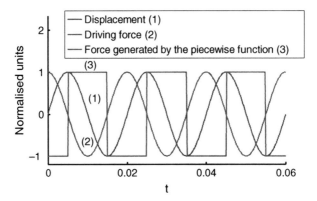

Fig. 8.5 SSD waveforms

Eq.(8.22). The last term corresponds to the transferred energy and can also be expressed as Eq. (8.23).

$$\int F\dot{u}\,dt = \frac{1}{2}M\dot{u}^2 + \frac{1}{2}K_E u^2 + \int C\dot{u}^2\,dt + \alpha\int V\dot{u}\,dt \qquad (8.22)$$

$$\alpha\int V\dot{u}\,dt = \frac{1}{2}C_0 V^2 + \int VI\,dt = \frac{1}{2}C_0 V^2 + \int V\,dQ \qquad (8.23)$$

For vibration-damping purposes, the principles of the damping consists of maximizing the transferred energy in order to reduce the kinetic and elastic energies (i.e., the two first terms of the second member of Eq. (8.22) respectively).

In steady state, the variation of kinetic energy, elastic energy, and electrostatic energy (first term of the second member of Eq. (8.23)) is a constant. Thus, the energy balance can be simplified as Eq. (8.24) and the transferred energy can be seen as the area of the cycles $\alpha V(\dot{u})$ or $V(Q)$ for one period. In the case of the SSD control, the transferred energy for one period is given as Eq. (8.25). Transferred energies as well as energy cycles for each technique are presented in Table 8.3 and Fig. 8.6, respectively. It can be noted that the SSDV voltage V_S and adaptive SSDV voltage coefficient β can be chosen such as the two techniques have to the same cycles. However, it is reminded that the SSDV suffers from instabilities problems.

$$\int F\dot{u}\,dt = \int C\dot{u}^2\,dt + \alpha\int V\,dQ \qquad (8.24)$$

$$(E_T)_{SSD} = \frac{\alpha^2}{C_0}\int_{1\ cycle} h\dot{u}\,dt = 4\frac{\alpha^2}{C_0}Hu_M \qquad (8.25)$$

Table 8.3 Synchronized Switch Damping: Transferred Energies

Method	Transferred energy
SSDS	$4\frac{\alpha^2}{C_0}u_M^2$
SSDI	$4\frac{\alpha^2}{C_0}\frac{1+\gamma}{1-\gamma}u_M^2$
SSDV	$4\frac{1+\gamma}{1-\gamma}\left(\frac{\alpha^2}{C_0}u_M^2 + \alpha V_S u_M\right)$
Adaptive SSDV	$4(\beta+1)\frac{\alpha^2}{C_0}\frac{1+\gamma}{1-\gamma}u_M^2$

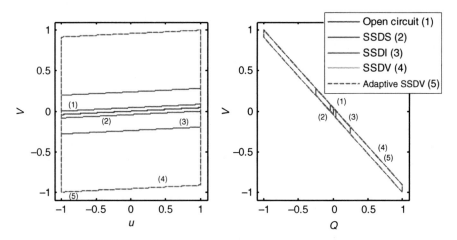

Fig. 8.6 SSD normalized energy cycles

8.2.1.4 Practical Consideration

The practical implementations of the switching devices are presented in Fig. 8.7. Typically, there are two ways of implementation.

The first one (Fig. 8.7 (a)) consists of two switches in parallel: one for positive switching and the other for negative switching. Diodes are inserted within the switching branches in order to ensure the stop of the switching process when the current sign changes. In other words, the diodes stop the switching process when the half of the pseudo-period of the LC_0 circuit is reached. Such a circuit presents the advantages of having a transistor command voltage independent of the switching electrical circuit. Indeed, the command is typically a square voltage that is equal to the sign of the speed. However, such a circuit necessitates two distinct switch branches (one for positive switching and the other for negative switching). Moreover, the diodes introduce an additional voltage gap that needs to be negligible when compared with the piezoelectric element voltage.

The second way of implementation (Fig. 8.7 (b)) uses only one switching branch, and operates both for positive and negative switchings. As well, the voltage gap

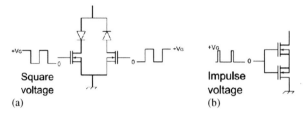

Fig. 8.7 Practical implementations of the switching device: (**a**) using diodes (**b**) diode-free

introduced by the transistors is usually less than the voltage gap introduced by the transistors and the diodes in the previous schematic. However, in this case, the control voltage needs to be pulse voltage with a high-state duration that has to be exactly equal to the inversion time period (i.e., half of the pseudo-period of the LC_0 circuit) to ensure a proper inversion with an optimal inversion coefficient. Thus, changing the switching inductor needs to modify the time duration of the command voltage high-state. It has to be noted that using the SSDV or adaptive SSDV requires either doubling the switching branch in the second case (i.e., one branch for positive switching with negative voltage and the other branch for negative switching with positive voltage) or using a tuneable voltage source (which is nevertheless necessary in the case of the adaptive SSDV), while there are still two branches with the first implementation technique.

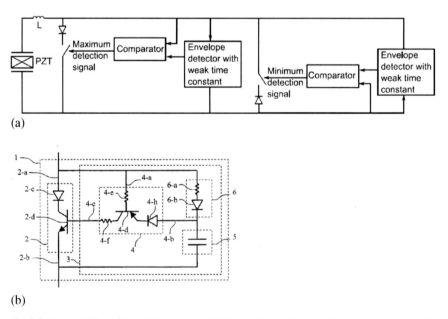

Fig. 8.8 (**a**) Principles of the self-powered switching device and (**b**) practical implementation for a single phase (Richard et al. 2007)

Figure 8.8 shows the implementation of the self-powered switching system (Richard et al. 2007). This system is more likely to be used with semi-passive methods (SSDS or SSDI), instead of semi-active methods (SSDV and adaptive SSDV), as these latter require external power supply. Typically, an extrema is detected when the absolute envelope detector voltage is greater than the absolute piezoelectric voltage.

As previously seen, the synchronized switch damping process creates a piecewise function. Particularly, in steady state, this function is a crenel function. Consequently, the voltage presents odd harmonics of the vibration frequency that could excite the structure on a higher mode. This could be overcome by the uses of a larger inductance[2] that would smooth the voltage during the switch process.

8.3 Energy Harvesting Using Nonlinear Techniques in Steady-State Case

Piezoelectric electrical generators (PEGs) are generally used to supply electronic circuits with low-average consumption (from a few microwatts to several hundred milliwatts) that usually require DC power supply, obtained by rectifying and filtering the AC voltage supplied by a piezoelectric insert. A DC–DC converter is sometimes placed between the filtering capacitor and the circuit to be supplied. According to case, this converter can be used as an impedance adapter to optimize energy transfer or as a voltage regulator.

The proposed technique is fully compatible with the structures of standard PEG. It consists of adding the SSDI switching system in parallel with the standard energy harvesting circuit, in order to artificially extend the electromechanical conversion cycles. This nonlinear energy harvesting technique is called SSHI[3] for synchronized switch harvesting on inductor. The results presented in this section were published in Guyomar et al. (2005) and Badel et al. (2006b).

The aim of this section is to describe the behavior of the SSHI technique in the case of sustained sinusoidal stress and compare this technique with standard energy harvesting technique. This section examines the case of an electromechanical structure excited in sinusoidal steady state. Under these conditions, the deformation of the piezoelectric elements is proportional to the displacement linked to the vibration. Thus, the extrema of deformation and displacement are mentioned indifferently.

Two cases will be considered, according to whether the energy harvesting process causes or does not lead to vibration damping. In each of the cases considered, the power and efficiency of the standard technique and the SSHI technique will be calculated.

---

[2] Even in this case, the inductance in SSD is still much smaller than the needed inductance in purely passive control

[3] SSHS in the case where no inductance is used. This case can be calculated from the SSHI results for $\gamma = 0$

8.3.1 Principles[4]

8.3.1.1 Standard Technique

The simplest technique for harvesting energy is to connect the electric circuit to be supplied directly to the piezoelectric elements through a bridge rectifier followed by a smoothing capacitor C_R. This setup is shown in Fig. 8.9(a). The voltage applied to the load R is continuous. When the piezoelectric voltage V is lower in absolute value than the rectified voltage V_{DC}, the bridge rectifier is blocked. The current I leaving the piezoelectric elements is therefore null so the voltage varies proportionally with the strain. When the absolute value of voltage V reaches V_{DC}, the bridge rectifier conducts, which stops V changing. The bridge rectifier stops conducting when the absolute value of the displacement u decreases. The waveforms of the considered signals are shown in Fig. 8.9(b).

8.3.1.2 Parallel Synchronized Switch Harvesting on Inductor (SSHI) technique

In comparison to the standard DC technique described earlier, the switching system is simply added in parallel with the piezoelectric elements for operation of the SSHI technique. This setup is shown in Fig. 8.10(a).

As long as the piezoelectric voltage V is lower in absolute value than the rectified voltage V_{DC}, the current I_P conducted through the bridge rectifier is null, and the voltage varies proportionally to the displacement. When the absolute value of V

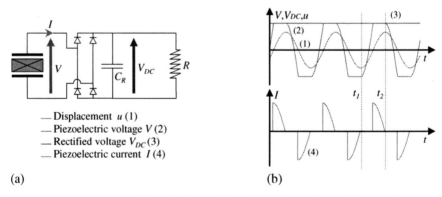

— Displacement u (1)
— Piezoelectric voltage V (2)
— Rectified voltage V_{DC} (3)
— Piezoelectric current I (4)

(a)   (b)

Fig. 8.9 Standard DC technique: (a) setup, (b) waveforms associated with sinusoidal steady state

---

[4] AC energy harvesting techniques (*i.e.* without AC-DC converter and smoothing capacitor) are not exposed here (as their applications are limited), but an analysis can be found in (Guyomar et al. 2005). As well, for clarity reasons, theoretical developments for the standard technique are not described, but are similar to those exposed for the SSH technique. These developments can also be found in (Guyomar et al. 2005).

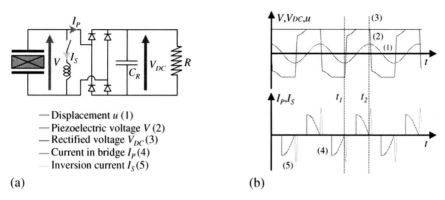

Fig. 8.10 DC SSHI technique: (a) setup, (b) waveforms associated with sinusoidal steady state

reaches V_{DC}, the bridge rectifier conducts, and the evolution of V is stopped. The bridge rectifier ceases to conduct when the displacement u decreases in absolute value, i.e., when a displacement extremum is reached, which coincides with the beginning of the voltage inversion through L_1. The cancelation of the current I_P absorbed by the bridge rectifier therefore corresponds to the release of the current I_S in the inversion inductor. The waveforms of the signals considered are shown in Fig. 8.10(b).

8.3.2 Analysis Without Induction of Vibration Damping

In this part, we study the power and efficiency of a PEG for which the energy harvesting process does not induce vibration damping. The latter can be very weakly coupled structures, structures excited out of the resonance, or structures for which displacement is imposed. The electromechanical structure can be modeled around a resonance frequency using the previously exposed lumped model and whose equations are given by Eq. (8.2). In the framework of a study in which a structure, whose displacement is not affected by energy harvesting is considered, only the constitutive electric equation in Eq. (8.2) is necessary.

8.3.2.1 SSHI Technique

This technique consists of simply adding the switching system in parallel with the piezoelectric elements in comparison to the standard rectified technique mentioned previously. The bridge rectifier is assumed to be perfect and the rectified voltage V_{DC} is constant. Here is considered a particular half-period corresponding to the interval $[t_1,\ t_2]$ represented in Fig. 8.10(b). t_1 and t_2 correspond to two consecutive inversions. t_1 is taken just before the first inversion and t_2 just before the second. The electric charge extracted from the piezoelectric elements between instants t_1 and t_2 is equal to the sum during this same half-period of the electric charge having flown through the resistor R and the electric charge having flown through the switching system, giving Eq. (8.26). The current flowing in the switching system is always null, except during the inversion phase just after instant t_1, where it is directly linked to the derivative of the piezoelectric voltage V. The electric charge having passed through the switching system is then given by Eq. (8.27). Using Eq. (8.26) and Eq. (8.27) and considering the constitutive electric equation (Eq. (8.2)), the rectified voltage V_{DC} can be expressed as a function of the magnitude of displacement, the load resistor, and parameters α and C_0 of the model. This expression is given by Eq. (8.28).

$$\int_{t_1}^{t_2} I \, dt = \frac{V_{DC}}{R} \frac{T}{2} + \int_{t_1}^{t_2} I_S \, dt \tag{8.26}$$

$$\int_{t_1}^{t_2} I_S \, dt = -C_0 \int_{t_1}^{t_1+t_i} dV = -C_0 V_{DC}(1+\gamma) \tag{8.27}$$

$$V_{DC} = \frac{2R\alpha}{RC_0(1-\gamma)\omega + \pi} \omega u_M \tag{8.28}$$

The harvested power is expressed by Eq. (8.29). It reaches a maximum P_{max} for an optimal resistance R_{opt}. The expressions of P_{max} and R_{opt} are given in Eq. (8.30). The maximum harvested power can then be expressed as a function of the maximum elastic potential energy in the structure (Eq. (8.31)) and the coupling coefficient, as shown in Eq. (8.32). The ratio of the maximum power with the SSHI technique over the maximum power with the standard technique is equal to $2/(1-\gamma)$. It is equal to 2 in the case of the SSHS technique ($\gamma = 0$) and can become very high if the electric inversion is good (e.g., 20 for $\gamma = 0.9$).

$$P = \frac{V_{DC}^2}{R} = \frac{4R\alpha^2}{(RC_0(1-\gamma)\omega + \pi)^2} \omega^2 u_M^2 \tag{8.29}$$

$$R_{opt} = \frac{\pi}{C_0(1-\gamma)\omega} \quad \text{and} \quad P_{max} = \frac{\alpha^2}{\pi C_0(1-\gamma)} \omega u_M^2 \tag{8.30}$$

$$E_e = \frac{1}{2} K_E u_M^2 \tag{8.31}$$

$$P_{\text{max}} = \frac{k^2}{1-k^2}\frac{2E_e}{\pi(1-\gamma)}\omega \tag{8.32}$$

The energy cycle corresponding to the transferred energy E_{Topt} in the case where the PEG supplies the resistance R_{opt} is shown in Fig. 8.11(b). In this case, the rectified voltage is given by Eq. (8.33). The maximum energy E_{Umax} consumed by the resistance R_{opt} during a period is given by Eq. (8.34) and the energy E_{Iopt} dissipated in the switching system by Eq. (8.35). The expression of the transferred energy E_{Topt} is given by Eq. (8.36). A graphic interpretation of energies E_{Umax} and E_{Iopt} is shown in Fig. 8.11(b).

$$V_{\text{DCopt}} = \frac{\alpha}{C_0(1-\gamma)}u_M \tag{8.33}$$

$$E_{\text{Umax}} = \frac{2\alpha^2}{C_0(1-\gamma)}u_M^2 = 2\alpha u_M V_{\text{DC}} \tag{8.34}$$

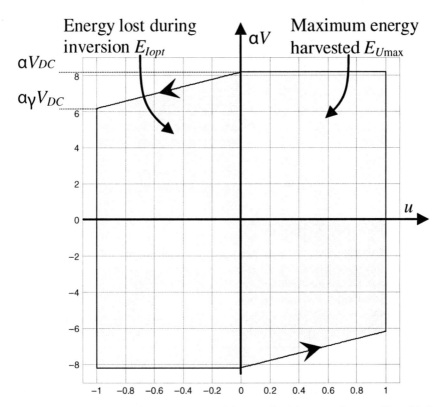

Fig. 8.11 Normalized optimal energy cycles. αV is normalized in comparison to αV_{DC} with the standard technique and u is normalized in comparison to u_M

$$E_{\text{Iopt}} = C_0 V_{\text{DC}}^2 (1 - \gamma^2) = \alpha u_{\text{M}} V_{\text{DC}} (1 + \gamma) \qquad (8.35)$$

$$E_{\text{Topt}} = E_{\text{Iopt}} + E_{\text{Umax}} = \frac{\alpha^2}{C_0} \frac{3 + \gamma}{1 - \gamma} u_{\text{M}}^2 = \frac{k^2}{1 - k^2} \frac{2 E_{\text{e}} (3 + \gamma)}{1 - \gamma} \qquad (8.36)$$

As it can be seen, the maximum energy harvested on R does not correspond to the maximum energy transferred. Indeed, the transferred energy is maximal when resistor R tends towards infinity, but in this case, the harvested energy is null. It can also be seen that the optimal ratio of the harvested energy over the extracted energy tends toward $^1/_2$ when γ tends toward 1 and equals $^2/_3$ with the SSHS technique (i.e., $\gamma = 0$).

If the vibration damping caused by the energy harvesting is negligible, then the energy supplied to the structure is given by Eq. (8.37). The efficiency of the system is therefore given by Eq. (8.38). This efficiency is maximal when the PEG supplies a resistance R_{opt}. In this case, the expression of the maximum efficiency is given by Eq. (8.39). This efficiency is necessarily very low in order to comply with the negligible damping effect assumption. However, it is $2/(1 - \gamma)$ times greater than the efficiency with the standard technique.

$$P_{\text{F}} = \frac{C \omega^2 u_{\text{M}}^2}{2} \qquad (8.37)$$

$$\eta = \frac{8 R \alpha^2}{C (R C_0 (1 - \gamma) \omega + \pi)^2} \qquad (8.38)$$

$$\eta_{\text{max}} = \frac{2 \alpha^2}{\pi C C_0 (1 - \gamma) \omega} = \frac{2 k^2 Q_{\text{m}}}{\pi (1 - \gamma)} \qquad (8.39)$$

8.3.2.2 Performance Comparison

Figure 8.12(a) represents the harvested power as a function of the load resistance R for the different techniques. This chart is normalized along the x- and y-axis in relation to the optimal resistance value and to the maximum harvested power, respectively, in the case of the standard technique. The advantage of this normalization is that this chart becomes totally independent of the model parameters. Only the electric inversion coefficient γ is required to plot the power in the SSHI case. For this chart, γ is set to 0.76, which corresponds to the experimental setup described further. This chart can also be interpreted as the representation of the PEG efficiency for the different techniques, normalized in comparison to the maximum efficiency with the standard technique.

For $\gamma = 0.76$, the power harvested with the SSHI technique is nearly eight times greater than the power harvested with the standard technique. Since the power gain

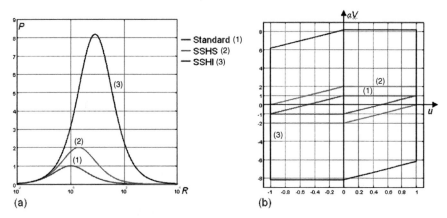

Fig. 8.12 Comparison of techniques at constant vibration magnitude: (**a**) harvested powers normalized as a function of normalized load resistor, (**b**) normalized optimal energy cycles

is given by $2/(1 - \gamma)$, the quality of the electric inversion is a crucial parameter. For the SSHS technique, this gain equals 2, and tends towards infinity when γ tends towards 1. In practice, it is very difficult to obtain inversion coefficients higher than 0.9, which corresponds to a power gain by a factor of 20.

The optimal energy cycles for the different techniques are shown in Fig. 8.12(b). The area of these cycles corresponds to the energy transferred in one period. The cycles are normalized by the magnitude of displacement (*x-axis*) and by αV_{DC} (*y-axis*) with the standard technique. Thus, the cycles no longer depend on the model parameters. Only the inversion parameter γ is required to plot the cycles corresponding to the SSHI techniques.

Figure 8.12(b) clearly shows the increase of the electromechanical conversion cycles generated by the nonlinear technique.

8.3.3 Damping Effect

In an electromechanical system, converting a part of the vibrational energy into electric energy is equivalent, from the mechanical point of view, to an increase in the internal losses. Therefore, when a sinusoidal driving force of constant magnitude is imposed, the energy harvesting process inevitably induces a vibration damping phenomenon. The latter is negligible in the case of systems for which the product $k^2 Q_m$ is very low or for systems excited far beyond their resonance frequencies. The value of $k^2 Q_m$ beyond which damping can be reasonably neglected will be precised in this section.

Naturally, the development presented in this part does not apply to systems with imposed vibration magnitude, as in the case of excitation using a shaker, for example.

In the literature, most of the analyses performed on PEGs use a fixed vibration magnitude that does not reflect a realistic excitation of electromechanical structures. Lesieutre et al. (2004) take damping effects into account in their analysis, but the global expression of the harvested power and the induced damping effect are not clearly defined. In this part, we study the behavior of a PEG excited at its resonance frequency by a sinusoidal force of constant magnitude. In the followings, a particular attention is paid to the power, efficiency, and damping induced by the PEGs for the two techniques described previously.

The squared coupling coefficient of most PEGs does not generally exceed a few percent. Consequently, we consider here that the angular resonance frequencies ω_E in short-circuit and ω_D in open-circuit are very close. The angular resonance frequency ω_r of the loaded PEG will be defined as the angular frequency for which the exciting force and the speed of vibration are in phase. This angular frequency must be between ω_E and ω_D and by assumption be assimilable with one or the other of these angular frequencies.

8.3.3.1 SSHI Technique

Here is considered the energy balance of the structure over a half-period corresponding to instants t_1 and t_2 defined in Fig. 8.10(b). The supplied mechanical energy is equal to the sum of the viscous losses in the structure, the energy harvested into the load and the energy lost during inversion, which is expressed in Eq. (8.40). At the resonance frequency, and by assuming that the displacement remains sinusoidal, the energy balance Eq. (8.40) can be simplified. This permits expressing the magnitude of the displacement as a function of the applied force magnitude, as shown in Eq. (8.41). The expression of the induced damping is then given by Eq. (8.42).

$$\int_{t_1}^{t_2} F\dot{u}\, dt = C \int_{t_1}^{t_2} \dot{u}^2\, dt + \frac{1}{2} C_0 V_{DC}^2 (1 - \gamma^2) + \frac{V_{DC}^2}{R} \frac{T}{2} \tag{8.40}$$

$$u_M = \frac{F_M}{C\omega_r + \frac{4R\omega_r\alpha^2}{\pi} \frac{(RC_0(1-\gamma^2)\omega_r+2\pi)}{(RC_0(1-\gamma)\omega_r+\pi)^2}} \tag{8.41}$$

$$A = 20 \log \left(\frac{C}{C + \frac{4R\alpha^2}{\pi} \frac{(RC_0(1-\gamma^2)\omega_r+2\pi)}{(RC_0(1-\gamma)\omega_r+\pi)^2}} \right) \tag{8.42}$$

The power supplied by the PEG is expressed by Eq. (8.43), as a function of the load R, the magnitude of the driving force, the electric inversion coefficient γ, and parameters α, C, and C_0 of the model. It is obtained from Eqs. (8.29) and (8.41). A numerical study of this function for realistic values of the parameters shows that it still supplies a single optimal resistor for which the power reaches a maximum.

$$P = \frac{4R\alpha^2}{(RC_0(1-\gamma)\omega_r + \pi)^2}$$

$$\times \frac{F_M^2}{\left(C + \frac{4R\alpha^2}{(RC_0(1-\gamma)\omega_r+\pi)^2} \frac{RC_0(1-\gamma^2)\omega_r+2\pi}{\pi}\right)^2} \qquad (8.43)$$

The power supplied by the PEG is given by Eq. (8.44) and the efficiency of the electromechanical conversion is given by Eq. (8.45). As for the power, a numerical study shows that the efficiency reaches a maximum for a single optimal resistor. However, this resistor is not the same as that giving the maximum power, except of course when $k^2 Q_m$ tends towards zero, which corresponds to the case of very weakly coupled structures.

$$P_F = \frac{F_M^2}{2\left(C + \frac{4R\alpha^2}{(RC_0(1-\gamma)\omega_r+\pi)^2} \frac{RC_0(1-\gamma^2)\omega_r+2\pi}{\pi}\right)} \qquad (8.44)$$

$$\eta = \frac{8R\alpha^2}{(RC_0(1-\gamma)\omega_r + \pi)^2}$$

$$\times \frac{1}{\left(C + \frac{4R\alpha^2}{(RC_0(1-\gamma)\omega_r+\pi)^2} \frac{RC_0(1-\gamma^2)\omega_r+2\pi}{\pi}\right)} \qquad (8.45)$$

The normalized power, efficiency, and damping generated by the SSHI technique are shown in Fig. 8.13. The loads are normalized in relation to the critical optimized resistor with the standard technique. The charts depicted are not functions of the model parameters, but only of the way in which $k^2 Q_m$ is made to evolve and of the electric inversion coefficient γ. For these charts, $\gamma = 0.76$, which corresponds to the experimental system that will be presented further.

The first curves represent the maximum power harvested, which tends towards power P_{limit} when product $k^2 Q_m$ increases. In this case, the efficiency tends towards 50% and the damping towards $-6\,dB$.

The second curves represent the maximum efficiency, which is an increasing function of product $k^2 Q_m$. When observing the maximum efficiency of the PEG, the harvested power is first an increasing function of $k^2 Q_m$, before reaching a maximum and then decreasing. The induced damping continues to increase and is higher than the damping corresponding to the maximum harvested power.

It should be noted that contrary to the standard technique, the maximum damping corresponds neither to the maximum efficiency nor to the maximum harvested

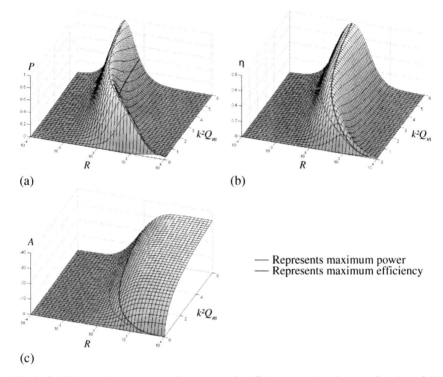

Fig. 8.13 SSHI technique: (**a**) normalized power, (**b**) efficiency, (**c**) damping, as a function of the normalized resistor and product $k^2 Q_m$

power. Damping is maximal when the load resistor tends towards infinity, which logically corresponds to the semi-passive SSDI vibration damping technique.

8.3.3.2 Performance Comparison

Figure 8.14(a) shows the progression of the normalized harvested power in comparison to P_{limit}, efficiency, and damping in the case where the load resistance is chosen to maximize the power supplied by the PEG. These magnitudes are plotted as a function of $k^2 Q_m$ for the standard, SSHS, and SSHI techniques. For the SSHI technique, γ was fixed at 0.76. The results would be very similar in the AC case. These charts confirm that the harvested power is limited by P_{limit}. This value is reached for $k^2 Q_m \geq \pi$ in the case of the standard technique, whereas for the SSH techniques, the power tends asymptotically towards this value. Likewise, the efficiency reaches or tends towards 50% and the damping reaches or tends towards $-6\,\text{dB}$.

The same amplitudes are shown in Fig. 8.14(b) when the load resistance is chosen in order to maximize the efficiency. In this case, the harvested electric power initially increases, and then it reaches a maximum before decreasing. In the case of the standard technique, the harvested power passes to a maximum when $k^2 Q_m = \pi$ and then tends towards P_{limit}. The SSH techniques reach a maximum

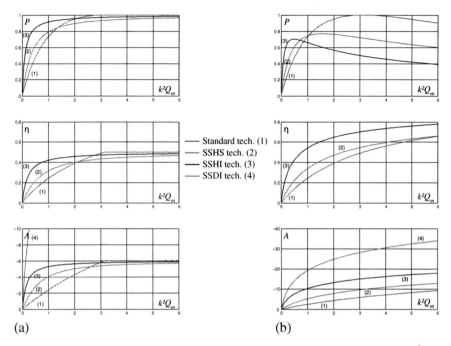

Fig. 8.14 Progression of the normalized power, efficiency, and damping as a function of $k^2 Q_m$: when maximizing the (**a**) power and (**b**) efficiency

power for $k^2 Q_m < \pi$, but this maximum power is lower than P_{limit}. Logically, the damping is greater when maximizing the efficiency than when maximizing the power of the PEG, though it is still less than the damping generated by the SSDI technique.

Figure 8.15 permits comparing the maximum power and efficiency of SSH techniques to the standard technique as a function of $k^2 Q_m$. The advantages of the SSH techniques can be seen clearly in the case of structures for which $k^2 Q_m$ is low. When $k^2 Q_m$ tends towards zero, the power and efficiency increase twofold in the case of the SSHS technique and by more than eightfold in the case of the SSHI technique. This gain corresponds to the power gain at a vibration of constant magnitude, when $k^2 Q_m$ tends to zero, the damping induced by energy harvesting also tends to zero. It should be noted that the gain obtained with the SSHI technique depends greatly on the electric inversion coefficient γ, which was fixed at 0.76. The evolutions of the maximum power and efficiency are shown in Fig. 8.16 for other values of γ.

If we consider that the quantity of used piezoelectric material is proportional to $k^2 Q_m$, then the obtained results mean that to recover a certain percentage of the maximum recoverable power P_{limit}, SSH techniques permit reducing dramatically the quantity of piezoelectric materials in comparison to the standard DC technique. For $\gamma = 0.75$, for example, the quantity of piezoelectric materials required

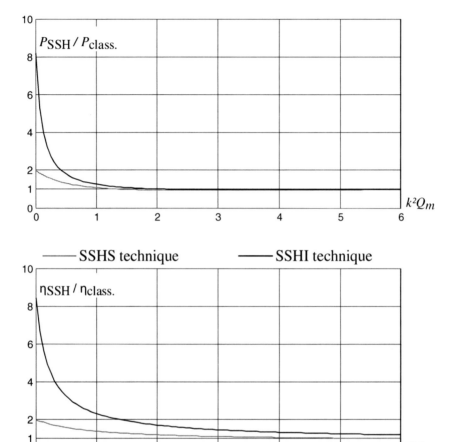

Fig. 8.15 Ratio of the maximum power and efficiency of SSH techniques over the maximum power and efficiency of the standard technique

to recover 75% of P_{limit} is four times less with the SSHI technique than with the standard technique.

Likewise, the SSH techniques permit significantly reducing the quantity of materials required to reach a certain efficiency. Thus, again for $\gamma = 0.75$, an efficiency of 60% was obtained with the SSHI technique for about four times less piezoelectric materials than with the standard technique.

8.3.4 Experimental Validation

The experimental setup is shown in Fig. 8.17. Sixty-eight small ceramic patches (type NAVY III) were arranged on both sides of the plate, distributed in four rows

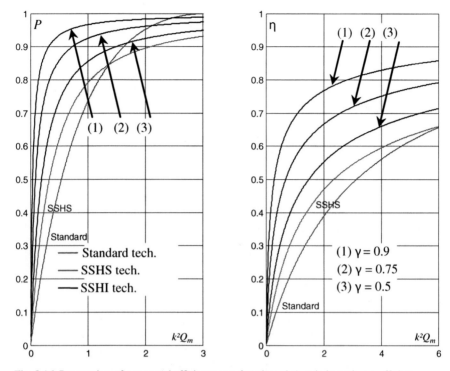

Fig. 8.16 Progression of power and efficiency as a function of electric inversion coefficient γ

of 17 (two rows above and two rows below). The geometric characteristics of these patches are given in Table 8.4. The patches were bonded close to the clamped end, and their direction of polarization was perpendicular to the beam. The coupling coefficient of the beam could be varied by linking more or fewer inserts in parallel with the energy harvesting circuit. The connection of the patches to the energy harvesting system was ensured by a box to enable the selection, via jumpers, of the patches to be used.

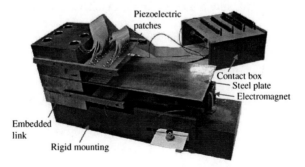

Fig. 8.17 Experimental setup

Table 8.4 Geometry of the piezoelectric patches

Number	68
Length	15 mm
Width	5 mm
Thickness	500 μm
Distance of embedding	14 mm

Table 8.5 Measurements and model parameters

f_E	60.18 Hz
f_D	60.46 Hz
Q_m	520
Λ	31,000 V/m
C_0	74.9 nF
A	0.0023 N/V
k^2	0.92%
K_E	7730 N m^{-1}
M	54 g
C	0.039 N m^{-1} s^{-1}

The structure is modeled with a simplified lumped model (second-order spring–mass–damper model), described in the first section. The measurements carried out to identify the parameters when all the patches were used are given in Table 8.5. For low levels of stress, Q_m remains relatively constant. In the case of the measurements presented here, the mechanical stress applied to the structure was deliberately limited so that the mechanical quality factor remained constant and high.

The beam is excited at about 65 Hz. Under these conditions (apart from resonance), the magnitude of displacement is quasi-constant whatever the load and the treatment applied to the voltage supplied by the inserts. A single range of patches is used. Under these conditions, the experimental inversion coefficient γ is equal to 0.76. Figure 8.18 represents the harvested power as a function of the load resistance R for the different techniques. This chart is normalized in the same way as for the theoretical study presented previously and shows that the experimental results are in good agreement with the predictions given by the model. According to the model, the ratio between the maximum powers for the standard and nonlinear techniques only depends on coefficient γ. Thus, the same ratio should be obtained between the maximum powers whatever the number of patches used. In practice, a deterioration of γ is observed with the decrease in the surface area of piezoelectric elements used. When the number of piezoelectric elements used decreases, the ratio between the piezoelectric capacitance C_0 (proportional to the number of patches connected) and the parasite capacitances of the electronic circuit ensuring the switching also decreases, affecting electric inversion quality. It should be underlined, however, that the development of better adapted electronics would probably limit this phenomenon.

When beam is excited at its resonance frequency, the magnitude of the displacement depends on the electric load connected to the inserts and the treatment applied

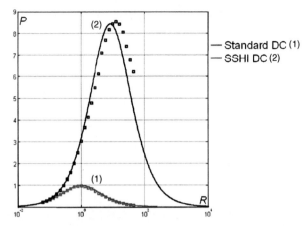

Fig. 8.18 Harvested powers normalized as a function of normalized load resistor. Comparison between the experimental measurements (*squares*) and theoretical results (*plain line*)

to the piezoelectric voltage. For each technique, the harvested power is measured as a function of the load applied, respectively, when all the inserts are connected ($k^2 = 0.92\%$), when a row of inserts is connected ($k^2 = 0.24\%$), when five inserts are connected ($k^2 = 0.072\%$) and when only two inserts are connected ($k^2 = 0.027\%$). Since the resonance frequency of the structure is influenced by the resistive load connected, the excitation frequency must be adjusted for each measurement.

The coupling coefficient is intended to vary proportionally to the surface area of the piezoelectric elements used. Under these conditions, parameters α and C_0 are proportional to the quantity of the patches connected, whereas dynamic mass M and the stiffness in open circuit K_D remain the same. The stiffness in short-circuit K_E varies slightly and can be determined from K_D, α, and C_0.

In reality, the coupling coefficient is not perfectly proportional to the number of patches connected, since the coupling of the piezoelectric elements making up the two rows closest to the clamped end is better. The variations of model parameters, α and C_0, are therefore performed proportionally to the coupling coefficient measured experimentally for each configuration and not proportionally to the number of patches connected. Stiffness K_D and mass M remain the same.

The inversion coefficient γ, which increases with the number of patches connected, is determined experimentally for each configuration. The comparisons of the theoretical and experimental results are presented for the standard DC technique in Fig. 8.19(a) and for the DC SSHI technique in Fig. 8.19(b). In these figures, the powers are normalized in relation to the maximum power P_{limit} and resistances in relation to the optimal resistance at which the power reaches P_{limit} for $k^2 Q_m = \pi$, which corresponds to the normalization performed in the previous theoretical studies.

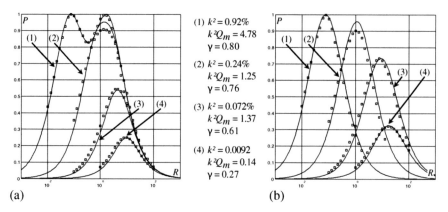

(1) $k^2 = 0.92\%$
    $k^2Q_m = 4.78$
    $\gamma = 0.80$

(2) $k^2 = 0.24\%$
    $k^2Q_m = 1.25$
    $\gamma = 0.76$

(3) $k^2 = 0.072\%$
    $k^2Q_m = 1.37$
    $\gamma = 0.61$

(4) $k^2 = 0.0092$
    $k^2Q_m = 0.14$
    $\gamma = 0.27$

Fig. 8.19 Normalized harvested powers as a function of normalized load resistor. Comparison between experimental measurements (*squares*) and theoretical results (*lines*) for different couplings: (**a**) standard technique and (**b**) SSHI technique

For these measurements, the maximum harvested power is in the region of 10 mW and corresponds to slight displacements of the structure. In reality, the maximum power that the experimental setup can supply is far greater. For example, for a displacement magnitude of 1.5 mm, the harvested power in SSDI is 80 mW when $\gamma = 0.76$.

8.4 Energy Harvesting in Pulsed Operation

In the previous section, it has been considered an electromechanical structure excited in steady state by a sinusoidal force. Under these conditions, energy is constantly supplied to the structure and the energy harvesting device continuously delivers electrical power. In this section, the case where energy is supplied punctually to the electromechanical structure is considered. This type of behavior corresponds, for example, to a structure excited by a force pulse, or by a static stress followed by a release. The aim of the energy harvesting device is, therefore, to convert the energy present in the structure after each mechanical stress and store it in a capacitor. The results presented in this section have been published in Badel et al. (2005).

Consequently, it is considered a structure with a certain initial internal energy that has to be transferred as electrostatic energy to a storage capacitor. During the energy harvesting process, the external mechanical excitation is assumed to be null and the structure is assumed to vibrate in pseudo-periodic state around one of its resonance frequencies.

It is considered that the lumped model is adequate for describing the electromechanical structure. This implies that the movement generated by the mechanical stress is pseudo-sinusoidal and thus that the response of the electromechanical structure can be assimilated with that of a second-order.

8.4.1 SSHI Technique

The theoretical results presented in the present and the following paragraphs are determined using a lumped model. The values of the used parameters are given in Table 8.5.

It is considered that a force pulse is applied to the structure at instant t_0. The energy supplied is, therefore, a step function, i.e., that for $t > t_0$, $E_F(t) = E_F(t_0)$. The function of the energy harvesting device is to transfer this energy to a capacitor C_R. The SSHI technique consists of connecting a bridge rectifier followed by the capacitor C_R to the terminals of the piezoelectric elements. Then the SSHI switching system is added in parallel with the piezoelectric elements, as shown in Fig. 8.20.

The constitutive equations of the model are recalled by Eqs. (8.46) and (8.47). At instant t_0, a force pulse is applied to the structure. At each instant $t > t_0$, the supplied energy E_F is equal to the sum of the energy E_R remaining in the electromechanical structure, the energy E_D dissipated in the form of viscous losses, the energy E_U harvested in the capacitor C_R, and the energy E_I dissipated in the switching system. The energy dissipated in the switching system corresponds to the losses during the voltage inversion process. These losses can be modeled by a resistor r_I in series with the inversion inductor L_I. The value of r_I can be determined from the quality factor Q_I of the electric inversion determined experimentally. The expression of r_I as a function of Q_I is given by Eq. (8.48), while Eq. (8.49) recalls the relationship between Q_I and inversion coefficient γ. The expressions of these energies are given in Table 8.6.

$$I = \alpha \dot{u} - C_0 \dot{V} \tag{8.46}$$

$$F = M\ddot{u} + K_E u + C\dot{u} + \alpha V \tag{8.47}$$

$$r_I = \frac{1}{Q_I} \sqrt{\frac{L_I}{C_0}} \tag{8.48}$$

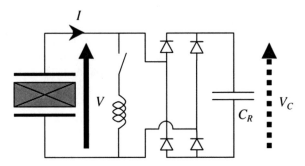

Fig. 8.20 Nonlinear energy harvesting device

Table 8.6 Definition of the energies

Energy remaining in the system	$E_R = \frac{1}{2}K_E u^2 + \frac{1}{2}M\dot{u}^2 + \frac{1}{2}C_0 V^2$
Energy dissipated by viscous losses	$E_D = \int_{t_0}^{t} C\dot{u}^2\, dt$
Energy dissipated in the switching system	$E_i = \frac{1}{2}C_0 V_i^2 (1 - \gamma^2)$
Harvested energy	$E_U = \frac{1}{2}C_R V_C^2$

$$\gamma = e^{-\pi/2Q_1} \tag{8.49}$$

It is assumed that the voltage of the capacitor C_R is null at the start of the energy harvesting process (i.e. $t = t_0$). When the absolute value of piezoelectric voltage V is lower than the rectified voltage V_C, the bridge rectifier is blocked and the current I flowing out the piezoelectric elements is null. The voltage V is then linked to the displacement by Eq. (8.50) and V_C remains constant until the absolute value of V reaches V_C. Then the bridge rectifier conducts. The voltage V is therefore linked to the displacement by Eq. (8.51) and the voltage V_C is equal to the absolute value of V. The bridge rectifier ceases to conduct when the displacement reaches an extremum. This instant also corresponds to the instant of closing of the electronic switch. During the electric inversion, the voltage in linked to the displacement by Eq. (8.52).

$$I = 0 \Rightarrow \alpha\dot{u} = C_0\dot{V} \tag{8.50}$$

$$I = C_R\dot{V} \Rightarrow \alpha\dot{u} = (C_0 + C_R)\dot{V} \tag{8.51}$$

$$V = L_1\dot{I} + r_1 I \Rightarrow \alpha L_1\ddot{u} + \alpha r_1\dot{u} = V + L_1 C_0\ddot{V} + r_1 C_0\dot{V} \tag{8.52}$$

The energy harvesting process ends when voltage V_C no longer increases, i.e., when the absolute value of V becomes lower than V_C. At this instant, the energy E_R remaining in the system is not null, but cannot be harvested.

Figure 8.21 shows the structure's electromechanical response obtained by the numerical integration[5] of Eqs. (8.47), (8.50), (8.51) and (8.52), when a force pulse is applied to the structure at instant t_0. For this simulation, the value of the capacitor C_R is fixed at 3.2 μF while that of electric inversion coefficient γ is 0.9. As shall be seen later on, the value of C_R is close to an optimum and γ corresponds to the experimental setup described earlier.

Figure 8.22 is the result of the same simulation. It shows the temporal evolution of the energies defined in Table 8.6 as well as the energy E_I lost in the inversion circuit. These energies are, once again, normalized in relation to supplied energy

---

[5] 4th order Runge Kutta numerical integration

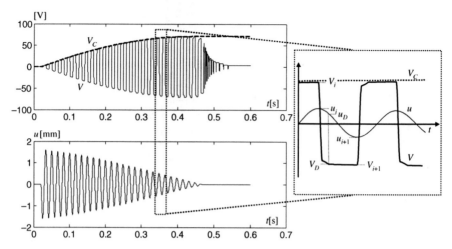

Fig. 8.21 Waveforms of the piezoelectric voltage V, voltage V_C on the capacitor, and displacement u in the case of the SSHI technique. Simulation by numerical integration: $\gamma = 0.9$, $C_R = 3.2\,\mu F$, $E_F = 14\,mJ$

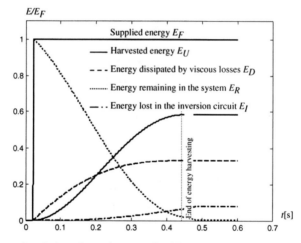

Fig. 8.22 Temporal evolution of energies normalized in relation to supplied energy E_F – Simulation by numerical integration: $\gamma = 0.9$, $C_R = 3.2\,\mu F$

E_F. For this simulation, the efficiency of the conversion reaches about 58%, which clearly shows the advantage of the nonlinear treatment of the voltage. The SSHI technique permits a better and faster extraction of mechanical energy, thereby limiting the viscous losses in the structure and reducing the energy remaining in the structure at the end of the energy harvesting process.

For a given electromechanical structure, it is interesting to know the harvested energy as a function of capacitance C_R. In the previous section, we made use of analytical formulae to directly determine the power harvested as a function of load resistance. Here, this type of analytical formula is not available to us, so several simulations have to be carried out successfully. In order to save time in comparison to simulations by numerical integration, another far faster tool has been developed based on iterative energetic analysis.

Two consecutive displacement extrema u_i and u_{i+1}, their instants of occurrence t_i and t_{i+1} and the two corresponding piezoelectric voltages V_i and V_{i+1} are considered. Also considered are displacement u_D and the corresponding voltage V_D when the bridge rectifier starts conducting. All these magnitudes are shown in Fig. 8.21. Equations (8.53) and (8.54) are, respectively, deduced from Eqs. (8.50) and (8.51), and equation Eq. (8.55) corresponds to the conduction condition of the bridge rectifier.

$$V_D + \gamma V_i = \frac{\alpha}{C_0}(u_D - u_i) \tag{8.53}$$

$$V_{i+1} - V_D = \frac{\alpha}{C_R + C_0}(u_{i+1} - u_D) \tag{8.54}$$

$$V_D = -V_i \tag{8.55}$$

On the basis of Eq. (8.53), Eqs. (8.54) and (8.55), u_{i+1} and V_{i+1} can be expressed as a function of u_i and V_i, respectively, as shown in Eq. (8.56).

$$V_{i+1} - \frac{\alpha}{C_r + C_0}u_{i+1} = -V_i\frac{C_r + \gamma C_0}{C_r + C_0} - \frac{\alpha}{C_r + C_0}u_i \tag{8.56}$$

The energy E_d dissipated by viscous losses between instants t_i and t_{i+1} can be approximated by considering that the displacement during this half pseudo-period is sinusoidal and its amplitude is equal to the average of the absolute values of u_i and u_{i+1}. These considerations are expressed by Eq. (8.57).

$$E_d = \int_{t_i}^{t_{i+1}} C\dot{u}^2 \, dt \approx \frac{1}{2}\pi C\omega \left(\frac{u_{i+1} - u_i}{2}\right)^2 \tag{8.57}$$

The energy E_i dissipated in the switching system between instants t_i and t_{i+1} corresponds to the energy lost during the voltage inversion phase. It is equal to the difference in electrostatic energy on the piezoelectric elements before and after voltage inversion, as shown by Eq. (8.58).

$$E_i = \frac{1}{2}C_0V_i^2(1 - \gamma^2) \tag{8.58}$$

Now it is considered the energy balance of the system composed by the electromechanical structure plus the energy harvesting device. The elastic energy plus the electrostatic energy in the system at instant t_i is equal to the elastic energy plus the electrostatic energy at instant t_{i+1} plus the dissipated energy E_d plus the energy E_i lost in the inversion circuit. This balance leads to Eq. (8.59), which links u_{i+1} and V_{i+1} to u_i and V_i, respectively.

$$\frac{1}{2}K_E u_i^2 + \frac{1}{2}(C_0 + C_r)V_i^2$$
$$= \frac{1}{2}K_E u_{i+1}^2 + \frac{1}{2}(C_0 + C_r)V_{i+1}^2 + \frac{1}{2}\pi C\omega\left(\frac{u_{i+1} + u_i}{2}\right)^2 + \frac{1}{2}C_0 V_i^2(1 - \gamma^2)$$

$$(8.59)$$

u_{i+1} and V_{i+1} can then be obtained from u_i and V_i using Eqs. (8.56) and (8.59). The system is initialized by forcing the first extremum u_1 in such a way as to fix the energy supplied to the electromechanical structure, as shown by Eq. (8.60). The simulation is terminated when $|V_{i+1}| < |V_i|$, we then obtain $V_C = |V_i|$.

$$\frac{1}{2}K_E u_1^2 = E_F \tag{8.60}$$

Figure 8.23(a) shows the different energies at the end of the energy harvesting process as a function of capacitance C_R. The energies are normalized in relation to E_F while capacitances C_R are normalized in relation to C_0. The calculations were performed with the iterative technique defined previously. The values of the model's parameters are those provided in Table 8.5, but the normalization of this figure only makes it a function of product $k^2 Q_m$ (which in this case is equal to 2.3)

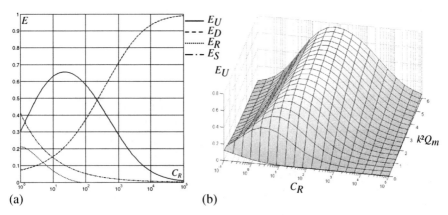

Fig. 8.23 SSHI technique with $\gamma = 0.9$ – Energies normalized in relation to E_F as a function of the capacitance C_R normalized in relation to C_0: (a) different energies for $k^2 Q_m = 2.3$, (b) harvested energy as a function of $k^2 Q_m$

for a given inversion coefficient γ (0.9 in this case). It can be clearly seen that the SSHI technique allows significantly reducing the energy E_R remaining in the system at the end of the energy harvesting process. This leads to a strong increase of the maximum harvested energy and a decrease in optimal capacity, amounting in this case to $20C_0 = 1.5\ \mu\text{F}$. It should be noted that this decrease in optimal capacitance permits considerably decreasing viscous losses in comparison to the standard technique. The SSHI technique thus permits more efficient and faster energy transfer.

Figure 8.23(b) shows the evolution of the harvested power as a function of the capacitance C_R and the product $k^2 Q_m$. The power and capacitance C_R are normalized as previously. The evolution of $k^2 Q_m$ is performed in the same way as in the previous section, i.e., by considering a variation of the surface of the piezoelectric elements connected to the energy harvesting device. The harvested energy normalized in relation to the energy supplied also corresponds to the efficiency of the electromechanical conversion. Logically, it is an increasing function of $k^2 Q_m$ and for a fixed product $k^2 Q_m$, a unique optimal capacity exists that maximizes the harvested energy. Normalized in this way and for a given inversion coefficient, Fig. 8.23(b) depends solely on the way in which $k^2 Q_m$ is made to vary. This figure shows that the amount of energy harvested using the SSHI technique is far greater than that harvested with the standard technique (Fig. 8.24), especially for the low values of $k^2 Q_m$.

Normalized in this way, Fig. 8.23(b) also shows the efficiency of the energy harvesting process. This efficiency is only linked to the electromechanical conversion itself and does not take into account the energy required to detect the extrema of the displacements and to control the switching system. It should be noted that in practice, these tasks are either performed using an external generator, or self-powered. The use of an external generator is very practical for experiments, but obviously does not represent real applications. The technical description of the self-powered

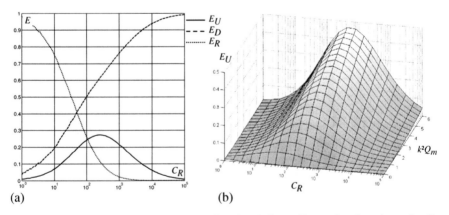

Fig. 8.24 Standard technique. Energies normalized in relation to E_F as a function of capacity C_R normalized in relation to C_0: (**a**) different energies for $k^2 Q_m = 2.3$ and (**b**) harvested energy as a function of $k^2 Q_m$

system is given in Richard et al. (2007). The energy required for the self-powering of the SSHI technique generally does not need more than 3–5% of the harvested energy.

8.4.2 Performance Comparison

The two methods can be compared by observing in each case the evolution of the maximal energy harvested as a function of product $k^2 Q_m$. Figure 8.25 shows the evolution of the maximum harvested energy for the standard, SSHS, and SSHI methods as well as the gain provided by the SSH methods in comparison to the standard technique. In the case of the SSHI technique, several curves corresponding to several values of electric inversion coefficient γ are plotted. We recall that the SSHS technique corresponds to the case where $\gamma = 0$.

Figure 8.25 is independent of the model parameters and is valid for any electromechanical structure that can be modeled by a second-order system. It shows that the performances of SSH methods are always better than those of the standard technique and that the gain provided by the SSH methods is as high as product $k^2 Q_m$ is low. These results are similar to those obtained in the previous section in the case

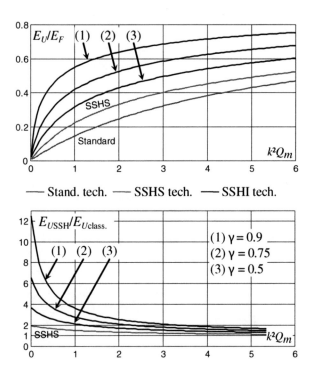

Fig. 8.25 Maximum energy harvested for the different methods and ratio of the energy harvested with the SSH methods over the energy harvested with the standard technique

of sustained sinusoidal stress. The crucial importance of inversion coefficient γ on the performances of the SSHI technique is clearly highlighted here.

8.4.3 Experimental Validation

The experimental setup is the same as that used in the previous section. Measurements performed to identify the model and the parameter values obtained are almost the same than those presented in Table 8.5. Although slight changes actually occurs between two parameter identifications (especially on the mechanical quality factor Q_m), the previous identification remains valid with an acceptable tolerance. The difference in the identification can be explained by changes regarding the limit conditions. As well, nonlinear effects such as the increase in aerodynamic losses with the amplitude of vibrations also alter the parameters, especially Q_m.

The beam is excited at its first resonance frequency so that its internal energy reaches 14 mJ, which corresponds to a displacement magnitude of the free end of the beam of 1.7 mm. The excitation source is then cut, and the energy harvesting device is connected to 68 piezoelectric elements ($t = 0$). Acquisitions of piezoelectric voltage V and voltage V_C on the capacitor and displacement u of the free end of the beam were carried out for the standard technique and for the SSHI technique, respectively, for $C_R = 18\,\mu F$ and $C_R = 3.2\,\mu F$. These results are shown in Figs. 8.26 and 8.27. The theoretical and experimental results are in good agreement. The only marked difference concerns the piezoelectric voltage after the end of the energy harvesting process in the case of the SSHI technique. Experimentally, when the displacement becomes very weak, the detection of the extrema is no longer

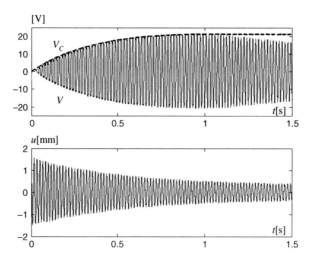

Fig. 8.26 Waveforms of the experimental piezoelectric voltage V and voltage V_C on the capacitor and displacement u in the case of the standard technique

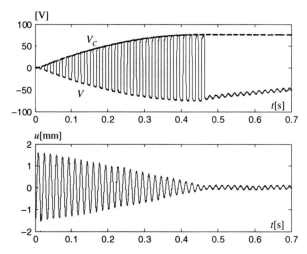

Fig. 8.27 Waveforms of the experimental piezoelectric voltage V and voltage V_C on the capacitor and displacement u in the case of the SSHI technique

operational and the voltage is no longer inverted. This can be seen in Fig. 8.27, but it does not affect the quantity of energy harvested.

For these measurements, the switching device is not self-powered and the value of the inversion inductor is 120 mH. The inversion coefficient γ is equal to 0.9, which corresponds to an electric quality factor Q_1 of 16. Regarding the measurements presented in this section, we used an air core inductor, which is larger than a classical magnetic core inductor, but whose losses are much lower.

The harvested energy using each of the two methods is measured as a function of capacitor C_R when all the patches are connected ($k^2 = 0.94\%$), when three quarters of the patches are connected (the two upper and the lower rows furthest from the clamped end – $k^2 = 0.62\%$), when half the patches are connected (the two rows furthest from the clamped end: $k^2 = 0.37\%$), and when a quarter of the patches are connected (the lower row furthest from the clamped end: $k^2 = 0.18\%$).

It can be observed that the coupling coefficient is not exactly proportional to the number of piezoelectric elements connected. Indeed, the electromechanical coupling of the patches located in the two rows closest to the clamped end is higher than that of the patches located in the two furthest rows.

For the same reasons as those mentioned in Section 8.4.2, the inversion coefficient is an increasing function of the quantity of piezoelectric elements connected. It is measured experimentally for each of the four configurations. The theoretical and experimental results are shown in Fig. 8.28 and are in very good agreement. The results were measured for a supplied energy of 14 mJ. The maximal energy harvested was 3.2 mJ with the standard technique and 8.7 mJ with the SSHI technique. In reality, the beam could be excited more strongly in order to harvest more energy. For example, for a displacement of the free end of the beam of 2.5 mm, the energy

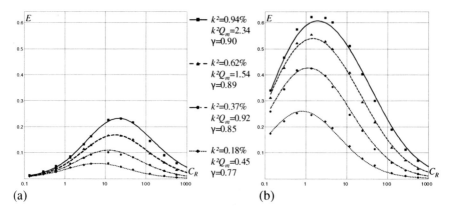

Fig. 8.28 Harvested energy normalized in relation to the energy supplied as a function of the capacitor C_R normalized in relation to C_0. Comparison between the experimental measurements (*dots*) and theoretical results (*solid lines*) for different couplings: (**a**) standard technique and (**b**) SSHI technique

harvested would be 18 mJ with the SSHI technique, by assuming that the system had a linear behavior.

The ratio of the maximum energy harvested using the SSHI technique over the maximal energy harvested with the standard technique is shown in Fig. 8.29 as a function of the product $k^2 Q_m$. This figure clearly highlights the gain provided by the SSHI technique, which varies from 250% for four rows to 450% for one row. Once again, the SSHI technique proved to be particularly interesting for structures with a low product $k^2 Q_m$, also meaning that for a given quantity of energy, the SSHI technique permits considerably reducing the required quantity of piezoelectric material.

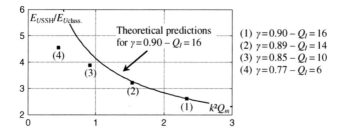

Fig. 8.29 Theoretical (*solid line*) and experimental (*squares*) comparison of the ratio of the energy harvested using the SSHI technique over the energy recovered using the standard technique as a function of $k^2 Q_m$

8.5 Other Nonlinear Energy Harvesting Techniques

This section presents two other energy harvesting methods that both draw advantage from the increased electromechanical conversion capacities brought about by the nonlinear treatment of the piezoelectric voltage. The first, called series SSH, is similar to the SSH technique presented previously, except for the configuration of the bridge rectifier which is connected in series to the switching system and not in parallel. The second technique presented in this section is called synchronous electric charge extraction (SECE). From the standpoint of the structure, this technique has the same effect as the SSDS damping method, but implements synchronous extraction of the electric charges generated on the electrodes of the piezoelectric elements. This section details the performances of these two new methods, by emphasizing their advantages and disadvantages in comparison to the parallel SSH technique. The methods will be studied in the case of a sinusoidal excitation. Two cases will be considered, according to whether the energy harvesting process leads to vibration damping or not.

8.5.1 Series SSHI Technique

8.5.1.1 Principles

The series SSHI technique was described under another name in Taylor et al. (2001). The system used by this technique is shown in Fig. 8.30(a). In this configuration, the voltage on the load resistor is continuous. The AC version of this technique, not studied here, consists of connecting the load resistance directly in series with an inversion inductor. In the series SSHS technique, the inversion inductor is replaced by a short circuit. The results obtained with the series SSHS technique can be deducted from the results obtained with the series SSHI technique by taking $\gamma = 0$.

The piezoelectric current I is always null, except during the voltage inversion phases. A perfect bridge rectifier is considered and the smoothing capacitor C_R is

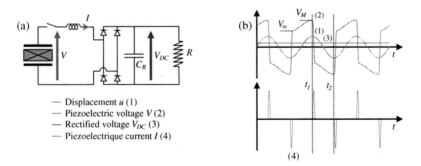

Displacement u (1)
Piezoelectric voltage V (2)
Rectified voltage V_{DC} (3)
Piezoelectrique current I (4)

Fig. 8.30 Series SSHI technique: (**a**) device and (**b**) associated waveforms in a steady sinusoidal state

assumed to be large enough to consider that the voltage V_{DC} is constant. During the voltage inversion phases, the voltage at the bridge rectifier input is equal to $+V_{DC}$ when the voltage switches to negative and $-V_{DC}$ when it switches to positive. From the structural point of view, the energy harvesting device behaves exactly like a semi-active damping device (SSDV) whose direct voltage is $-V_{DC}$ (Badel et al. 2006a; Lefeuvre et al. 2006a). The waveforms of the considered signals are shown in Fig. 8.30(b).

8.5.1.2 Theoretical Development Without Damping Effect

A structure animated by a sinusoidal movement of constant magnitude u_M is considered. It is recalled that it could be either a very weakly coupled structure, or a structure excited outside the resonance frequency, or a structure for which the displacement is imposed, in order to fulfil the assumption of negligible damping effect.

The inversion of the piezoelectric voltage around voltage V_{DC} is expressed by Eq. (8.61), whereas Eq. (8.62) expresses the evolution of the voltage between two switching instants. The corresponding half-period is considered as the interval $[t_1,\ t_2[$ defined in Fig. 8.30(b). The energy E_U harvested during this half-period corresponds indifferently to the energy absorbed by the bridge rectifier during the inversion of the voltage and to the energy consumed by load resistance R. This is summarized in Eq. (8.63).

$$V_m + V_{DC} = \gamma\,(V_M - V_{DC}) \tag{8.61}$$

$$V_M = V_m + \frac{2\alpha}{C_0} u_M \tag{8.62}$$

$$E_U = V_{DC} \int_{t_1}^{t_2} I\,dt = C_0 V_{DC}\,(V_M + V_m) = \frac{\pi}{\omega}\frac{V_{DC}^2}{R} \tag{8.63}$$

The expression of the voltage V_{DC} can be obtained from Eqs. (8.61), (8.62) and (8.63), as shown in Eq. (8.64). The harvested power is expressed by Eq. (8.65). It has a maximum P_{max} for an optimal resistance R_{opt}. The expressions of P_{max} and R_{opt} are given in Eq. (8.66). The maximum power can be expressed as a function of the maximum elastic potential energy E_e in the structure, defined by Eq. (8.31), and the coupling coefficient, as set out in Eq. (8.67).

$$V_{DC} = \frac{2\alpha R\,(1+\gamma)}{2RC_0\omega\,(1+\gamma) + \pi\,(1-\gamma)}\omega u_M \tag{8.64}$$

$$P = \frac{4R\alpha^2 (1+\gamma)^2}{(2RC_0\omega(1+\gamma) + \pi(1-\gamma))^2}\omega^2 u_M^2 \tag{8.65}$$

$$R_{opt} = \frac{\pi}{2C_0\omega}\frac{1-\gamma}{1+\gamma} \quad \text{and} \quad P_{max} = \frac{\alpha^2}{2\pi C_0}\frac{1+\gamma}{1-\gamma}\omega u_M^2 \tag{8.66}$$

$$P_{max} = \frac{k^2}{1-k^2}\frac{E_e}{\pi}\frac{1+\gamma}{1-\gamma}\omega \tag{8.67}$$

Figure 8.31(a) represents the harvested electric power as a function of the load resistance R for the series and parallel SSH methods as well as for the standard technique. This chart is normalized along x- and y-axis in relation to the optimal resistance and the maximum power harvested in the case of the standard technique, respectively. Normalized in this way, the chart is solely a function of the inversion coefficient γ, set to 0.76 in this case.

The ratio of the maximum harvested power with the series SSHI technique to the maximum harvested power with the SSHI parallel technique is equal to $(1+\gamma)/2$. It is equal to $^1/_2$ in the case of the SSHS technique ($\gamma = 0$) and tends towards one when the electric inversion improves. The series SSHS technique provides exactly the same performances as the standard technique. For a sufficiently large inversion coefficient, the performances of the series and parallel SSHI methods are very close and their respective optimal resistances are distributed symmetrically on a logarithmic scale of R, in comparison to the optimal resistance with the standard technique, as expressed in Eq. (8.68).

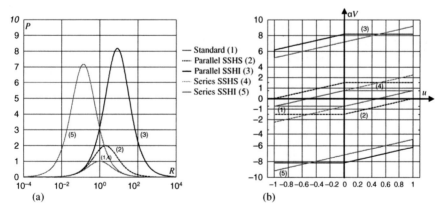

Fig. 8.31 Comparison of series and parallel SSH methods at constant vibration magnitude: **(a)** harvested power normalized as a function of normalized load resistance and **(b)** normalized optimal energetic cycles

$$\text{For } \gamma \to 1, \sqrt{\left(\frac{\pi}{2C_0\omega}\frac{1-\gamma}{1+\gamma}\right)\left(\frac{\pi}{C_0(1-\gamma)\omega}\right)} \approx \frac{\pi}{2C_0\omega} \qquad (8.68)$$

The series SSHI technique therefore permits harvesting the energy at much lower levels of voltage. In the case shown in Fig. 8.31(a), the optimal resistance is 58 times weaker with the series SSHI technique than with the parallel SSHI technique, meaning that for the same harvested power, the voltage is 7.6 times lower. For the experimental device presented previously, for example, the optimal voltage for a 1 mm displacement of the free end of the beam is 120 V in the parallel SSHI technique whereas it is only 15.5 V in the series SSHI technique. This reduction of the optimal resistance therefore provides an advantage for piezoelectric systems with high-output impedances and high-voltage levels. The expression of the optimal transferred energy, corresponding to the case where the PEG supplies the resistance R_{opt}, is given by Eq. (8.69). This energy is twice as high as the energy consumed by the resistance R_{opt}, meaning that half of the transferred energy is dissipated in the switching circuit. The optimal energetic cycles corresponding to series SSH methods are shown in Fig. 8.31(b), as well as the energetic cycles corresponding to parallel SSH methods and to the standard technique. It can be verified that in the case of the series SSHS technique, the energy transferred is twice as high as in the case of the standard technique, even though the maximum harvested power is the same.

$$E_{\text{Topt}} = \frac{2\alpha^2}{C_0}\frac{1+\gamma}{1-\gamma}u_M^2 = \frac{4k^2}{1-k^2}\frac{1+\gamma}{1-\gamma}E_e \qquad (8.69)$$

A structure excited at its resonance frequency by a sinusoidal force F is considered. It is assumed that this structure is sufficiently weakly coupled so that the energy harvesting process does not affect the vibration magnitude. The power supplied to the structure by the exciting force F is given by Eq. (8.70). It is by hypothesis not a function of the energy harvesting device. The efficiency of the PEG is equal to the power dissipated in the resistor over the power supplied, as expressed in Eq. (8.71). The efficiency reaches a maximum η_{max} for a resistance R_{opt} defined previously. It can be expressed very simply as a function of the structure's coupling coefficient, its mechanical quality factor, and electric inversion coefficient, as set out in Eq. (8.72).

$$P_F = \frac{C\omega^2 u_M^2}{2} \qquad (8.70)$$

$$\eta = \frac{8R\alpha^2(1+\gamma)^2}{C(2RC_0\omega(1+\gamma)+\pi(1-\gamma))^2} \qquad (8.71)$$

$$\eta_{max} = \frac{\alpha^2}{\pi C C_0 \omega} \frac{1+\gamma}{1-\gamma} = \frac{k^2 Q_m}{\pi} \frac{1+\gamma}{1-\gamma} \qquad (8.72)$$

8.5.2 Theoretical Development with Damping Effect

In the following, the behavior of a PEG excited at its resonance frequency by a sinusoidal force of constant magnitude is studied, taking into account the damping effect. Focus is given on the PEG's power, efficiency, and damping effect.

As with the study of the parallel SSH technique presented in the second section, we focus on the structure's energy balance for a half period corresponding to the interval $[t_1, t_2[$ as shown in Fig. 8.30(b). The energy supplied to the structure is equal to the sum of the viscous losses, the energy lost during inversion, and the energy consumed by the load resistor, which is expressed by Eq. (8.73). At the resonance frequency, and by assuming that the displacement remains sinusoidal, this energy balance can be simplified and used to obtain the expression (Eq. (8.74)) of the magnitude of displacement. The damping is then given by Eq. (8.75).

$$\int_{t_1}^{t_2} F\dot{u}\, dt = C \int_{t_1}^{t_2} \dot{u}^2\, dt + \frac{1}{2} C_0 (V_M - V_{DC})^2 \left(1 - \gamma^2\right) + \frac{V_{DC}^2}{R} \frac{T}{2} \qquad (8.73)$$

$$u_M = \frac{F_M}{C\omega_D + \frac{4\alpha^2}{C_0} \frac{1+\gamma}{2RC_0(1+\gamma)\omega_D + \pi(1-\gamma)}} \qquad (8.74)$$

$$A = 20 \log \left(\frac{C\omega_D}{C\omega_D + \frac{4\alpha^2}{C_0} \frac{1+\gamma}{2RC_0(1+\gamma)\omega_D + \pi(1-\gamma)}} \right) \qquad (8.75)$$

The power supplied by the PEG is expressed by Eq. (8.76), as a function of load R, the magnitude of the exciting force, the electric inversion coefficient γ, and the parameters α, C, and C_0 of the model. It is obtained by substituting Eq. (8.74) in Eq. (8.65).

$$P = \frac{4R\alpha^2 (1+\gamma)^2}{(2RC_0\omega_D (1+\gamma) + \pi(1-\gamma))^2}$$
$$\times \frac{F_M^2}{\left(C + \frac{4\alpha^2}{C_0\omega_D} \frac{1+\gamma}{2RC_0(1+\gamma)\omega_D + \pi(1-\gamma)}\right)^2} \qquad (8.76)$$

The mechanical power supplied to the PEG is given by Eq. (8.77) and the efficiency of electromechanical conversion by Eq. (8.78). The efficiency reaches a maximum for a single optimal resistance that is different from the resistance that maximizes the power delivered by the PEG.

$$P_F = \frac{F_M^2}{2\left(C + \frac{4\alpha^2}{C_0\omega_D}\frac{1+\gamma}{2RC_0(1+\gamma)\omega_D+\pi(1-\gamma)}\right)} \tag{8.77}$$

$$\eta = \frac{8R\alpha^2(1+\gamma)^2}{(2RC_0\omega_D(1+\gamma)+\pi(1-\gamma))^2}$$
$$\times \frac{1}{\left(C + \frac{4\alpha^2}{C_0\omega_D}\frac{1+\gamma}{2RC_0(1+\gamma)\omega_D+\pi(1-\gamma)}\right)} \tag{8.78}$$

The normalized power, efficiency, and damping generated by the series SSHI technique are shown in Fig. 8.32. The evolution of the product $k^2 Q_m$ is performed in the same way as in the previous sections, i.e., by considering a variation of the surface of the piezoelectric materials used. The load resistances are normalized in

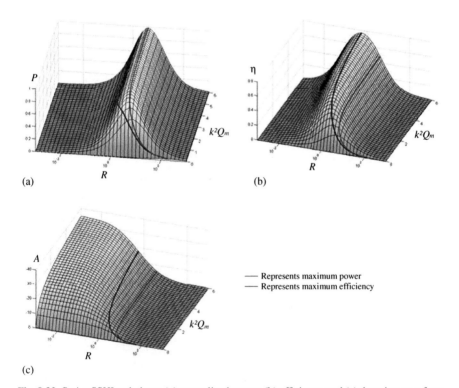

(a)

(b)

(c)

Represents maximum power
Represents maximum efficiency

Fig. 8.32 Series SSHI technique: (a) normalized power, (b) efficiency, and (c) damping, as a function of the normalized resistance and product $k^2 Q_m$

relation to the critical optimized resistance with the standard technique (corresponding to the case where P_{max} reaches P_{limit}) and the power in relation to power P_{limit}. The first curves are related to the maximum power harvested, while the second curves are related to the maximum efficiency.

Contrary to the parallel SSHI technique, maximum damping is obtained for a null load resistance and not for an infinite load resistance. When the bridge rectifier is short circuited, the series SSHI technique is exactly the same as the SSDI damping technique. This difference in behavior explains why, contrary to the parallel SSHI technique, the maximum efficiency is reached for a resistance lower than that of the one which maximizes the power. It can also be observed that the value of the optimal resistance, whether in terms of harvested power or efficiency, is less sensitive to variations of product $k^2 Q_m$ than in the case of the parallel SSHI technique.

Figure 8.33(a) shows the evolution of the harvested power, efficiency, and damping in the case where a load resistance is chosen to maximize the power supplied by the PEG, while Fig. 8.33(b) shows these same magnitudes in the case where a load resistance is chosen to maximize the efficiency. These magnitudes are plotted as a function of $k^2 Q_m$ for the standard, series SSH, and parallel SSH methods. The harvested power is normalized in relation to P_{limit} and γ has been fixed at 0.76 for the SSH methods.

The performances obtained with the series SSHS technique are far less interesting than those obtained with the parallel SSHS technique. The series SSHS tech-

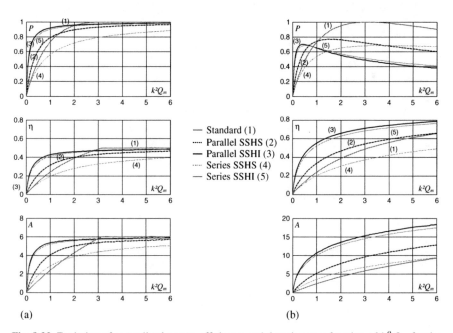

(a)   (b)

Fig. 8.33 Evolution of normalized power, efficiency, and damping as a function of $k^2 Q_m$ for the series SSH, parallel SSH, and standard methods: when maximizing the (**a**) power and (**b**) efficiency

nique is even less efficient than the standard technique (except for very low values of $k^2 Q_m$ where they are similar). On the other hand, for usual values of the inversion coefficient γ, the series SSHI technique is almost as efficient as the parallel SSHI technique. In reality, the performances of the parallel SSHI technique are always higher, but the curves of Fig. 8.33 representing these two methods tend to superpose when γ tends to 1.

8.5.3 Synchronous Electric Charge Extraction (SECE) Technique

8.5.3.1 Principles

A device able to extract the electrostatic energy on the piezoelectric elements within a very short period of time is considered. The device used to implement the SECE technique is shown in Fig. 8.34(a). The electrostatic energy of the piezoelectric elements is extracted at each extremum of displacement, bringing the voltage of the piezoelectric element back to zero. The waveforms associated with this technique are shown in Fig. 8.34(b). For a positive alternation of voltage between instants t_1 and t_2, V_M represents the maximum voltage on the piezoelectric elements.

This energy harvesting method leads to the voltage waveforms similar to those exhibited when using the SSDS technique.

8.5.3.2 Theoretical Development Without Damping Effect

It is considered a structure animated by a sinusoidal movement of constant magnitude u_M. The relation (Eq. (8.62)) expressing the evolution of the voltage between the two switching instants is still valid, but the cancelation of the voltage leads to a harvested energy at each half-period given by Eq. (8.79). The power expression Eq. (8.80) is obtained from Eq. (8.79). It corresponds to the expression of the power harvested as a function of the magnitude of displacement and the model parameters α and C_0. P can also be expressed as a function of the coupling coefficient and the maximum elastic energy in the piezoelectric elements, as shown in Eq. (8.81).

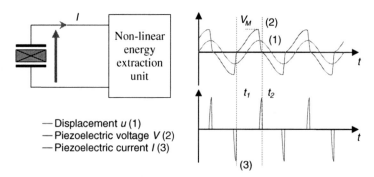

Fig. 8.34 SECE technique: (**a**) device and (**b**) waveforms associated in sinusoidal steady state

$$E_U = \frac{1}{2} C_0 V_M^2 \tag{8.79}$$

$$P = \frac{2\alpha^2}{\pi C_0} \omega u_M^2 \tag{8.80}$$

$$P = \frac{k^2}{1-k^2} \frac{4E_e}{\pi} \omega \tag{8.81}$$

Figure 8.35(a) shows the power harvested as a function of load resistance R for the SECE, parallel SSH, and standard harvesting methods. This chart is normalized as previously.

The power extracted with the SECE techniques is not a function of the input impedance of the device to be supplied, since by assumption, the piezoelectric voltage is not a function of the electric load but only of the quantity of energy extracted before each extraction process.

The ratio of the maximum power harvested with the SECE technique over the maximum power harvested with the parallel SSHS techniques is equal to 2. The SECE technique is therefore twice more efficient than the parallel SSHS technique.

The energy transferred during a period when the PEG supplies the resistance R_{opt} is given by Eq. (8.82) and the ratio of the harvested energy from the transferred energy is therefore equal.

$$E_{Topt} = \frac{4\alpha^2}{C_0} u_M^2 = \frac{8k^2}{1-k^2} E_e \tag{8.82}$$

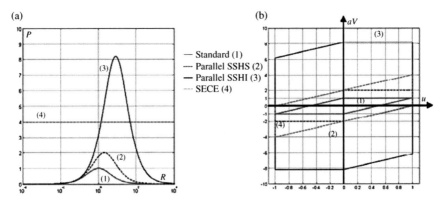

Fig. 8.35 Comparison of standard, SECE, and parallel SSH methods at constant vibration magnitude: (**a**) normalized harvested powers as a function of the normalized load resistance and (**b**) normalized optimal energetic cycles

The energetic cycles corresponding to the SECE method is compared in Fig. 8.35(b) to the energetic cycles corresponding to the parallel SSH methods and the standard technique.

A PEG excited at its resonance frequency is considered. It is assumed that the energy harvesting leads to negligible vibration damping, which amounts to saying that the electromechanical structure is characterized by a very weak product $k^2 Q_m$. The efficiency of this PEG when the SECE technique is implemented is given by Eq. (8.83).

$$\eta = \frac{4\alpha^2}{\pi C C_0 \omega} = \frac{4 k^2 Q_m}{\pi} \tag{8.83}$$

8.5.3.3 Theoretical Development with Damping Effect

Here, an electromechanical structure excited at one of its resonance frequencies is considered with particular attention paid to the power and efficiency of the device by taking into account the vibration damping induced by the extraction and conversion of the mechanical energy in the structure.

From the standpoint of the electromechanical structure, the SECE technique is equivalent to the SSDS semi-passive vibration damping technique. The magnitude of displacement is therefore given by Eq. (8.84) and the expression of the damping by Eq. (8.85).

$$u_M = \frac{F_M}{C\omega_D + \frac{4\alpha^2}{\pi C_0}} \tag{8.84}$$

$$A = 20 \log \left(\frac{C\omega_D}{C\omega_D + \frac{4\alpha^2}{\pi C_0}} \right) = 20 \log \left(\frac{1}{1 + \frac{4}{\pi} k^2 Q_m} \right) \tag{8.85}$$

The harvested power is given by Eq. (8.86) as a function of the parameters of the model and by Eq. (8.87) as a function of more physical characteristics. It is calculated by substituting Eq. (8.84) in Eq. (8.80). It allows a maximum for $k^2 Q_m = \pi/4$ and thus reaches P_{limit}, as stated in Eq. (8.88).

$$P = \frac{2\alpha^2}{\pi C_0 \omega_D} \frac{F_M^2}{\left(C + \frac{4\alpha^2}{\pi C_0 \omega_D} \right)^2} \tag{8.86}$$

$$P = \frac{2\pi k^2 Q_m}{\left(\pi + 4 k^2 Q_m \right)^2} \frac{F_M^2}{C} \tag{8.87}$$

$$\frac{dP}{d\left(k^2 Q_m\right)} = 0 \Rightarrow \begin{cases} k^2 Q_m = \frac{\pi}{4} \\ P = P_{limit} = \frac{F_M^2}{8C} \end{cases} \tag{8.88}$$

For $k^2 Q_m \geq \pi/4$, the maximum harvested power is given by Eqs. (8.89) and (8.90). The study of its derivative in relation to $k^2 Q_m$ shows that the maximum power is a decreasing function of $k^2 Q_m$, that it is maximal for $k^2 Q_m = \pi/4$ where it reaches P_{limit}. The continuity of P as a function of $k^2 Q_m$ is therefore verified.

$$P_{max} = \frac{2\alpha^2}{\pi C_0 \omega_D} \frac{F_M^2}{\left(C + \frac{4\alpha^2}{\pi C_0 \omega_D}\right)^2} \tag{8.89}$$

$$P_{max} = \frac{2\pi k^2 Q_m}{\left(4k^2 Q_m + \pi\right)^2} \frac{F_M^2}{C} \tag{8.90}$$

The power supplied to the PEG is given by Eq. (8.91) and the efficiency of the electromechanical conversion by Eq. (8.92) and Eq. (8.93), which is an increasing function of $k^2 Q_m$.

$$P_F = \frac{F_M^2}{2\left(C + \frac{4\alpha^2}{\pi C_0 \omega_D}\right)} \tag{8.91}$$

$$\eta = \frac{4\alpha^2}{\pi C_0 \omega_D} \frac{1}{\left(C + \frac{4\alpha^2}{\pi C_0 \omega_D}\right)} \tag{8.92}$$

$$\eta = \frac{4k^2 Q_m}{\left(\pi + 4k^2 Q_m\right)} \tag{8.93}$$

Figure 8.36 shows the comparison of the performances of standard, parallel SSH, and SECE methods as a function of product $k^2 Q_m$. For the SSHI method, γ is fixed to 0.76. Fig. 8.36(a) shows the evolution of the harvested power, efficiency, and damping in the case where the load resistance is chosen to maximize the power supplied by the PEG. It clearly shows that the curves corresponding to the SECE method reaches the power P_{limit} for $k^2 Q_m = \pi/4$, i.e., for product $k^2 Q_m$ four times lower than that required to reach P_{limit} with the standard technique. This means that from a theoretical standpoint, SECE method permits obtaining power P_{limit} for four times less piezoelectric material. For low $k^2 Q_m$ values (lower than 1), the SECE technique is the most efficient. In practice, the efficiency of the charge extracting device is not perfect and so the power harvested is lower, but the SECE technique is

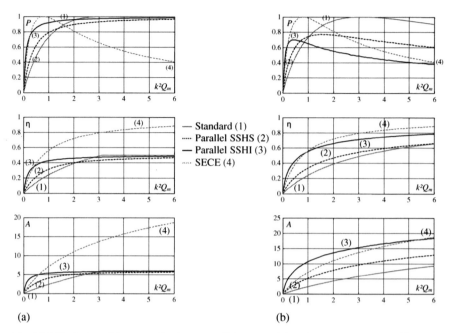

Fig. 8.36 Evolution of the normalized power, efficiency, and damping as a function of $k^2 Q_m$ for the SECE, parallel SSH, and standard methods: when maximizing (**a**) power and (**b**) efficiency

certainly worthy of interest, especially because the energy transfer is not *a priori* a function of the input impedance of the device supplied.

Figure 8.36(b) represents the harvested power, the efficiency, and damping in the case where the load resistance is chosen to maximize the efficiency. It can be seen that the efficiency of the SECE method is greater than that of parallel SSH methods.

8.5.4 Experimental Validation

The experimental setup is the same as that used for the experimental results previously obtained. The measurements performed to identify the model, as well as the values of the parameters obtained are roughly the same than those summarized in Table 8.5 (the drifts are explained by the same reasons than those in the previous section).

8.5.4.1 Series SSH Technique

Figure 8.37 shows the power harvested at constant vibration magnitude for the series and parallel SSH methods as a function of load resistance. The power and resistances are normalized as usual, in relation to the maximum power and optimal resistance to the standard technique, respectively.

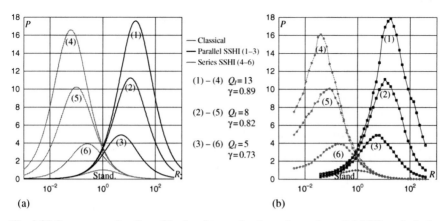

Fig. 8.37 Powers as a function of load resistance for the series and parallel SSH methods for different values of γ: (**a**) theoretical and (**b**) experimental results

Three series of measurements were performed, for $Q_1 = 13$, $Q_1 = 8$, and $Q_1 = 5$, respectively, in order to highlight the crucial influence of electric inversion quality on the power harvested. For $Q_1 = 8$ and $Q_1 = 5$, a resistor was added in series with the inversion inductor in order to reduce the electric inversion quality. The theoretical results agree very well with the experimental results, and confirm that the series SSHI technique is almost as efficient as the parallel SSHI technique, particularly when the electric inversion quality factor is high.

For these experimental measurements, the magnitude of displacement at the free end of the beam was fixed at about 1.7 mm. The power harvested with the parallel SSHI technique when $Q_1 = 13$ is therefore about 140 mW on an optimal resistance of 490 kΩ, which corresponds to a DC voltage of 260 V. With the series SSHI technique, the power harvested is slightly lower, 132 mW, but the optimal resistance is also lower, 56 kΩ, which permits limiting the voltage to 86 V.

8.5.4.2 SECE Technique

A possible setup for a load extraction circuit corresponding to the SECE technique is shown in Fig. 8.38. This circuit is composed of a bridge rectifier and a DC–DC switching mode converter. The electric load is represented by resistance R. In this section, the flyback-type converter will be used.

The converter is controlled by the voltage gate V_G of the MOSFET transistor T. This voltage is determined by a control circuit that detects maxima and passages at zero of rectified voltage V_R. When voltage V_R reaches a maximum, a voltage of 15 V is applied to the transistor gate. The transistor is then in on-state and permits the transfer to inductance L of the electrostatic energy of the piezoelectric elements. When all the electric charges present in the electrodes of the piezoelectric patches have been extracted, the control circuit detects the cancelation of rectified voltage V_R and applies a null voltage to the transistor gate. This operation blocks the transistor

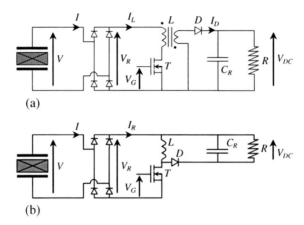

(a)

(b)

Fig. 8.38 Examples of wiring diagram of the SECE device: (**a**) flyback architecture and (**b**) buck-boost architecture

and the piezoelectric elements are once again in open circuit and the energy present in the coupled inductor L flows to the capacitor C_R. The waveforms of the displacement, voltages, and currents mentioned previously are shown in Fig. 8.39. According to these timing diagrams, the piezoelectric elements are in open circuit for 8.3 ms, whereas the load extraction lasts about 10 µs. The theoretical efficiency of the ideal flyback converter is one. The output power is therefore equal to the input power of the converter and is not a function of load resistance R. This also means that the output voltage V_{DC} of the DC–DC converter is only determined by load resistance R.

In practice, the efficiency of the converter is obviously not perfect and is a function of load resistance R Fig. 8.40(a) shows the power at the converter input and the power harvested as a function of the load resistance at a fixed vibration magnitude. The power and resistances are normalized as before. The power at the converter input also corresponds to the power extracted from the piezoelectric elements. In its normal operating range, the efficiency of the converter is about 70%. The power harvested is therefore 2.8 greater than with the standard technique, for a much wider range of resistance, almost two decades for this non-optimized set-up.

With the standard technique, the maximum power reaches 4.3 mW for an optimal resistance of 55 kΩ and a displacement magnitude of the free end of the beam of about 0.9 mm. For this same displacement magnitude, the power extracted from the piezoelectric elements with the SECE technique is four times higher whatever the load resistance, which corresponds to 17.2 mW. The flyback converter therefore delivers a useful power of 12.3 mW for a load resistance between 500 Ω and 55 kΩ, corresponding to an output voltage between 2.5 and 25 V. This operating range corresponds to a converter efficiency of 70%. The consumption of the flyback converter is 4.9 mW. Three quarters of this energy are dissipated in the power circuit shown in Fig. 8.38, and about a quarter is consumed by the self-supplied control circuit (not

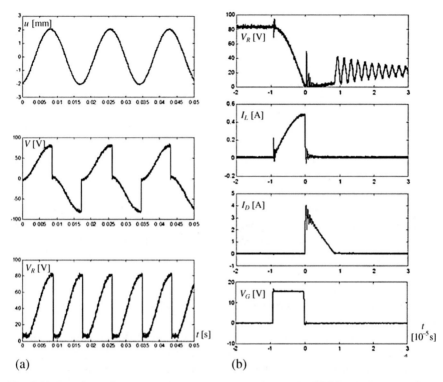

Fig. 8.39 Experimental voltage and currents corresponding to the SECE technique: (a) global waveforms and (b) waveforms during load extraction

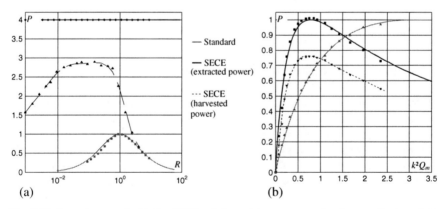

Fig. 8.40 Power harvested with the SECE technique: (a) at constant vibration magnitude and (b) at the resonance frequency, as a function of $k^2 Q_m$

shown). The technology of the converter could probably be optimized in order to obtain a better efficiency and an even wider operating range.

Figure 40(b) shows the extracted and the harvested powers with the SECE technique as a function of the product $k^2 Q_m$. These powers are normalized with respect to power P_{limit} and are compared with the maximum power harvested with the standard technique. The extracted power corresponds very well with the theoretical predictions. The theoretical harvested power is equal to the extracted power. In reality, as the efficiency of the converter is 70%, the harvested power decreases by as much. The power harvested with the SECE technique is, however, higher than that harvested with the standard technique, as long as $k^2 Q_m < 1$. When $k^2 Q_m$ is very low, the gain in power is 2.8.

The design of the SECE device required the use of a DC–DC converter to supply the electric load. In practice, the use of any other technique (standard, series SSH, and parallel SSH) also requires implementing such a converter to ensure realistic use of the PEG. A converter must be used to adapt the voltage level and/or impedance between the PEG and the load. Assuming that the efficiency of the converter is of the same order as that used for the SECE technique, it is reasonable to postulate that the SSDS extraction technique permits a reduction of 75% of the quantity of materials required to obtain power P_{limit} (or 70% of P_{limit} by taking into account the efficiency).

In addition, the higher the voltage delivered by the PEG, the poorer the efficiency of the converter intended to reduce the voltage to a level suitable for supplying a standard electronic circuit. This means, for example, that in the case of a realistic application making use of a DC–DC converter, the series SSHI technique would perhaps be more efficient than the parallel SSHI technique, since it supplies its maximum power with a lower voltage.

8.6 Energy Harvesting Techniques under Broadband Excitation

The case of sinusoidal or pulsed excitation is practically unachievable in real applications. Realistic systems would be more likely excited by a random force. In this case, ambient vibrations are neither of constant amplitude nor on a single frequency. It is, therefore, difficult to achieve a proper tuning of the load resistance.

Thus, in this case, the self-adaptive nature of the synchronous electric charge extraction technique (as shown Fig. 8.35) is particularly well adapted to the random distribution of the excitation. This section proposes to compare the performances of the standard energy harvesting technique to the SECE technique under various broadband, random vibrations. In this section, the implementation of the SECE is done using a buck-boost type DC–DC converter as shown in Fig. 8.38(b). The parallel and series SSHI techniques will not be exposed here, as these methods present the same drawbacks than the standard technique in the case of broadband excitation. The results exposed in this section have been published in Lefeuvre et al. (2007).

8.6.1 Multimodal Vibrations

The previous development considered separated harmonic excitation (i.e., monochromatic excitation). In the case of a combination of vibrations at different frequencies, the amount of power that can be harvested is different from the sum of the harvested power for each mode. Experimental results are depicted in Fig. 8.41.

Although the SECE technique still outperforms the standard technique and offers a rather constant value of the harvested power over a wide range of resistance value, the ratio of the harvested powers becomes smaller. This ratio can be increased by a proper selection of the switching extrema, as for vibration damping (Guyomar and Badel 2006; Guyomar et al. 2007a). It is interesting to note that the SECE still exhibits a constant value of the extracted power over the full range of resistance value.

8.6.2 Random Vibrations

Figure 8.42 shows the average extracted power with driving forces on different frequency bands from a 200 s recording of the instantaneous powers. The first frequency band includes the first three bending modes (56, 334, and 915 Hz), the second, the two first modes (56 and 334 Hz), and finally, the last frequency band only corresponds to the first bending mode (56 Hz). The performances of the SECE technique over the standard are still evident, but the gain is limited to 2, while it can reach a factor of 3 in harmonic operation.

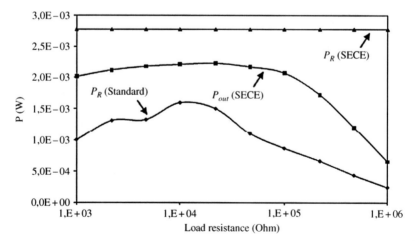

Fig. 8.41 Performance of the SECE technique over the standard DC technique with three modes mixed

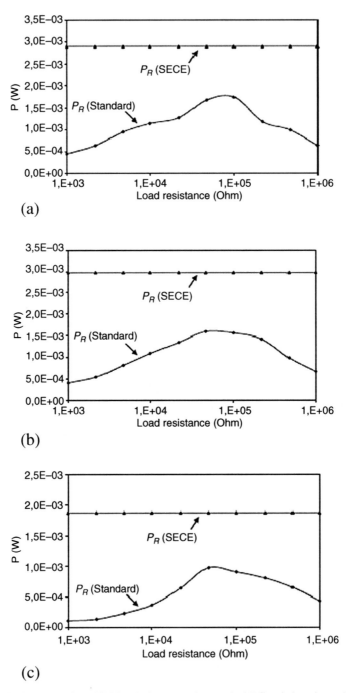

Fig. 8.42 Performance of the SECE technique over the standard DC technique in random vibration: (**a**) random force in [10 Hz, 2 kHz], (**b**) random force in [10 Hz, 500 Hz], and (**c**) random force in [10 Hz, 150 Hz]

8.7 Conclusion

Even if the energy delivered by a piezoelectric generator is usually in the range of a few microwatts to a few milliwatts, this power is now compatible of many advanced functions that can be implemented by low-power electronic devices (Guyomar et al. 2007b). In most of the applications, the electromechanical device including the piezoelement always has to cope with size and mechanical constraints resulting in a low-global electromechanical coupling factor. The question of optimizing the extracted power for a given device, while making a proper usage of the piezoelement, appears therefore of major importance. The SSHI (series or parallel) techniques discussed in this paper appear to be effective techniques allowing a strong increase of the total extracted power. SSHI applied to a non-resonant steady-state structure motion allows extracting 10 times more energy than a conventional technique for a given strain. Applied on resonance mode or in pulsed mode, the nonlinear techniques allow the extraction of a given energy with much smaller devices exhibiting much lower electromechanical coupling factors. The SECE technique which mixes nonlinear technique with a buck converter topology allows extracting four times more energy than a conventional technique along with a total independence of the energy extraction performance with the load impedance or voltage matching. The extension of these techniques to complex or broadband signal emphasizes the universal aspect of this approach which can be summarized as a better usage of a piezoelectric element as an electromechanical converter.

References

Badel A, Guyomar D, Lefeuvre E, Richard C (2005) Efficiency Enhancement of a Piezoelectric Energy Harvesting Device in Pulsed Operation by Synchronous Charge Inversion, J. Intell. Mater. Syst. Struct. 16(10):889–901. doi: 10.1177/1045389X05053150

Badel A, Sebald G, Guyomar D, Lallart M, Lefeuvre E, Richard C, Qiu J (2006a) Wide Band Semi-Active Piezoelectric Vibration Control by Synchronized Switching on Adaptive Continuous Voltage Sources, J. Acoust Soc. Am. 119(5):2815–2825. doi: 10.1121/1.2184149

Badel A, Guyomar D, Lefeuvre E, Richard C (2006b) Piezoelectric Energy Harvesting using a Synchronized Switch Technique, J. Intell. Mater. Syst. Struct. 17(8–9):831–839.

Badel A, Benayad A, Lefeuvre E, Lebrun L, Richard C, Guyomar D (2006c) Single Crystals and Nonlinear Process for Outstanding Vibration Powered Electrical Generators, IEEE Trans. Ultrason. Ferr. 53(4):673–684.

Badel A, Lagache M, Guyomar D, Lefeuvre E, Richard C (2007) Finite Element and Simple Lumped Modeling for Flexural Nonlinear Semi-passive Damping. J. Intell. Mater. Syst. Struct. 18(7):727–742. doi: 10.1177/1045389X06069447

Guyomar D, Badel A, Lefeuvre E, Richard C (2005) Toward Energy Harvesting using Active Materials and Conversion Improvement by Nonlinear Processing, IEEE Trans. Ultrason. Ferr. 52(4):584–595. doi: 10.1109/TUFFC.2005.1428041

Guyomar D, Badel A (2006) Nonlinear Semi-Passive Multimodal Vibration Damping: An Efficient Probabilistic Approach, J. Sound Vib. 294(1–2):249–268. doi:10.1016/j.jsv.2005.11.010

Guyomar D, Richard C, Mohammadi S (2007a) Semi-Passive Random Vibration Control Based on Statistics, J. Sound Vib. 307(3–5):818–833. doi:10.1016/j.jsv.2007.07.008

Guyomar D, Jayet Y, Petit L, Lefeuvre E, Monnier T, Richard C, Lallart M (2007b) Synchronized Switch Harvesting Applied to Self-Powered Smart Systems: Piezoactive Microgenerators for Autonomous Wireless Transmitters, Sensor Actuat. A-Phys. 138 (1):151–160. doi:10.1016/j.sna.2007.04.009

Lallart M, Badel A, Guyomar D (2008) Non-Linear Semi-Active Damping Using Constant or Adaptive Voltage Sources: A Stability Analysis, J. Intell. Mater. Syst. Struct 19(10):1137–1142.

Lefeuvre E, Badel A, Petit L, Richard C, Guyomar D (2006a) Semi Passive Piezoelectric Structural Damping by Synchronized Switching on Voltage Sources, J. Intell. Mater. Syst. Struct. 17 (8–9):653–660.

Lefeuvre E, Badel A, Richard C, Guyomar D (2007 – *available online*) Energy Harvesting using Piezoelectric Materials: Case of Random Vibrations, J. Electroceram. doi: 10.1007/s10832-007-9051-4

Lesieutre GA, Ottman GK, Hofmann HF (2004) Damping as a Result of Piezoelectric Energy Harvesting, J. Sound Vib. 269(3–5):991–1001. doi:10.1016/S0022-460X(03)00210-4

Richard C, Guyomar D, Audigier D, Ching G (1998) Semi Passive Damping Using Continuous Switching of a Piezoelectric Device, Proceedings of SPIE International Symposium on Smart Structures and Materials: Damping and Isolation 3672:104–111.

Richard C, Guyomar D, Audigier D, Bassaler H (2000) Enhanced Semi Passive Damping Using Continuous Switching of a Piezoelectric Device on an Inductor, Proceedings of SPIE International Symposium on Smart Structures and Materials: Damping and Isolation 3989:288–299.

Richard C, Guyomar D, Lefeuvre E (2007) Self-Powered Electronic Breaker with Automatic Switching by Detecting Maxima or Minima of Potential Difference Between Its Power Electrodes, patent # PCT/FR2005/003000, publication number: WO/2007/063194

Roundy S and Wright PK (2004) A Piezoelectric Vibration Based Generator for Wireless Electronics, Smart Mater. Struct. 13:1131–1142.

Taylor GW, Burns JR, Kammann SA, Powers WB, Welsh TR (2001) The Energy Harvesting Eel: a small subsurface ocean/river power generator, IEEE J. Oceanic Eng. 26(4):539–547. doi: 10.1109/48.972090

Chapter 9
Power Sources for Wireless Sensor Networks

Dan Steingart

Abstract Many environmental and industrial monitoring scenarios require wireless instrumentation with a small form factor and a long service life, a combination that forces designers to move beyond batteries and into energy harvesting techniques. This chapter considers the average requirements of wireless sensor networks, and assesses the suitability of modern thermal, photonic, and vibration-harvesting methods to power such networks across various application spaces.

9.1 Introduction

Wireless sensor nodes are tiny computers with sensors and radios (Fig. 9.1) capable of measuring aspects of their environment from simple quantities such as temperature, pressure, humidity, and insolation to complex recordings of images, audio, and video. These devices transmit this data to either a central aggregation point or other peer nodes on the network that can act on this data. These networks can form intricate topologies (Fig. 9.2), and there are multiple routing protocols and standards, which attempt to minimize power consumption while maintaining network and data integrity.

Applications for wireless sensor networks to date have focused on environmental monitoring and industrial automation on the milli-Hertz scale. Measurements are taken every few seconds to hours to provide a time-lapse view of a given process or phenomenon.

The focus in powering these nodes lies in long-term stability rather than short-term resolution, and as such these nodes provide an interesting challenge to the power and energy research community.

These nodes tend to:

1. Have low-duty cycles
2. Have low-power sleep states (10 to 300 μW)

D. Steingart (✉)
Department of chemical Engineering, City College of New York, 140th Street at Convent Avenue, New York, NY 10031
e-mail: steingart@che.ccny.cuny.edu

S. Priya, D.J. Inman (eds.), *Energy Harvesting Technologies*,
DOI 10.1007/978-0-387-76464-1_9 © Springer Science+Business Media, LLC 2009

Fig. 9.1 An example of a small wireless galvanostat produced by the author which uses wireless mesh network communication

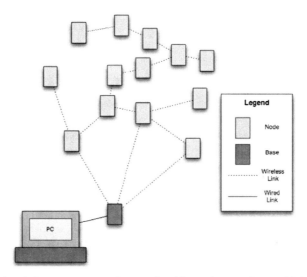

Fig. 9.2 Simple wireless sensor network example with one base station, multiple nodes, and a logging PC

3. Have relatively high power on states (where polling, sending, and receiving is done). These on states can be anywhere from 500 μW to 60 mW, and if a radio is involved, is typically at last 2 or 3 mW. The pulse duration is rather short, 500 ms or less.

Table 9.1 lists the four main states of a wireless sensor node, sleeping, polling, sending, and receiving. Note that a "high-power" sensor node may always be in a listening state, but truly low-power nodes use either a lower power reactive radio that wakes up a main receiver (and is included as an interrupt in the power budget),

Table 9.1 Table of Wireless Sensor States and Associated Power Consumption

Function	Duration (s)	Power (mW)	Description
Sleep	10^{-1} to 10^{4}	10^{-1} to 10^{-3}	A minimal power state allowing the device to "wake up" on event interrupts and power an internal clock
Polling	10^{-5} to 1	10^{-4} to 100	The power required to interact with the node's environment, from sensing to actuating, and whatever data processing are required on node
Transmitting	10^{-6} to 1	1 to 100	The power and time required to send a data packet to another node or a base station. This packet includes sensor ID, routing information, and sensor data value. The power required scales with the distance the data need to travel and the rate at which the information is sent.
Receiving	10^{-3} to 2	10^{-1} to 1	The power and time required to listen for a data packet note that listening takes more time than sending to ensure data are received

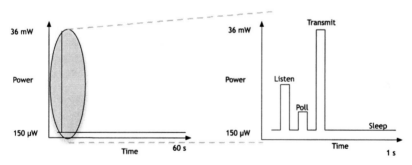

Fig. 9.3 Representative power trace for a wireless sensor node

or a time synchronized protocol to turn the receiver on when a package in expected (Pletcher 2004). Figure 9.3 shows a common power trace, the most simple state flow for a transceiver equipped wireless sensor node. Robust wireless sensor networks generally poll multiple times before triggering a send event, and when data are exchanged rather than a single send–receive pair nodes will often send an acknowledgment packet. These techniques increase reliability but tend to decrease the power budget (Polastre et al. 2004).

Wireless sensor nodes, at a circuit level, are simply traditional micrcontrollers with serial communication routed through a dedicated digital radio, as opposed to conventional communication approaches such as RS-232, ethernet, USB, and so on. Generally, the microcontroller provides the necessary digital and analog inputs and outputs, but in cases where enhanced accuracy or extensive channels are required "daughter" chips may be added.

Figure 9.4 shows, perhaps, the most simple block diagram for a wireless sensor network, discounting passives such as resistors, diodes, and capacitors, as well as any timing crystals. Figure 9.5 shows a more complicated node requiring digital communication on board with varied chips. As the number elements for digital communication increase on chip, so does the power consumption, *unless* the additional elements are treated as interrupts to use external control signals (temperature change or reactive radio signal (Pletcher 2004) through a comparator, for an example) to trigger an input. Generally, a board such as the one in Fig. 9.5 is used as an environmental characterization tool for a few hours or days, and is generally fairly large (any device with 24 sensor inputs is going to be bulky), and is not powered through environmental harvesting. For the purposes of this chapter, we will consider devices closer to Fig. 9.4.

There are extensive reviews on applications and details about the nature of the power consumption; the following thoughts are provided to give the reader a boundary on the expected values. Overall, these networks are used for simple data quantities that require low data rate sampling (typically milli-Hertz) and infrequent communications, at most a few times a minute per node. It must be considered that these are generally *multi-hop* networks, where each node acts as both a sensor and a relay. As the network density increases, the power consumption of each node increases as its likelihood of relaying a packet increases (Shnayder et al. 2004).

A low-power node is generally in a "sleep" state for most of its life. However, the nature of that sleep depends on the configuration of the network. If the network is

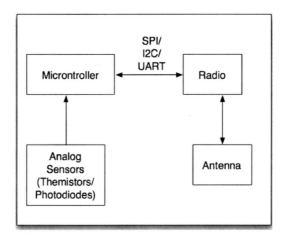

Fig. 9.4 Simple wireless sensor node block diagram

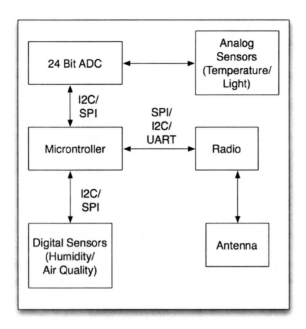

Fig. 9.5 Complex wireless sensor node block diagram

simply a star topology, and thee nodes are simply transmitters, each node can enter a low-power state that floats a simple clock and the basic volatile memory of a node. Most advanced nodes, however, function in a mesh topology, and therefore require receivers as well.

Once a receiver is required, it has to be turned on in order to be useful. The general power requirements of a receive state can be on the order of $500\,\mu W$ to $2\,mW$, a few orders of magnitude greater than the previously mentioned sleep state. The degree to which this receive state can be duty cycled is a balance between overall power consumption and reliability. Clock timing is temperature and humidity sensitive, so if the receiver-on state is too brief packets will be missed at an unacceptable rate.

Transmit cycles can be based on a clock cycle or an event trigger. The power required to transmit goes as $1/R^2$ as described later. Thus, decreasing the transmit distance only increases the need for a mesh network, reinforcing the receiver timing issue previously mentioned.

9.2 Primary Batteries

Take a low-power node with an average sleep state of $80\,\mu W$ and receive-poll-transmit cycle of five to ten times an hour for an overall average power draw of $100\,\mu W$. This device would require $876\,mW$ h to run for 1 year. In 2007, it is arguable whether such a device, reliable enough for commercial consumption, exists, but chipsets and algorithms exist that, at least in research environments, are capable of realizing this power budget in a sparse network configuration. Radio frequency identification (RFID) tags meet these power specifications, but due to the need for a large, power intensive reader node such a setup cannot be considered autonomous.

A small lithium cell with an open-circuit potential of 3 V and a capacity of 300 mA h would meet this goal, discounting self-discharge and the poor high-current pulse response common to such cells. Packaged, this battery weighs under 5 g, is roughly 5 mm^3, and is a highly reliable solution. Any on-chip energy-harvesting, storage and management system must surpass this benchmark to be a competitive solution. Not only is this the benchmark for density, but for shelf life as well. Lithium–sulfur and lithium–manganese dioxide primary cells have a useful shelf life of more than 10 years.

Beyond size and energy density, economics must considered. Alkaline batteries deliver on the order of 10 W h per dollar. Thus if size was not a constraint, this provides a good benchmark price. As of this writing, the battery described in the previous paragraph costs roughly $3 USD. Discounting cost, three requirements are used to filter an application's suitability for use of environmental energy harvesting in place of batteries.

1. Lifetime
2. Size
3. Environment

The first principle approaches to battery degradation are difficult to generalize, but qualitative boundaries can be used to frame the comparison. Lifetime and size are balanced about the energy density of a cell in the short term (1–18 months), beyond two years the self-discharge of a cell becomes a complicating factor. The environmental concerns are superimposed on top of this calculation. In general:

- Depending on the battery chemistry, monthly self-discharge[1] is on the order of 0.1 to 5% a month. A typical $LiMnO_2$ coin cell loses 0.1% of its charge while in storage at room temperature, so this should be taken into account when designing a long mission cell.
- Repeated high-current pulses diminish the capacity of a cell disproportionally. Based on the "C"[2] methodologies ratings, rates above C/5 for a Li–MnO$_2$ cell will hasten cell depletion. For large cells this is not a problem, but for thin film cells (Bates et al. 2000), this may be a concern as peak currents may reach 20 C.
- High-temperature exposure (temperatures consistently in excess of 40 °C) tend to increase the rate of self-discharge for batteries. The severity of this relationship is also chemistry dependent.

---

[1] Self-discharge is the degree to which batteries lose charge without any load placed across the cells.

[2] "C" rates are short-hand for describing the charge or discharge current of a cell based on the capacity of cell. For a 1000 mA h cell, C/5 indicates a rate of 200 mA, C/2 a rate of 500 mA, and C a rate of 1000 mA.

Thus, if an application can be flexible in terms of size and is in moderate climate, primary batteries are the most cost-effective choice for wireless sensor nodes. Beyond cost, batteries:

- require little or no environmental calibration
- do not influence the placement of sensor nodes (no need for thermal, vibrational, or photonic exposure beyond the sensor goals)
- do not place exotic packaging requirements on sensor nodes
- do not depend on environmental variables
- have predictable behavior
- require less energy conversion overhead

Particularly for academic research, where environmental studies tend to range from a few weeks to a year, batteries allow researchers to quickly implement systems in packaging that is protected from the environment and in locations where it is not obtrusive the environment to be monitored or its habitat (Werner-Allen et al. 2005).

9.3 Energy harvesting

Despite the ease of use and cost-effectiveness of batteries, certain applications can simply not depend on primary cells as the sole source of power. Such applications may include:

- *Industrial installations*: Industrial sensors often operate in harsh environments on a continual basis. Lifetime expectancies for almost all continuously operating industrial processes exceed 10 years, and most industries would not rely on technicians to service batteries for such sensors.
- *Biomedical devices*: Perhaps, the "holy grail" of micropower harvesting devices such as defibrillators and pacemakers that require routine invasive surgeries simply to replace the power source could be run "indefinitely" from a device that converted a small fraction of the body's 120 W (Starner 1996) of continuous power.
- *Environmental sensors for regulatory purposes*: The use of smart dust in forests to prevent forest fires and pollution detection.

9.3.1 Energy Harvesting versus Energy Scavenging

Energy scavenging refers to environments where the ambient sources are unknown or highly irregular, whereas energy harvesting refers to situations where the ambient energy sources are well characterized and regular.

Though this difference seems pedantic, the author feels the semantic nuances reveal a deep need of *all* existing environmental power generating solutions: an extensive understanding of the nature of environment to be "mined" for energy. As of 2007, it is a pipe dream to be able to drop a sugar cube-sized node into an unknown environment without a primary battery and expect any useful information to speak forth. Each of the sources of environmental generation covered in this chapter (photonic, vibrational, and thermal) require both energy harvesting devices tailored for specific applications, and unfortunately, a hard limit on the energy budget for an existing application. As a farmer must understand and prepare the land to harvest a predictable and useful crop, an engineer must understand the boundaries and gradients of an environment to harvest enough energy to produce a useful application.

With this subtle but important distinction considered, a brief survey of energy-harvesting technologies examined with respect to use in wireless sensor networks follows.

9.3.2 Photonic Methods

Traditional silicon solar cells have been widely demonstrated as a power source for wireless sensor networks. Roundy et al. (2003) demonstrated a simple, battery-free transmitting only matchbox-sized device that sent a simple radio packet (through a custom, low power 1.9 GHz radio) with a temperature reading when an attached solar panel charged a capacitor back to a threshold potential (Table 9.2). In the same vein of simplicity, Warneke et al. (2002) demonstrated a small node that used optical communications and was also solar powered.

Jiang et al. (2005) created a more complex system that integrated a small solar panel with a combination of super-capacitors and lithium–ion–polymer batteries which powered a fully functional Berkeley Telos B (Polastre et al. 2005) mode. A DC–DC converter with a comparator determined whether the lithium–ion–polymer should be used rather than the solar cells for power. The device could automatically change between power sources to account for diurnal cycles. Studies have been undertaken that demonstrate the effectiveness of systems inclusive of this complexity range (Dutta et al. 2006).

All of the aforementioned examples used silicon solar cells, typically amorphous silicon (a-Si), which are able to convert at most 11% of incident radiation into useful energy, and generally at an open-circuit potential of 700 mV. Since most of the shelf wireless sensor nodes require at least 2 V, three cells must either be placed in series, or a DC–DC converter must be used to boost the potential by diminishing current (discussed later).

The cost-per-area considerations for photovoltaics are secondary as most nodes have at most a few square centimeters of exposed surface. This makes multiple band-gap systems attractive. Multi-gap systems have multiple excitation states

Table 9.2 Power available for a variety of lighting conditions (Roundy et al. 2003)

Condition	Power incident (mW/cm^2)
Mid-day, no clouds	100
Outdoors, overcast	5
10 ft from an incandescent bulb	10
10 ft from a CF bulb	1

allowing them to make more use of the solar spectrum than traditional solar cells (Barnett et al. 2007). These solar cells have yet to reach market: however, if the results achieved in laboratory testing can be maintained though commercial scaling, the size of a solar cell for a given application can decrease by almost 60%.

Organic systems (Hoppe and Sariciftci 2004) and copper indium gallium selenide (CIGS) (Dherea and Dhere 2005) systems have demonstrated flexibility and durability innate to their design with minimal packaging, however, their conversion efficiency is on par with that of a-Si. While clever "wrap" designs have been demonstrated, it has to be considered that a solar cell face touching a surface provides little or no power. Thus, one can only assume, realistically, only a twofold performance gain in using all sides of a wireless sensor node for harvesting, not a sixfold gain. What multiple angled cells do allow, however, is longer high exposure to the sun given during a diurnal cycle and, perhaps, arbitrary placement of a device (air dropping nodes, for example) (Table 9.3) (Fig. 9.6). However, assuming that node packaged in a solar panel will see any light when blindly positioned neglects effects of dirt, animals, and seasonal vegetation change. Wrap-around cells for indoor nodes, to be powered from ambient electrical light sources, ease placement but do not increase energy production substantially, again, due to angle of incidence concerns.

Table 9.3 Various photovoltaic technologies and reported maximum conversion efficiencies (Green et al. 2005)

Technology	Best reported conversion efficiency (%)
a-Si	11
p-Si	18
SC-Si	25
Dye-sensitized	11
Organic	5
CdTe	15
CIGS	19
Multi-gap	35

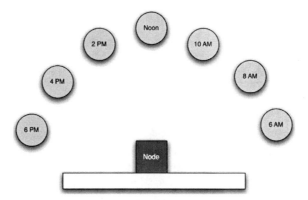

Fig. 9.6 Example of a cubic node covered with solar panels with respect to position of the sun

Unless data are only required at night, all outdoor solar methods need a secondary cell or super capacitor for energy storage. In indoor situations, incident light intensity decreases significantly (90%), requiring a backup battery/capacitor and an extremely low-duty cycle or a relatively large area to compensate (roughly 10 times more area for the harvesting unit than an outdoor solution).

Complicating factors aside, photovoltaic-harvesting methods remain the most widely available harvesting resource for both research and commercial applications, and the most promising method for long-term (10 years and beyond) installations given proper energy-handling components and placement.

Depending on the nature of the environment and the area available is a DC–DC converter may not be necessary. For example, if a single 1 in.^{2} CdTe cell provides 13 mA at 700 mV, and the application only needs to run during the day, five of the cells could be strung in series to provide the 3.3 V most low-power microcontrollers require. The most efficient DC–DC converters available have a 5% cost, but provide optimal regulation.

9.3.3 Vibrational Methods

While photovoltaic methods are a natural fit for providing energy for powering an outdoor wireless sensor network, ambient indoor light intensity is difficult to harvest in footprints of 2 cm^2 or less. Vibrational energy sources, particularly in industrial settings, appear quite promising. Common methods for harvesting vibrations include:

- *Piezoelectric materials*: Systems in which mechanical strains across a material layer generate a surface charge, and when an oscillating load is placed on the structure an AC power source results.
- *Inductive systems*: Systems in which a magnet moving through a wound coil induces a current through the coil, akin to an electric motor.
- *Capacitive systems*: Systems in which a charge on a capacitor is "pumped" by varying the distance between the plates of the capacitor. In this case, the harvester always requires a voltage source from which to pump.

Williams and Yates (1996) developed an equation that gives a good first-order approximation of the theoretical power available from a vibrating surface:

$$P = \frac{m\zeta_e A^2}{4\omega(\zeta_e + \zeta_m)^2} \tag{9.1}$$

A is the acceleration on a proof mass m, ω is the frequency at which the system is vibrating, ζ_e is the electrical damping coefficient, and ζ_m is the mechanical damping coefficient. From a cursory analysis, it can be surmised that power increases with increasing mass and acceleration, and decreases with increasing frequency and dampening coefficients. This model is valid regardless of the harvester geometry or scavenging mechanism.

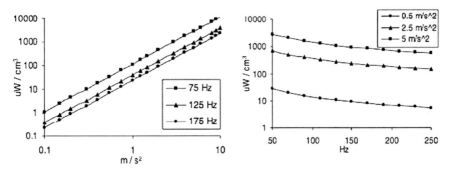

Fig. 9.7 Power per cubic centimeter for varying accelerations and frequencies given a constant mass and damping factor based on Roundy and Wright (2004)

Roundy et al. (2004) demonstrated available power from such vibrations given a constant mass and damping coefficients (Fig. 9.7). Given varying input vibrations, it is almost always better to choose a larger acceleration.

Lead zirconia titanate has been demonstrated by Roundy and Wright (2004) and Beeby among others. The use of polyvinylidene fluoride (PVDF), a piezoelectric polymer, has been proposed by Starner and Paradiso (2004). Inductive designs have been developed (Beeby et al. 2007a) and commercialized (Beeby et al. 2007b). Capacitive designs tailored for energy harvesting have been proposed and developed by various groups (Miyazaki et al. 2004, Yaralioglu et al. 2003).

Overall, the energy to be harvested mechanically is limited by kinetic initiator (the moving sources), and practically only a few percent of the initial source can be converted into electricity with these methods. Most designs are sensitive to driving frequencies, they can provide peak power only in a narrow band. Of course, multiple bands can be accounted for by multiplexing generators, but this decreases the overall power density of the device. As applied to wireless sensor networks, this constrains the design of nodes: each harvester must be matched to a particular application. Tunable designs have been demonstrated (Leland and Wright 2006), but these require a technician to manually set the frequency of the device at the time of installation. Table 9.4 lists a brief comparison of achievable power from various techniques.

Beyond power, *lifetime* of these benders must considered. As power increases with mass and acceleration, the fatigue stresses on these devices becomes significant

Table 9.4 Comparison of various vibrational-harvesting technologies for use with Wireless sensor networks

Technology	Power	Conditions	Size	Source
PZT	0.375 mW	9.1 g, 2.25 m/s², 85 Hz	1 cm³	Roundy (2005)
Electromagnetic	3 mW	50 g, 0.5 m/s², 50 Hz	41.3 cm³	Beeby et al. (2007)
Capacitive	3.7 µW	1.2 mg, 10 m/s², 800 Hz	0.75 cm³	Mitcheson et al. (2003)

as power density increases. Techniques such as magneto-restriction (Li et al. 2004) to optimize strain of piezoelectric devices while minimizing fatigue are currently being investigated, but it is always important to optimize instantaneous power in light of device lifetime.

Since these harvesting systems produce an inherently AC signal, it needs to be rectified and conditioned for use in any circuit (as opposed to the DC signals produced by photovoltaic and thermoelectric elements). This makes piezoelectric conversion perhaps the most difficult case for power circuitry. Figure 9.8 shows a basic example. It is important to note that photonic and thermal harvesting are generally steady-state phenomena, while vibrations may be transient or steady state. In the case of transient inputs, it is necessary to build a buffering circuit (generally, a bank of capacitors) to store accumulation of charge over a period of time that will then fire a burst of current into a secondary battery or directly into the device. Such circuitry is cumbersome, but adds flexibility (Roundy et al. 2003). This scheme is not to be confused with buffering schemes for *overnight* storage such as Jiang's work (Jiang et al. 2005). In the vibrational case, this buffering may be critical to have the device work in any situation, while in the photonic case the buffering is only required if the device is to run overnight. If multiple structures are used in a harvesting array, it is critical to independently rectify *each* independent unit to avoid compensation and wave cancelation.

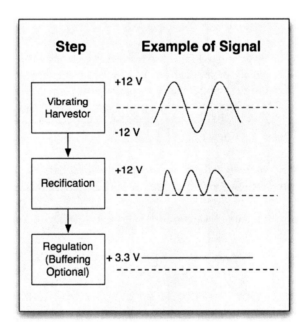

Fig. 9.8 Example of steps necessary to condition a vibrational-harvesting signal

9.3.4 Thermal Methods

Conventional Bi_2Te_3 thermoelectric devices have successfully been applied to wireless sensor nodes. Schnieder et al. (2006) demonstrated a system that functioned purely from a temperature gradient across an exhaust duct of a Hall–Héroult aluminum smelter, without any secondary storage whatsoever. This design was later commercialized. Other successful implementations of thermoelectrics for low power include a wrist watch developed by Seiko (Kishi et al. 1999) capable of operating from skin gradients, and other body-centric systems Paradiso (2006).

Like photovoltaics, thermoelectric device requirements for micropower are nearly identical to those of their larger counterparts in terms of materials. Also similar is the constraint of a minimal footprint. For small devices, it may be difficult to achieve the necessary temperature gradient.

Overall, a balance must be struck between the leg area of thermoelectrics, which determines current and the number of legs in series which determine potential. If this balance is met, DC–DC conversion may be unnecessary, which would allow more volume for energy generation and improve the overall power-train efficiency. The maximum energy that can be harvested is capped by the available gradient, as described by the Carnot efficiency, η:

$$\eta_{Carnot} = \frac{T_{High} - T_{Low}}{T_{High}} \tag{9.2}$$

where the maximum power attained from the heat flux, q' is determined by conductivity k, gradient ΔT and distance L.

$$P_{max} = q' = k\frac{\Delta T}{L} \tag{9.3}$$

Thus

$$P_{actual} = P_{max}\eta_{Carnot} \tag{9.4}$$

Roundy et al. (2004) calculated that based on a k of 140 W/mK and a length of 1 cm, a gradient of 5 K above room temperature would produce a heat flux of 7 W/cm². Given the Carnot efficiency, this would produce a maximum of 117 mW/cm². Unfortunately, most thermoelectric devices only realize a fraction of that power. Microscale heat engines have been proposed and designed (Epstein 2004, Frechette et al. 2003) but none have been realized at a scale useful for wireless sensor networks.

When using thermoelectrics, unless there are sufficient leg-pairs to create requisite potential at the temperature gradient in use, a DC–DC converter is necessary, as with a photovoltaic system. Unlike piezoelectric harvesters, these devices produce DC signals, and are generally not subject to the same cycling stresses. Most industrial applications operate at temperature well above room temperature night and

Table 9.5 Reported performance of various thermoelectric systems

System	Power (mW)	Conditions	Source
Bismuth telluride	60	20 °C above RT, 16 cm^2	Schneider et al. (2006)
Bismuth telluride	0.67	5 °C above RT, 1 mm^2	Bottner et al. (2004)
Bismuth telluride	45	5 °C above RT, 28 mm^2	Stordeur and Stark (1997)

day and therefore provide large, constant gradients from which to generate energy. These sensor locations also tend to be at high temperatures, often above 60 °C, are are therefore environments which shorten the working life of a primary battery.

On the other hand, maintaining a useful temperature gradient requires separation of the physical plates, which adds bulk to packaging. While industrial gradients are readily available, human gradients are just under the threshold for useful sensors currently, but they are perhaps useful for next-generation ultra-low-power nodes.

9.4 Alternative Methods

While these methods fall outside the scope of harvesting technologies, they are nonetheless interesting as a basis of comparison.

9.4.1 RF Power

A cousin to wireless sensor networks, RFID tags are nodes that are either powered entirely a received RF transmission (passive tags), or activated by said RF transmission and then powered by an internal battery or power source (semi-active tags). It is problematic to classify passive tags as a harvesting technology because the power source for these nodes exists solely to power these nodes and communicate. To approximate the range from a large power source, these nodes would work, a simple Gaussian model (Smith 1998):

$$P_{RF} = \frac{P_0 \lambda^2}{4\pi R^2} \tag{9.5}$$

where P_{RF} is the power incident on the node, P_0 is the radiative power, λ is the wavelength, and R is the distance between the reader and the node. Assume that the maximum distance between the power transmitter and any sensor node is 5 m, and that the power is being transmitted to the nodes is Zigbee protocol, so the frequency is 2.4 GHz. Regulations limit P_0 to 1 W, so the maximum power 5 m away would be on the order of 50 μW. This falls below our stated 100 μW requirement, and is generally illustrative of the difference in application space between RFID tags and the active wireless sensor nodes discussed here.

9.4.2 Radioactive Sources

Power sources for wireless sensor nodes that exploit radioactive decay have been explored (Lal and Li 2005, Duggirala et al. 2006) mostly with ^{63}Ni. Radioactive sources have an enormous potential energy density, generally greater than 40,000 W h/cm^3. The power density of such a cell is inversely proportional to the half-life of the given isotope. ^{32}Si has an energy density of roughly 100,000 W h/cm^3, and a half-life of 172 years, for an average power density of 60 mW/cm^3. If only 10% of this energy could be utilized, existing nodes would be able to operate for over a century. Safety and environmental issues aside, it is clear that radioactive sources are by far the most effective source of energy per unit mass or volume, providing an effective method to harness the decay. This utilization remains a challenge to be overcome in the laboratory, let alone a field installation.

9.5 Power Conversion

For all of the aforementioned techniques, a DC–DC converter is required. For most situations outside the thermal–industrial case, there are elements of randomness or cycling for the power, such as vibrational harvesting on humans as well as diurnal solar cycles. These cases require a super-capacitor at minimum, or more likely a secondary battery.

DC–DC conversion is generally 90–95% efficient assuming a low-drop out (LDO) boost circuit. These designs require large inductors, which require a relatively large area. This prevents such DC–DC converters from being used in wireless sensor nodes smaller than 1 cm^3. Switched capacitor designs, which are smaller (as they lack large inductors) and can run from lower initial potentials (due to the lack of a diode) run, at best, at 70% efficiency and generally can only obtain 50–60% efficiency (Maksimovic and Dhar 1999).

9.6 Energy Storage

The generally low-duty cycles of wireless sensor node cause energy density and cycle life requirements take precedent over power density. Thin-film batteries have shown promise for the past decade, and are on the brink of mass commercialization, but low capacity per area and high-processing temperatures are still engineering obstacles to be overcome (Bates et al. 2000). Thick-film approaches will improve capacity, but a compatible, effective solid-polymer electrolyte has yet to be deployed commercially (Arnold et al. 2007, Steingart et al. 2007). While liquid-phase electrolytes are an option for wireless sensor nodes, the packaging costs and environmental restraints required are generally prohibitive as nodes approach 1 cm^3 or smaller.

9.7 Examples

In this section, we will consider a few examples of how to determine an optimal power source for a wireless sensor network application.

9.7.1 Sensors in a Cave

Given the recent human loss in mining accidents in the US, and the continued reliance on mined ores for energy and materials production, the use of wireless sensor networks within mining operations is of great interest, particularly for location tracking and environmental monitoring.

Unfortunately, caves provide very little in the way of harvestable energy. There is very little light, vibrations are sporadic, and thermal gradients exist but over distance not amenable to node placement. In this environment, primary batteries or line power with secondary battery reserves are most feasible.

Fortunately, modern mines require significant power and internal infrastructure for operations, and line power is generally readily available. In an emergency, when line power is cut, the secondary batteries can kick in and depending on the size of the node support activity for a matter of hours to a few months.

9.7.2 Sensors in an Industrial Plant

Industrial plants provide a rich variety of sources from which to harvest energy. Most high-temperature operations, even when insulated, create large areas with gradients of at least 10 °C with the ambient environment. These operations include:

- Smelters
- Blast furnaces
- Roasting operations
- Crackers
- Distillation columns
- Evaporation chambers

In large-scale operations, most of the aforementioned units are run continuously, and therefore can be run completely without any type of secondary power source. The previously mentioned work (Schneider et al. 2006) obtained 60 mW from a 4 in^2 bismuth telluride thermoelectric unit at a gradient of 25 °C. For most low-power units, the size of the area is 1/20th would be sufficient.

While the area required for thermoelectric harvesting can decrease with size needs, there must always be adequate thickness to the entire device (thermoelectric and convective/conductive surface) to ensure a sufficient physical barrier between the heat source and the sink.

Beyond thermal gradients, regular vibrations are also widely available in these environments. Motors, conveyors, and other mechanical assembly are often, as a side benefit of tolerance required for high yield, design to operate with a particular resonant frequency. As long as the attached harvesting device is of a negligible mass when compared with the vibrating source (as a guide less than 1/100th of the mass of the source) the additional mass of the sensor should not alter the frequency of the vibrating object.

While it may be counter-intuitive to consider energy harvesting for sensing on units that are receiving (massive) amounts of power while operating, remember:

1. These units are quite large and wiring may be an impediment to proper operation
2. The areas where these units are implement are quite hostile, and wires running more than a meter have an increase chance of failure
3. In the case of high-current and/or high-voltage units, wires are a safety issue for workers

When we consider the sheer number of sensors that may be required in a given operation, it becomes clear that the added liability of wires for data transmission can quickly outweigh the added complication and cost of energy harvesting for each sensor.

9.7.3 Sensors in Nature

Perhaps, the most demonstrated use of wireless sensor networks to date is outdoor monitoring. This includes:

- Vineyards
- Animal habitats
- Volcanos
- Water resources

The majority of these applications use photovoltaic harvesting because the sun, of all harvestable quantities, is the most reliable and least costly. What must be determined is the necessity of data at night: if this is required, a system capable of buffering overnight must be used (Jiang et al. 2005).

Particularly difficult in outdoor applications away from the relative constancy of the roofs in which most solar panels lie are the accumulative effects of moisture and dust and the sporadic influences of flora and fauna. While the sun may be more reliable than a spinning motor or a smelter, industrial environments tend to be more predictable and regularly maintained. Placement must consider these factors, and software design of the sensors should include alarms which are tripped when the harvesting mechanism are compromised. An unforeseen benefit, however, of this addition is that the harvesting mechanism becomes a crude sensor!

9.8 Conclusion

Primary batteries provide a drop-in solution for wireless sensor networks. In this case, placement of nodes can be based purely on the sensing and actuation mission of the units. The author believes if the mission is within 3 years, and a size of $\sim$50 cm^3 is acceptable, most wireless applications, providing moderately long sense and send intervals, would be most easily met by such a solution. Given the robustness of current multi-hop algorithms, and the decreasing cost of nodes, it may be more cost effective when considering labor to simply replace an entire node than to change its batteries.

If a mission is of sufficient duration, duty cycle, and environment that primary cells do not meet an investigator's needs, energy harvesting methods should be considered. Given an outdoor application, photovoltaic methods, supplemented with a battery backup for nocturnal operation, are currently the most cost effective choice. In industrial operations, where thermal gradients exist persistently, thermoelectric modules may be used without secondary energy storage. Vibrational energy scavenging remains attractive in theory, but mechanical coupling issues and narrow-band frequency response remain obstacles to widespread implementation. In all cases, the energy harvesting solutions *do not yet provide significantly smaller form factors* than their primary battery counterparts.

Ultimately, there is no one correct solution for powering wireless sensor nodes. The system integrator must survey the application space and choose the correct solution based on fabrication and maintenance budgets, lifetime, duty cycle, and node density.

References

Arnold CB, Serra P, Piqué A (2007) Laser Direct-Write Techniques for Printing of Complex Materials. Mrs Bulletin 32:24–30

Barnett AM, Kirkpatrick D, Honsberg CB, Moore D (2007) Milestones Toward 50% Efficient Solar Cell Modules. 22nd European Photovoltaic Solar Energy Conference

Bates JB, Dudney NJ, Neudecker B, Ueda A, Evans CD (2000) Thin-film lithium and lithium-ion batteries. Solid State Ionics 135:33–45

Beeby SP, Torah RN, Tudor MJ, Glynne-Jones P, O'Donnell T, Saha CR, Roy S (2007) A micro electromagnetic generator for vibration energy harvesting. Journal of Micromechanics and Microengineering 17:1257–1265

Beeby SP, Tudor MJ, Torah RN, Roberts S (2007) Experimental comparison of macro and micro scale electromagnetic vibration powered generators. Microsystem Technologies 13, pp. 1647–1653

Bottner H, Nurnus J, Gavrikov A, Kuhner G, (2004) New thermoelectric components using microsystem technologies. Microelectromechanical Systems 13:414–420

Dherea NG, Dhere RG (2005) Thin-film photovoltaics. Journal of Vacuum Science & Technology A 23:1208–1214

Duggirala R, Lal A, Polcawich RG, Dubey M (2006) CMOS compatible multiple power-output MEMS Radioisotope μ-Power generator. IEEE Electron Devices Meeting: 1–4

Dutta P, Hui J, Jeong J, Kim S, Sharp C, Taneja J (2006) Trio: enabling sustainable and scalable outdoor wireless sensor network deployments. International Conference on Information Processing in Sensor Networks 407–415

Epstein AH (2004) Millimeter-Scale, Micro-Electro-Mechanical Systems Gas Turbine Engines. Journal of Engineering for Gas Turbines and Power 126:205–226

Frechette L, Lee CS, Arslan S, Liu YM (2003) Design of a Microfabricated Rankine Cycle Steam Turbine for Power Generation. Proceedings of ASME IMECE

Green MA, Emery K, King DL, Igari S, Warta W (2005) Solar cell efficiency tables (Version 25). Progress in Photovoltaics: Research and Applications 13:49–54

Hoppe H, Sariciftci NS (2004) Organic solar cells: an overview. Journal of Materials Research:1924–1946

Jiang X, Polastre J, Culler D (2005) Perpetual environmentally powered sensor networks. Proceedings of the 4th International Symposium on Information Processing in Sensor Networks

Kishi M, Nemoto H, Okano H (1999) Thermoelectric device. US Patent 5,982,013

Lal A, Li RH (2005) Pervasive power: a radioisotope-powered piezoelectric generator. Pervasive Computing

Leland ES, Wright P (2006) Resonance tuning of piezoelectric vibration energy scavenging generators using compressive axial preload. Smart Materials & Structures 15:1413–1420

Li YQ, O'Handley RC, Dionne GF, Zhang C (2004) Passive solid-state magnetic field sensors and applications therefor. US Patent 6,809,515

Maksimovic D, Dhar S (1999) Switched-capacitor DC-DC converters for low-power on-chip applications. IEEE Power Electronics Specialists Conference 1:54–59

Mitcheson PD, Stark BH, Miao P, Yeatman EM, Holmes A, Green T (2003) Analysis and Optimisation of MEMS Electrostatic On-Chip Power Supply for Self-Powering of Slow-Moving Sensors Proc. Eurosensors 17:48–51

Miyazaki M, Tanaka H, Ono G, Nagano T, Ohkubo N, Kawahara T (2004) Electric-energy generation through variable-capacitive resonator for power-free LSI. IEICE Transaction on Electronics:549–555

Paradiso JA (2006) Systems for human-powered mobile computing. ACM IEEE Design Automation 645–650

Pletcher NM (2004) Micro Power Radio Frequency Oscillator Design. Thesis

Polastre J, Hill J, Culler D (2004) Versatile low power media access for wireless sensor networks. Sensys:95–100

Polastre J, Szewczyk R, Culler D (2005) Telos: enabling ultra-low power wireless research. Information Processing in Sensor Networks 364–369

Roundy S (2005) On the effectiveness of vibration-based energy harvesting. J Intel Mat Syst Str:809–823

Roundy S, Otis B, Chee YH, Rabaey J, Wright P (2003) A 1.9 GHz RF transmit beacon using environmentally scavenged energy. IEEE Int. Symposium on Low Power Elec and Devices

Roundy S, Steingart D, Frechette L, Wright P (2004) Power sources for wireless sensor networks. Lecture Notes in Computer Science 2920:1–17

Roundy S, Wright P (2004) A piezoelectric vibration based generator for wireless electronics. Smart Materials & Structures 13:1131–1142

Schneider MH, Evans JW, Wright PK, Ziegler D (2006) Designing a thermoelectrically powered wireless sensor network for monitoring aluminium smelters. P I Mech Eng E-J Pro:181–190

Shnayder V, Hempstead M, Chen B, Allen GW, Welsh M (2004) Simulating the power consumption of large-scale sensor network applications. 2nd international conference on Embedded networked sensors

Smith A (1998) Radio frequency principles and applications: the generation, propagation, and reception of signals and noise, Wiley-IEEE Press Hobken, NJ 236pp

Starner T (1996) Human-powered wearable computing. IBM Systems Journal vol. 35, pp 618–629

Starner T, Paradiso JA (2004) Human generated power for mobile electronics. Low Power Electronics Design, CRC Press, Boston

Steingart D, Ho CC, Salminen J, Evans JW, et al. (2007) Dispenser printing of solid polymer-ionic liquid electrolytes for lithium ion cells. Polymers and Adhesives in Microelectronics and Photonics 2007:261–264

Stordeur M, Stark I (1997) Low power thermoelectric generator-self-sufficient energy supply for micro systems. Thermoelectrics 1997:575–577

Warneke B, Scott MD, Leibowitz BS, Zhou L (2002) An autonomous 16 mm 3 solar-powered node for distributed wireless sensor networks. Proceedings of Sensors' 02

Werner-Allen G, Johnson J, Ruiz M, Lees J, Welsh M (2005) Monitoring volcanic eruptions with a wireless sensor network. Wireless Sensor Networks, Proceedings of the Second European Workshop on Wireless Sensor Networks, pp. 108–120

Williams CB, Yates RB (1996) Analysis of a micro-electric generator for microsystems. Sensors & Actuators: A. Physical 52:8–11

Yaralioglu GG, Ergun AS, Bayram B, Haeggstrom E, Khuri-Yakub BT (2003) Calculation and measurement of electromechanical coupling coefficient of capacitive micromachined ultrasonic transducers. IEEE Transaction on Ultrasonic Ferroelectrics and Frequency:449–456

Chapter 10
Harvesting Microelectronic Circuits

Gabriel A. Rincón-Mora

Abstract This chapter focuses on how to most efficiently transfer and condition harvested energy and power with emphasis on the imposed requirements of microscale dimensions. The driving objective is to maximize operational life by reducing all relevant power losses. The chapter therefore briefly reviewes the electrical characteristics and needs of available harvesting sources and the operational implications of relevant conditioners. It then discusses the energy and power losses associated with transferring energy and useful schemes for maximizing system efficiency. It ends by presenting a sample microelectronic harvester circuit.

Increasingly dense microelectronic solutions spark an expanding new wave of non-invasive wireless microsensor and bio-implantable devices in commercial, military, space, and medical applications for monitoring, control, and surveillance. The driving features of this technology are self-power, long life, wireless telecommunication, and sensing-control functionality. Integration, however, in spite of its progress over the years, hampers the growth of such applications because energy and power are scarce when conformed to microscale dimensions. The fact is, even with the most energy-dense state-of-the-art battery, the operational life of a miniaturized system capable of sensing, storage, and wireless telemetry is relatively short, maybe on the order of minutes or less, depending on its needs, requiring periodic maintenance by personnel, which is costly and in many cases prohibitive and/or dangerous.

Harvesting energy from the surrounding environment is of growing interest to the research community because of its ability to not only replenish energy into an otherwise easily exhaustible reservoir but also substitute more expensive and perhaps less environmentally friendly alternatives. The ultimate promise of successful harvesting technologies is indefinite life. Unfortunately, practical realities challenge and limit its viability and ability to penetrate the market.

G.A. Rincón-Mora (✉)
School of Electrical and Computer Engineering, Georgia Institute of Technology, Atlanta, GA 30332-0250, USA
e-mail: rincon-mora@ece.gatech.edu

S. Priya, D.J. Inman (eds.), *Energy Harvesting Technologies*,
DOI 10.1007/978-0-387-76464-1_10 © Springer Science+Business Media, LLC 2009

At a fundamental level, the output power of state-of-the-art harvesting technologies is not yet commensurate with power-hungry high-performance loads such as transmitting power amplifiers (PAs) and their associated antennae whose propagated energy in the far field space decreases by the square of the distance traveled, increasing its energy demands with telecommunication distance. The fact the power is not in a usable state has adverse compounding effects. For one, harvesting does not necessarily increase its output power in response to transient load (power) dumps. What is more, the power delivered is not necessarily in the form required, as loading applications often demand specific voltage and impedance characteristics. Transferring energy and conditioning power also require energy and power, substantially reducing the already limited capabilities of the system.

10.1 Harvesting Sources

10.1.1 Energy and Power

Before embarking on the means of transferring and storing harvested energy, it is important to understand the implications of limited energy and power in a microscale system. To extend operational life, for instance, power-hungry functions are predominantly off to conserve energy while subjecting the system to deep low-power conditions. The power the system demands may therefore extend to five to six decades, from maybe 100 nW (e.g., 1 V and 100 nA of load power) during idle conditions to 20 mW (e.g., 2 V and 10 mA) when fully loaded. Under these variable conditions, an energy-constrained environment limits the life of the system, even in its low-power mode, while power deficiency impedes the operation of specific power-demanding features like transmission. As a result, self-contained microsystems demand both high-energy and high-power densities.

Energy-dense storage devices, however, often exhibit low-power densities, and vice versa, as shown in the Ragone plot of Fig. 10.1. Atomic betavoltaic batteries and fuel-constrained fuel cells, on one hand, are probably most notable for their high-energy densities, or similarly, high-specific energies (Joules per kilogram). They consequently yield long operational lives, but only when lightly loaded because their power densities are substantially low. Capacitors, on the other hand, provide instantaneous and seemingly unbounded current and power, but only momentarily, because their energy densities are low. Inductors may store considerable energy, but they must continue to receive, release, or lose it, making them ideal for transferring energy but not storing it for longer periods. Lithium–ion batteries (Li Ion) are popular because they yield a happy compromise between these two extremes, providing moderate energy and power levels.

The most appealing attribute of the harvester is that it draws energy from the surrounding environment, from a practically boundless supply, and its most debilitating drawback is low-power density. Like the fuel cell, it is an energy-conversion device, but unlike the fuel cell, its fuel supply is not constrained to a small reservoir.

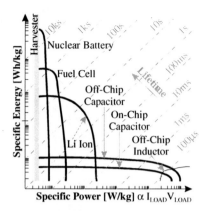

Fig. 10.1 Ragone plot: comparative energy-power performance of various energy-sourcing technologies

Its seemingly infinite energy density, in fact, is the driving motivation behind this book and all related research. Its most useful metric is therefore power, but only when coupled and normalized to practical ambient conditions. Note the harvester cannot replace the power source in a microscale system, only the energy source, because it can extend operational life under idling conditions but not supply power to seldom-occurring but power-demanding functions like transmission.

10.1.2 Energy Sources

DC sources: In designing a suitable power conditioner, it is important to ascertain and model the impedance, voltage, and current levels a particular harvesting technology produces because the conditioner, in effect, matches the impedance of the load to its source. A photovoltaic solar cell, for instance, is a pn-junction diode that generates photon-induced steady-state current i_{PH} when exposed consistently to light so its model can take the form of a dc Norton-equivalent current source, as shown in Fig. 10.2(a). Not all harvested current i_{PH} reaches the target load, however, because (1) charge carriers recombine before reaching their intended destination and (2) diode current i_D diverts a portion of i_{PH} to ground, as illustrated in the model. The periphery of the photovoltaic cell also presents a load to the current harvested near the edges of the device, effectively losing current to parallel resistance R_R. Lastly, as with any practical device, the medium through which the carriers traverse offers resistance and series source resistor R_S models that effect.

Applying a temperature gradient across p- and n-type segments joined at their high-temperature terminals (i.e., thermocouple) results in a heat gradient that induces high-energy charge-carrier diffusion to the low-temperature ends of the device. The electron and hole carriers flow in opposite directions and therefore constitute a net current, whose ionizing effects on the terminals of the thermocouple produce voltages across each segment, the series combination of which is shown in

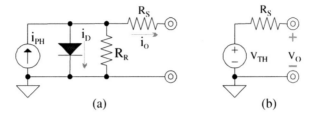

Fig. 10.2 (a) Photovoltaic and (b) thermoelectric circuit-equivalent models

the model of Fig. 10.2(b) by Thevenin-equivalent voltage v_{TH}. This voltage and its associated power are directly proportional to the Seebeck coefficient of the thermoelectric material (Seebeck refers to the flow of charge carriers in response to heat) and the temperature differential across the thermocouple. As with the photovoltaic cell, the material offers thermal impedance and electrical resistance to heat and charge-carrier flow, respectively, the linear effects of which series source resistor R_S models.

AC sources: Unlike solar and thermal energies, which produce a steady flow of charge carriers, kinetic energy in vibrations, much like a pendulum, stresses harvesters in alternating directions, the result of which are ac voltages and ac currents. Unfortunately, rectifying and/or synchronizing the harvester circuit to vibrations requires energy in the form of higher intrinsic losses through series asynchronous diodes or more work on the part of the conditioner and the synchronous switches it controls. The peak–peak voltages that these technologies produce is important because, on one hand, they must be high enough to drive metal-oxide-semiconductor (MOS) switches, bipolar-junction transistors (BJTs), and diodes (e.g., 0.4–0.8 V), and on the other hand, low enough to avoid incurring breakdown events (e.g., less than 1.8 V).

Harvesting kinetic energy in vibrations may take one of several forms based on piezoelectric, electrostatic, and/or electromagnetic means. A piezoelectric device, for example, produces a voltage in response to strain or deformation caused by the movement an attached mass produces under vibrating conditions. The voltage is proportional to the mechanical–electrical coupling of the material, that is, its piezoelectric properties, and the stress applied.

The harvester, as a system, is typically a two-layer bending bimorph fixed on one end to a stationary base and attached to a free-moving mass on the other end. The dielectric shim used to separate the piezoelectric bars couples the open-circuit voltages the rods produce (v_{PE1} and v_{PE2}, or combined, v_{PE}) electrostatically, as depicted by coupling capacitor C_{Shim} in the model shown in Fig. 10.3; open-circuit voltage v_{PE} can be on the order of Volts. Series resistor R_S, as before, models the series resistance through the device.

Microscale electrostatic harvesters convert mechanical to electrical energy by allowing the separation distance or overlap area of the plates in a micro-electromechanical systems (MEMS) capacitor to vary. By keeping the charge (Q) or voltage (V) constant, because charge and voltage relate through capacitance (C),

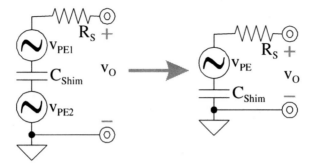

Fig. 10.3 Circuit-equivalent model for the two-layer piezoelectric bending bimorph

$$Q = CV, \tag{10.1}$$

changes in capacitance produce open-circuit voltage variations (Δv_{ES}) or current (i_{ES}):

$$\Delta v_{ES} = \frac{Q_{CONST}}{\Delta C}, \tag{10.2}$$

or

$$i_{ES} = \frac{dQ}{dt} = \left(\frac{dC}{dt}\right) V_{CONST}. \tag{10.3}$$

Practical electrostatic voltage variations reach 25–200 V with initial voltages of 0.125–1 V and under 1–200 pF variations. Modeling these effects, because the harvester is already an electrical component, amounts to using a vibration-dependent variable capacitor (i.e., varactor C_{VAR}), as shown in Fig. 10.4(a). Capacitors $C_{PAR.A}$, $C_{PAR.B}$, and $C_{PAR.C}$ describe the parasitic capacitances present at each terminal of the device and across it, while resistors $R_{S.A}$ and $R_{S.B}$ model the parasitic series resistance each plate and terminal offers to moving charges.

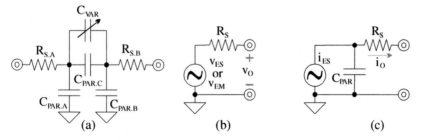

Fig. 10.4 Equivalent circuit models of a variable parallel-plate capacitor (varactor) with (**a**) variable capacitance C_{VAR}, (**b**) charge-constrained voltage source v_{ES} (and equivalent electromagnetic circuit model), and (**c**) voltage-constrained current source i_{ES}

Under charge-constrained conditions, the model may simplify to Thevenin-equivalent open-circuit voltage v_{ES} and series resistance R_S, as depicted in the model of Fig. 10.4(b), although loading the model with a resistor violates charge-constrained conditions, changing the characteristics of the harvester. A voltage-constrained capacitor, on the other hand, conforms best to Norton-equivalent current i_{ES}, and as before, series resistance R_S. Note the model includes parasitic capacitor C_{PAR} because it shunts and diverts some of the electrostatic current generated to ground. From an electrical engineering perspective, the purely passive model first described is appealing because it conveys more physical information about the system, applies to either charge- or voltage-constrained scenarios, and circumvents the need for transposing the voltage or current generated into source equivalents v_{ES} or i_{ES}.

Electromagnetic harvesters harness kinetic energy in vibrations by moving a coil across the magnetic field established by a stationary magnet (or moving a magnet through a stationary coil), thereby inducing, as Faraday's law dictates, a voltage across the coil. The Thevenin-equivalent model shown in Fig. 10.4(b) also aptly describes electromagnetic open-circuit voltage v_{EM} and the series resistance the coil offers to moving charges (R_S). Given the microscale dimensions of the system, however, the voltages induced across the coil are on the order of micro-Volts, which fall below the driving requirements of most semiconductor devices.

Conditioning notes: Although the underlying mechanism through which solar and thermal harvesters generate electrical energy is moving charges in one direction and producing a dc current, the voltages they induce preclude them from directly channeling the charges to low-impedance secondary (rechargeable) batteries or loads. Photovoltaic cells, for one, are more practical for outdoor applications, where its power levels for a square centimeter are on the order of milli-Watts; indoor lighting produces power levels on the order of micro-Watts under similar area constraints. As to thermoelectric harvesters, conforming them to microscale dimensions produce substantially low power levels because the temperature differential a typical semiconductor-based microchip sustains without extraneous heat sinks is relatively low, maybe around 5–10 °C, ultimately yielding considerably low voltages, which are also difficult to manage.

One of the most challenging aspects of harnessing kinetic energy in vibrations is the oscillating or pendulum-like (ac) nature of the resulting voltages and currents. To scavenge reasonable amounts of energy, the conditioner must therefore tune and synchronize itself to the vibrations. Asynchronous full- or half-wave rectifiers, that is, diode-based ac–dc converters, circumvent the need for synchronization, but at the cost of Ohmic power losses and voltage drops, which electromagnetic devices are fundamentally less able to accommodate. Alternatively, synchronizing the conditioner to the vibration so as to only drive or supply energy in the positive cycles also requires work on the part of the controller circuit.

Using a semiconductor microelectronic conditioner imposes additional parasitic devices and restrictions on the foregoing harvesters. Mainstream complementary metal-oxide-semiconductor (CMOS), bipolar, and mixed bipolar-CMOS (BiCMOS) technologies, for example, suffer from limited breakdown voltages. The fact is finer photolithographic dimensions necessarily expose semiconductor junctions to more intense electric fields, inducing breakdown conditions at relatively low voltages, at around 1–5 V, which are incompatible, for instance, with charge-constrained electrostatic harvesters whose voltages exceed this limit by a large margin.

Another technological restriction is leakage current because highly doped pn junctions and tunneling and impact-ionization effects resulting from high-electric fields, which are prevalent in high-density technologies, shunt and divert harvested energy to ground, reducing the efficacy of the harvester; a resistor in parallel with the harvester models this energy loss. Additionally, voltage headroom also presents another limitation because diode and bipolar-junction transistors (BJTs) conduct current when the voltages across their respective forward-biased pn junctions are approximately 0.6–0.7 V. Similarly, MOS field-effect transistors (FETs) conduct current when their gate–source voltages are on the order of 0.4–1 V, even under sub-threshold conditions. As a result, thermoelectric and electromagnetic voltages, which are considerably below 1 V, impose additional challenges, as mentioned earlier, because the conditioner requires higher voltages to operate.

10.2 Power Conditioning

A power conditioner massages voltage and current so as to supply the needs of a loading application. This typically amounts to converting and regulating a voltage (or current) by supplying whatever current (or voltage) is necessary to sustain it. When considering wireless microsensors, however, the field narrows because the electrical requirements of most loads are moderate-to-low dc voltages (e.g., 0.5–2 V) demanding anywhere from 100 to maybe 20 mA. As such, only conditioners that can supply dc voltages are considered in this chapter. Additionally, since the source can be dc or ac, this section describes dc–dc and ac–dc converters, all of which, in one way or another, conform to linear and/or switching strategies.

The purpose of this section is not to present the general discipline of power conditioning, which can easily monopolize an entire textbook, if not more, but illustrate how it generally relates and applies to the foregoing harvesting technologies. As such, for brevity and purpose, this section discusses some devices, physics, feedback control, and circuit details not intrinsic to harvesting in broad terms. To better evaluate what is and not important and consequently avoid tangential deliberations, the section starts with a system block-level overview of the harvesting microsystem, incorporating many of the comments from the previous section. The discussion then leads to an overview and comparison of various converter topologies, evaluating them according to the needs of the general microsystem presented.

10.2.1 Microsystem

As noted earlier, energy-dense sources do not supply the most power, and oversizing a sourcing technology to compensate for its deficits is in direct conflict with

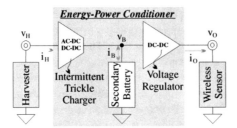

Fig. 10.5 Harvesting microsystem

integration and the general goals of non-invasive microsystems. As a result, the most space-efficient solution is a hybrid structure that exploits the complementary advantages of its constituent parts, as shown in Fig. 10.5. The harvester, for instance, because of its boundless energy supply and deficiency in power, is the energy source of the system. The secondary battery yields higher output power but stores less energy so it functions as a power cache (i.e., source), supplying power when required but otherwise constantly receiving charge from the harvester.

The harvester, battery, and wireless sensor load all have different voltage and current characteristics. The battery voltage (v_B) across a thin-film Li Ion, for instance, is around 2.7 V when discharged and 4.2 V when charged, and overloading or overcharging it past its fully discharged or charged states, respectively, is destructive, appreciably decreasing the capacity of the device. The load, on the other hand, may not survive supply voltage (v_O) variations exceeding 0.25 V around a nominal value of maybe 1–1.5 V. The harvester's output voltage (v_H) and current also differs vastly from that of the battery and load. As a result, an intermittent dc–dc or ac–dc trickle-charger circuit and dc–dc voltage regulator combination is used to transfer and condition energy from the harvester, through the battery, and to the load, effectively matching the impedances of the various components in the system and suppressing the effects of voltage and current variations in the harvester and battery from affecting the load.

Although the art of designing voltage regulators is not new, as most systems today derive their energy and power from a Li Ion, its requirements are more stringent in microsystems because light loading conditions are pervasive and any power losses associated with the circuit, irrespective of its magnitude, decreases operational life. These performance requirements worsen when considering integration into microscale spaces limits the number of low-loss in-package capacitors (e.g., 0.1–1 μF) and inductors (e.g., 1–10 μH) to maybe one or two. On-chip inductors and capacitors, unfortunately, do not typically surpass 20–50 nH and 100–200 pF, forcing the accompanying circuits to perform more work and consequently consume more energy and power.

The difference between the trickle charger and the voltage regulator is typically output impedance because the latter regulates a voltage and the former delivers a charging current. The nature of the feedback loop used to control the circuit helps define this, as shunt negative feedback at the output regulates voltage and series negative feedback regulates current. The input of the charger may mimic that of the voltage regulator if harvester voltage v_H is at a reasonably high steady-state dc level, which is normally not the case. Motion harvesters, for instance, source characteristically low ac voltages and photovoltaic and thermoelectric devices low dc voltages. Nonetheless, the general strategies for transferring energy and conditioning power efficiently (i.e., without incurring considerable losses), be it linear or switching, amplifying or attenuating, apply to both.

10.2.2 Linear DC–DC Converters

Perhaps, the most straightforward means of converting high to low dc voltages is through a linear regulator, which is nothing more than an operational amplifier (op amp) in non-inverting configuration, as shown in Fig. 10.6(a). The differences between an op amp and a regulator are loading conditions and requirements. For one, the variable and unpredictable nature of the load imposes stringent stability requirements on the regulator, which often lead to vastly different filter-compensation strategies. Additionally, the current demands of a regulator substantially exceed those of an op amp.

Within the context of supply circuits, the most attractive features of a linear regulator are low noise and relatively fast response times. The fact is no systematic ac variations or noise appears in the output because there are no periodic switching events polluting it. Additionally, its response time is relatively fast because its loop, when compared with more complicated switching converters, is simpler and consequently quicker. However, even with fast response times, low-value output capacitors (e.g., 150 pF), when exposed to substantial load dumps (e.g., 0–5 mA), still droop considerably (e.g., 0.8–1 V), as shown with the experimental results of Fig. 10.6(b). Higher bandwidth (i.e., a faster loop), higher output capacitance, or a lower load step reduce this voltage excursion.

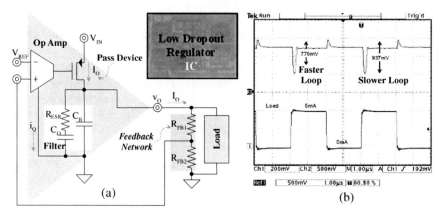

Fig. 10.6 Linear regulator (**a**) circuit and (**b**) typical transient response to quick load dumps

The general strategy for stabilizing the feedback loop of a linear regulator is to establish a dominant low-frequency pole either at the output, where a larger output capacitor (C_O) also helps suppress transient load-dump-induced variations, or inside the loop, at a high-gain node, which typically offers high resistance. Ensuring a single pole predominates over all others is often difficult, however, because the load is unpredictable, C_O is typically large to suppress noise, the tolerance of C_O's equivalent series resistance (ESR) R_{ESR} is relatively high, and quiescent current must remain low to reduce drain current drawn from the battery, which means the parasitic poles in the op amp reside at low-to-moderate frequencies. As a result, designers often sacrifice loop gain and bandwidth to accommodate all these ill-fated variations and scenarios, trading off dc- and ac-accuracy (i.e., conditioning) performance for electrical stability.

Ultimately, within the context of harvesting circuits, linear regulators suffer from considerable power losses and therefore shorter lifespan performance because output current I_O flows directly from input V_{IN} to output v_O through the regulator (i.e., through a pass device). The regulator, therefore, dissipates an average power that is equivalent to the product of the average voltage dropped across the regulator (i.e., pass device) and I_O, in addition to losses induced by quiescent current I_Q through the op amp and across the input:

$$P_{LIN} = (V_{IN} - V_O)\,I_O + V_{IN}\,I_Q, \qquad (10.4)$$

where V_O is the steady-state value of output voltage v_O. Because this power is strongly dependent on input–output voltages V_{IN} and V_O, respectively, neither of which is typically a free design variable, linear regulators do not appeal to applications that require moderate-to-high voltage drops and high-power efficiency. The fact that the regulator can only down-convert (or buck) a voltage further limits its use, given intermittent trickle chargers often amplify (or boost) harvested voltages.

10.2.3 Switching DC–DC Converters

Switching supplies fall under two broad categories: electrostatic and magnetic. Electrostatic converters or charge pumps, as they are typically called, use only capacitors to transfer energy and condition power. Magnetic-based switching supplies, on the other hand, use capacitors only for storage and inductors for transferring energy. Fundamentally speaking, though, irrespective of the category, switching power supplies enjoy higher power-efficiency performance than linear regulators (i.e., they yield lower power losses) because the voltages impressed across the power transistors (i.e., switches), unlike the pass devices in linear regulators, are small in steady state (on the order of 50–150 mV), irrespective of input voltage V_{IN} and target output voltage V_O. They are sometimes exposed to higher voltages, but only momentarily, limiting the total energy lost during those instances.

Electrostatic switchers (charge pumps): Charge pumps work in phases, charging capacitors and connecting them to the output in alternate cycles. The basic strategy

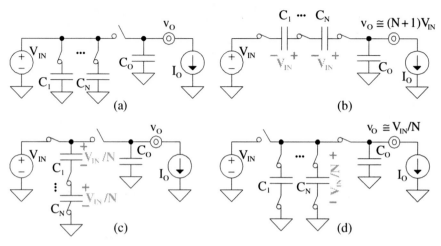

Fig. 10.7 Basic alternating cycles of (a)–(b) boosting (*step-up*) parallel-series and (c)–(d) bucking (*step-down*) series-parallel charge pumps

in step-up or boosting charge pumps, for instance, is to charge one or several capacitors to input voltage V_{IN}, as shown in Fig. 10.7(a)–(b), and then connect them in series with V_{IN} to output voltage v_O. Because each capacitor holds an initial voltage equivalent to V_{IN} across it, their series combination with V_{IN} sums to $(N + 1)V_{IN}$, where N refers to the number of pumping capacitors. Bucking or step-down switchers are similar in principle but reversed, because they charge a series combination of capacitors to V_{IN} (Fig. 10.7(c)–(d)), initializing each capacitor to V_{IN}/N, and then connect all capacitors in parallel with v_O, producing an initial voltage equivalent to V_{IN}/N at v_O.

To increase the amplification across N capacitors in the boosting case, designers often connect several low-gain (i.e., $2\times$), one-capacitor charge pumps in series, as shown in Fig. 10.8, to use the boosted output of one to charge the next stage, and so on. Each stage consequently amplifies the pre-charge voltage of the subsequent stage, ultimately achieving a higher overall gain of $2^N V_{IN}$. The pre-charge and pump cycles of each stage complements the next to ensure the pre-charge voltage is ready for the subsequent stage. As a result, the clock signals between adjacent charge pumps are out of phase.

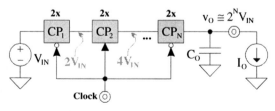

Fig. 10.8 Cascaded charge-pump configuration

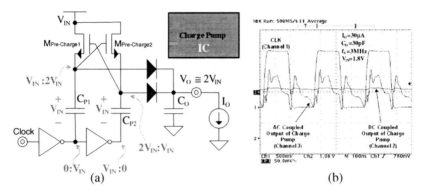

Fig. 10.9 Practical 2× CMOS boosting charge-pump (**a**) schematic and (**b**) resulting measurement results

To avoid disconnecting the charge-pumping capacitors from v_O for long, which would otherwise further exhaust output capacitor C_O, designers also add complementary parallel stages so that, while one capacitor charges, the other pumps the output in complementary fashion. Figure 10.9 shows a practical CMOS realization and the resulting measurement results of such a case, of a 2× CMOS boosting charge pump with 30 μA, 30 pF, 3 MHz, and 1.8 V of I_O, C_O, switching frequency f_S, and V_{IN}, respectively. When the clock is at V_{IN}, n-channel field-effect transistor (FET) $M_{Pre-Charge1}$ charges pump capacitor C_{P1} to V_{IN} while its bottom terminal is at ground. As the clock cycles to ground, the bottom plate of C_{P1} connects to V_{IN}, raising the top plate to twice V_{IN}, since C_{P1} holds V_{IN} across its plates. At this point, the top asynchronous diode switch forward biases and conducts charge to v_O. While C_{P1} pumps v_O, C_{P2} charges to V_{IN} so that it can then pump v_O when C_{P1} cycles back to its charging state. Note the top-plate voltages of C_{P1} and C_{P2} also provide sufficient gate drive for switches $M_{Pre-Charge1}$ and $M_{Pre-Charge2}$ to conduct and fully charge their counter capacitors.

The fundamental drawback of a charge pump is that a considerable fraction of the energy delivered is lost across the connecting switches. The initial voltage across the pumping capacitors (C_P) during the charging phase, for example, is below its target value of V_{IN} (Fig. 10.10(a)), because the load partially discharged it in the previous cycle. As a result, fully charging C_P to V_{IN} incurs an energy loss in V_{IN} that is directly and linearly proportional to voltage difference ΔV, V_{IN}, and C_P:

$$\Delta E_{IN} = V_{IN} i_C t = V_{IN} Q_C = V_{IN} C_P \Delta V, \qquad (10.5)$$

where i_C is the capacitor current, t is time, and Q_C is the charge delivered to C_P. Considering the energy in C_P increased from $0.5 C_P (V_{IN} - \Delta V)^2$ to $0.5 C_P V_{IN}^2$ by

$$\Delta E_C = 0.5 C_P [V_{IN}^2 - (V_{IN} - \Delta V)^2] = C_P V_{IN} \Delta V - 0.5 C_P \Delta V^2, \qquad (10.6)$$

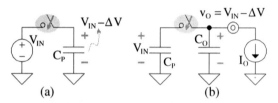

Fig. 10.10 Initial charge-pump equivalent (**a**) charge and (**b**) transfer circuits

the energy lost in the switch (ΔE_{SW}) is

$$\Delta E_{SW} = \Delta E_{IN} - \Delta E_C = 0.5 C_P \Delta V^2 = 0.5 Q_C \Delta V, \tag{10.7}$$

which is directly and linearly proportional to Q_C and ΔV, respectively.

Similarly, transferring energy from pump capacitors C_P to C_O also incurs energy losses (Fig. 10.10(b)). When connected, C_P shares its charge with C_O and consequently produces a final combined voltage that is between the initial pre-charged value of C_P and the loaded value of C_O. Much like in the previous scenario, the difference in energy lost in C_P (ΔE_{CP}) and energy gained in C_O (ΔE_O), which is the loss incurred by the switch, is directly proportional to charge transferred Q_C and the initial voltage difference between C_P and C_O (ΔV),

$$\Delta E_{SW} = \Delta E_{CP} - \Delta E_O = 0.5 Q_C \Delta V = 0.5 (C_P \oplus C_O) \Delta V^2, \tag{10.8}$$

or equivalently, linearly proportional to the series combination of C_P and C_O (i.e., $C_P \oplus C_O$) and the square of ΔV.

Generally, basic power losses in a charge pump decrease with output current I_O, or more specifically, with the lower voltage droop a smaller I_O causes in v_O. Partially discharging C_O with I_O forces the circuit to transfer charge from V_{IN} to C_O via charge-pump capacitors C_P, losing energy in every operation of the charge-pumping sequence. Note I_O not only includes the load current but also any and all parasitic leakage currents at v_O, which are more pervasive and therefore more troublesome in fined-pitched modern CMOS technologies. Also note regulating v_O to a target value with negative feedback amounts to decreasing v_O from its highest possible charge-pumped value, thereby increasing voltage difference ΔV and decreasing the overall efficiency of the circuit. Ultimately, because of these foregoing reasons, charge pumps tend to be most practical and useful in unregulated (i.e., open loop and therefore low accuracy) light-to-no load applications, which exist but not to the extent regulated and loaded applications do.

Magnetic switchers: Magnetic-based dc–dc converters circumvent these basic switch-related losses by transferring energy through a quasi-lossless inductor rather than a switch. The idea is to energize and de-energize an inductor from input voltage V_{IN} to output voltage v_O in alternate cycles, as shown in Fig. 10.11(a)–(b). The voltages across the energy-steering switches are substantially low because the inductor, to prevent its current from varying instantaneously, allows its voltage to

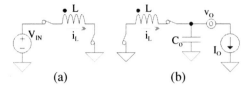

Fig. 10.11 (a) Energizing and (b) de-energizing phases in magnetic-based dc–dc converters

change instantly, in the process cutting the time the switch undergoes any voltage drop to nearly zero. Generally speaking, as inferred from the particular switching configuration shown in the figure, the average value of v_O is independent of V_{IN} and can therefore reach above or below V_{IN}, depending on the duty cycle of the alternating phases.

Energizing and de-energizing an inductor do not require the voltage drop across L to be any particular value with respect to V_{IN} or v_O, as long as the voltage across the inductor reverses in the de-energizing state relative to its energizing counterpart. As such, a step-down or bucking magnetic-based dc–dc converter, because V_{IN} is higher than v_O, energizes inductor L from V_{IN} to v_O, instead of V_{IN} to ground, as accomplished by switch S_1 or its p-channel FET realization in Fig. 10.12(a)–(b). The de-energizing cycle, as in Fig. 10.11(b), de-energizes from v_O to ground via switch S_2 or its asynchronous diode realization. Considering the average voltage across L is zero, v_O's average voltage equals v_{PH}'s average voltage, which is V_{IN} times the fraction of time S_1 applies V_{IN} to v_{PH} (i.e., S_1's duty cycle):

$$v_O(\text{avg}) = V_O = v_{PH}(\text{avg}) = V_{IN}D_{S1}, \qquad (10.9)$$

where V_O is the average value of v_O and D_{S1} is S_1's turn-on duty cycle, whose value is necessarily less than one. Applying negative feedback, as shown in Fig. 10.12(b), ultimately modulates D_{S1} with pulse-width modulated (PWM) signal v_{PWM} to regulate v_O about reference voltage V_{REF}.

The use of diodes for switches carries both advantages and disadvantages. The appeal, when possible and practical, is no synchronizing signal is necessary to drive

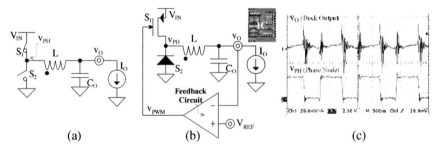

Fig. 10.12 (a) Conceptual and (b) circuit-level magnetic-based buck (*step-down*) dc–dc converter, and (c) corresponding waveforms

a diode because it switches on its own, asynchronously, whenever the forward-bias voltage exceeds approximately 0.6–0.8 V. Because v_{PH} switches from V_{IN} to ground during the energizing and de-energizing cycles, respectively, a diode in place of S_2 naturally conducts and clamps v_{PH} when it decreases below ground, as S_2 should, and shuts off when S_1 raises v_{PH} above ground. The drawback of the diode is its relatively high switch-on voltage, which translates to power losses and consequently lower overall system efficiency and shorter operational life. As a result, designers often replace S_2 with a synchronous NFET switch, which like S_1, requires a synchronizing signal to switch.

The experimental v_O and v_{PH} waveforms shown in Fig. 10.12(c) correspond to a CMOS buck converter with synchronous switches for both S_1 and S_2. Although not directly shown, the current square-wave voltage v_{PH} induces through L is triangular because current ramps up with constant positive inductor voltages ($v_L = L di/dt$) and similarly ramps down with negative inductor voltages. The output shown (v_O) mimics this behavior because the difference in inductor current i_L and output current I_O is the ac ripple current driven into output capacitor C_O, whose equivalent series resistance (ESR, not shown in Fig. 10.12(a)–(b)) translates into the triangular voltage shown in Fig. 10.12(c). The output voltages produced by low-ESR (i.e., high-frequency) capacitors are more sinusoidal in nature because the ESR voltage is negligible with respect to the charging and discharging voltage across C_O that results from driving said triangular ripple current into C_O.

The buck regulator exploits the fact V_{IN} is greater than v_O to energize L directly from V_{IN} to v_O and reduce the number of switches associated with the v_O side of L. Taking advantage of this characteristic, however, limits the circuit to bucking applications only. Nevertheless, a boosting or step-up converter can also similarly exploit v_O is greater than V_{IN} by de-energizing L from v_O to V_{IN}, rather than ground, in the process reducing the number of switches associated with the V_{IN} side of L, as shown in Fig. 10.13(a). In this case, S_1 or its NFET equivalent, as shown in Fig. 10.13(b), energizes L from V_{IN} to ground and S_2 or its diode equivalent de-energizes it from v_O to V_{IN}.

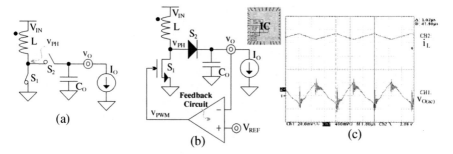

Fig. 10.13 (a) Conceptual and (b) circuit-level magnetic-based boost (*step-up*) dc–dc converter, and (c) corresponding waveforms

Because the average voltage across L is zero, average phase node voltage V_{PH} is V_{IN}, which means v_{PH} swings below and above V_{IN}. As such, V_{PH} is its peak voltage (V_{PK}) times the fraction of time S_2 applies V_{PK} to v_{PH} (i.e., S_2's duty cycle):

$$v_{PH}(\text{avg}) = V_{PH} = V_{IN} = V_{PK}D_{S2}. \tag{10.10}$$

Switch S_2 or its diode equivalent and C_O constitute a peak detector, which means the average value of v_O (or V_O), assuming the voltage drop across S_2 is negligibly small, is V_{PK} and therefore related to V_{IN} and D_{S2}, or equivalently, one minus the turn-on duty cycle of S_1 (or $1\text{-}D_{S1}$) by

$$v_{O(\text{avg})} = V_O \approx V_{PK} = \frac{V_{IN}}{D_{S2}} = \frac{V_{IN}}{1 - D_{S1}}, \tag{10.11}$$

Since D_{S1} is necessarily less than one, V_O is greater than V_{IN}.

As in the buck case, PWM feedback signal v_{PWM} modulates D_{S1} to regulate v_O about target voltage V_{REF}. Similarly, synchronous FET switches can implement S_2 to decrease the turn-on voltages associated with the diodes to 50–150 mV levels and consequently improve the energy and power efficiency of the system. As in the buck case, v_{PH} remains a square wave, except it now swings from ground to v_O (i.e., V_{PK}). Because the circuit also applies a square wave across L, the resulting inductor current (i_L) is also a ripple (Fig. 10.13(c)). Unlike the buck case, however, not all of i_L flows to v_O, only the fraction S_2 conducts. As a result, v_O discharges with I_O when S_2 is off (i.e., L energizes and i_L increases) and charges with the difference of i_L and I_O when S_2 conducts (i.e., L de-energizes to v_O and i_L decreases), the main result of which is a rippling v_O. As before, the magnitude of C_O's ESR relative to its capacitance dictates the exact nature of the output ripple voltage.

In addition to the difficulties present in the linear regulator with respect to stabilizing the negative feedback loop, such as load variations, ESR tolerance, the presence of additional parasitic poles, and so on, magnetic-based converters suffer from the complex-conjugate poles L and C_O produce, which tend to fall at relatively low-to-moderate frequencies. The general compensation strategy is to convert them into a single pole by regulating the inductor current with one loop up to higher frequencies so that L behaves like a current source at lower frequencies. Another loop then regulates v_O about V_{REF} with a bandwidth that falls below that of the current loop to ensure L remains a current source throughout its bandwidth, in the process turning the complex-conjugate pole pair into a single RC pole, as prescribed by C_O and the equivalent resistance present at v_O.

This dual-loop approach, however, limits system bandwidth (i.e., the speed of the voltage loop) and increases circuit complexity. Unfortunately, the only other means of compensating the loop is to add a substantially lower frequency pole, limiting the bandwidth even further. As a result, magnetic-based converters tend to respond more slowly to transient load dumps than linear regulators. Note charge pumps do not normally regulate v_O because efficiency degrades with regulation so the variation its output experiences under load-dump events is even greater.

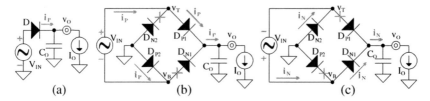

Fig. 10.14 (a) Half- and (b)–(c) full-wave rectifying ac–dc converters

10.2.4 Switching AC–DC Converters

Switching ac–dc converters, or as they are often called, inverters or rectifiers *invert* ac signals to dc or *rectify* ac power into dc form by appropriately channeling charge into a holding capacitor. The basic strategy, as shown in Fig. 10.14(a), is to transfer energy in the form of positive current i_P from input voltage source V_{IN} into output capacitor C_O only when V_{IN} is positive with respect to ground. Diode D and capacitor C_O, as in the magnetic-based boost converter, constitute a peak and hold detector; that is to say, D captures V_{IN}'s peak voltage and C_O holds it until V_{IN} is again ready, as it cycles through, to channel more charge and therefore energy back into C_O. Typically, ac–dc converters are open-loop architectures, as they do not regulate their outputs about a reference, so designers often cascade dc–dc converters to their outputs as a means of regulating v_O.

The single diode-capacitor combination, however, only rectifies the positive phase of the cycle, which means C_O must be sufficiently large to sustain output current I_O without allowing the voltage across C_O to droop considerably. Unfortunately, larger capacitors are bulky and expensive, which impede integration and market penetration, respectively. Figure 10.14(b)–(c) illustrates how to combine two half-wave rectifiers to implement a full-wave rectifier, otherwise called a Wheatstone Bridge, easing the holding requirements of C_O and consequently reducing the total variation of v_O. Considering the positive half of the cycle, when top voltage v_T is positive with respect to ground, D_{P1} and C_O constitute a simple rectifying circuit, except D_{P2} is in series to source current and complete the circuit. During this time, because v_T and bottom voltage v_B are above and below zero, respectively, D_{N2} and D_{N1} are reverse biased and therefore off. Conversely, when v_T and v_B are below and above zero, respectively, during the other half of the cycle, D_{N1} and C_O half-wave rectify V_{IN}, D_{N2} sources the current from ground and completes the circuit, and D_{P1} and D_{P2} are off.

Asynchronous diodes may ease the design of the circuit but only at the expense of power losses, given they drop approximately 0.6–0.8 V across them every time they conduct current to v_O. As such, designers normally replace them with synchronous MOSFET switches, which drop considerably lower voltages, maybe on the order of 50–150 mV. The drawback to MOSFETs, as before, is that they require synchronizing signals to switch when desired, which means the overall circuit is more complex.

Also, as in charge pumps, switches channel input energy to the output capacitor, incurring similar charge-sharing energy losses as electrostatic-based converters do.

Another problem with diodes is that V_{IN} must exceed 0.6–0.8 V for them to conduct, which is possible with piezoelectric harvesters but not necessarily so with electromagnetic devices. MOSFETs also help in this regard, but only by shifting the problem to their control input terminals (i.e., gates). For a MOSFET to conduct current, its control voltage must exceed its low-voltage terminal (i.e., source) by a technology-defined threshold voltage, which is typically on the order of 0.5–0.7 V. In these cases, however, much like a car uses a battery to start the engine, a precharged Li Ion may be used to start the switches and allow the harvester to sustain the charge in the battery during operation; a self-starting, low ac voltage harvester is difficult to realize.

10.2.5 Comparison

Table 10.1 presents a comparison of the foregoing power-conditioners with respect to their fundamental performance limits. To start, when considering the most important metric of microscale harvesters is output power, magnetic- or L-based converters are the most appealing structures because they incur, in theory, no power losses, allowing most of the sourced power to reach the load. Electrostatic- or C-based converters (or charge pumps) and linear regulators, on the other hand, dissipate finite power across their connecting switches, and the latter loses more energy than the former because the power lost depends strongly on V_{IN} and the targeted V_O, irrespective of the load.

L- and C-based converters enjoy the flexibility of their linear regulator and ac–dc counterparts do not, which is the ability to buck or boost V_{IN}. The drawback to L-based regulators with respect to linear regulators is ac accuracy because the negative feedback loops of the former are more difficult to compensate and consequently lead to slower response times. Charge pumps enjoy the benefits of an all-capacitor solution, which perceivably achieves the goals of integration, but not necessarily so in practice. The fact is accurate charge pumps require substantially large capacitors, on the order of micro-Farads, which are as difficult to integrate as micro-Henry inductors. Co-packaging one or two off-chip components, however, be it capacitors, inductors, or a combination thereof, is not necessarily prohibitive because the

Table 10.1 Fundamental comparison table of power-conditioning circuits

	DC–DC linear	DC–DC C-based	DC–DC L-based	AC–DC
Power losses	High	Moderate	Low	Moderate
DC accuracy	High	Low	High	Low
AC accuracy (speed)	High	Low	Moderate	Low
Control	One loop	Open loop	Two loops	Open loop
Buck V_{IN}	Yes	Yes	Yes	No
Boost V_{IN}	No	Yes	Yes	No

dimensions of state-of-the-art off-chip capacitors and inductors are on the order of $2 \times 2 \times 1 \text{ mm}^3$, conforming to the packaging dimensions and integration objectives of targeted system-in-package (SiP) solutions. The challenge, in these cases, would be to constrain the solutions to only a few off-chip power devices.

10.3 Power Losses

In practice, magnetic-based converters are not completely lossless, charge pumps incur more losses than just the energy lost in connecting switches, and linear regulators require quiescent current to operate. Because microharvesters generate necessarily low power levels, any and all power losses in the conditioner constitute a considerable portion of its output power, if not all of it, leaving little to no power left for the system. The objective of this section is to therefore understand the power losses associated with practical energy- and power-supply circuits.

By and large, power losses result because switches, inductors, and capacitors, in real life, present parasitic diodes, resistances, and capacitances, and controllers require quiescent current. To illustrate these losses in a general manner, it is useful to use a sample circuit that is comprehensive enough to include all power-consuming mechanisms, yet sufficiently simple to visualize, understand, and evaluate, drawing relevant and perhaps far-reaching conclusions. The magnetic-based buck converter shown in Fig. 10.15 is one such example. To start, it uses switches (e.g., MOS switches MP and MN), just as charge pumps and linear regulators do, the latter in the form of a power pass device. These high-power switches present finite series resistances to the conduction path, parasitic capacitances to their incoming control signals, and pn-junction diodes across their conduction terminals. Output capacitor C_O and power inductor L, the former of which is present in all conditioners and the latter applies only to L-based converters, present equivalent series resistances (ESRs) to incoming charges (e.g., $R_{C.ESR}$ and $R_{L.ESR}$ for the capacitor and inductor, respectively). Finally, the feedback loop in linear regulators and L-based topologies demand quiescent current to operate (e.g., i_Q).

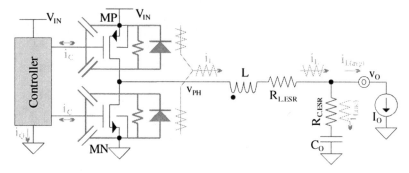

Fig. 10.15 Magnetic-based buck dc–dc converter and its parasitic devices

The approach adopted in this section is to categorize the additional power losses these parasitic devices induce into (1) dc and ac conduction, (2) switching, and (3) quiescent losses. Conduction losses refer to the power dissipated in resistors in the power-flowing path, switching losses to the power lost during clock-like transient transitions, and quiescent losses, as before, to the power required to operate the controller. To better evaluate the system ramifications of inductor current i_L, it is also helpful to decompose it into its average value $i_{L(avg)}$ or I_L and its ac or ripple component $i_{L(ac)}$ or i_l, both of which share some but not all of their current-flowing path, as seen in Fig. 10.15.

10.3.1 Conduction Losses

Because i_L cannot change instantaneously, given the basic nature of L, i_L flows linearly and continuously. As such, in the buck converter shown in Fig. 10.15, high- and low-side switches MP and MN conduct in alternate phases and each therefore supplies a fraction of i_L and combined continuously supply all of i_L. The collective equivalent resistance MP and MN present to i_L, as depicted by R_{SW} in the equivalent conduction-loss circuit of Fig. 10.16, is roughly the sum of their respective turn-on resistances R_{MP} and R_{MN} when weighted by the fraction of time they each conduct i_L. Although not necessarily so but often the case, R_{MP} and R_{MN} are on the same order and, for the purposes of the foregoing analysis, assumed equal (i.e., $R_{MN} \approx R_{MP} \equiv R_{ON}$). As such, because MP and MN conduct with complementary duty cycles D_{MP} and D_{MN} (i.e., $D_{MN} \approx 1 - D_{MP}$), D_{MP} and D_{MN} add to one and series resistance R_{SW} approximately equals switch turn-on resistance R_{ON}:

$$R_{SW} = R_{MP}D_{MP} + R_{MN}D_{MN}$$
$$\approx R_{MP}(D_{MP} + D_{MN}) \approx R_{MP} \approx R_{MN} \equiv R_{ON}. \qquad (10.12)$$

The average or dc portion of i_L ($i_{L(avg)}$), which equals output current I_O, consequently flows through equivalent switch resistance R_{SW}, L, and L's ESR $R_{L.ESR}$. As such, R_{SW} and $R_{L.ESR}$ incur dc conduction losses proportional to the square of I_O:

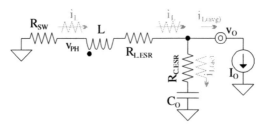

Fig. 10.16 Equivalent conduction-loss circuit of the buck dc-dc converter

$$P_{C.DC} = i_{L(avg)}^2 (R_{SW} + R_{L.ESR})$$
$$= I_O^2 (R_{SW} + R_{L.ESR}) \equiv I_O^2 R_{C.DC}, \tag{10.13}$$

where $R_{C.DC}$ represents the equivalent dc conduction resistance. The ripple or ac part of i_L ($i_{L(ac)}$) also flows through R_{SW}, L, and $R_{L.ESR}$; but instead of reaching the load, it flows to ground through C_O and C_O's ESR $R_{C.ESR}$. Note L suffers from skin effects at higher frequencies, so if the switching frequency is sufficiently high, current crowds on the outer edges of the inductor coil, decreasing the equivalent conductance of the coil and presenting a larger ESR.

All resistances in the conduction path incur ac power losses proportional to the square of the root-mean-square (RMS) of $i_{L(ac)}$, which increases with peak-to-peak inductor current ΔI_L. The magnitude of the voltage across L(i.e.,v_L), inductance L, and switching frequency f_{SW} determine the peak-to-peak value of triangular current i_L, as ΔI_L depends on the rising (or falling) rate of i_L (i.e., $di_L/dt = v_L/L$) and the time i_L continues to rise (or fall), when high-side switch MP conducts from V_{IN} to v_O (or low-side switch MN conducts from v_O to ground). Voltage v_L, which is the difference between phase voltage v_{PH} and v_O, in turn, depends on V_{IN}, ground, and predominantly the average value of v_O. In other words, V_{IN}, average output voltage V_O, L, and f_{SW} dictate the value of ΔI_L:

$$\Delta I_L = \frac{(V_{IN} - v_O)t_{MP}}{L} \approx \frac{(V_{IN} - V_O)t_{MP}}{L} = \frac{(V_{IN} - V_O)d_{MP}}{Lf_{SW}}, \tag{10.14}$$

where duty cycle d_{MP} refers to the on-duty cycle of MP (i.e., $d_{MP} = t_{MP}f_{SW}$). As a result, ac conduction losses $P_{C.AC}$ increase linearly with series conduction resistances and decrease with L and the square of f_{SW}:

$$P_{C.AC} = i_{L.AC(rms)}^2 (R_{SW} + R_{L.ESR}' + R_{C.ESR})$$
$$= \left(\frac{\Delta I_L}{\sqrt{12}}\right)^2 (R_{SW} + R_{L.ESR}' + R_{C.ESR}') \equiv \left(\frac{(V_{IN} - V_O)d_{MP}}{L\sqrt{12}}\right)^2 \left(\frac{R_{C.AC}}{f_{SW}^2}\right), \tag{10.15}$$

where $R_{L.ESR}'$ is L's ESR at the ripple frequency ($R_{L.ESR}'$ is slightly larger than $R_{L.ESR}$), $\Delta I_L/\sqrt{12}$ is the root-mean-square current of ac triangular current $i_{L(ac)}$ about zero and $R_{C.AC}$ represents the combined ac-conduction resistance of the circuit.

These ac-conduction losses, because they do not decrease with I_O, incur prohibitively detrimental energy losses in micropower harvesting applications, monopolizing most if not all available power. Decreasing these power losses amounts to decreasing the ac ripple in i_L by increasing inductance, which contradicts the limits of integration; increasing switching frequency, which increases switching losses, as will be noted later; or entering discontinuous-conduction mode (DCM), a condition where L is allowed to deplete all its energy before the end of the switching period.

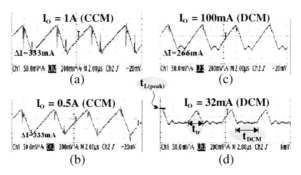

Fig. 10.17 Inductor current i_L for a magnetic-based buck converter as it transitions from (**a**)–(**b**) continuous-conduction mode (CCM) to (**c**)–(**d**) discontinuous-conduction mode (DCM) (i.e., as output current I_O decreases from 1 A to 32 mA)

Figure 10.17 shows i_L as a buck converter transitions from continuous-conduction mode (CCM) to DCM, the latter of which is characterized by a delay discontinuity (t_{DCM} in Fig. 10.17(d)) between triangular ripple events.

The onset of DCM is natural in buck converters that use a diode for low-side switch MN because the diode prevents i_L from reversing direction and flowing back to ground, which would have otherwise occurred in a bidirectional CMOS switch like MN. To be more specific, before the onset of DCM, as before, $i_{L(avg)}$ flows to v_O to supply I_O and i_L ripples about $i_{L(avg)}$ or I_O (Fig. 10.17(a)–(b)); but when I_O falls below half the peak-to-peak value of the ripple current ($0.5\Delta I_L$), MN allows i_L to fall below zero, drawing current to ground (instead of from ground), and therefore losing energy through the conducting Ohmic switches. A diode in place of MN blocks this negative current flow and consequently clamps i_L's negative peak to roughly 0 A (Fig. 10.17(c)–(d)), while slowly decreasing the positive peak and increasing delay discontinuity t_{DCM} with decreasing I_O. This type of bucking switching supply is said to be asynchronous because the diode used is essentially an asynchronous switch.

Following the same methodology as in CCM, dc conduction losses $P_{C.DC}$ remain valid in DCM because average current continues to flow through $R_{C.DC}$ to I_O. With switching bidirectional transistor MN, rippling current i_L momentarily falls below zero for low I_O values, keeping the circuit in CCM so ac conduction losses $P_{C.AC}$, as before, apply. In DCM, however, ac current $i_{L(ac)}$ is different and $P_{C.AC}$ is therefore no longer valid. As such, following the prescribed approach, subtracting I_O losses from all i_L losses equates to ac conduction losses in DCM, now an abstract but equally valid variable.

Because $i_{L(avg)}$ necessarily equates to I_O, peak inductor current $i_{L(peak)}$ depends on the duration of each triangular current (t_{tr}) event and I_O:

$$I_O = i_{L(avg)} = \frac{0.5 t_{tr}\, i_{L(peak)}}{T_{SW}} \text{ or } i_{L(peak)} = \frac{2I_O}{t_{tr}\, f_{SW}}. \qquad (10.16)$$

Peak current $i_{L(peak)}$, as in ΔI_L, increases with the rising rate of i_L (or $V_{IN} - V_O)/L$) and rising time $t_{L(rise)}$, except $t_{L(rise)}$ differs in DCM:

$$i_{L(peak)} = \frac{t_{L(rise)}(V_{IN} - v_O)}{L} \approx \frac{t_{L(rise)}(V_{IN} - V_O)}{L}$$

or

$$t_{L(rise)} \approx \frac{Li_{L(peak)}}{V_{IN} - V_O}. \tag{10.17}$$

Time $t_{L(rise)}$ relates to t_{tr} and high-side duty cycle d_{MP} in DCM as it does to switching period T_{SW} and d_{MP} in CCM because the energy V_{IN} delivers to L must equal the energy released to v_O in steady-state:

$$t_{L(rise)} = \left(\frac{t_{MP}}{T_{SW}}\right) t_{tr} = d_{MP} t_{tr}. \tag{10.18}$$

Combining the last three relations to solve for t_{tr} reveals t_{tr} is directly proportional to $I_O^{0.5}$ and $L^{0.5}$ and inversely proportional to $f_{SW}^{0.5}$:

$$t_{tr} = \frac{t_{L(rise)}}{d_{MP}} \approx \frac{i_{L(peak)} L}{d_{MP}(V_{IN} - V_O)}$$

$$= \frac{2I_O L}{t_{tr} f_{SW} d_{MP}(V_{IN} - V_O)} = \sqrt{\frac{2I_O L}{f_{SW} d_{MP}(V_{IN} - V_O)}}. \tag{10.19}$$

Ultimately, subtracting I_O^2- from i_L^2-conduction (i.e., resistor) losses amounts to subtracting I_O^2 from the square of RMS current $i_{L(rms)}$, which by using the previous relation, yields an equivalent ac conduction-loss current ($i_{L(ac).DCM}$) with one term proportional to $I_O^{1.5}/f_{SW}^{0.5}$ and the other, by definition, to $-I_O^2$:

$$i_{L(ac).DCM}^2 = i_{L(rms)}^2 - i_{L(avg)}^2 = \left(\frac{i_{L(peak)}^2}{3}\right)\left(\frac{t_{tr}}{T_{SW}}\right) - I_O^2 = \left(\frac{2I_O}{f_{SW} t_{tr} \sqrt{3}}\right)^2$$

$$(t_{tr} f_{SW}) - I_O^2 = \frac{4}{3}\sqrt{\frac{d_{MP}(V_{IN} - V_O) I_O^3}{2Lf_{SW}}} - I_O^2 \equiv A\sqrt{\frac{I_O^3}{f_{SW}}} - I_O^2. \tag{10.20}$$

In other words, DCM ac conduction losses $P_{C.AC.DCM}$ increases with $I_O^{1.5}$ and decreases with $f_{SW}^{0.5}$ and $L^{0.5}$:

$$P_{C.AC.DCM} = i_{L.AC.DCM}^2(R_{SW} + R'_{L.ESR} + R_{C.ESR})$$

$$= \left(\frac{4}{3}\sqrt{\frac{d_{MP}(V_{IN} - V_O) I_O^3}{2Lf_{SW}}} - I_O^2\right) R_{C.AC}. \tag{10.21}$$

10.3.2 Switching Losses

There are three basic types of switching power losses: (a) shoot-through and dead-time losses, (b) gate-drive, and (c) current–voltage overlap. All of these, as the category implies, are directly proportional to frequency because, as frequency increases, more power-consuming instances occur. As a result, switching frequency, to start, is a critical design parameter, and the higher the frequency, the higher the switching losses, generally.

Shoot through and dead time: Shoot-through current through conducting switches across V_{IN} and ground (in this case, MP and MN) incur considerable power losses because they momentarily short-circuit V_{IN} to ground. This loss is not specific to power conditioners, but is especially troublesome because the switches involved are typically large, thereby conducting considerably higher currents and doing so for relatively longer periods. The power lost by V_{IN} during a shoot-through event ($P_{SW.ST}$) through turn-on resistances R_{MP} and R_{MN} occurs twice per period and is proportional to the square of V_{IN} and the fraction of time they both conduct (ratio of shoot-through time t_{ST} and switching period T_{SW}):

$$P_{SW.ST} = 2\left(\frac{V_{IN}^2}{R_{MP} + R_{MN}}\right)\left(\frac{t_{ST}}{T_{SW}}\right) = \frac{2V_{IN}^2 t_{ST} f_{SW}}{R_{SW.ST}}, \tag{10.22}$$

where $R_{SW.ST}$ models the combined equivalent shoot-through resistance (i.e., $R_{SW.ST} = R_{MP} + R_{MN} = 2R_{OW}$).

Shoot-through losses in the power stage are often prohibitive and therefore eliminated by introducing dead time (t_{DT}) between low- and high-side switching instances. Inductor L, however, prevents i_L from dropping to zero during dead time, swinging phase voltage v_{PH} until one of the parasitic diodes MP and MN present forward biases and conducts i_L. The diodes therefore incur power losses during dead time that are approximately proportional to diode voltage v_D (or 0.6–0.8 V) and the value of i_L at those times, that latter of which refers to i_L's positive peak value in one transition (i.e., $i_{L+} = i_{L(avg)} + 0.5\Delta I_L$) and negative peak in the other (i.e., $i_{L-} = i_{L(avg)} - 0.5\Delta I_L$). The average switching dead-time losses ($P_{SW.DT}$) that result are therefore directly proportional to v_D, $i_{L(avg)}$ or I_O, and f_{SW} and independent of peak-to-peak current ΔI_L, since i_{L+} exceeds $i_{L(avg)}$ by the same amount i_{L-} falls below $i_{L(avg)}$:

$$P_{SW.DT} \approx v_D(i_{L+} + i_{L-})\left(\frac{t_{DT}}{T_{SW}}\right)$$
$$= 2v_D\, i_{L(avg)}\, t_{DT}\, f_{SW} = 2v_D\, I_O\, t_{DT}\, f_{SW}. \tag{10.23}$$

Designers often place off-chip Schottky diodes across power switches to shunt current away from the parasitic diode, which would otherwise induce current flow through the substrate of the IC. In doing so, they also reduce the voltage drops across the switches during dead time and therefore its related dead-time power losses. Note the circuits driving the switches must also conduct considerable current to charge and discharge the parasitic capacitors the power switches present so they are also large and incur additional shoot-through losses. Dead time is often included in the stage driving the power switches, but it is impractical and costly in terms of speed and silicon real estate to do so for all remaining driver stages.

Gate drive: Gate-drive losses refer to the energy used to charge and discharge the parasitic capacitors (**e.g.,** i_C **in Fig. 10.15**) the power switches (e.g., MP and MN) introduce to their respective driving pulse-width modulated signals. This gate-drive power (P_{GD}) is directly proportional to parasitic switch capacitance C_G, the square of the voltage transitioned ΔV_C, and switching frequency f_{SW}:

$$P_{GD} = v_C i_C \left(\frac{\Delta t_C}{T_{SW}}\right) = \Delta V_C \left(C_G \frac{\Delta V_C}{\Delta t_C}\right)\left(\frac{\Delta t_C}{T_{SW}}\right) = \Delta V_C^2 C_G f_{SW}, \qquad (10.24)$$

where Δt_C is the time required to charge C_G ΔV_C Volts. As it turns out, most parasitic capacitors MP and MN in the buck converter shown in Fig. 10.15 transition to and from V_{IN} and ground so ΔV_C reduces to V_{IN}. This is not normally true, though, for MP's parasitic capacitor to v_{PH} because, as its driving gate signal transitions from V_{IN} to ground, v_{PH} transitions from ground to V_{IN}, forcing a total capacitor–voltage variation of twice V_{IN} ($\Delta V_C = V_{IN} - (-V_{IN}) = 2V_{IN}$). Nevertheless, the difference in voltage variation is a constant and can therefore be absorbed, for the purposes of this discussion, into total equivalent gate-drive capacitance $C_{G.EQ}$. As a result, switching gate-drive losses $P_{SW.GD}$ increase with the square of V_{IN}, $C_{G.EQ}$, and f_{SW}:

$$P_{SW.GD} = V_{IN}^2 C_{G.EQ} f_{SW}, \qquad (10.25)$$

where $C_{G.EQ}$ increases with the number of switches used, their respective dimensions, and increasing gate-drive voltages.

IV overlap: As the voltage across a switching transistor transitions, current flows through the device, dissipating in the process what is typically called IV-overlap power. Consider top-side switch transistor MP in Fig. 10.15, for instance, just before it conducts. First, immediately after low-side switch MN stops conducting, dead time t_{DT} prevents MP from turning on, which means L, because current must flow continuously, swings phase voltage v_{PH} below ground until MN's parasitic diode conducts i_L, which at this point is at its minimum value of $i_{L(avg)} - 0.5\Delta I_L$. Concurrently, the voltage across MP is approximately V_{IN} and its current zero, as in the initial conditions shown in Fig. 10.18.

After dead time t_{DT}, as MP's gate drive increases, MP's current i_{MP} rises, but v_{PH} does not rise until i_{MP} exceeds i_L by enough to charge the parasitic capacitance present at that node. As such, MP's voltage v_{MP} does not fall until i_{MP} reaches $i_{L(avg)} - 0.5\Delta I_L$, after which v_{PH} rises to V_{IN} and v_{MP}, therefore, falls near zero. Inductor L subsequently energizes for the duration MP is on until its current reaches its peak value of $i_{L(avg)} + 0.5\Delta I_L$, at which point the reverse process occurs: as MP's gate drive decreases and I_{MP} drops slightly, L quickly swings v_{PH} low until MN's parasitic diode conducts the difference, after which I_{MP} transitions to zero.

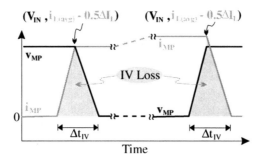

Fig. 10.18 IV-overlap power losses in high-side switching transistor MP

Low-side switch MN undergoes the same sequence of events, except the voltage across the device, instead of swinging to and from V_{IN} and ground, swings to and from a diode below ground and ground, as dictated by L during dead time t_{DT}. As a result, IV-overlap losses normally apply to one of the two switches in a buck converter, not both. In any case, the switching IV-overlap power lost ($P_{SW.IV}$) is the energy associated with the overlap of i_{MP} and v_{MP} and is equal to the two areas highlighted in Fig. 10.18, when normalized to switching period T_{SW}, the sum of which is directly proportional to the total voltage variation ΔV_{MP} or in this case peak voltage V_{IN}, average peak current $i_{L(avg)}$ or I_O, average overlap time $\Delta t_{IV(avg)}$, and switching frequency f_{SW}:

$$P_{SW.IV} = \frac{0.5 \Delta V_{mp} i_{L(avg)}(t_{I.R} + t_{V.F})}{T_{SW}} \approx V_{IN} I_O \Delta t_{IV(avg)} f_{SW}, \qquad (10.26)$$

where $t_{I.R}$ and $t_{V.F}$ correspond to the rise and fall times of i_{MP} and v_{MP}, respectively.

10.3.3 Quiescent Losses

Feedback control, protection, and other functions normally attached to what is termed the controller, while not directly supplying output current I_O, which the power stage does, require quiescent current to operate, as generally depicted by ground current i_Q in Fig. 10.15. Input voltage V_{IN} normally supplies this current and therefore incurs a quiescent power loss (P_Q) proportional to V_{IN} and the average value of i_Q:

$$P_Q = V_{IN} i_{Q(avg)} = V_{IN} I_Q, \qquad (10.27)$$

where I_Q is the average value of i_Q. Note V_{IN} also supplies gate-drive energy to the switches via the controller.

10.3.4 Losses Across Load

One of the most difficult challenges to conquer in microscale harvester systems is to produce a net power gain, that is to say, a net profit because the energy and power costs of channeling and conditioning scavenged power can easily surpass the limited mechanical revenue produced by a harvesting device. The power levels under consideration, to provide some perspective, given volumes under $5 \times 5 \times 5 \, \text{mm}^3$, are on the order of micro-Watts, which means the losses associated with the harvester circuit must fall below those levels to achieve a net gain. Fortunately, dc and ac conduction in discontinuous-conduction mode (DCM) and IV-overlap and dead-time switching losses decrease with the square and linearly with respect to output current I_O, respectively, as discussed earlier and shown in Fig. 10.19(a). The main problem at these micro-Watt power levels is reducing I_O-independent losses such as quiescent and shoot-through and gate-drive switching losses, which do not normally scale with I_O and consequently lead to power efficiencies on the order of 20–40% or less (Fig. 10.19(b)). If a harvesting electromechanical device, for example, generates $50 \mu \text{W}$, a magnetic-based step-down conditioner, given the possible but somewhat optimistic conditions defined in Fig. 10.19, produce $15 \mu \text{W}$ of conditioned output power.

The conditions identified in Fig. 10.19, in actuality, apply to a theoretical 2–1 V buck converter with an off-chip $2 \times 2 \times 1 \, \text{mm}^3$ $25 \mu \text{H}$ inductor and a fast sub-threshold-operated 10-MHz CMOS controller, all of which describes a relatively simple case with hopeful attributes. Many harvesting conditioners, to start, are more complex than a buck converter, requiring more switches and drivers and therefore

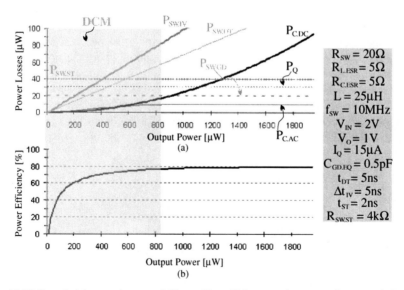

Fig. 10.19 Sample (a) power losses and (b) resulting efficiency performance of a magnetic-based buck dc–dc converter at low power levels

incurring more I_O-independent losses. Scavenging kinetic energy from vibrations electromagnetically, for instance, demands an ac–dc rectifier or synchronizer with a series boost converter, which requires five to six more power switches and the same number of relevant drivers. The same example just cited would also present a larger input–output voltage spread, considering electromagnetic voltages from microharvesters are on the order of micro- and milli-Volts and the Li–Ion batteries they are meant to charge (Fig. 10.5) are at 2.7–4.2 V. Additionally, replacing the off-chip device with a 20–100 nH on-chip inductor increases ac conduction losses by an order of magnitude, and increasing the number of off-chip inductors violates the integration objectives of these microsystems. Lastly, designing a 15μA, 10 MHz CMOS controller is optimistic, given present-day technologies, the driving requirements of the switches in the power stage, and the functionality requirements of the system.

Research is currently underway to achieve a net power gain, even if at low power efficiencies. One way of increasing efficiency is to decrease the switching frequency with decreasing I_O, which trades off ac conduction losses (because of larger ripple inductor currents) for switching losses. Another approach is designing nano-Watt converters and decreasing the switching frequency to accommodate the speed of the circuit, trading off ac conduction losses for quiescent losses. There is also the issue of operating in discontinuous-conduction mode (DCM) in more complicated topologies, and controlling its onset without compromising the electrical stability of the system. Low-power and high-performance control-loop architectures are also under scrutiny in research circles.

From a system viewpoint, as described in this text, the harvester-charger system presented in Fig. 10.5 is generally under investigation, but so are harvester-supply systems, the latter of which attempt to condition the power directly from the harvester to the output, incurring additional losses in the process. Charge-pump approaches have been proposed but overcoming the need for multiple off-chip capacitors to suppress power losses and maintaining performance remains a challenge. Ultimately, the power-management system must supply the energy and power needs of wireless microsensors, which demand milli-Watts during transmission events and micro-Watts when idling. The operational needs, in fact, are often sporadic and diverse, requiring more functionality in the controller in the form of power modes, all the while attempting to increase idling conditions so as to decrease energy use.

10.4 Sample System: Electrostatic Harvester

The objective of this section is to illustrate by way of an example how to apply some of the teachings derived from the previous discussions. Of the energy harvesters discussed in Section II, drawing power from the kinetic energy in vibrations is probably the most practical alternative because vibrations are abundant, and in many cases, consistent and systematic. In considering microscale integration, though, thermoelectric devices exhibit small thermal gradients and therefore, much like their electromagnetic counterparts, produce low-voltage levels, which are challenging to process. Piezoelectric devices are difficult to integrate on chip, and they also supply power in the form of an ac voltage signal, requiring a synchronizer or an ac–dc converter and incurring the additional losses associated with the same. Variable parallel-plate micro-electromechanical systems (MEMS) capacitors are on-chip electrostatic harvesters that produce charge when constrained to a particular voltage. The integration qualities and voltage flexibility of these latter devices echo the needs of a CMOS microsystem, albeit at the cost of synchronizing hardware. The foregoing example, because of its integration features, is on an electrostatic harvesting system.

Because the power demanded by a wireless microsensor load exceeds what a microharvester can generate, it is perhaps more practical to target a harvester-charger scheme, like the one shown in Fig. 10.5, as opposed to a harvester supply. In conditioning the incoming scavenged power, the trickle-charger (i.e., harvesting) circuit must not dissipate much energy so the least power-consuming converter topology is best. Although linear regulators are faster and less noisy and charge pumps moderately low power and inductor-free, magnetic-based converters are fundamentally more efficient and therefore adopted in the example offered.

10.4.1 Harvesting Current

Changing the distance between the parallel plates of a capacitor, when constraining the voltage across its terminals, induces charge flow. If variable capacitor C_{VAR} decreases from 101 to 1 pF (i.e., $\partial C/\partial t$ is negative) in 1ms, for instance, when constrained to a capacitor voltage of 3.6 V, charges flow out (i.e., $\partial Q_C/\partial t$ is negative) of the capacitor at a rate of 360nA:

$$i_H = \frac{\partial Q_C}{\partial t} = \frac{\partial(C_{VAR}v_C)}{\partial t} = \left(\frac{\partial C_{VAR}}{\partial t}\right)V_C, \qquad (10.28)$$

where i_H is the harvesting current and V_C is the dc value of capacitor voltage v_C; similarly, charges flow into the capacitor (i.e., $\partial Q_C/\partial t$ is positive) when C_{VAR} increases (i.e., $\partial C/\partial t$ is positive). When considering a harvester-charger system, however, only current flowing out of the capacitor (when capacitor decreases) charges a battery.

10.4.2 Trickle Charging Scheme

Since only decreasing capacitances harvest current, the circuit must detect and synchronize to vibrations so as to harvest only when varactor C_{VAR} decreases. Detecting when C_{VAR} reaches its maximum value of $C_{VAR(max)}$ (Fig. 10.20(a)) therefore marks the onset of the harvesting sequence, the first phase of which is to charge C_{VAR}

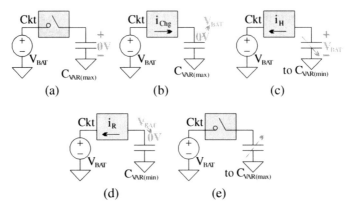

Fig. 10.20 (a) Onset of the voltage-constrained electrostatic-harvesting sequence (at $C_{VAR(max)}$) and its subsequent (b) pre-charge, (c) harvest, (d) recover, and (e) reset phases

with charge current i_{Chg} to battery voltage V_{BAT} (Fig. 10.20(b)). Pre-charging to this target voltage paves the way for harvesting current i_H to flow back to V_{BAT} when C_{VAR} decreases (Fig. 10.20(c)) because C_{VAR} remains clamped to V_{BAT}, that is, voltage-constrained to V_{BAT}.

Once C_{VAR} reaches its minimum value of $C_{VAR(min)}$, the harvesting phase ends and the battery recovers the energy left in C_{VAR} by discharging it from voltage V_{BAT} to ground with recover current i_R (Fig. 10.20(d). Once the circuit completely drains C_{VAR} of charge, the reset cycle begins (Fig. 10.20(e)), as C_{VAR} again reaches $C_{VAR(max)}$. Note that, because practical vibrations are most abundant at 50–300 Hz and microelectronic circuit delays are on the order of micro- to milliseconds, the pre-charge and recovery phases, which are under the control of the harvester circuit, occur substantially faster than the harvesting and reset phases, whose frequency is set by the vibrations.

The charging phase (with i_{Chg}) actually constitutes an investment (drain) in energy from V_{BAT}, so the objective for the trickle charger is to harvest sufficient energy to exceed this amount, ultimately increasing the battery's state of charge (SOC) in steady state. Assuming the trickle charger incurs no losses in the charging process, the energy lost by the battery (ΔE_{Invest}) to charge C_{VAR} to V_{BAT} is

$$\Delta E_{Invest} = 0.5 C_{VAR(max)} \Delta V_c^2 = 0.5 C_{VAR(max)} V_{BAT}^2. \tag{10.29}$$

The energy harvested ($\Delta E_{Harvest}$) is the difference in energy in V_{BAT} as a result of C_{VAR}'s harvesting current i_C as C_{VAR} transitions from $C_{VAR(max)}$ to $C_{VAR(min)}$ or

$$\Delta E_{Harvest} = \int V_{BAT} i_C \, dt = V_{BAT}^2 \int \left(\frac{\partial C_{VAR}}{\partial t} \right) dt = V_{BAT}^2 \Delta C_{VAR}, \tag{10.30}$$

where ΔC_{VAR} is the total variation in capacitance. At this point, however, C_{VAR} is at V_{BAT} and recovering the energy associated with that charge adds energy back to the battery ($\Delta E_{\text{Recover}}$) equivalent to

$$\Delta E_{\text{Recover}} = 0.5 C_{\text{VAR(min)}} \Delta V_c^2 = 0.5 C_{\text{VAR(min)}} V_{\text{BAT}}^2, \qquad (10.31)$$

which means the net energy gain of the system is

$$\begin{aligned}\Delta E_{\text{Gain}} &= \Delta E_{\text{Harvest}} - \Delta E_{\text{Invest}} + \Delta E_{\text{Recover}} - E_{\text{Losses}} \\ &= 0.5 \Delta C_{\text{VAR}} V_{\text{BAT}}^2 - E_{\text{Losses}}, \qquad (10.32)\end{aligned}$$

where E_{Losses} refers to the conduction, switching, and quiescent power losses associated with the circuit and all energy transfers, from charging C_{VAR} and harvesting current back into V_{BAT} to recover the energy left in C_{VAR}.

In this scheme, like a car, the battery must have an initial charge to start or ignite the system (i.e., invest energy). Starting the system with a fully discharged battery is difficult with a single source and research is underway to discover new ways of doing so. Hybrid solutions may address this by using another harvester whose output may more easily charge C_{VAR} during a start-up sequence, like a piezoelectric device in combination with an ac–dc converter. Besides acting as a starter, the second harvester can also harvest energy during regular operation, except its output must exceed the power required to harvest and mix with the primary harvester for the effort to merit consideration.

10.4.3 Harvesting Microelectronic Circuit

All pre-charge, harvest, and recovery phases cost energy and power. To reduce the investment in pre-charging variable capacitor C_{VAR}, a magnetic-based pre-charger is chosen, as shown in the switch-level circuit of Fig. 10.21. The basic idea is to first energize inductor L with battery V_{BAT} by connecting it between V_{BAT} and ground via energizing switches S_{E1} and S_{E2} and then de-energize and channel its energy to C_{VAR} by connecting it from ground to C_{VAR} via de-energizing switches S_{D1} and S_{D2}. In implementing the harvesting function, it is possible to invert the energy flow in the pre-charger by reversing the energizing and de-energizing source–load

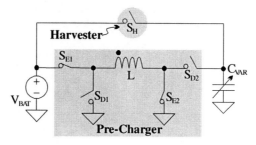

Fig. 10.21 Switch-level harvesting circuit

combination (i.e., energize with C_{VAR} and de-energize to V_{BAT}), except C_{VAR} must remain charged to V_{BAT} to achieve voltage-constrained conditions, which the charger cannot do. As a result, harvesting switch S_H short-circuits V_{BAT} to C_{VAR} while allowing harvesting current i_H to flow to V_{BAT}; little power is lost across S_H because the voltage across it, in theory, nears zero.

Harvesting switch S_H would have also worked for the pre-charge function but the power losses would have been prohibitively high (on the order of the harvested power) because the initial voltage difference between the fully discharged C_{VAR} and V_{BAT} is substantial (e.g., 3.6 V) and the energy loss associated with charging the capacitor is proportional to C_{VAR} at its maximum value and the square of the aforementioned voltage difference ($0.5\,C_{VAR(max)}\Delta V_C^2$ or $0.5\,C_{VAR(max)}V_{BAT}^2$). Switch S_H does not work for the recovery phase either, when C_{VAR} is at $C_{VAR(min)}$, because S_H cannot discharge C_{VAR} to ground while connected to V_{BAT}. It is possible, however, for the pre-charger to reverse its operation and energize L with C_{VAR} and de-energize it to V_{BAT}. The only problem is the energy left in C_{VAR} is relatively low because its capacitance is at its minimum value ($0.5C_{VAR(min)}V_{BAT}^2$) and the switching and quiescent energy losses of the charger can easily exhaust it, negating the benefits of a recovery phase.

Figure 10.22 shows the practical implementation of the microelectronic harvesting circuit, including all relevant parasitic devices. Varactor C_{VAR} presents parasitic capacitance C_{PAR} to ground, slightly increasing the minimum capacitance from $C_{VAR(min)}$ (e.g., 1 pF) to the sum of $C_{VAR(min)}$ and C_{PAR} (e.g., 1.5 pF). Contacting C_{VAR} also introduces equivalent series resistance (ESR) $R_{C.ESR}$ in the power-conduction path, dissipating additional power during both the pre-charge and the harvesting phases as a result. Referring to Fig. 10.4, because one of the C_{VAR}'s terminals is at ground, $C_{PAR.A}$ and $C_{PAR.B}$ roughly reduce to C_{PAR}, $R_{C.ESR}$ includes the effects of $R_{S.A}$ and $R_{S.B}$, and $C_{PAR.C}$ incurs inconsequential effects. Battery

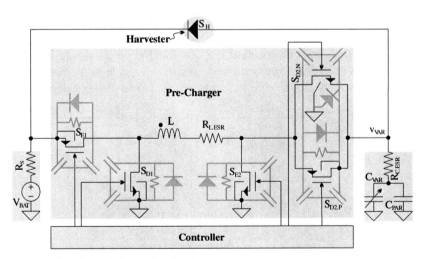

Fig. 10.22 Microelectronic harvesting circuit

V_{BAT} also offers series source resistance R_S, again incurring losses during both said phases and reducing the overall efficiency of the circuit.

N- and p-type transistors implement energizing and de-energizing switches S_{E1}–S_{E2} and S_{D1}–S_{D2} in the charger. They all introduce power-consuming series Ohmic resistances in the conduction path equivalent to their switch-on resistances (R_{ON}), which are in turn a function of their size (i.e., width-to-length ratio) and the peak voltage their controlling gates receive from the controller (i.e., gate drive); they also present parasitic diodes some of which incur dead-time losses. Switches S_{E1}, S_{E2}, and S_{D1} are all single-transistor devices because their gate-driving voltages are with respect to V_{BAT} and ground and therefore sufficiently high to produce reasonably low turn-on resistances. The gate-drive voltage of S_{D2}, on the other hand, is at first between V_{BAT} (as supplied from the controller) and ground (initial voltage across C_{VAR} or v_{VAR}), which is enough to drive NMOS transistor $S_{D2.N}$, but then between V_{BAT} and a rising v_{VAR} voltage whose ultimate target value is V_{BAT}, which is not enough to drive $S_{D2.N}$. As a result, PMOS device $S_{D2.P}$ is added in parallel to help conduct and charge C_{VAR} when v_{VAR} nears V_{BAT}; note $S_{D2.P}$ has similar gate-drive problems when v_{VAR} is near ground, which is why a complementary switch (i.e., transmission gate) is used.

A PMOS transistor or the asynchronous diode shown in Fig. 10.22 can implement harvesting switch S_H. The diode allows current to flow back to V_{BAT} during the charging phase, though, if overcharged, but this is acceptable because that energy would have otherwise constituted a larger than needed investment on the part of V_{BAT} and channeling it back to V_{BAT} restores some of it (the conducting diode loses a power equivalent to the product of the induced overcharging current and diode voltage V_D). The parasitic diode, PMOS transistor introduces, by the way, has an equivalent effect.

Using a PMOS device, however, incurs an additional power loss because of the parasitic pnp bipolar-junction transistor (BJT) it presents when fabricated on a standard n-well, p-substrate technology. To ensure the pn-junction diode is not in the direction where it forward-biases when v_{VAR} is below V_{BAT}, which would otherwise bypass the pre-charger during the pre-charge phase and charge C_{VAR} with a lossy diode, the back gate (n-well) of the PMOS must connect to V_{BAT}, as shown in Fig. 10.23(a), where the emitter-base terminals of the parasitic BJT shown in Fig. 10.23(b) and modeled in Fig. 10.23(a) includes said pn junction. As a result, in events when L de-energizes more energy than required to charge C_{VAR} to V_{BAT}, capacitor voltage v_{VAR} may exceed V_{BAT} by diode voltage V_D before the onset of the harvesting phase, allowing the parasitic pn junction to conduct what amounts to base current i_B. The problem with this scenario is allowing this pn junction to conduct current effectively forward-biases the parasitic vertical pnp BJT, conducting current gain β times higher current to ground than to V_{BAT}, where β may be 40–100, losing considerable power (i.e., $\beta i_B(V_{BAT} + V_D)$) and energy across the BJT instead of recovering it back to V_{BAT}.

Energizing L with just enough energy to charge C_{VAR} to V_{BAT} and ensuring the charging current and $R_{C.ESR}$ do not produce a voltage higher than V_D altogether circumvents the substrate-BJT scenario just described. In practice, though, C_{VAR}'s

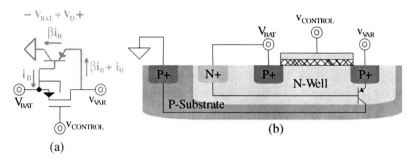

Fig. 10.23 (a) Equivalent schematic and (b) physical-profile view of a p-channel metal-oxide-semiconductor (MOS) field-effect transistor (FET) in an n-well, p-type substrate

initial voltage may not always be at ground, given its dependence to possibly inconsistent vibrations, offsetting its energy-investment requirement from what the system supplies through L; and the voltage across $R_{C.ESR}$ compounds this effect, although not necessarily by much. Moreover, the tolerance and temperature-drift variations of the circuit also vary the total energy transferred during the pre-charge phase, and regulating the energy with a loop requires more work (quiescent current), which may defeat the purpose of saving energy.

A solution to the PMOS dilemma is to realize the switch with a more complex structure involving at least another series PMOS device with its n-well connected in the opposite direction. The motivation to pursue this path is the diode incurs losses in channeling harvesting current i_H back to V_{BAT} equivalent to the product of i_H and V_D and the shunting effect of a parallel PMOS transistor decreases this voltage and consequently the power it incurs. However, justifying the additional complexity and gate-drive switching losses the more complicated structure introduces in practice over the harvesting losses associated with the simple diode shown in Fig. 10.22 is not apparent, which is why the foregoing example, for the time being, adopts the diode.

Figure 10.24 shows the relevant single- and multiple-cycle waveforms generated by the electrostatic voltage-constrained harvesting microelectronic circuit shown in Fig. 10.22 when subjected to a manually varied 60–250 pF trim capacitor. The pre-charger circuit is in discontinuous-conduction mode (DCM) because it charges C_{VAR} in one sequence for microseconds and remains off for the remainder of the switching period, which is over 700 ms. The long switching period implies switching gate-drive, IV-overlap, and shoot-through losses are relatively low. The harvesting phase lasted over 250 ms, collecting roughly 10 nJ per cycle with 10–30 nA by investing 1.7 nJ per cycle and harnessing over 60 nJ in six cycles by investing 10.2 nJ. The experimental results shown, however, do not include the power losses of the controller because part of it was off chip and not ideally suited for the experiment, especially with respect to steady-state quiescent power.

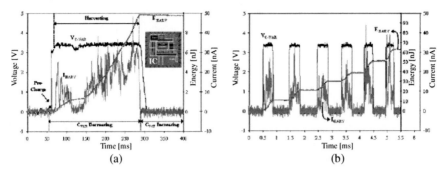

Fig. 10.24 Sample (**a**) single- and (**b**) multiple-cycle response of an electrostatic voltage-constrained CMOS harvesting circuit when sourced with a manual 60–250 pF trim capacitor

10.5 Summary

Irrespective of the source, electromechanical harvesters generate characteristically low-power levels. While linear regulators are relatively easier to implement and charge pumps are inductor free, magnetic-based converters are fundamentally more efficient, if switched at low frequencies and operated in discontinuous-conduction mode (DCM). On-chip inductors, however, are low, on the order of 50–100 nH, and present challenges in the form of instantaneous peak power. Fortunately, microscale off-chip inductors conform to $2 \times 2 \times 1$ mm^3, and possibly smaller dimensions in the future, making single-chip system-in-package (SiP) solutions viable, only if no more than one inductor is used, as another off-chip capacitor is already necessary to suppress the ac variations load dumps produce.

Magnetic-based converters may be fundamentally more efficient but not completely lossless in practice. Although dc- and ac-conduction and dead-time and IV-overlap losses are important, they decrease with output current (if allowed to operate in DCM) so their impact on efficiency is often less pronounced than quiescent, gate-drive, and shoot-through losses; and decreasing these latter losses amounts to operating at low switching frequencies and in sub-threshold operation. Practical microelectronic circuits, however, impose additional challenges in the form of deficient gate drive for the switches, power losses through parasitic current paths, start-up requirements, and functionality in synchronizing to vibrations, among others.

Research continues, nonetheless, exploring and improving both the mechanical and the electrical performance of the transducing agent and accompanying harvesting microelectronic integrated circuit (IC). Ultimately, the motivating force behind microscale harvesters rests on the potential ramifications of realizing self-powered wireless microsensors and microimplantable technologies, which can range from optimizing and monitoring the use of energy in rooms, factories, and cities to reanimating limbs and managing debilitating biological functions and life-threatening deficiencies in vivo, in situ, and in tempo.

Part III
Thermoelectrics

Chapter 11
Thermoelectric Energy Harvesting

G. Jeffrey Snyder

Abstract Temperature gradients and heat flow are omnipresent in natural and human-made settings and offer the opportunity to harvest energy from the environment. Thermoelectric energy harvesting (or energy scavenging) may one day eliminate the need for replacing batteries in applications such as remote sensor networks or mobile devices. Particularly, attractive is the ability to generate electricity from body heat that could power medical devices or implants, personal wireless networks or other consumer devices. This chapter focuses on the design principles for thermoelectric generators in energy harvesting applications, and the various thermoelectric generators available or in development. Such design principles provide good estimates of the power that could be produced and the size and complexity of the thermoelectric generator that would be required.

11.1 Harvesting Heat

Environments that naturally contain temperature gradients and heat flow have the potential to generate electrical power using thermal-to-electric energy conversion. The temperature difference provides the potential for efficient energy conversion, while heat flow provides the power. Even with large heat flow, however, the extractable power is typically low due to low Carnot and material efficiencies. In addition, limited heat availability will also limit the power produced. Nevertheless, for systems with exceptionally low-power requirements, such as remote wireless sensors, thermoelectric energy harvesting has shown to be a viable technology and promise to become more prevalent as the power requirements for such applications drop (Paradiso and Starner, 2005).

A good example of thermoelectric energy harvesting is the thermoelectric wristwatch that converts body heat into the electrical power that drives the watch. At least two models have been built, one by Seiko and another by Citizen. The Seiko watch

G.J. Snyder (✉)

Materials Science, California Institute of Technology, 1200 East California Boulevard, Pasadena, CA 91125, USA

e-mail: jsnyder@caltech.edu

S. Priya, D.J. Inman (eds.), *Energy Harvesting Technologies,*
DOI 10.1007/978-0-387-76464-1_11 © Springer Science+Business Media, LLC 2009

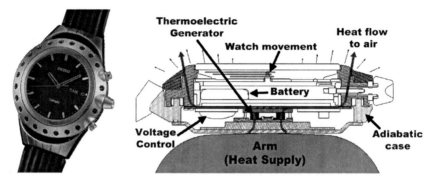

Fig. 11.1 Seiko Thermic, a wristwatch powered by body heat using a thermoelectric energy harvester. (*left*) the watch, (*right*) cross-sectional diagram. Copyright by Seiko Instruments Inc., reprinted with permission

(Kishi et al. 1999) under normal operation produces $22\,\mu\text{W}$ of electrical power (Fig. 11.1). With only a $1.5\,\text{K}$ temperature drop across the intricately machined thermoelectric modules, the open-circuit voltage is $300\,\text{mV}$, and thermal-to-electric efficiency is about 0.1%.

11.2 Thermoelectric Generators

Thermoelectric generators are solid-state devices with no moving parts. They are silent, reliable, and scalable, making them ideal for small, distributed power generation, and energy harvesting.

The thermoelectric effects arise because charge carriers in metals and semiconductors are free to move much like gas molecules while carrying charge as well as heat.(Snyder and Toberer, 2008) When a temperature gradient is applied to a material, the mobile charge carriers at the hot end preferentially diffuse to the cold end. The buildup of charge carriers results in a net charge (negative for electrons, e^- and positive for holes, h^+) at the cold end, producing an electrostatic potential (voltage). An equilibrium is thus reached between the chemical potential for diffusion and the electrostatic repulsion due to the build up of charge. This property, known as the Seebeck effect, is the basis of thermoelectric power generation.

Thermoelectric devices contain many thermoelectric couples (Fig. 11.2a) consisting of n-type (containing free electrons) and p-type (containing free holes) thermoelectric elements wired electrically in series and thermally in parallel (Fig. 11.2b). The best thermoelectric materials are heavily doped semiconductors.

A thermoelectric generator utilizes heat flow across a temperature gradient to power an electric load through the external circuit. The temperature difference provides the voltage ($V = \alpha\Delta T$) from the Seebeck effect (Seebeck coefficient α), while the heat flow drives the electrical current, which therefore determines the power output. The rejected heat must be removed through a heat sink. The thermoelectric

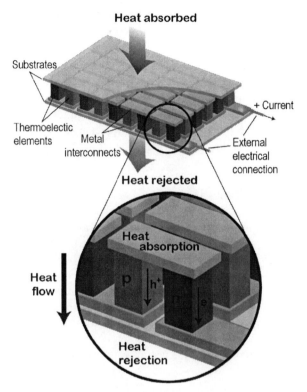

Fig. 11.2 Schematic of a thermoelectric generator. Many thermoelectric couples (inset *bottom*) of n- and p-type thermoelectric materials are connected electrically in series and thermally in parallel to make a thermoelectric module (*top*) or thermopile. The height of the thermoelectric elements and the area of the substrates are used to determine the thermal resistance of the module (see Eq. 11.14). Copyright Nature Publishing Group (Snyder and Toberer, 2008), reprinted with permission

figure of merit of the materials (zT) depends on the Seebeck coefficient (α), absolute temperature (T), electrical resistivity (ρ), and thermal conductivity (κ) of the material:

$$zT = \frac{\alpha^2 T}{\rho \kappa} \qquad (11.1)$$

The maximum efficiency of a thermoelectric device is determined by its figure of merit (ZT), which is largely an average of the component materials' zT values.

For the past 40 years, solid-state thermoelectric generators have reliably provided power in remote terrestrial and extraterrestrial locations, most notably on deep space probes such as *Voyager*. One key advantage of thermoelectrics is their scalability to small sizes, making them the most appropriate thermal-to-electric technology for energy harvesting.

11.3 Design of a Thermoelectric Energy Harvester

Energy harvesting systems, because they use natural chemical-potential gradients, are best designed for their particular environment and are typically not amenable to a one-size-fits-all solution. This is in stark contrast to batteries where the chemical-potential gradient is known, stable, and well regulated as it is an integral part of the power source. Thus, a battery can supply similar voltage and power in many different environments, but the power output of an energy harvester could vary by orders of magnitude depending on its location and use.

11.4 General Considerations

Viable energy harvesting systems need to outperform a battery solution in terms of energy density, power density, and/or cost. Typically, the niche for energy harvesting is in long-lived applications where energy density is critical and routine maintenance (replacing batteries) is not an option. A likely scenario for use of an energy harvester is as a means of recharging a battery. In this case, the battery supplies high power (mW or W) during a short period of time (e.g., sensing and communications for few seconds or milliseconds), while the majority of the time the energy harvester trickle charges the battery (μW).

While heat is a form of energy, the useful work content of heat is limited by the Carnot factor,

$$\eta_{\text{Carnot}} = \frac{\Delta T}{T_h} \qquad (11.2)$$

where $\Delta T = T_h - T_c$ is the temperature difference across the thermoelectric. This puts thermoelectric energy harvesting at a distinct disadvantage when compared with the other forms of energy harvesting that are not Carnot limited. Visible light, for example, has high useful work content that enables photovoltaics to outperform thermoelectrics when sunlight or bright lighting is available. Photovoltaics can produce $100\,\text{mW/cm}^2$ in direct sunlight and about $100\,\mu\text{W/cm}^2$ in a typically illuminated office—significantly more than the wristwatch in Fig. 11.1.

11.5 Thermoelectric Efficiency

A thermoelectric generator converts heat (Q) into electrical power (P) with efficiency η.

$$P = \eta Q \qquad (11.3)$$

The maximum efficiency of a thermoelectric converter depends heavily on the temperature difference ΔT_{TE} across the device. This is because the thermoelectric generator, like all heat engines, cannot have an efficiency greater than that of a Carnot cycle (Eq. 11.2).

$$\eta = \Delta T_{TE} \frac{\eta_r}{T_h} \tag{11.4}$$

Here η_r is the reduced efficiency, the efficiency relative to the Carnot efficiency.

While the exact thermoelectric materials' efficiency is complex (Snyder, 2006), the constant properties approximation (Seebeck coefficient, electrical conductivity, and thermal conductivity independent of temperature) leads to a simple expression for efficiency:

$$\eta = \frac{\Delta T}{T_h} \cdot \frac{\sqrt{1 + ZT} - 1}{\sqrt{1 + ZT} + T_c/T_h} \tag{11.5}$$

Here, ZT is the thermoelectric *device figure of merit*. Fig. 11.3 shows this is quite a good approximation for a typical commercial thermoelectric device made from bismuth telluride alloys. The efficiency of an actual thermoelectric device should be about 90% of this value due to losses from electrical interconnects, thermal and electrical contact resistances, and other thermal losses.

The efficiency of a thermoelectric generator increases nearly linearly with temperature difference (Fig. 11.3), indicating η_r/T_h (Eq. 4) is fairly constant. In energy-harvesting applications, where the temperature difference ΔT is small the efficiency is, to a good approximation, directly proportional to the ΔT across the thermoelectric. For good bismuth telluride devices, the efficiency is approximately 0.04% for each 1 K of ΔT.

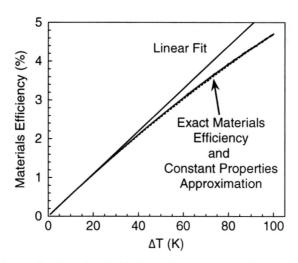

Fig. 11.3 Efficiency of a bismuth telluride-based thermoelectric module (cold side at 300 K, assuming no additional losses)

$$\eta = \eta_1 \cdot \Delta T \tag{11.6}$$

$$\eta_1 \approx 0.05\%/K$$

11.6 Matched Thermal Resistance

If the temperature across the thermoelectric, ΔT_{TE}, could truly be kept constant, the power output, P, could be made arbitrarily large by engineering the harvester to conduct more heat. Unfortunately, most real heat sources cannot supply arbitrarily large heat fluxes without loss of temperature difference (the heat exchange to the thermoelectric has non-zero thermal impedance).

Figure 11.4 shows a simple thermal model of a thermoelectric energy harvester. ΔT_{supply} is the temperature difference between the heat source and the cold bath, Θ_{TE} is the thermal resistance of the thermoelectric, and Θ_{Hx} is the combined thermal resistance (impedance) of the hot and cold side heat exchangers in series with the thermoelectric. Heat exchangers used as sinks and sources are often well characterized by a thermal resistance Θ_{Hx}, according to:

$$\Delta T_{Hx} = Q\Theta_{Hx}\Delta T_{Hx} = Q\Theta_{Hx} \tag{11.7}$$

Note that this implies that the heat supplied to the thermoelectric is proportional to the temperature drop across the heat exchanger.

If all the available ΔT is across the thermoelectric with no ΔT across the heat exchangers ($\Delta T_{Hx} = 0$), the heat exchangers supply no heat ($Q = \Delta T_{Hx}/\Theta_{Hx}$) and therefore no power is produced. Since the ΔT across the heat exchanger increases, the heat flux supplied increases, increasing the power output, but the temperature drop across the thermoelectric generator will decrease ($\Delta T_{TE} = \Delta T_{supply} - \Delta T_{Hx}$), reducing efficiency. Even if the heat source has low-thermal impedance, the same consideration applies to the cold-side heat sink. Invariably, some of the ΔT will be needed to transport heat through the heat exchangers to the heat source and/or sink which will reduce ΔT_{TE} to less than ΔT_{supply}.

For maximum power, it can be shown (see below) that the thermal resistance of the heat source and sink should be designed to match (Stevens, 2001). If either the heat source or heat sink has a large thermal resistance, the thermoelectric must

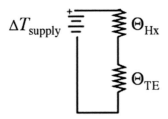

Fig. 11.4 Typical thermal circuit in a thermoelectric energy harvesting device

have a large thermal resistance to build a significant ΔT_{TE} across the thermoelectric. If both the heat source and the sink have low thermal resistance, then the low-thermal resistance for the thermoelectric will enable large heat flow and therefore high power.

From this thermal circuit, one can show that the temperature difference across the thermoelectric is given by

$$\Delta T_{TE} = \Delta T_{supply} \frac{\Theta_{TE}}{\Theta_{Hx} + \Theta_{TE}} \tag{11.8}$$

and the heat flow through the circuit is given by

$$Q = \frac{\Delta T_{supply}}{\Theta_{Hx} + \Theta_{TE}} \tag{11.9}$$

Combining these with the linear relationship between efficiency and ΔT_{TE} (Eq. 11.6) gives power:

$$P = \eta_1 \Delta T_{supply}^2 \frac{\Theta_{TE}}{(\Theta_{Hx} + \Theta_{TE})^2} \tag{11.10}$$

From this analysis, it is clear that larger heat sinks (with smaller heat sink impedance, Θ_{Hx}) provide higher power.

Once the largest heat sink viable for a particular application has been selected (Θ_{Hx} now a constant), the highest power is provided when the thermal resistance of the thermoelectric is designed to match that of the heat exchangers ($\Theta_{Hx} = \Theta_{TE}$) (Stevens, 1999).

$$P_{max} = \frac{\eta_1 \Delta T_{supply}^2}{4\Theta_{Hx}} \tag{11.11}$$

Under the matched condition, the temperature difference across the thermoelectric is exactly half of the total temperature difference across the cold and hot heat baths.

$$\Delta T_{TE} \approx \frac{\Delta T_{supply}}{2} \tag{11.12}$$

$$\eta \approx \frac{\eta_1 \Delta T_{supply}}{2}$$

From this simple analysis, the efficiency (and even the number of couples needed for a particular voltage) can be estimated for a given application with only the knowledge of ΔT_{supply}.

Thus, to at least a first approximation, the thermal resistance of the thermoelectric should be designed to be equal to that of the heat exchangers. This is particularly true in energy harvesting where power and size are more important than efficiency, and heat exchangers typically limit size.

11.7 Heat Flux

Typically, power and size are the primary concerns in energy harvesting. Obviously, a bigger device that utilizes more heat Q will produce more power P. Similarly, the use of twice as many power converters will naturally produce twice the power and consume twice the heat.

Thus, without a specific application in mind, it is natural to focus on power per unit harvested area (P/A) produced and heat flux (Q/A) rather than absolute power and heat consumed. This is particularly convenient for thermoelectric power generation because the systems are so easily scalable: a large system can simply be an array of smaller systems.

$$\frac{P}{A} = \eta \frac{Q}{A}$$
(11.13)

For maximum power flux (P/A), it is necessary to maximize both heat flux (Q/A) and efficiency.

At a constant temperature difference across the thermoelectric (ΔT_{TE}) and the thermal conductance ($K = \kappa A/l$), the inverse of thermal resistance and therefore, the heat/area absorbed into the thermoelectric generator can be modified by adjusting its height, l (**Fig. 11.2**), and therefore engineered for thermal impedance matching with the heat exchangers.

$$\frac{Q}{A} = \frac{\kappa_{eff}\Delta T_{TE}}{l}$$
(11.14)

This relation holds because, to a good approximation, the thermoelectric device acts as a simple heat conductor. The effective thermal conductivity, κ_{eff}, of the thermoelectric module depends not only on the thermal conductivity of the n- and p-type materials but also on the thermoelectric material filling fraction, parallel heat losses within the module, and Peltier and Thomson effects.

11.8 Matching Thermoelectrics to Heat Exchangers

Heat exchangers also scale with size, larger heat exchangers carry more heat and have lower thermal resistance. In this way, the product of thermal resistance times the area, ΘA, is a relatively constant quantity ($\Theta A = 1/h$, where h is the heat transfer coefficient). A low but achievable ΘA value for air cooled heat sinks is 5 K cm^2/W. For a forced fluid heat exchanger a typical value is 0.5 K cm^2/W.

Heat exchangers are likely to be the physically largest component in an energy-harvesting system (Fig. 11.5), and thermoelectrics are typically small in comparison. Because of this, even when space is at a premium, it is usually best to size the thermoelectric to achieve the most power from a given allowed size of heat exchanger.

Even before the exact size and design of the heat exchanger is selected, the general size and type of thermoelectric module can be determined from available heat flux from the heat exchangers. Specifically, because Θ_{TE} and Θ_{Hx} should be matched for maximum power, the size-independent quantities $\Theta_{TE}A$ and $\Theta_{Hx}A$ should also be matched.

The ΘA value for a thermoelectric is derived using the effective thermal conductivity from Eq. (11.14),

$$\Theta_{TE}A = \frac{\Delta T_{TE}}{Q/A} = \frac{1}{\kappa_{\text{eff}} f} \tag{11.15}$$

but now including a filling factor, f, in case the heat exchangers also act as heat spreaders ($f < 1$). The filling factor is the ratio of the area of the thermoelectric to the area of the heat exchangers. In Fig. 11.5, $f = 1$.

Matching $\Theta_{TE}A$ and $\Theta_{Hx}A$ for a given type of heat exchanger (e.g., natural convection air cooled, forced convection, or water cooled) then determines the length, l, of the thermoelectric module. This is because the heat transfer coefficient, $h = 1/\Theta A$, will be relatively constant, which gives

$$l = \frac{\kappa_{\text{eff}} f}{h} \tag{11.16}$$

for an impedance-matched thermoelectric generator.

Typical bulk thermoelectric modules have $0.1\,\text{cm} < l < 0.5\,\text{cm}$. Custom modules and thermopiles can be made outside this range. Thin bulk devices routinely have l as small as $0.02\,\text{cm}$ (Semenyuk, 2001), while unicouples and miniature thermopiles (Snyder et al., 2002) can be made with $l > 2\,\text{cm}$. A typical densely packed thermoelectric module (thermopile) has κ_{eff} of about $0.02\,\text{W/cm K}$. From such val-

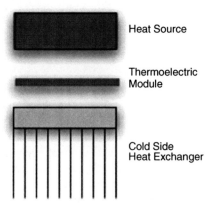

Fig. 11.5 Schematic of a thermoelectric energy harvesting system consisting of a hot and cold side heat exchanger and a thermoelectric module

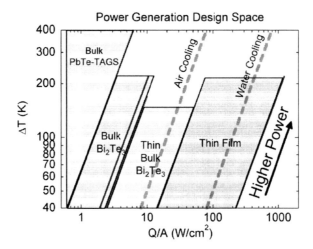

Fig. 11.6 Available design space for energy harvesting from commercial thermoelectric generators. Approximate limits of air and water coolings are shown. Traditional bulk devices are most
appropriate for low heat flux (Q/A), passive air-cooled heat exchangers. High performance air-
cooled heat exchangers require higher heat flux than is available with thin bulk modules. To utilize
the very high heat fluxes at the limit of water cooling, thin film devices are needed. Highest power
is achieved from both large heat flux and ΔT, in the upper right corner of the figure

ues of l and κ_{eff} in available devices, one can map out the available design space
(Fig. 11.6) for thermoelectric generators.

11.8.1 Thin Film Devices

Recently developed thin film devices have very thin thermoelectric material, ranging
from about 0.0005 to 0.004 cm. In out-of-plane devices (Snyder et al., 2003; Böttner
et al., 2004, 2006), this provides a very small value for l, which allows exceptionally high-heat fluxes and low-thermal resistances. These thin films have the great-
est advantage when the heat exchangers are nearly ideal, having very low-thermal
resistances, such as in forced water cooling (Fig. 11.6). So far, these devices have
lower efficiency due to the larger fraction of electrical and thermal contact resistance
losses.

Thin film thermoelectrics used in the in-plane direction have the capability of
producing a much greater number of higher thermal impedance couples (Stark and
Stordeur, 1999). The large number of couples produces significantly higher voltage
and the higher thermal impedance is more appropriate for low-heat flux energy-
harvesting applications. The inherent disadvantage of in-plane thermoelectrics is
that the substrate used to deposit the thermoelectrics acts as a thermal short, reducing
the efficiency.

11.9 Additional Considerations

Power conditioning of a thermoelectric energy harvester should also be a consideration. Even in steady-state, maximum-power operation, the output voltage of a generator is small due to the low voltage per couple, typically 0.2 mV/K. Modules have many couples in series, but even these may use a DC–DC transformer to achieve the few Volts necessary for most applications (e.g., the Seiko watch in Fig. 11.1). When the energy harvester is recharging a battery, the power conditioning also protects the battery from overcharging.

When the temperature changes, power conditioning is needed to stabilize the output voltage. If no battery is present then the power conditioning circuitry may itself require a minimum voltage to start, which will be a challenge for a thermoelectric harvester when the temperature difference is small.

Since high power and efficiency of a thermoelectric system require optimizing the thermal impedance of the harvester with respect to the environment, changes in environmental conditions will adversely affect performance. Devices are typically designed for a steady thermal supply. If the heat and temperatures drop dramatically, the power and voltage produced will rapidly fall. Increases in temperatures may damage contacting solders or the thermoelectric materials themselves. Altering the thermal impedance of the harvester or redirecting the heat to only a portion of the thermoelectric generator would increase power, but methods for such thermal switching are not common.

11.10 Summary

Natural thermal gradients are omnipresent sources of available energy for powering remote, low-power electronics. Thermoelectric generators are the ideal converter for such applications because of their small size and solid-state, maintenance-free operation. The development of several micro-fabrication techniques to build small thermoelectric modules ensures that devices will be available for even the smallest applications.

The most power is provided when the thermal resistance of the thermoelectric device is matched with that of the heat sinks. Under the matched condition, the temperature difference across the thermoelectric is exactly half of the total temperature difference across the cold and hot heat baths. The power output can then be easily estimated using the performance of the available heat exchangers.

References

H. Böttner, G. Chen, and R. Venkatasubramanian, *MRS Bulletin* **31**, 211 (2006).
H. Böttner, J. Nurnus, A. Gavrikov, et al., *Journal of Microelectromechanical Systems* **13**, 414 (2004).

M. Kishi, H. Nemoto, T. Hamao, M. Yamamoto, S. Sudou, M. Mandai, and S. Yamamoto, in *Eighteenth International Conference on Thermoelectrics. Proceedings, ICT'99*, 301 (1999).

J. A. Paradiso and T. Starner, *IEEE Pervasive Computing* **4**, 18 (2005).

V. Semenyuk, *Proceedings ICT2001. 20 International Conference on Thermoelectrics (Cat. No.01TH8589)*, 391 (2001).

G. J. Snyder, "Thermoelectric Power Generation: Efficiency and Compatibility", in *Thermoelectrics Handbook Macro to Nano*, edited by D. M. Rowe (CRC, Boca Raton, 2006), p. 9.

G. J. Snyder and E. S. Toberer, *Nature Materials* **7**, 105 (2008).

G. J. Snyder, A. Borshchevsky, A. Zoltan, et al., in *Twentyfirst International Conference on Thermoelectrics*, 463 (2002).

G. J. Snyder, J. R. Lim, C.-K. Huang, and J.-P. Fleurial, *Nature Materials* **2**, 528 (2003).

I. Stark and M. Stordeur, in *Eighteenth International Conference on Thermoelectrics. Proceedings, ICT'99*, 465 (1999).

J. Stevens, in *34th Intersociety Energy Conversion Engineering Conference*, (1999).

J. W. Stevens, *Energy Conversion and Management* **42**, 709 (2001).

Chapter 12
Optimized Thermoelectrics For Energy Harvesting Applications

James L. Bierschenk

Abstract This chapter highlights the design characteristics of thermoelectric generators (TEGs), electronic devices capable of harvesting power from small temperature differences. Power produced from these generators is low, typically in the microwatt to low milliwatt range. Due to the TEG's unique nature, not only does the TEG's electrical resistance have to be matched to the connected electrical load, the TEG also needs to be thermally matched to the attached heat sink, which is used to dissipate heat to the surrounding ambient. Proper TEG-to-heat sink thermal matching is required to produce sufficient power and voltage to continuously power small wireless sensors, switches, and other wireless devices from temperature differences as small as $5-10\,°C$. Traditional bulk thermoelectric devices and thin film thermoelectric devices are not well suited for these low ΔT, low heat flux applications. When used with small, natural convection heat sinks, TEGs containing hundreds of thermocouples with extreme length-to-area ratios are necessary. New TEG device structures which incorporate thin, adhesive-filled gaps to separate the TEG elements are the best TEG device configuration for small ΔT energy harvesting applications.

12.1 Introduction

Bismuth telluride thermoelectric (TE) technology is well established in many small-scale cooling applications, including picnic boxes, auto seats, telecommunications lasers, military smart munitions, and satellite applications. However, applications using thermoelectrics in the power generation mode, which harvest power from small temperature differences, are much less mature and not as well understood. This chapter highlights the TE design characteristics for low ΔT energy harvesting thermoelectric generators (TEGs) and utilizes the term "energy harvesting" to describe small-scale power production scavenged or harvested by thermoelectrics from small temperature differences. Typically, harvested power is in the micro to milliwatt power range at sufficient voltages to power electronic circuits. The uses for

J.L. Bierschenk (✉)
Marlow Industries, Inc., 10451 Vista Park Road, Dallas TX 75238, USA
e-mail: jbierschenk@marlow.com

S. Priya, D.J. Inman (eds.), *Energy Harvesting Technologies*,
DOI 10.1007/978-0-387-76464-1_12 © Springer Science+Business Media, LLC 2009

energy harvesting TEGs include powering wireless sensors for structural buildings or bridge health monitoring, sensors for engine health monitors, sensors for battlefield surveillance, and reconnaissance, as well as medical sensors and implants. Traditional bulk thermoelectric devices and thin film thermoelectric devices are not well suited for these low ΔT, low heat flux applications due to the mismatch between the TEG and the heat sink thermal resistances. This chapter will focus on the optimization of thermoelectrics specifically for energy harvesting applications. Performance comparisons between bulk TEG devices and new thin film thermoelectric approaches will also be provided.

12.2 Basic Thermoelectric Theory

Thermoelectric generators join dissimilar p and n semiconductor materials electrically and thermally to produce electrical power from the input of thermal energy. The amount of useful power produced, P_{out}, divided by the heat flowing through the TE device, q_h, defines the efficiency of a thermoelectric generator:

$$\eta = \frac{P_{out}}{q_h} \qquad (12.1)$$

where P_{out} is defined by the electrical load resistance, R_{Load}, times the square of the internal current, I:

$$P_{out} = I^2 R_{Load} \qquad (12.2)$$

and the heat flow through the TEG being a function of the thermoelectric material properties and TE element geometries:

$$q_h = N(\alpha_p - \alpha_n)I T_h + K(T_h - T_c) - \frac{1}{2}I^2 R \qquad (12.3)$$

In the above equations, the following relationships for current (I), TEG thermal conductance (K), and TEG electrical resistance (R) are defined:

$$I = \frac{V_{oc}}{R_t} = \frac{N(\alpha_p - \alpha_n)(T_h - T_c)}{(R + R_{Load})} \qquad (12.4)$$

$$K = N\left[\frac{\lambda_p A_p}{L_p} + \frac{\lambda_n A_n}{L_n}\right] = 2N\lambda\frac{A}{L} \qquad (12.5)$$

$$R = N\left[\frac{L_p \rho_p}{A_p} + \frac{L_n \rho_n}{A_n}\right] = 2N\rho\frac{L}{A} \qquad (12.6)$$

where T_h and T_c are the TE element hot and cold side temperatures, respectively, and α_p is the p-type and α_n is the n-type Seebeck coefficients.

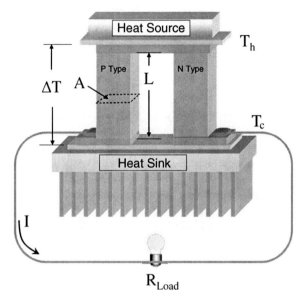

Fig. 12.1 TE power generator device

The relationships can be further simplified assuming the p- and n- type materials have the same element cross-sectional area (A) and length (L), and similar electrical resistivity (ρ), and thermal conductivity (λ). R_{load} is the electrical resistance of the component or circuit (load) which connects to the thermoelectric as shown in Fig. 12.1. In Fig. 12.1, only a single thermoelectric couple (comprised of a single p-type and single n-type thermoelectric element) is shown. Most thermoelectric devices contain many couples wired electrically in series and configured thermally in parallel. In each of the above equations, N represents the number of thermocouples wired electrically in series.

The efficiency of a TEG is influenced by the electrical load attached to the device. Therefore, it becomes useful to define the ratio of the electrical load resistance, R_{Load}, to the TE internal electrical resistance, R, with the variable m:

$$m = \frac{R_{Load}}{R} \tag{12.7}$$

By substituting Eq. (12.4) for current into Eq. (12.2) and putting it in terms of the load ratio, m, the efficiency can be defined as a function of the Carnot efficiency, η_c, load resistance, and a TE material parameter, ZT_h (Cobble, 1995):

$$\eta = \eta_c \frac{m}{\frac{(m+1)^2}{ZT_h} + (m+1) - \frac{\eta_c}{2}} \tag{12.8}$$

The Carnot efficiency defines the theoretical maximum efficiency for a heat engine operating between the temperatures T_h and T_c and is given by:

$$\eta_c = \frac{T_h - T_c}{T_h} \tag{12.9}$$

TE energy harvesting devices using the best bulk materials and operating near room temperature typically run approximately 14% of the Carnot efficiency. It is also useful to evaluate two special thermoelectric efficiency conditions: the efficiency when a TEG produces maximum power output and the maximum theoretical thermoelectric efficiency.

For a given thermoelectric device, maximum power output is achieved when the electrical load matches the TEG electrical resistance (Cobble, 1995), thus $m = 1$. The resulting efficiency can be determined by substituting $m = 1$ into Eq. (12.8):

$$\eta_{mp} = \eta_c \frac{1}{2 + \frac{4}{ZT_h} - \frac{\eta_c}{2}} \tag{12.10}$$

In the design of a TEG operating between temperature difference $T_h - T_c$, when the TE element length is a design variable, the maximum thermoelectric efficiency, is determined by differentiating Eq. (12.8) with respect to m, setting the resulting equation equal to 0, and solving for m (Ioffe, 1956):

$$m = \frac{R_{Load}}{R} = \sqrt{1 + ZT_M} \tag{12.11}$$

Substituting this value of m back into Eq. (12.8), the maximum efficiency is shown to be:

$$\eta_{max} = \frac{(T_h - T_c)}{T_h} \frac{\sqrt{1 + ZT_M} - 1}{\sqrt{1 + ZT_M} + \frac{T_c}{T_h}} = \eta_c \frac{m - 1}{m + \frac{T_c}{T_h}} \tag{12.12}$$

It appears that the resistance ratio for maximum power is in conflict with the ratio for maximum efficiency and thus a conflict between η_{mp} and η_{max}. For a given TEG with an electrical resistance, R, operating between two temperatures T_h and T_c, maximum power output is achieved when R is equivalent to R_{Load} ($m = 1$). The efficiency associated with this operating point is η_{mp}. There also exists a single optimum TEG, one having different TE element dimensions (different R), for each respective combination of T_c and T_h operating points, that can be connected to the same electrical load R_{Load}, that will produce the same power output but at a lower heat flow and higher efficiency. TEGs with the optimum element dimensions can operate at the maximum efficiency for these conditions, η_{max}. When designing a thermoelectric system with heat exchangers, η_{max} from Eq. (12.12) should always be used (Snyder, 2006).

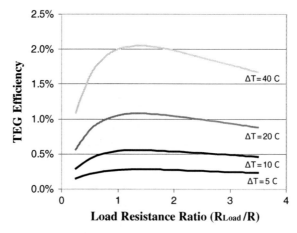

Fig. 12.2 TEG efficiency as a function of load ratio for different TEG ΔTs. $T_c = 27\,^\circ$C for all cases, $ZT = 1$

In each of the above equations, the dimensionless thermoelectric *Figure of Merit* *(ZT)* is a measure of the efficiency of a TE material comprised of temperature (T) in Kelvin, material Seebeck coefficient (α) in V/$^\circ$C, thermal conductivity (λ) in W/m K, and electrical resistivity (ρ) in Ω m^2. In the present day, bulk materials have a ZT value of approximately 1 near room temperature. The higher the ZT, the better the material is.

$$ZT = \frac{\alpha^2}{\lambda\rho}T \qquad (12.13)$$

As stated previously, the electrical load attached to the TEG impacts the TEG's efficiency. This is shown in Fig. 12.2, where TEG efficiency is plotted against the load resistance ratio, m, for several different temperature differences and a $ZT = 1$ material. As can be seen from the curves, the maximum efficiency, $\eta_{\max}$, occurs at a load ratio of approximately 1.4. Efficiency drops off considerably for load ratios less than 1, while the losses with load ratios greater than 1 are less severe.

12.3 Device Effective ZT

ZT is normally a measure of the material efficiency. However, as any material is incorporated into an actual working device, passive losses within the device derate this material efficiency. Therefore, it becomes useful to think in terms of an effective device ZT, Z_eT (Bierschenk and Miner, 2007; Miner, 2007), that can substituted into the theoretical equations in this paper in order to provide improved estimates of device performance. This metric takes the material performance, ZT, and includes the effect of all the passive losses that degrade the thermoelectric performance from the classical theory. This effective device Z_eT includes the impact of electrical contact resistance, interconnect resistance, exterior ceramic conduction losses, and substrate-to-substrate conduction losses in the gaps that separate the TE elements. If the material is very thin, the electrical losses can be quite significant. Likewise, if the spaces between TE elements are filled with thermally conductive polymers and those gaps are not minimized, these device losses can also be appreciable.

Equation (12.14) defines this effective device ZT in terms of the material ZT times the degradation factors. In this relation, K_b is the thermal conductance of parasitic heat flow paths from the hot to the cold side that bypass the elements: K is the thermal conductance through the elements (Eq. 12.5); R_i is the series sum of all electrical interface resistances (electrical contact and interconnect resistances); and R is the series sum of all element electrical resistances given in Eq. (12.6).

$$Z_eT = ZT \frac{1}{\left(1 + K_b/K\right)\left(1 + R_i/R\right)} \qquad (12.14)$$

With most bulk material devices used for energy harvesting, the losses K_b and R_i are small and have historically been ignored. However, for thin film materials, these losses become significant and cannot be ignored. Some of the highest performing thin film TE materials are the superlattice materials with material ZT values in excess of 2 (Venkatasubramanian et al., 2001). The electrical losses, R_i, associated with fabricating a working couple from these materials reduces this $ZT = 2$ material to an effective ZT of only 0.65 (Venkatasubramanian et al., 2006). This value is further reduced when the thermal conduction losses associated with these extremely high heat flux elements are also taken in account. These conduction losses further reduce the device performance to an effective ZT less than 0.4 (Bierschenk and Miner, 2007), i.e., device $Z_eT < 0.4$. For these thin film devices, the derated

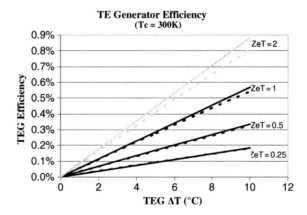

Fig. 12.3 TEG load-matched efficiency (η_{max}: *solid lines* and η_{mp}: *dashed lines*) as a function of TEG Z_eT and ΔT

device effective Z_eT values should be used instead of ZT with the equations in this chapter in order to estimate device and system performance.

Using Eqs. (12.10 and 12.12), Fig. 12.3 shows the impact of TE generator efficiency as a function of TEG ΔT for different device Z_eT values ranging from 0.25 to 2. As Fig. 12.2 indicates, TEG efficiency increases with increasing ΔT and with increasing device Z_eT. For most low ΔT energy harvesting applications, TEG efficiency will be quite low, likely less than 1%. Figure 12.3 also shows the differences between η_{mp} and η_{max} for the same Z_eT values. For the present day devices, operating over small temperature differences, there is little difference between the η_{mp} and η_{max} given in Eqs. (12.10 and 12.12), respectively.

12.4 System Level Design Considerations

The previous sections demonstrated the electrical performance of the TEG for various TEG temperature differences. In the design of actual energy harvesting systems, when the TEG is connected thermally with a small natural convection heat sink, one cannot assume that the desired temperature difference can be achieved across the TEG. The TEG must be matched thermally to the available heat sink in order to maximize the power output and voltage.

Many energy harvesting applications can be represented thermally by the following simple analogous electrical circuit (Fig. 12.4). For small temperature differences existing between a heat source and the local ambient, the temperature difference is divided between the TEG and the heat sink.

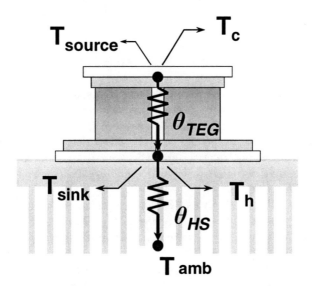

Fig. 12.4 Electrical representation of the thermal circuit of a simple energy harvesting system

The thermal resistance of the heat sink is characterized by:

$$\theta_{HS} = \frac{T_{HS} - T_{amb}}{q_h} \tag{12.15}$$

Likewise, the thermal resistance of the TEG is the inverse of the TEG thermal conductance. For typical bulk thermoelectric devices, the thermal resistance is dominated by the resistance of the TE elements. In the cases where thin film materials are used, or devices where the gaps between TE elements are filled with thermally conductive polymers, it may be necessary to adjust this TEG thermal resistance to account for additional conductive losses:

$$\theta_{TE} = \frac{1}{K} = \frac{1}{2N\lambda\frac{A}{L}} \tag{12.16}$$

Ignoring the impact of bolts or other fasteners that may be used to secure the TEG between the heat sink and the heat source, the heat flow through the system will be given by:

$$q_h = \frac{T_{source} - T_{amb}}{\theta_{TE} + \theta_{HS}} \tag{12.17}$$

Since these resistances are in series, Eq. (12.17) can be broken down and combined with Eq. (12.1).

$$q_h = q_{HS} = \frac{T_{source} - T_{HS}}{\theta_{TE}} = \frac{T_{HS} - T_{amb}}{\theta_{HS}} = \frac{P_{out}}{\eta} \tag{12.18}$$

12.5 System Optimization for Maximum Power Output

In the following example, the impact of the heat sink and TEG thermal resistances on the output power of an energy harvesting system is analyzed. It is assumed that the TEG electrical resistance is optimally matched to the attached load thus the maximum efficiency and power output is produced for the given heat flow and TEG ΔT. In addition, the thermal interface losses between the heat sink and the TEG cold side, and between the heat source and the TEG hot side are assumed to be negligible ($T_c \sim T_{HS}$ and $T_{source} \sim T_h$).

Using Eqs. (12.12 and 12.18), the heat flow through the system, TEG efficiency, and the resulting power output can be plotted as functions of the TEG to heat sink resistance ratio, θ_{TEG}/θ_{HS}, as shown in Fig. 12.5. If the thermal resistance of the TEG is greater than the thermal resistance of the heat sink, a larger proportion of the source-to-ambient ΔT is across the TEG. As a result, the TEG efficiency increases. In this case, however, the amount of heat flow through the TEG and heat sink becomes limited. Even though the heat flowing through the TEG is converted at

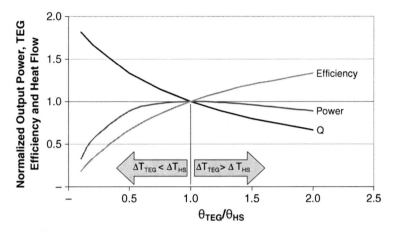

Fig. 12.5 TEG power output, efficiency and heat flow as a function of TEG and heat sink thermal resistance ratio

a higher efficiency, the total power output is limited. Similarly, if the TEG thermal resistance is much lower than the heat sink thermal resistance, the larger proportion of the source-to-ambient ΔT is across the heat sink. Since the TEG ΔT decreases, the resulting TEG efficiency suffers, as does the output power. Peak power output occurs when the TEG and heat sink thermal resistances are equivalent. Therefore, for maximum power output, TE element length and cross sections should be tailored so that the TEG thermal resistance matches that of the heat sink. With the matched thermal resistance, the system ΔT is equally split between the heat sink and the TEG (Mayer and Ram, 2006). It can also be shown when heat exchangers are placed on both sides of the TEG, the sum of all the thermal resistances including the heat sinks and thermal interface resistances should equal the thermal resistance of the TEG for maximum power output (Stevens, 2001).

Today, several types of thermoelectric materials exist, ranging from bulk to thin film, even materials with superlattice structures. These materials vary significantly in thickness (and thus in element length), ranging from several millimeters thick for bulk materials to as low as 5–10 μm thick for the thin film superlattice materials. Material thickness influences not only the electrical resistance of the TEG (Eq. 12.5), but also the thermal resistance (and conductance) as shown in Eqs. (12.6 and 12.16). It should be noted that the electrical resistance is a function of the length-to-area ratio (L/A) of the TEG. Ignoring the electrical contact resistance and interconnect losses, different thickness TE elements will have the same electrical resistance as long as the length-to-area ratio is constant. Likewise, different thickness elements can have the same thermal conductance if the area-to-length ratio is maintained.

The impact of material thickness on the design of a TEG system is best demonstrated by the following example. Consider an ideal TEG energy harvesting system having a 10 °C heat source-to-ambient temperature difference where the TEG is attached to a small 2.5×2.5 cm natural convection heat sink having a thermal resistance of $20\,°C/W$. In this example, $1.0\,V$ is needed to power a low-voltage circuit. Assume the following parameters will be held constant as the material thickness is varied and the resulting TEG size is determined:

- Bi_2Te_3-based materials with $\alpha_p = |\alpha_n| = 220\,\mu V/°C$
- $\lambda_p = \lambda_n = 1.4\,W/mK$
- matched electrical load for maximum efficiency (Eqs. (12.11 and 12.12))
- 35% packing factor (percentage of TEG cross-sectional area occupied by TE elements)
- $Z_eT = 0.8$

Since maximum power is produced when the ratio of TEG to heat sink thermal resistances is 1.0 (Fig. 12.5), the $10\,°C$ system ΔT is evenly split between the heat sink and the TEG and the resulting TEG ΔT is $5\,°C$. In addition, with a matched TE electrical resistance and electrical load and the 1 V output voltage requirement, the TEG open-circuit voltage is $2V_{oc}$. Using Eq. (12.4), approximately 900 couples are needed to produce the $2\,V_{oc}$ and using Eq. (12.16), the TE element area-to-length ratio (A/L) of 0.02 mm is required to achieve $\theta_{TE} = \theta_{HS} = 20\,°C/W$. The impact of material thicknesses ranging from 10 mm to 10 μm on the TEG size, heat flux, and element dimensions is summarized in Table 12.1. In addition, assuming the Z_eT of the device is constant for all material thickness, the power output for all material thickness is approximately $600\,\mu W$.

Several items are noteworthy from Table 12.1

- With the 35% fill factor, the 10 mm thick material device exceeds the heat sink area by almost 60%. At the other extreme, the 30 and 10 μm thick material devices cover only a very small fraction of the heat sink, 0.47 and 0.16%, respectively.
- TEGs that are too small to mechanically support the heat sink mass will require fasteners to secure the TEG between the heat sink and the heat source. These fasteners, made from materials many times more thermally conductive than the TE materials, will provide additional thermal paths for the system heat flow. If the heat passes through the fasteners and not through the TEG, it cannot be converted

Table 12.1 TEG element and device dimensions needed to produce maximum power output and 1 V for a $20\,°C/W$ heat sink with $10\,°C\,\Delta T$

TE Element Height (Thickness) mm	10	3	1	0.1	0.03	0.01
TE Element Width (mm)	0.443	0.243	0.140	0.044	0.024	0.014
TEG Width (mm)	31.9	17.5	10.1	3.2	1.7	1.0
% coverage on 2.5 (cm) sq. heat sink	158	47	16	2	0.47	0.16
TEG Heat Flux (W/cm^2)	0.025	0.082	0.245	2.45	8.17	24.50

to electricity, and the system efficiency drops. Likewise, these parallel heat paths to the TEG will decrease the effective TEG thermal resistance, reducing the TEG ΔT and thus further decreasing the system efficiency.

- For the natural convection heat sink in this example, the heat flux dissipated would be approximately $0.04 \, W/cm^2$. As noted in Table 12.1, as the TE material thickness is decreased to below $100 \, \mu m$, the mismatch in TEG versus heat sink heat flux becomes extreme. With these large heat fluxes through the TEG, thermal interface losses, ceramic conduction losses, and spreading losses can become appreciable further decreasing the TEG ΔT resulting in decreased efficiency and power output.

12.6 Design Considerations for Maximizing Voltage Output

When typical manufacturing design guidelines are taken into account for the different material thicknesses, many of the TE element dimensions that are required to achieve the 1 V output at maximum power output are not achievable. Therefore, it is useful to determine for each of the material thicknesses, how many couples could be incorporated into the device and what the resulting voltage would be. These values are summarized in Table 12.2.

In Table 12.2, the material thicknesses are labeled to represent typical device types. Superlattice materials are typically no more than approximately $10 \, \mu m$ thick. Thin film sputtered devices have materials that are approximately $30 \, \mu m$ thick. Miniature bulk devices, as used in the Telecom Industry, have TE elements limited to approximately 1.4 mm tall and elements in traditional large bulk devices are around 2 mm thick. The compact bulk devices represent a variety of devices produced with adhesive-filled gaps between the elements, enabling very long aspect ratio elements, and very high fill factors (Anatychuk and Pustovalov, 2006; Elsner et al., 1999; Kishi et al., 1999). For each of these device types, the minimum TE element cross

Table 12.2 Typical TE element limitations for different device types and resulting couples and voltage for $20 \, °C/W$ heat sink with $10 \, °C \, \Delta T$

	Compact Bulk	Traditional Bulk	Miniature Bulk	Sputtered	Superlattice
Max TE Element Height (Thickness) mm (est)	20 3	2	1.4	0.03	0.01
Minimum TE Element Width (mm) (est)	0.30 0.15	1.0	0.25	0.035	0.10
Length to Area Ratio (L/Acm^{-1})	2200 1330	20	225	245	10
Max # Couples for $\theta_{TE} = \theta_{HS}$	3968 2381	36	400	437	17
Output Voltage $\theta_{TE} = \theta_{HS}$	4.4 2.6	0.04	0.44	0.48	0.02
Device ZeT	0.60 0.70	0.75	0.86	0.32	0.37

section is estimated. The resulting *A/L* of the TE elements then represents the minimum thermal resistance elements for each device technology. As can be noted from Table 12.2, for the bulk devices and the thin film devices (sputtered and superlattice), the number of thermocouples needed to match the thermal resistance of the heat sink are quite limited. As a consequence, the resulting output voltage is very low, too low in many cases to power even a voltage multiplier circuit.

Since these voltages are very low for most CMOS circuits, deviating from the "max power" assumption is necessary. It is possible to design for less than the theoretical maximum output power in order to achieve higher voltages as demonstrated below. Keeping the same assumptions as before (20 °C/W heat sink, source-to-ambient $\Delta T = 10$ °C, $\alpha_p = |\alpha_n| = 220\,\mu V/$°C, $\lambda_p = \lambda_n = 1.4\,W/m\,K$) and a matched electrical output load condition, Fig. 12.6 summarizes the output power versus voltage capabilities of the various TEG configurations and material thicknesses. The approach taken is to maximize the number of high-thermal resistance elements (longest length/smallest cross-sectional area) in the TEG device. As the number of couples is increased, voltage and power output increase. Peak power output occurs when the heat sink and TEG thermal resistances are equal. As additional couples are added, the TEG thermal resistance decreases, becoming increasingly less than the heat sink thermal resistance and the TEG ΔT decreases. The voltage increase that accompanies the additional couples is partially offset by the decreased TEG ΔT. As can be seen from Fig. 12.6, higher voltages can be achieved, but only at the expense of a rapid drop-off in TEG output power. In addition, a point of diminishing returns is quickly reached in which adding additional thermocouples to the TEG only makes very marginal increases in voltage, but significantly greater decreases in TEG output power.

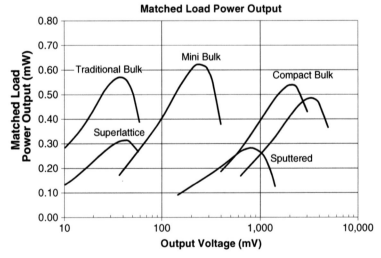

Fig. 12.6 Matched load power output versus voltage for different TEG device types for 20 °C/W heat sink with 10 °C ΔT

Key items to note from Fig. 12.6:

- The impact of the device Z_eT is clearly seen in the peak output power for each configuration. Since the superlattice TEGs and sputtered material TEGs have lower device Z_eT (Bierschenk and Miner, 2007), output power is considerably lower than that of achievable with the bulk materials.
- Figure 12.6 ignores any bolt or fastener thermal loads used to sandwich the heat sink to the heat source. For many of the bulk configurations, it is possible to use the TEG as a structural member and "glue", the heat sink directly to the TEG. However, as noted in the previous example, for the much smaller thin film devices these bolt conductive loads will be appreciable, significantly decreasing both the power output and the voltage as the available heat partially bypasses the TEG in favor of the lower thermal resistance bolts.
- Both traditional bulk TEGs and superlattice TEGs have very low output voltage for the small ΔT systems being analyzed, but for different reasons. TE element aspect ratios, limited by the cross-sectional area in the traditional bulk device and by the thickness of the superlattice device, limit the thermal resistance of each individual TE element. Low-device thermal resistance severely reduces the TEG ΔT and voltage output.

Clearly, the best performance is achievable from devices that can be built with the lowest A/L ratio. To achieve these extreme area-to-length ratios on the TE devices, various approaches have been implemented, all of which incorporate adhesives to provide structural rigidity to the TE elements. Similar constructions were utilized for harvesting energy from body heat to power watches. Figure 12.7 shows one of the many devices produced by Marlow Industries that incorporate very long aspect ratio TE elements with adhesive filled spaces between TE elements, creat-

Fig. 12.7 Example of marlow compact bulk energy harvester

ing a structurally sound energy harvesting TEG that is designed to match the thermal resistance of the heat sink while providing structural support and high-voltage output.

12.7 Conclusions

The optimization of TEG energy harvesting systems requires proper matching of the TEG with the heat sink to be used in the application. Maximum power output is achieved when the temperature difference between the heat source and the ambient is equally split between the TEG and the heat sink. To achieve reasonable output voltages, large numbers of thermocouples are required. These thermocouples must collectively have a thermal resistance equaling the thermal resistance of the heat sink. With traditional bulk and thin film devices, it is difficult to obtain the TE element area-to-length aspect ratios needed to achieve the desired device thermal resistance. In addition, the thin film energy harvesting devices will inherently be very small devices with drastically different heat fluxes than the small natural convection heat sinks. As such, these devices will incur higher thermal interface losses and likely require the use of fasteners to support the mass of the heat sink, both of which further degrade the TEG system performance. Device configurations that enable TEG devices with long aspect ratio TE elements provide the highest power and voltage outputs. These are typically devices with low conductivity adhesive fillers to support the long, small cross-section thermoelectric elements. These device designs can easily produce sufficient power and voltage to power a variety of low-power sensors, even when harvesting energy from temperature differences as low as 5–10 °C. Traditional bulk TEGs and the thin film TEGs will only be applicable in energy harvesting applications when large temperature differences exist and when means of heat sinking beyond natural convection are available.

Acknowledgments I would like to thank Michael Gilley, Joshua Moczygemba, Robin McCarty, and Jeff Sharp for their insightful comments and technical discussion. I would also like to thank Alicia Saddock, Amanda Bierschenk, and Christine Taylor for their proofreading and editing.

References

Anatychuk LI, Pustovalov AA, "Thermoelectric Microgenerators with Isotope Heat Sources", Chapter 53, Thermoelectrics Handbook: Macro to Nano, D.M. Rowe, editor, CRC Press, Baco Raton, FL, 2006

Bierschenk JL, Miner, A, "Practical Limitations of Thin Film Materials for Typical Thermoelectric Applications," International Conference on Thermoelectrics, 2007

Cobble MH, "Calculations of Generator Performance", Chapter 39, CRC Handbook on Thermoelectrics, **CRC** Press, New York, 1995

Elsner NB, et al, "Fabrication of Milliwatt Modules", 18th International Conference on Thermoelectrics, 1999

Ioffe AF, Semiconductor Thermoelements and Thermoelectric Cooling. Infosearch Limited, London, 1956

Kishi M, et al, "Micro Thermoelectric Modules and Their Application to Wrist Watches as an Energy Source", Eighteenth International Conference Thermoelectrics, 1999

Mayer PM, Ram RJ, "Optimization of Heat Sink Limited Thermoelectric Generators", Nanoscale and Microscale Thermophysical Engineering, Vol. 10, No. 2, 2006

Miner A, "The Compatibility of Thin Film and Nanostructures in Thermoelectric Cooling Systems," ASME Journal of Heat Transfer, Vol. 129, No. 7, pp. 805–812, 2007

Snyder GJ, "Thermoelectric Power Generation: Efficiency and Compatibility", Chapter 9, Thermoelectrics Handbook: Macro to Nano, DM Rowe, editor, CRC Press, Baco Raton, FL, 2006

Stevens JW, "Optimal Design of Small ΔT Thermoelectric Generation Systems", Energy Conversion and Management, Vol. 42, pp 709–720, 2001

Venkatasubramanian R, Siivola E, Colpitts T, O'Quinn B, "Thin-Film Thermoelectric Devices with High Room Temperature Figures of Merit", Nature, Vol. 413, pp. 597–602, 2001

Venkatasubramanian, R, Siivola E, O'Quinn B, "Superlattice Thin-Film Thermoeletric Materials and Device Technologies", Chapter 49, Thermoelectrics Handbook: Macro to Nano, D.M. Rowe, editor, CRC Press, Baco Raton, FL, 2006

Part IV
Microbatteries

Chapter 13
Thin Film Batteries for Energy Harvesting

Nancy J. Dudney

Abstract Batteries are one solution for charge accumulation and storage of energy from harvesting and scavenging devices. Because many harvesting devices capture low levels of ambient energy, only very small batteries are required for most applications requiring energy storage and intermittent use. This chapter highlights the fabrication and performance of research batteries and recently commercialized thin film batteries (TFB) including the energy and power densities, charging requirements, cycle-life and shelf-life, and also oerformance at both high and low temperatures. With flexible charging models and excellent cycle life, thin batteries are very promising for paring with a variety of energy harvesting devices including solar cells and piezoelectrics.

13.1 Introduction

Batteries are one solution for charge accumulation and storage of energy from harvesting and scavenging devices. Because many harvesting devices capture low levels of ambient energy, only very small batteries are required for energy storage and intermittent use. Particular applications include self-powered wireless sensors or wearable electronics. Very small batteries are commonly referred to as microbatteries, although common units for power, energy, and capacity are in the milliwatt (mW), milliwatt-hour (mW h), and milliamp-hour (mA h) range.

This chapter will present the results for thin film batteries (TFB) based on Oak Ridge National Laboratory's (ORNL) materials and designs which have been developed and recently commercialized by several manufacturers (Cymbet Corp, Excellatron, Front Edge Technologies, Infinite Power Solutions, Oak Ridge Micro-Energy). No attempt is made to review all the literature and advances on thin film battery design or materials, rather the focus is to highlight the performance characteristics of these batteries that is of particular interest for energy harvesting. Issues where

N.J. Dudney (✉)

Material Science and Technology Division, Oak Ridge National Laboratory, Oak Ridge, TN, USA
e-mail: dudneynj@ornl.gov

S. Priya, D.J. Inman (eds.), *Energy Harvesting Technologies*,
DOI 10.1007/978-0-387-76464-1_13 © Springer Science+Business Media, LLC 2009

the performance differs substantially from larger and more common rechargeable batteries will also be emphasized.

Surprisingly, there are only a few publications demonstrating the use of harvesting devices with different conversion circuits for charging small batteries and none that we know of with TFBs. Sodano et al. (2005) demonstrated partial charging of a 40 mA h Ni metal hydride battery with a piezoelectric plate and diode bridge circuit and, in a later publication, demonstrated charging at higher power with thermoelectric generators. At the time, these investigators avoided the available lithium–ion batteries because they would have required additional components for voltage regulation. Although unpublished, TFBs have been successfully recharged using power from solar cells and piezoelectric devices (personal communication).

13.2 Structure, Materials, and Fabrication of TFB

As constructed by ORNL (Bates et. al. 2000a, Dudney 2005), a thin film battery is all solid state and fabricated by a sequential series of physical vapor deposition processes. A cross section for a common battery layout supported on an insulating substrate is shown in Fig. 13.1. Each of the component layers is 0.3 to several micrometers thick. Ideally, the substrate is also a component of the device; otherwise even with very thin substrates, the weight and volume of the battery is largely determined by the inactive support. Thin film batteries are being developed using a variety of supports, including silicon, mica, alumina, polymers, and metal foils. As shown in Fig. 13.2, many of the batteries are not only thin, but also quite flexible. As yet, these TFBs are limited to a single layer on a substrate surface, as there are engineering hurdles to depositing a multicell stack. Typical active areas are several square centimeter, but cells with active areas from $< 1\ mm^2$ to $25\ cm^2$ have been fabricated. Capacities are in the range of < 1 to 10 mAh.

As demonstrated by extensive work at Oak Ridge, a variety of materials may be used for the active cathode and anode of the batteries. Being intercalation or insertion compounds, each of these materials can reversibly incorporate and release atomic lithium with each charge and discharge of the battery. Choice of the electrode materials determine the cell voltage profile, and to some extent the capacity, energy, and power. Both "lithium batteries," utilizing metallic lithium as the anode, and

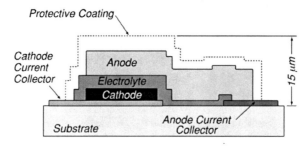

Fig. 13.1 Schematic cross section of TFB

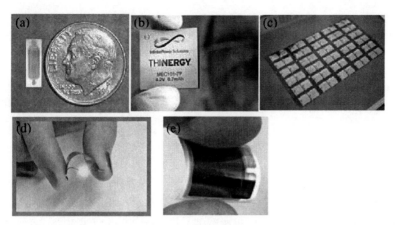

Fig. 13.2 Examples of prototype batteries manufactured by (**a**) Oak Ridge Micro-Energy, (**b**) Infinite Power Solutions, (**c**) Excellatron, (**d**) Front Edge Technology, and (**e**) Cymbet Corporation

"lithium-ion batteries," with a compound or alloy for the anode, have been investigated (Neudecker and Zuhr 2000). In addition, a "lithium-free battery" lacks an anode material until lithium is electrodeposited at the current collector during the first charge (Neudecker et al. 2000).

With the exception of lithium metal, the active anode and cathode materials are deposited as single-phase films by magnetron sputter deposition processing onto unheated substrates. The resulting films are typically X-ray amorphous or composed of very fine crystallites. Subsequent annealing at temperatures of 300–800 °C gives a crystalline microstructure capable of more rapid lithium diffusion and hence a higher power battery. Many of the active compounds and alloys for the TFBs are also used in conventional lithium–ion batteries, but because the films have smaller and well-connected grains, the materials are typically more robust when cycled over a wider voltage range than conventional lithium–ion battery materials (Jang et. al. 2002). Stresses due to volume changes upon cycling limit the thicknesses of the anode and cathode films.

The solid electrolyte film is the most critical battery component. For batteries described here, a glassy lithium phosphorus oxynitride, now known as Lipon, is sputtered deposited over the cathode film. Lipon has been shown to have a wide voltage stability window, 0–5.5 V versus Li^+/Li, an ionic conductivity due to mobile Li^+ ions of 1–2 $\mu S/cm$, and an electronic resistivity in excess of $10^{14}\Omega$ cm (Yu et. al. 1997). Furthermore, because there are no grain boundaries and the Lipon is mechanically rigid, short-circuit paths and lithium dendrites do not penetrate the electrolyte film.

Work to improve the manufacturability of these batteries has resulted in a number of improvements and variations over the original research from the group at ORNL. Advanced materials and faster, more efficient deposition practices have produced batteries that are being used or evaluated for a variety of applications (Cymbet Corp., Excellatron, Front Edge Technologies, Infinite Power Solutions, Oak Ridge Micro-Energy). Researchers at NASA's Jet Propulsion Laboratory have developed processes for lithographic patterning of the battery components for true micropower batteries (West et. al. 2002). Because of the very reactive materials, the batteries must be protected from air. A variety of protective layers have been evaluated. Ideally, a conformal coating would provide a hermetic barrier. Multilayers of Parylene[TM] and titanium, with a total thickness of 7 μm have been used at ORNL, but this only provides protection of the battery for several months at room temperature and humidity. More effective barriers are provided by manufacturers.

13.3 Performance of TFBs

13.3.1 Energy and Power

Typical discharge and charge curves for different lithium batteries (i.e., with a lithium metal anode) are shown in Fig. 13.3. These were measured at low-current densities, so that there is little loss due to polarization of the battery materials. The horizontal axis is the capacity in microampere hour normalized by the active area of the battery and thickness of the cathode. The cathode limits the capacity of these TFBs and may be as much as 2–4 μm thick (Dudney and Jang Y-I 2003). Much thicker cathode films are under development (personal communication). The batteries with the crystalline cathodes show distinct constant voltage plateaus and steps in the discharge/charge profiles reflecting two- and single-phase composition ranges of the cathode material, respectively. Where changes in the host lattice are small, the charge curve nearly overlaps the discharge curve at low current. In contrast, batteries with cathodes which are either amorphous or poorly crystalline have gradually sloping voltage profiles. For these cathodes, there is often a large voltage gap between the discharge and the charge curves which reduces the energy efficiency of

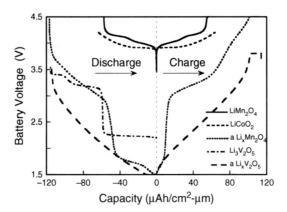

Fig. 13.3 Battery voltage for a low-current discharge and charge cycle of five different batteries with metallic lithium as the anode. The capacity is normalized by the active battery area and the thickness of the cathode. The cathodes are identified in the legend. X-ray amorphous cathodes are denoted as a; other cathodes are crystalline

the battery cycles. Voltage profiles for lithium–ion batteries (i.e., with an alternative anode) differ somewhat from those shown in the figure for the lithium batteries as the voltage reflects the electrochemical potential of lithium in *both* the anode and the cathode. Damage to the electrode materials may occur from over charge or discharge when the lithium content is forced beyond the highly reversible composition range. Some lithium–ion batteries are less susceptible to these conditions and can recover from dead shorts (Oak Ridge Micro-Energy).

When batteries are discharged at higher current densities, increased polarization within the electrolyte and electrodes reduces the capacity that can be utilized before the cell voltage drops below a useful cutoff. When discharged at higher power, the effective energy is reduced. This information is best presented as a Ragone plot of the energy versus power, such as shown in Fig. 13.4. Each of the curves shows the performance of a different battery normalized to the active area. Often these curves cross each other highlighting the tradeoff between batteries with thinner electrodes that can be discharged faster versus batteries with thick electrodes and hence more available energy. The TFBs, and any battery, should be optimized for the particular duty cycle. For thin film batteries fabricated at ORNL, the maximum power realized for batteries under continuous discharge is about $10 \, \text{mW/cm}^2$, and the maximum energy at lower power is about $1 \, \text{mW h/cm}^2$. Pulsed discharge duty cycles can utilize the battery capacity most efficiently, as polarization can relax during the rest periods. The best performance by far is for crystalline $LiCoO_2$ cathodes; this layered crystal structure facilitates a particularly high-lithium intercalation rate. Figure 13.4 also includes axes showing the volumetric energy and power densities. For this estimate, the substrate for the TFB has been excluded, but all other components and protective coating have been included. This represents the ideal case where the

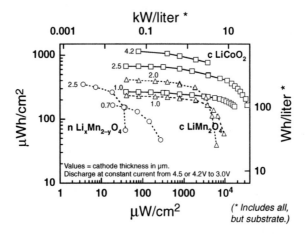

Fig. 13.4 Energy and power per active area of seven different lithium thin film batteries. The cathodes are: crystalline $LiCoO_2$ (□), crystalline $LiMn_2O_4$(Δ), or nanocrystalline $Li_x Mn_{2-y}O_4$ (○). Calculations of the energy and power densities include all battery components and protective coating, but exclude the substrate

TFB is integrated with the harvesting device such that substrates or packaging are shared. Compared with the other lithium–ion batteries and various ultracapacitors, these energy and power densities are very attractive. However, unless the active area of the TFB is very large, the energy provided is small.

13.3.2 Charging

Most of the published performance of the TFBs emphasizes the discharge performance, however for energy harvesting applications, charging the battery is equally or more important. For maximum efficiency and battery life, the harvesting circuit and battery storage voltage must be well matched (Guan and Liao 2007).

The voltage requirements for charging the TFBs are stringent, while the current can vary widely. The voltage profiles in Fig. 13.3 indicate the minimum voltage that must be applied in order to charging the cell continuously, while the maximum is the voltage of the fully charged cell, $V_{charged}$. Higher voltages can be tolerated, but only for short times. This gives a strict voltage target for most TFBs and control is needed to prevent discharge when the charging voltage is low or interrupted. The variety of battery materials offers some flexibility to match the battery with the voltage output of the harvesting device, and in addition, nearly identical batteries can be connected in series to achieve higher storage voltages. Cycle tests of thin film batteries connected in series have given the expected performance and are not degraded by small variations in the cell-to-cell properties (Bates et. al. 2000a,b).

The charge current, on the other hand, can vary by orders of magnitude from a trickle charge to about $10\,mA/cm^2$. The high-current densities obviously give the most rapid charge. For thin film batteries, rapid charge is achieved by simply applying $V_{charged}$ delivering a high-initial current to the battery which decays as the cell approaches full charge. An example of such a rapid continuous charge is shown in Fig. 13.5. Pulsed charge is also acceptable and the maximum applied voltage may be extended by the ohmic voltage so the $V_{max.} = V_{charged} + i\,R_{cell}$, where i is the current and R_{cell} is the internal cell resistance. A stepwise charge with this extended voltage is also shown in Fig. 13.5.

For some energy harvesting devices, the charging current may be very low, referred to as a trickle charge. Currents used for the battery charging shown in Fig. 13.1 are quite low, only $5–20\,\mu A/cm^2$, but from tests of self-discharge it is clear that effective charge can be realized at any currents greater than $\sim 1\,nA/cm^2$.

13.3.3 Cycle-Life and Shelf-Life

Most studies of TFBs report deep cycles, from full charge to complete discharge, at continuous or pulsed currents. Here several thousand cycles are realized with minimal degradation. For use with energy harvesting, actual cycles may be shallow and/or more variable. Although not specifically evaluated, shallow cycles present

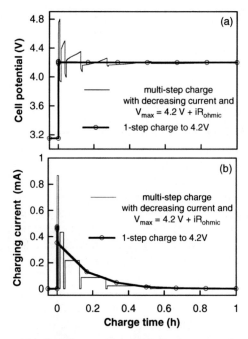

Fig. 13.5 Comparison of (**a**) the cell potential and (**b**) charge current of a small area Li–LiCoO₂ thin film battery charged by applying: (1) a constant 4.2 V charge potential (*bold line*) and (2) high-current steps until the cell voltage exceed a maximum (*thin line*)

less stress to the battery and may extend the cycle life. Interruptions in cycling also cause no degradation and if the battery is at open circuit, there is an almost negligible selfdischarge of the cell. While most often tested with a fully charged cell, results in Fig. 13.6 demonstrate negligible self-discharge over 80 days for a discharged

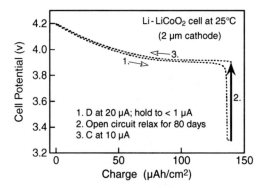

Fig. 13.6 Self-life test of a Li–LiCoO₂ TFB. The battery was first discharged (1) so that the cell voltage is very sensitive to the state of charge. The cell potential was monitored for 80 days at open circuit (2). The subsequent charge capacity equals the initial discharge capacity (3)

battery. Here, the cell voltage increases from 3.3 V continuously over at least 80 days with equilibration of the cell overwhelming any self-discharge. This and other tests indicate that self-discharge is < 1 μA h/cm^2/year for most cells with Lipon electrolyte and metallic Li anode. Obviously, it is essential that the leakage current through the control electronics be minimized to preserve the battery charge.

13.3.4 High and Low Temperature Performance

Because of the stability of the materials, TFBs can withstand high and low temperature extremes (Bates et. al. 2000a). At $-25\,°C$, the lithium transport in both the electrolyte and the electrodes becomes sluggish, so discharge and charge powers are reduced. Batteries have been cycled for long periods at $100\,°C$ (Dudney et. al. 1999), yet for some electrode materials, the cycle life is compromised. Also, lithium-free and lithium–ion batteries can withstand short times at very high temperatures, $\sim 250°C$, needed for solder-bonding processes (Neudecker et al. 2000). Several manufacturers (Excellatron, Front Edge Technologies, Oak Ridge Micro-Energy) have presented results demonstrating good performance at more extreme conditions, including: cycles at $-40\,°C$, and hundreds of cycles at 150 or $175\,°C$. Further, batteries suffer no degradation in performance from treatments of 20 min over $220\,°C$ with excursions to $265\,°C$.

13.4 Outlook and Summary

With robust performance and excellent cycle life, thin film batteries are very promising for pairing with energy harvesting devices. The best scenario is for the TFB to be full integrated with the harvesting device, both in shared components and also in a shared fabrication line. This will complicate manufacturing, but ultimately greatly reduce the weight or volume of the device compared to add-on components. Perhaps, the biggest challenge is design of electronics to efficiently match the output of the harvesting device with the voltage requirements of the TFB to ensure the most efficient energy storage from ambient sources.

Acknowledgments Research leading to invention of these thin film materials and study of their electrochemical performance has been sponsored by the Division of Materials Sciences and Engineering and by the Laboratory Technology Transfer Research Program, U.S. Department of Energy under contract with UT-Battelle, LLC.

References

Bates JB, Dudney NJ, Neudecker BJ, Wang B (2000a) Thin film lithium batteries. In: Osaka T, Datta M (eds) New Trends in Electrochemical Technology: Energy Storage System for Electronics, Gordon and Breach, pp. 453–485.

Bates JB, Dudney NJ, Neudecker BJ, Ueda A, Evans CD (2000b) Thin-film lithium and lithium-ion batteries. Solid State Ionics 135:33–45.

Cymbet Corporation (2007) http://www.cymbet.com/

Dudney NJ, (2005) Solid-state thin-film rechargeable batteries. Mater. Sci. Eng. B. 116:245–249; and http://www.ms.ornl.gov/researchgroups/Functional/BatteryWeb/index.htm

Dudney NJ, Bates JB, Zuhr RA, Young S, Robertson JD, Jun HP, Hackney SA (1999) Nanocrystalline $Li_xMn_{2-y}O_4$ cathodes for solid-state thin-film rechargeable lithium batteries J. Electrochem. Soc. 146:2455–2464.

Dudney NJ, Jang Y-I (2003) Analysis of thin-film lithium batteries with cathodes of 50 nm to 4 μm thick $LiCoO_2$. J. Power Sources 119–121:300–304

Excellatron (2007) http://www.excellatron.com/

Front Edge Technologies (2007) http://www.frontedgetechnology.com/

Guan MJ, Liao WH (2007) On the efficiencies of piezoelectric energy harvesting circuits towards storage device voltages. Smart Mater. Struct. 16:498–505.

Infinite Power Solutions (2007) http://www.infinitepowersolutions.com/

Jang Y-I, Dudney NJ, Blom DA, Allard LF (2002) High-voltage cycling behavior of thin-film $LiCoO_2$ cathodes. J. Electrochem. Soc. 149:A1442–1447.

Neudecker BJ, Dudney NJ, Bates JB (2000) "Lithium-free" thin-film battery with in situ plated Li anode. J. Electrochem. Soc. 147:517–523.

Neudecker BJ, Zuhr RA (2000) Li-ion thin-film batteries with tin and indium nitride and subnitride anodes MeN_x. In: Nazri G-A, Thackeray M, Ohzuku T (eds) Intercalation Compounds for Battery Materials, The Electrochemical Society Proceedings, Pennington, NJ PV 99–24, pp. 295–304.

Oak Ridge Micro-Energy (2007) http://www.oakridgemicro.com/

Sodano HA, Inman DJ, Park GJ (2005) Comparison of piezoelectric energy harvesting devices for recharging Batteries. Intell. Mater. Sys. Struct. 16: 799–807.

Sodano HA, Simmers GE, Dereuz R, Inman DJ, (2007) Recharging batteries using energy harvested from thermal gradients. J. Intell. Mater. Sys. Struct. 18:3–10.

West WC, Whitacre JF, White V, Ratnakumar BV (2002) Fabrication and testing of all solid-state microscale lithium batteries for microspacecraft applications. J. Micromech. Microeng. 12: 58–62.

Yu X, Bates JB, Jellison GE, Hart FX (1997) A stable thin-film lithium electrolyte: Lithium phosphorus oxynitride. J. Electrochem. Soc. 144:524–532.

Chapter 14
Materials for High-energy Density Batteries

Arumugam Manthiram

Abstract Lithium-ion batteries have emerged as the choice of rechargeable power source as they offer much higher energy density than other systems. However, their performance factors such as energy density, power density, and cycle life depend on the electrode materials employed. This chapter provides an overview of the cathode and anode materials systems for lithium-ion batteries. After providing a brief introduction to the basic principles involved in lithium-ion cells, the structure-property-performance relationships of cathode materials like layered $LiMO_2(M = Mn, Co,$ and $Ni)$ and their soiled solutions, spinel $LiMn_2O_4$, and olivine $LiFePO_4$ are presented. Then, a brief account of the carbon, alloy, oxide, and nanocomposite anode materials is presented.

14.1 Introduction

Batteries are the main energy storage devices used in modern society. They power invariably the portable devices we use in our daily life. They are also being pursued and developed intensively for widespread automobile applications. With respect to the topic of this book, batteries are also critical to store the energy harvested from various sources like solar, wind, thermal, strain, and inertia and to use the harvested energy efficiently when needed. In all of these applications, the energy density of the battery, which is the amount of energy stored per unit volume (W h/L) or per unit weight (W h/kg), is a critical parameter. The amount of energy stored depends on the capacity (amount of charge stored) per unit volume (A h/L) or per unit weight (A h/kg) and the voltage (V) each cell can deliver. In addition, rechargeability, charge–discharge cycle life, and the rate at which the cell can be charged and discharged are also important parameters. Moreover, cost and environmental impact considerations need to be taken into account. All of these parameters and properties are related to the battery chemistry and materials involved.

A. Manthiram (✉)

Electrochemical Energy Laboratory, Materials Science and Engineering Program, The University of Texas at Austin, Austin, TX 78712, USA

e-mail: rmanth@mail.utexas.edu

S. Priya, D.J. Inman (eds.), *Energy Harvesting Technologies*,
DOI 10.1007/978-0-387-76464-1_14 © Springer Science+Business Media, LLC 2009

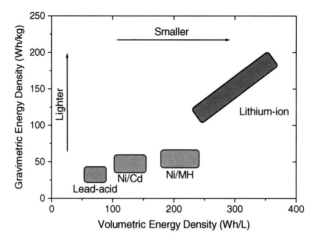

Fig. 14.1 Comparison of the gravimetric and volumetric energy densities of lithium–ion batteries with those of other rechargeable systems

Among the various rechargeable battery chemistries known to date, lithium–ion batteries offer the highest energy density when compared with the other rechargeable battery systems such as lead–acid, nickel–cadmium, and nickel–metal hydride batteries as shown in Fig. 14.1. The higher volumetric and gravimetric energy densities of the lithium–ion cells are due to the higher cell voltages ($\sim$4 V) achievable by the use of non-aqueous electrolytes in contrast to $<$2 V achievable with most of the aqueous electrolyte-based cells. This chapter, after briefly providing the basic principles involved in lithium–ion cells, focuses on the various materials systems employed or being currently pursued to maximize the energy density, power density, or both, while keeping in mind the other parameters like cycle life, cost, and environmental concerns.

14.2 Principles of Lithium–Ion Batteries

Lithium–ion batteries involve a reversible insertion/extraction of lithium ions into/from a host matrix during the discharge/charge process as shown in Fig. 14.2. The host matrix is called as a lithium insertion compound and serves as the electrode material in the cell. The present generation of lithium–ion cells mostly uses graphite and layered $LiCoO_2$ as the lithium insertion compounds, which serve, respectively, as the anode and cathode materials. A lithium-containing salt such as $LiPF_6$ dissolved in a mixture of aprotic solvents like ethylene carbonate (EC) and diethyl carbonate (DEC) is used as the electrolyte. During the charging process, the lithium ions are extracted from the layered $LiCoO_2$ cathode, flow through the electrolyte, and get inserted into the layers of the graphite anode, while the electrons flow through the external circuit from the $LiCoO_2$ cathode to the graphite anode in order to maintain charge balance. Thus, the charging process is accompanied by

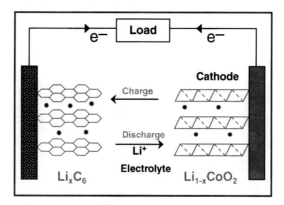

Fig. 14.2 The charge/discharge process involved in a lithium–ion cell consisting of graphite as an anode and layered $LiCoO_2$ as a cathode

an oxidation reaction (Co^{3+} to Co^{4+}) at the cathode and a reduction reaction at the anode. During discharge, exactly the reverse reactions occur at the anode and cathode with the flow of lithium ions (through the electrolyte) and electrons from the anode to the cathode. The free energy change involved in the relevant chemical reaction 14.1 is taken out as electrical energy during the discharge process:

$$LiCoO_2 + C_6 \rightarrow Li_{1-x}CoO_2 + Li_xC_6 \qquad (14.1)$$

The open-circuit voltage V_{oc} of such a lithium–ion cell shown in Fig. 14.2 is given by the difference in the lithium chemical potential between the cathode ($\mu_{Li(c)}$) and the anode ($\mu_{Li(a)}$) as follows:

$$V_{oc} = \frac{\mu_{Li(c)} - \mu_{Li(a)}}{F} \qquad (14.2)$$

where F is the Faraday constant. Figure 14.3 shows a schematic energy diagram of a cell at open circuit. The cell voltage V_{oc} is determined by the energies involved in both the electron transfer and the Li^+ transfer. While the energy involved in electron transfer is related to the work functions of the cathode and the anode, the energy involved in Li^+ transfer is determined by the crystal structure and the coordination geometry of the site into/from which Li^+ ions are inserted/extracted (Aydinol and Ceder, 1997). Although the cell voltage is a function of the separation between the redox energies of the cathode (E_c) and anode (E_a), thermodynamic stability considerations require the redox energies E_c and E_a to lie within the bandgap E_g of the electrolyte, as shown in Fig. 14.3, so that no unwanted reduction or oxidation of the electrolyte occurs during the charge–discharge process. Thus, the electrochemical stability requirement imposes a limitation on the cell voltage as follows:

$$FV_{oc} = \mu_{Li(c)} - \mu_{Li(a)} < E_g \qquad (14.3)$$

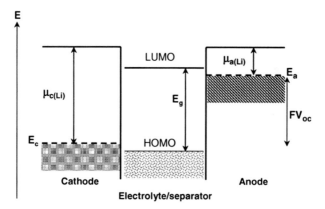

Fig. 14.3 Schematic energy diagram of a lithium cell at open circuit. HOMO and LUMO refer, respectively, to the highest occupied molecular orbital and lowest unoccupied molecular orbital in the electrolyte

Several criteria should be satisfied in order for a lithium insertion compound to be successful as a cathode or an anode material in a rechargeable lithium cell. Some of the most important criteria are listed below.

- The cathode should have a low-lithium chemical potential ($\mu_{Li(c)}$) and the anode should have a high-lithium chemical potential ($\mu_{Li(a)}$) to maximize the cell voltage (V). This implies that the transition metal ion M^{n+} in the lithium insertion compound $Li_xM_yX_z$ should have a high oxidation state to serve as a cathode and a low oxidation state to serve as an anode.
- The lithium insertion compound $Li_xM_yX_z$ should allow an insertion/extraction of a large amount of lithium per unit weight or per unit volume to maximize the cell capacity (A h/L or A h/kg). This depends on the number of lithium sites available in the lithium insertion/extraction host for reversible lithium insertion/extraction and the accessibility of multiple valences for M in the lithium insertion/extraction host. A combination of high capacity and cell voltage can maximize the energy density (W h/L or W h/kg), which is given by the product of the cell capacity and cell voltage.
- The lithium insertion compound $Li_xM_yX_z$ should support a reversible insertion/extraction of lithium with no or minimal changes in the host structure over the entire range of lithium insertion/extraction in order to provide good cycle life for the cell. This implies that the insertion compound $Li_xM_yX_z$ should have good structural stability without breaking any M–X bonds.
- The lithium insertion compound should support both high-electronic conductivity (σ_e) and high-lithium–ion conductivity (σ_{Li}) to facilitate fast charge/discharge (rate capability) and offer high-power capability, i.e., it should support mixed ionic–electronic conduction. This depends on the crystal structure, arrangement of the MX_n polyhedra, geometry, and interconnection of the lithium sites, nature, and electronic configuration of the M^{n+} ion, and the relative positions of the M^{n+} and X^{n-} energies.

- The insertion compound should be chemically and thermally stable without undergoing any reaction with the electrolyte over the entire range of lithium insertion/extraction.
- The redox energies of the cathode and anode in the entire range of lithium insertion/extraction process should lie within the bandgap of the electrolyte as shown in Fig. 14.3 to prevent any unwanted oxidation or reduction of the electrolyte.
- The lithium insertion compound should be inexpensive, environmentally benign, and lightweight from a commercial point of view. This implies that the M^{n+} ion should preferably be from the 3d transition series.

In addition to the criteria outlined earlier for the cathode and the anode materials, the electrolyte should also satisfy several criteria. The electrolyte should have high lithium–ion conductivity, but should be an electronic insulator in order to avoid internal short-circuiting. A high-ionic conductivity in the electrolyte is essential to minimize IR drop or ohmic polarization and realize a fast charge–discharge process (high rate or power capability). With a given electrolyte, the IR drop due to electrolyte resistance can be reduced and the rate capability can be improved by having a higher electrode interfacial area and thin separators. The electrolyte should also have good chemical and thermal stabilities without undergoing any direct reaction with the electrodes. It should act only as a medium to transport efficiently the Li^+ ions between the two electrodes (anode and cathode). Additionally, the engineering involved in cell design and fabrication plays a critical role in the overall cell performance including electrochemical utilization, energy density, power density, and cycle life (Linden, 1995). For example, although ideally high-electronic conductivity and lithium–ion diffusion rate in the electrodes are preferred to minimize cell polarizations, the electronic conductivity of the electrodes can be improved by adding electronically conducting additives such as carbon. However, the amount of additive should be minimized to avoid any undue sacrifice in cell capacity and energy density. Finally, cell safety, environmental factors, and raw materials and processing/manufacturing costs are also important considerations in materials selection, cell design, and cell fabrication.

14.3 Cathode Materials

Intensive materials research during the past three decades has led to the identification and development of several lithium insertion compounds as cathodes for lithium–ion batteries (Whittingham and Jacobson, 1982; Gabano, 1983; Venkatasetty, 1984; Pistoia, 1994; Julien and Nazri, 1994; Wakihara and Yamamoto, 1998; Nazri and Pistoia, 2003). Although initial efforts were focused on transition metal chalcogenides (sulfides and selenides), their practical use was limited as it is difficult to achieve high-cell voltages with the chalcogenide cathodes. The cell voltage is limited to <2.5 V versus metallic lithium anode with the chalcogenides due to an overlap of the higher valent M^{n+} : d band with the top of the nonmetal:p band

and the inability to stabilize higher oxidation states of the transition metal ions in chalcogenides. For example, an overlap of the M^{n+} : 3d band with the top of the S^{2-} : 3p band results in an introduction of holes into the S^{2-} : 3p band to form S_2^{2-} ions rather than oxidizing the transition metal ions and accessing higher valent M^{n+}. Recognizing this difficulty with chalcogenides, Goodenough's group at the University of Oxford focused on oxide cathodes during the 1980s (Mizushima et al., 1980; Goodenough et al., 1980; Thackeray et al., 1983). The location of the top of the O^{2-} : 2p band much below the top of the S^{2-} : 3p band and a larger raising of the M^{n+} : d energies in an oxide when compared with that in a sulfide due to a larger Madelung energy make the higher valent states accessible in oxides. For example, while Co^{3+} can be readily stabilized in an oxide, it is difficult to stabilize Co^{3+} in a sulfide since the $Co^{2+/3+}$ redox couple lies within the S^{2-} : 3p band.

Accordingly, several transition metal oxide hosts crystallizing in different structures have been identified as cathode materials during the past 25 years. Among them, oxides with a general formula $LiMO_2$ (M = Mn, Co, and Ni) having a layered structure, $LiMn_2O_4$ having the spinel structure, and $LiFePO_4$ having the olivine structure have become appealing cathodes for lithium–ion batteries. These three systems of cathodes are discussed in the sections below.

14.3.1 Layered Oxide Cathodes

$LiMO_2$ with M = Co and Ni crystallize in a layered structure in which the Li^+ and M^{3+} ions occupy the alternate (111) planes of the rock salt structure so that the MO_2 sheets alternate with Li^+ layers along the c-axis as shown in Fig. 14.4. The structure consists of a cubic close-packed oxygen array with the Li^+ ions occupying the octahedral interstitial sites and three MO_2 sheets per unit cell. Accordingly, this structure is designated as the O3 layer structure in which the letter "O" refers to the presence of the alkali metal ions in the octahedral sites and the number "3" refers to the number of MO_2 sheets per unit cell. While the interconnected lithium–ion sites through the edge-shared LiO_6 octahedral arrangement between the strongly

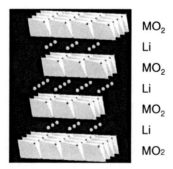

MO$_2$
Li
MO$_2$
Li
MO$_2$
Li
MO$_2$

Fig. 14.4 Crystal structure of layered $LiMO_2$ (M = Co or Ni) showing the arrangement of Li^+ ions between the strongly bonded MO_2 sheets

bonded MO_2 layers provide high lithium–ion conductivity σ_{Li}, the edge-shared MO_6 octahedral arrangement with a direct M–M interaction provides good electronic conductivity σ_e, which are critical to realize fast charge–discharge rates. A large charge and ionic radii differences between the Li^+ and Co^{3+} ions offer a good ordering without any mixing between the cations in the lithium and transition metal layers. Also, a strong preference of the low spin Co^{3+} : $3d^6$ ($t_{2g}^6 e_g^0$) ions for the octahedral sites prevents the migration of the Co^{3+} ions from the transition metal layer to the lithium layer via the neighboring tetrahedral sites during the charge–discharge process unlike in the case of other layered oxides like $LiMnO_2$. On the other hand, the highly oxidized $Co^{3+/4+}$ redox couple with a large work function offers a high-discharge voltage of around 4 V versus Li/Li^+ (Mizushima et al., 1980). Moreover, with a direct Co–Co interaction and a partially filled t_{2g} orbital for Co^{4+} ($t_{2g}^5 e_g^0$) ions, $Li_{1-x}CoO_2$ becomes a metallic conductor on partially charging it. As a result $LiCoO_2$ has become an attractive cathode, and most of the lithium–ion cells presently use $LiCoO_2$ as the cathode.

However, only 50% of the theoretical capacity of $LiCoO_2$, which corresponds to a reversible extraction of 0.5 lithium ions per $LiCoO_2$ formula and around 140 A h/kg, could be utilized in practical cells. Although early studies attributed this limitation in practical capacity to an ordering of lithium ions and consequent structural distortions (hexagonal to monoclinic transformation) around $x = 0.5$ in $Li_{1-x}CoO_2$ (Reimers and Dahn, 1992), more recent characterizations of chemically delithiated samples suggest that the limitation is primarily due to the chemical instability at deep charge for $(1 - x) < 0.5$ in $Li_{1-x}CoO_2$ (Venkatraman et al., 2003; Choi et al., 2006). This is supported by the fact that the reversible capacity of $LiCoO_2$ has been increased significantly close to 200 A h/kg with a reversible extraction of 0.7 lithium ions per formula unit on modifying the surface of $LiCoO_2$ by other oxides like Al_2O_3, ZrO_2, and TiO_2 (Cho et al., 2001a, b). The surface modification prevents the direct contact of the highly oxidized $Co^{3+/4+}$ with the electrolyte and thereby improves the chemical stability. The chemical instability of $Li_{1-x}CoO_2$ at deep charge also leads to safety concerns.

The chemical instability of $Li_{1-x}CoO_2$ is due to a significant overlap of the redox active $Co^{3+/4+}$: t_{2g} band with the top of the O^{2-} : 2p band as shown in Fig. 14.5 and the consequent introduction of significant amount of holes into the O^{2-} : 2p band at deep charge (i.e. $(1 - x) < 0.5$). Although the analogous layered $Li_{1-x}MnO_2$ and $Li_{1-x}NiO_2$ offer better chemical stability than $Li_{1-x}CoO_2$ as the redox active $Mn^{3+/4+}$: e_g and $Ni^{3+/4+}$: e_g bands lie well above the O^{2-} : 2p band or barely touches the top of O^{2-} : 2p band (Fig. 14.5), $Li_{1-x}MnO_2$ suffers from a migration of the Mn^{3+} ions from the transition metal plane to the lithium plane to form a spinel-like phase while $Li_{1-x}NiO_2$ suffers from phase changes during the charge–discharge process. As a result, both $LiMnO_2$ and $LiNiO_2$ are not promising cathodes.

Although both $LiMnO_2$ and $LiNiO_2$ by themselves could not be used as cathodes, solid solutions among $LiMnO_2$, $LiCoO_2$, and $LiNiO_2$ have become attractive cathodes. For example, layered $LiNi_{1/3}Mn_{1/3}Co_{1/3}O_2$ has emerged as a replacement for $LiCoO_2$ as it offers a lower cost and better safety when compared with

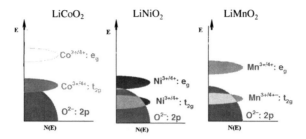

Fig. 14.5 Comparison of the energy diagrams of LiCoO$_2$, LiNiO$_2$, and LiMnO$_2$

LiCoO$_2$. While the lower cost arises from a lower raw material cost of Mn and Ni when compared with that of Co, the better safety is due to the lying of the Mn$^{3+/4+}$: e$_g$ and Ni$^{3+/4+}$: e$_g$ bands above the O^{2-} : 2p band when compared with the Co$^{3+/4+}$: t$_{2g}$ band as seen in Fig. 14.5. The better chemical stability also allows Li$_{1-x}$Ni$_{1/3}$Mn$_{1/3}$Co$_{1/3}$O$_2$ to be charged to higher cutoff charge voltages with a higher reversible capacity of around 180 A h/kg when compared with Li$_{1-x}$CoO$_2$ (Choi and Manthiram, 2004).

More recently, solid solutions between layered Li[Li$_{1/3}$Mn$_{2/3}$]O$_2$, which is commonly designated as Li$_2$MnO$_3$, and layered Li[Ni$_{1-y-z}$Mn$_y$Co$_z$]O$_2$ have also become interesting as they exhibit much higher capacities of around 250 A h/kg, which is nearly two times higher than that found with LiCoO$_2$ (Lu et al., 2002; Armstrong et al., 2006; Thackeray et al., 2007; Arunkumar et al., 2007a, b). These layered solid solutions between Li[Li$_{1/3}$Mn$_{2/3}$]O$_2$ and Li[Ni$_{1-y-z}$Mn$_y$Co$_z$]O$_2$ exhibit an initial sloping region A, which corresponds to the oxidation of the transition metal ions to 4+ state, followed by a plateau region B, which corresponds to an oxidation of the O^{2-} ions and an irreversible loss of oxygen from the lattice, during the first charge as seen in Fig. 14.6. After the first charge, the material cycles with a sloping discharge–charge profile involving a reversible reduction–oxidation of the transition metal ions.

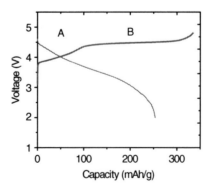

Fig. 14.6 First charge–discharge profiles of solid solutions between layered Li[Li$_{1/3}$Mn$_{2/3}$]O$_2$ and Li[Ni$_{1-y-z}$Mn$_y$Co$_z$]O$_2$

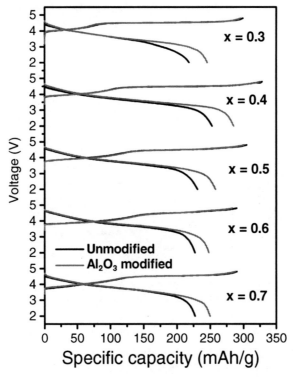

Fig. 14.7 First charge–discharge profiles of the layered $(1 - x)$ Li[Li$_{1/3}$Mn$_{2/3}$]O$_2$ − xLi[Ni$_{1/3}$Mn$_{1/3}$Co$_{1/3}$]O$_2$ solid solutions before and after surface modification with Al$_2$O$_3$ followed by heating at 400 °C

However, these layered solid solution cathodes tend to exhibit a large irreversible capacity loss in the first cycle, i.e., a large difference between the first charge capacity and the first discharge capacity as seen in Fig. 14.6. This large irreversible capacity loss is generally believed to be due to the extraction of lithium as "Li$_2$O" in the plateau region B in Fig. 14.6 and an elimination of the oxide ion vacancies formed to give an ideal composition "MO$_2$" without any oxide ion vacancies existing in the lattice at the end of first charge, resulting in a less number of lithium sites available for lithium insertion/extraction during the subsequent discharge–charge cycles (Armstrong et al., 2006; Thackeray et al., 2007). However, a careful analysis of the first charge and discharge capacity values recently with a number of compositions suggests that part of the oxide ion vacancies should be retained in the layered lattice to account for the high-discharge capacity values observed in the first discharge (Wu et al., 2008).

The irreversible capacity loss observed during the first cycle could also be reduced by a modification of the cathode surface with other oxides like Al$_2$O$_3$ (Wu et al., 2008; Park et al., 2001). For example, Fig. 14.7 shows the first charge–discharge profiles of a series of solid solutions between layered Li[Li$_{1/3}$Mn$_{2/3}$]O$_2$ and Li[Ni$_{1/3}$Mn$_{1/3}$Co$_{1/3}$]O$_2$ before and after surface modification with Al$_2$O$_3$, while

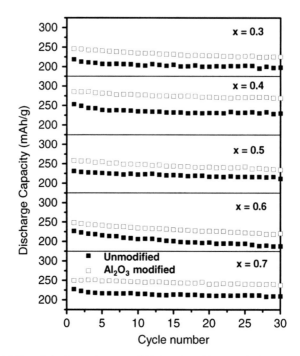

Fig. 14.8 Cyclability of the layered $(1 - x)$ Li[Li$_{1/3}$Mn$_{2/3}$]O$_2$ − xLi[Ni$_{1/3}$Mn$_{1/3}$Co$_{1/3}$]O$_2$ solid solutions before and after surface modification with Al$_2$O$_3$ followed by heating at 400 °C

Fig. 14.8 shows the corresponding cyclability data. Clearly, the surface modified samples exhibit lower irreversible capacity loss and higher discharge capacity values than the pristine, unmodified samples. This improvement in the surface modified samples has been explained on the basis of the retention of more number of oxide ion vacancies in the layered lattice after the first charge when compared with that in the unmodified samples. It is remarkable that the surface modified $(1 - x)$ Li[Li$_{1/3}$Mn$_{2/3}$]O$_2$ − xLi[Ni$_{1/3}$Mn$_{1/3}$Co$_{1/3}$]O$_2$ composition with $x = 0.4$ exhibit a high-discharge capacity of 280 mA h/g, which is two times higher than that of LiCoO$_2$. These oxides have high potential to increase the energy density values significantly. However, these layered oxides require charging up to about 4.8 V and more stable, compatible electrolyte compositions need to be developed. Moreover, oxygen is lost irreversibly from the lattice during the first charge, and it may have to be vented appropriately during cell manufacturing before sealing the cells. Furthermore, the long-term cyclability of these cathodes under more aggressive environments such as elevated temperatures need to be fully assessed.

14.3.2 Spinel Oxide Cathodes

LiMn$_2$O$_4$ crystallizes in the normal spinel structure in which the Li$^+$ and the Mn$^{3+/4+}$ ions occupy, respectively, the 8a tetrahedral and 16d octahedral sites of the cubic close-packed oxygen array (Fig. 14.9) to give $(Li)_{8a}[M_2]_{16d}O_4$. While the interconnection of the 8a tetrahedral sites via the neighboring empty 16c octahedral sites offers a fast 3D lithium–ion diffusion (high σ_{Li}) within the covalently bonded $[Mn_2]O_4$ framework, the edge shared MnO_6 octahedra with a direct Mn–Mn interaction provide good electronic conductivity σ_e needed for high rate capability. Unlike the layered $LiMO_2$ cathodes that could suffer from the migration of the transition metal ions from the transition metal layer to the lithium layer, the 3D $[Mn_2]O_4$ spinel framework provides excellent structural stability, supporting high rate capability. Additionally, the lying of the $Mn^{3+/4+} : e_g$ band well above the $O^{2-}:2p$ band as seen in Fig. 14.5 offers excellent chemical stability unlike the $Co^{3+/4+}$ couple. Moreover, Mn is inexpensive and environmentally benign. As a result, $LiMn_2O_4$ with a discharge voltage of about 4 V (Thackeray et al., 1983) has become appealing as a cathode, particularly for high-power applications such as hybrid electric vehicles (HEV).

However, only about 0.8 lithium ions per $LiMn_2O_4$ formula unit could be reversibly extracted, which limit the practical capacity to <120 A h/kg. Although an additional lithium could be inserted into the empty 16c sites of $(Li)_{8a}[M_2]_{16d}O_4$ around 3 V versus Li/Li^+ to give the lithiated spinel $\{Li_2\}_{16c}[M_2]_{16d}O_4$, it is accompanied by a transformation of the cubic $(Li)_{8a}[M_2]_{16d}O_4$ into tetragonal $\{Li_2\}_{16c}[M_2]_{16d}O_4$ due to the Jahn–Teller distortion associated with the high spin Mn^{3+} : $3d^4$ $(t_{2g}^3 e_g^1)$ ion. This results in poor capacity retention during cycling in the 3 V region due to huge volume changes, so the 3 V region cannot be utilized in practical cells. In addition to the lower capacity ($<120A$ h/kg) when compared with that of the layered oxides, the $LiMn_2O_4$ spinel cathode encounters severe capacity fade even in the 4 V region particularly at elevated temperatures. This has been attributed

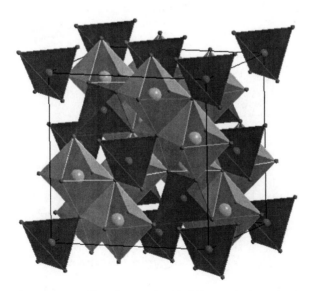

Fig. 14.9 Crystal structure of $LiMn_2O_4$ spinel with LiO_4 tetrahedra and edge-shared MnO_6 octahedra

to a disproportionation of Mn^{3+} ions into Mn^{4+} and Mn^{2+} ions in the presence of trace amounts of hydrofluoric acid generated by a reaction of the $LiPF_6$ salt with the parts per million levels of water present in the electrolyte solution; while the Mn^{4+} ion remains in the solid, Mn^{2+} leaches out into the electrolyte and poisons the anode, resulting in capacity fade.

Several efforts including cationic and anionic substitutions and surface modifications (Park et al., 2001; Kannan and Manthiram, 2002) have been pursued to overcome the capacity fade problem. Among them, optimized substitutions like $Li[Mn_{1.8}Ni_{0.1}Li_{0.1}]O_4$ suppresses Mn dissolution drastically and improves the capacity retention significantly as seen in Fig. 14.10 (Shin and Manthiram, 2003; Choi and Manthiram, 2006, 2007). However, such substitutions increase the manganese valence and decrease the capacity to around 80 A h/kg. Nevertheless, the capacity values could be increased to around 100 A h/kg without increasing the manganese dissolution by a partial substitution of fluorine for oxygen. For example, oxyfluoride compositions like $Li[Mn_{1.8}Ni_{0.1}Li_{0.1}]O_{3.8}F_{0.2}$ exhibit excellent capacity retention at ambient and elevated temperatures (Fig. 14.10) along with the high-rate capability (Choi and Manthiram, 2006, 2007). Although the capacity values

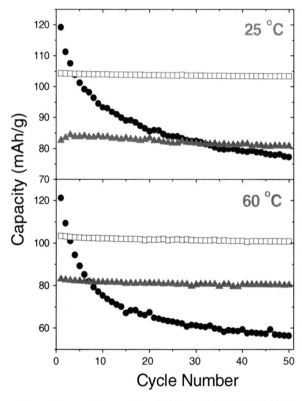

Fig. 14.10 Comparison of the capacity retention of $LiMn_2O_4$ (*solid circles*), $Li[Mn_{1.8}Ni_{0.1}Li_{0.1}]O_4$ (*solid triangles*), and $Li[Mn_{1.8}Ni_{0.1}Li_{0.1}]O_{3.8}F_{0.2}$ (*open squares*) spinel cathodes at 25 and 60 °C

of the spinel cathodes are lower when compared with that of the layered oxides discussed in the previous section, the low cost of Mn and the high charge–discharge rate capabilities make the optimized oxyfluoride spinel cathodes appealing for high-power applications such as power tools and hybrid electric vehicles. Interestingly, the capacity values of the spinel cathodes could also be increased by blending with appropriate amounts of a layered oxide like $LiNi_{1/3}Mn_{1/3}Co_{1/3}O_2$ without sacrificing the cost and rate capability benefits.

One interesting way to increase the energy density of $LiMn_2O_4$ spinel is to increase the operating voltage. For example, substitution of other transition metal ions like Ni^{2+} for $Mn^{3+/4+}$ to give $LiMn_{1.5}Ni_{0.5}O_4$ has been found to increase the operating voltage to 4.7 V with a capacity of about 130 A h/kg (Zhong et al., 1997). In $LiMn_{1.5}Ni_{0.5}O_4$, Mn is present as Mn^{4+} and Ni is present as Ni^{2+}, and the oxidation of Ni^{2+} to Ni^{3+} and Ni^{4+} during the charging process and the operation of the $Ni^{2+/3+}$ and $Ni^{3+/4+}$ couples along with an extraction/insertion of lithium from/into the tetrahedral sites offers a voltage of 4.7 V when compared with 4 V for the $Mn^{3+/4+}$ couple. It is interesting to note that while the $Ni^{3+/4+}$ couple with an octahedral site lithium operates around 4 V in the layered oxides like $LiNiO_2$, the same couple with a tetrahedral site lithium operates around 4.7 V in the spinel $LiMn_{1.5}Ni_{0.5}O_4$. This is a direct reflection of the contributions of the energies involved in both the electron transfer (work function) and the Li^+ ion transfer (site energy) to the cell voltage as discussed in Section 14.2. Although the synthesis of $LiMn_{1.5}Ni_{0.5}O_4$ encounters the formation of a small amount of NiO impurity, appropriate cationic substitutions to give compositions like $LiMn_{1.42}Ni_{0.42}Co_{0.16}O_4$ and $LiMn_{1.5}Ni_{0.42}Zn_{0.08}O_4$ eliminate the impurity phase with a significant improvement in capacity retention (Arunkumar and Manthiram, 2005). However, the major issue with this class of spinel cathodes is the electrolyte stability at 4.7 V. Development of more stable, compatible electrolytes could make these cathodes attractive for high power applications with a significantly increased energy density when compared with the $LiMn_2O_4$ cathode.

14.3.3 Olivine Oxide Cathodes

One disadvantage with layered oxide cathodes containing highly oxidized redox couples like $Co^{3+/4+}$ and $Ni^{3+/4+}$ is the chemical instability at deep charge and the associated safety problems. Recognizing this, oxides like $Fe_2(XO_4)_3$ that contain the polyanion $(XO_4)^{2-}$ (X = S, Mo, and W) were initiated as lithium insertion/extraction hosts in the late 1980s. Although the lower valent $Fe^{2+/3+}$ couple in a simple oxide like $LiFeO_2$ would be expected to offer a lower discharge voltage of <3 V, the covalently bonded groups like $(SO_4)^{2-}$ lower the redox energies of $Fe^{2+/3+}$ through inductive effect and increase the cell voltage to >3 V (Manthiram and Goodenough, 1987, 1989). Following this, $LiFePO_4$ crystallizing in the olivine structure (Fig. 14.11) and offering a flat discharge profile around 3.4 V with a theoretical capacity of around 170 A h/kg was identified as a cathode (Padhi et al., 1997) in late1990s. Despite a higher capacity, the lower discharge voltage when compared

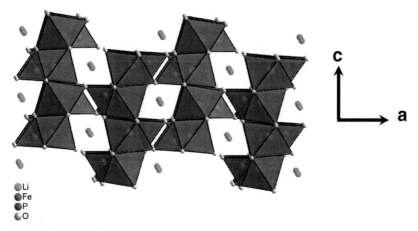

Li
Fe
P
O

Fig. 14.11 Crystal structure of olivine LiFePO$_4$ with LiO$_6$ and FeO$_6$ octahedra and PO$_4$ tetrahedra

with that of the layered and spinel oxides together with a less dense olivine structure reduces particularly the volumetric energy density. Nevertheless, as Fe is inexpensive and environmentally benign and the covalently bonded PO$_4$ groups together with the chemically more stable Fe$^{2+/3+}$ couple offer excellent safety, LiFePO$_4$ has become an appealing cathode.

However, the major drawback with LiFePO$_4$ is the poor lithium–ion conductivity resulting from the 1D diffusion of Li$^+$ ions along the chains (b-axis) formed by edge-shared LiO$_6$ octahedra and poor electronic conductivity resulting from the little solubility between LiFePO$_4$ and FePO$_4$ (lack of mixed valency) and the corner shared FeO$_6$ octahedra with highly localized Fe^{2+} or Fe^{3+} ions. These problems have been overcome in recent years by cationic doping, decreasing the particle size through solution-based synthesis, and coating with electronically conducting agents like carbon (Chung et al., 2002; Ellis et al., 2007; Kim and Kim, 2006; Vadivel Murugan et al., 2008). Among them, a microwave-irradiated solvothermal approach offers nanorod morphologies with controlled particle size as seen in Fig. 14.12 within a short reaction time of 5 min at temperatures as low as 300 °C, offering important savings in manufacturing cost (Vadivel Murugan et al., 2008). The nanorod dimension could be controlled by altering the reaction conditions such as the reactant concentrations. The transmission electron microscopic (TEM) data in Fig. 14.12 reveal the highly crystalline nature of the products. The samples obtained by the microwave-solvothermal method followed by coating with a mixed ionic–electronic conducting polymer at ambient temperatures offers close to theoretical capacity with excellent capacity retention as seen in Fig. 14.13. The small lithium diffusion length also offers excellent rate capability.

Although LiFePO$_4$ offers a lower voltage (3.4 V), the analogous LiMnPO$_4$, LiCoPO$_4$, and LiNiPO$_4$ offer much higher voltages of, respectively, 4.1, 4.8, and 5.2 V. These materials could also be prepared as nanorods with controlled dimensions by the microwave-solvothermal approach. However, the extremely low-electronic conductivity and the Jahn–Teller distortion associated with LiMnPO$_4$ lead to

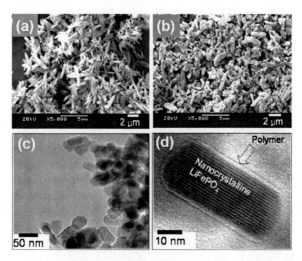

Fig. 14.12 (a) SEM image of the LiFePO$_4$ nanorods prepared by the microwave-solvothermal method. (b) SEM image of the LiFePO$_4$ nanorods after coating with a mixed ionic–electronic conducting polymer, p-toluene sulfonic acid (p-TSA) doped poly(3,4-ethylenedioxythiophene) (PEDOT). (c) TEM image of LiFePO$_4$ nanorods (25 ± 6 nm width and up to 100 nm length). (d) High resolution TEM image of LiFePO$_4$ nanorods after coating with p-TSA doped PEDOT

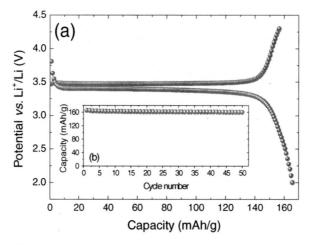

Fig. 14.13 (a) First charge–discharge profiles and (b) cyclability of LiFePO$_4$ nanorods (25 ± 6 nm width and up to 100 nm length) after coating with p-toluene sulfonic acid (p-TSA) doped poly(3,4-ethylenedioxy thiophene) (PEDOT)

poor performance, while the lack of stable electrolytes at higher voltages leads to poor performance for LiCoPO$_4$ and LiNiPO$_4$. Development of alternate electrolytes that are stable around 5 V could make these olivine cathodes attractive for both high-energy density and high-power applications.

14.4 Anode Materials

Although the concept of rechargeable lithium batteries was initially pursued with metallic lithium anodes (Whittingham and Jacobson, 1982; Gabano, 1983; venkata-setty, 1984; Pistoia, 1994; Julien and Nazri, 1994), the dendrite formation during cycling and the associated safety concerns and capacity fade forced the replacement of metallic lithium anode by lithium insertion compounds. The use of lithium insertion compounds both as anode and as cathode led to the lithium–ion cell concept discussed in Section 14.2 (Fig. 14.2). With this strategy, carbon with a low atomic weight and a redox energy close to that of metallic lithium has become an attractive anode in the present generation of lithium–ion cells (Wakihara and Yamamoto 1998; Winter et al., 1998; Megahed and Scrosati, 1994). Graphite exhibits a high theoretical capacity of 372 A h/kg, which corresponds to an insertion of one lithium per six carbon atoms ($x = 1$ in Li_xC_6). With a discharge potential close to $0 V$ versus Li/Li^+, coupling of graphite anode with the layered $LiMO_2$ or spinel $LiMn_2O_4$ cathodes offers around $4 V$ per cell. In addition to graphite, highly disordered hard carbons have also attracted much attention as they offer higher capacities than graphite (Sato et al., 1994; Dahn et al., 1995). Several mechanisms such as adsorption of lithium in the microcavities, on both the sides of single graphene sheets, and on internal surfaces of the nanopores formed by monolayer and multilayer graphene sheets have been proposed (Sato et al., 1994; Dahn et al., 1995; Mabuchi et al., 1995; Zheng et al., 1996). However, one of the drawbacks with the carbon anodes is the occurrence of a significant amount of irreversible capacity loss during the first discharge–charge cycle due to unwanted, irreversible side reactions with the electrolyte. The reaction with the electrolyte produces a solid-electrolyte interfacial (SEI) layer, which could also pose safety concerns with large cells.

The difficulties with carbon anode have generated a lot of interest in alternative anode materials. Among them, alloying with other elements like Si, Sn, and Ge has become appealing as some of them exhibit much higher theoretical capacities ($>1,000$ A h/kg) when compared with carbon (Chan et al., 2008a, b; Kim et al., 2007). However, the major drawback with such alloys is the huge volume change occurring during the discharge–charge process, resulting in cracking and crumbling of the electrodes and a severe capacity loss during cycling. Several approaches are being pursued to overcome this problem. One of them is an active/inactive nanocomposite concept in which while one material reacts with lithium, the other material acts as an electrochemically inactive matrix to buffer the volume variations during the alloying process (Tarascon and Armand, 2001; Hassoun et al., 2007). With this approach, the use of the nanosize metallic clusters as lithium insertion hosts suppresses the associated strains considerably and improves the reversibility of the alloying reaction. Another approach is to use 1D Si or Ge nanowires that can accommodate large strain without pulverization while providing good electrical contact (Chan et al., 2008a, b). The nanowire strategy has been found to improve the cyclability significantly with high capacities, but it could lead to a significant reduction in overall volumetric energy density and difficulties for cost-effective manufacturing of large volume of materials.

More recently, there has been considerable interest in a variety of nanoarchitectures as anode hosts, employing novel synthesis approaches (Wen et al., 2007; Deng and Lee, 2008; Zhang et al., 2008). For example, mesoporous SnO_2 grown on multiwalled carbon nanotubes, hollow core–shell mesospheres of SnO_2 and carbon, and tin particles encapsulated in hollow carbon spheres have been pursued. They exhibit high capacities and further work is needed to realize their full potentials.

Another class of anodes is the metal oxides that undergo displacement reactions during the charge–discharge process and exhibit high capacities. For example, oxides like SnO_2 react with lithium to form Li_2O and nanosize tin particles (Idota et al., 1997; Courtney and Dahn, 1997; Brousse et al., 1998; MacKinnon and Dahn, 1999). The Sn particles remain finely dispersed in the Li_2O matrix, which by surrounding the tin particles accommodates the mechanical stresses occurring during the alloy formation–decomposition process. This greatly improves the cycling performance although there is a significant irreversible capacity loss during the first cycle and agglomeration of tin particles tend to occur during prolonged cycling. Other oxides like FeO, NiO, and CoO involving displacement reactions with lithium to form Li_2O and Fe, Ni, or Co have also been pursued as anode materials, but they exhibit higher voltages versus Li/Li^+ when compared with carbon anodes, resulting in a lower cell voltage (Poizot et al., 2000).

In addition, several other materials have been pursued over the years as anode hosts, but they tend to exhibit higher voltages versus Li/Li^+ and lower capacities than carbon, resulting in a significantly reduced voltage and energy density. For example, nanocrystalline $Li_4Ti_5O_{12}$ crystallizing in the spinel structure exhibits excellent lithium insertion/extraction properties with very little volume change, and it is often referred to as a zero strain material. More importantly, unlike carbon anodes, it does not suffer from solid-electrolyte interfacial (SEI) layer problems or any serious safety problems (Ferg et al., 1994). Unfortunately, it exhibits a much lower capacity of 175 A h/kg with a much higher voltage of 1.5 V versus Li/Li^+. Thus, coupling of $Li_4Ti_5O_{12}$ with the layered $LiMO_2$ or spinel $LiMn_2O_4$ cathodes results in much reduced cell voltages of around 2.5 V and lower energy densities. Other materials that have been pursued over the years are MoO_2 (Auborn and Barbeiro, 1987), WO_2 (Abraham et al., 1990), MnP_4 (Souza et al., 2002), intermetallic compounds like Cu_6Sn_5 (Kepler et al., 1999) and Li_2CuSn (Vaughey et al., 1999), and nitrides like $Li_{7-x}MnN_4$ and $Li_{3-x}FeN_2$. However, they could not compete with carbon anode in terms of overall performance.

14.5 Conclusions

This chapter has provided an overview of the cathode and anode materials systems for lithium–ion batteries. With respect to maximizing the energy density and other performance factors, layered $LiCoO_2$, spinel $LiMn_2O_4$, and olivine $LiFePO_4$ have become attractive cathodes, while carbon has become an attractive anode. Each cathode material has its own advantages and disadvantages with respect to the various performance factors, and their adoption will depend on the energy versus power needs, cost and safety considerations, and type of applications (e.g., portable electronic devices versus automobile versus stationary energy storage). More recent work suggests that it is possible to achieve capacities as high as 300 A h/kg with some layered oxide cathodes when compared with the $\sim$150 A h/kg achievable with the present generation of cathode materials. Similarly, the use of olivine cathodes like $LiNiPO_4$ and $LiCoPO_4$ as well as spinel cathodes like $LiMn_{1.5}Ni_{0.5}O_4$ can increase the energy density significantly due to their higher operating voltages ($\sim$5 V). However, compatible, robust electrolytes that can operate at the higher voltages of 5 V need to be developed to realize the full potential of these cathodes. On the other hand, although carbon is the choice of anode in the present generation of lithium–ion cells, other materials like Si and Sn offer great promise due to their higher capacities. Novel synthesis and processing approaches including nanotechnology to overcome some of the persistent problems (huge volume changes) with these anodes could make them commercially viable. Moreover, search for entirely new class of cathode and anode materials as well as compatible electrolytes could advance the field.

Acknowledgments Financial support by the Welch Foundation grant F-1254 is gratefully acknowledged.

References

Abraham KM, Pasquariello DM, Willstaedt EB (1990) Preparation and characterization of some lithium insertion anodes for secondary lithium batteries. J Electrochem Soc 137: 743–749.

Armstrong R, Holzapfel M, Novak P, Johnson,CS, Kang, SH, Thackeray MM, Bruce PG (2006) Demonstrating oxygen loss and associated structural reorganization in the lithium battery cathode $Li[Ni_{0.2}Li_{0.2}Mn_{0.6}]O_2$. J Am Chem Soc 128: 8694–8698.

Arunkumar TA, Alvarez E, Manthiram A (2007a) Structural, chemical, and electrochemical characterization of layered $Li[Li_{0.17}Mn_{0.33}Co_{0.5-y}Ni_y]O_2$ cathodes. J Electrochem Soc 154: A770–A775.

Arunkumar TA, Manthiram A (2005) Influence of lattice parameter differences on the electrochemical performance of the 5 V spinel $LiMn_{1.5-y}Ni_{0.5-z}M_{y+z}O_4$ (M = Li, Mg, Fe, Co, and Zn). Electrochem Solid State Lett 8: A403–A405.

Arunkumar TA, Wu Y, Manthiram A (2007b) Factors influencing the irreversible oxygen loss and reversible capacity in layered $Li[Li_{1/3}Mn_{2/3}]O_2 - Li[M]O_2$ (M = $Mn_{0.5-y}Ni_{0.5-y}Co_{2y}$ and $Ni_{1-y}Co_y$) solid solution. Chem Mater 19: 3067–3073.

Auborn JJ, Barbeiro YL (1987) Lithium intercalation cells without metallic lithium. J Electrochem Soc 134: 638–641.

Aydinol MK, Ceder G (1997) First-principles prediction of insertion potentials in Li-Mn oxides for secondary Li batteries. J Electrochem Soc 144: 3832–3835.

Brousse T, Retoux R, Herterich U, Schleich DM (1998) Thin-film crystalline SnO_2-lithium electrodes. J Electrochem Soc 145: 1–4.

Chan CK, Peng H, Liu G, McIlwrath K, Zhang XF, Cui Y (2008a) High-performance lithium battery anodes using silicon nanowires. Nat Nanotechnol 3: 31–35.

Chan CK, Zhang XF, Cui Y (2008b) High capacity Li ion battery anodes using Ge nanowires. Nano Lett 8: 307–309.

Cho J, Kim YJ, Kim TJ, Park B. (2001a) Zero-strain intercalation cathode for rechargeable Li-ion cell. Angew Chem Int Ed Engl 40: 3367–3369.

Cho J, Kim YJ, Park B (2001b) LiCoO$_2$ cathode material that does not show a phase transition from hexagonal to monoclinic phase. J Electrochem Soc 148: A1110–A1115.

Choi J, Alvarez E, Arunkumar TA, Manthiram A (2006), Proton insertion into oxide cathodes during chemical delithiation. Electrochem Solid State Lett 9: A241–A244.

Choi J, Manthiram A (2004) Comparison of the electrochemical behaviors of stoichiometric LiNi$_{1/3}$Co$_{1/3}$Mn$_{1/3}$O$_2$ and lithium excess Li$_{1.03}$(Ni$_{1/3}$Co$_{1/3}$Mn$_{1/3}$)$_{0.97}$O$_2$. Electrochem Solid State Lett 7: A365–A368.

Choi W, Manthiram A (2006) Superior capacity retention spinel oxyfluoride cathodes for lithium ion batteries. Electrochem Solid State Lett 9: A245–A248.

Choi W, Manthiram A (2007) Factors controlling the fluorine content and the electrochemical performance of spinel oxyfluoride cathodes. J Electrochem Soc 154: A792–A797.

Courtney IA, Dahn JR (1997) Electrochemical and *in situ* X-ray diffraction studies of the reaction of lithium with tin oxide composites. J Electrochem Soc 144: 2045–2052.

Chung SY, Bloking JT, Chiang YM, (2002) Electronically conductive phospho-olivines as lithium storage electrodes. Nature Mater 1: 123–128.

Dahn JR, Zheng T, Liu Y, Xue JS (1995) Mechanisms for lithium insertion in carbonaceous materials. Science 270: 590–593.

Deng D, Lee JY (2008) Hollow core-shell mesospheres of crystalline SnO$_2$ nanoparticle aggregates for high capacity Li$^+$ ion storage. Chem Mater 20: 1841–1846.

Ellis B, Kan WH, Makahnouk WRM, Nazar LF (2007) Synthesis of nanocrystals and morphology control of hydrothermally prepared LiFePO$_4$. J Mater Chem 17: 3248–3254.

Ferg E, Gummow RJ, de Kock A, Thackeray MM (1994) Spinel anodes for lithium-ion batteries. J Electrochem Soc 141: L147–L150.

Gabano JP (1983) Lithium batteries. Academic Press, London.

Goodenough JB, Mizushima K, Takeda T (1980) Solid-solution oxides for storage-battery electrodes. Jap J Appl Phys Suppl 19–3: 305–313.

Hassoun J, Reale P, Scrosati B (2007) Recent advances in liquid and polymer lithium-ion batteries. J Mater Chem 17: 3668–3677.

Idota Y, Kubota T, Matsufuji A, Maekawa Y, Miyasaka T (1997) Tin-based amorphous oxide: a high-capacity lithium-ion-storage material. Science 276: 1395–1397.

Julien C, Nazri, GA (1994) Solid state batteries: materials design and optimization. Kluwer, Boston, MA.

Kannan AM, Manthiram A (2002) Surface/chemically modified LiMn$_2$O$_4$ cathodes for lithium-ion batteries. Electrochem Solid State Lett 5: A167–A169.

Kepler KD, Vaughey JT, Thackeray MM (1999) Li$_x$Cu$_6$Sn$_5$ (0<x<13): an intermetallic insertion electrode for rechargeable lithium batteries. Electrochem Solid State Lett 2: 307–309.

Kim DH, Kim J (2006) Synthesis of LiFePO$_4$ nanoparticles in polyol medium and their electrochemical properties. Electrochem Solid State Lett 9: A439–A442.

Kim DW, Hwang IS, Kwon SJ, Kang HY, Park KS, Choi YJ, Choi KJ, Park JG (2007) Highly conductive coaxial SnO$_2$-In$_2$O$_3$ heterostructured nanowires for Li ion battery electrodes. Nano Lett 7: 3041–3045.

Linden D (ed) (1995) Handbook of batteries. 2nd ed., McGraw Hill, New York.

Lu Z, Beaulieu LY, Donaberger RA, Thomas CL, Dahn JR (2002) Synthesis, structure, and electrochemical behavior of Li[Ni$_x$Li$_{1/3-2x/3}$Mn$_{2/3-x/3}$]O$_2$. J Electrochem Soc 149: A778–A791.

Mabuchi A, Tokumitsu K, Fujimoto H, Kasuh T (1995) Charge-discharge characteristics of the mesocarbon miocrobeads heat-treated at different temperatures. J Electrochem Soc. 142: 1041–1046.

MacKinnon MR, Dahn JR (1999) On the aggregation of tin in SnO composite glasses caused by the reversible reaction with lithium. J Electrochem Soc 146: 59–68.

Manthiram A, Goodenough JB (1987) Lithium insertion into $Fe_2(MO_4)_3$ frameworks: comparison of M = W with M = Mo. J Solid State Chem 71: 349–360.

Manthiram A, Goodenough JB (1989) Lithium insertion into $Fe_2(SO_4)_3$-type frameworks. J Power Sources 26: 403–406.

Megahed S, Scrosati B (1994) Lithium-ion rechargeable batteries. J Power Sources 51: 79–104.

Mizushima K, Jones PC, Wiseman PJ, Goodenough, JB (1980) Li_xCoO_2 ($0 < x < -1$): a new cathode material for batteries of high energy density. Mat Res Bull 15: 783–789.

Nazri GA, Pistoia G (eds) (2003) Science and technology of lithium batteries. Kluwer Academic Publishers, Boston, MA.

Padhi AK, Nanjundasawamy KS, Goodenough JB (1997) Phospho-olivines as positive-electrode materials for rechargeable lithium batteries. J Electrochem Soc 144: 1188–1194.

Park S, Han Y, Kang Lee PS, Ahn S, Lee H, Lee J (2001) Electrochemical properties of $LiCoO_2$-coated $LiMn_2O_4$ prepared by solution-based chemical process. J Electrochem Soc 148: A680–A686.

Pistoia G (ed) (1994) Lithium batteries: new materials, developments and perspectives. Vol. 5, Elsevier, Amsterdam.

Poizot P, Laruelle S, Grugeon S, Dupont L, Tarascon JM (2000) Nano-sized transition-metal oxides as negative-electrode materials for lithium-ion batteries. Nature, 407: 496–499.

Reimers JN, Dahn JR (1992) Electrochemical and *in situ* X-ray diffraction studies of lithium intercalation in Li_xCoO_2. J Electrochem Soc 139: 2091–2097.

Sato K, Noguchi M, Demachi A, Oki N, Endo M, (1994) A mechanism of lithium storage in disordered carbons. Science 264: 556–558.

Shin Y, Manthiram A (2003) High rate, superior capacity retention $LiMn_{2-2y}Li_yNi_yO_4$ spinel cathodes for lithium ion batteries. Electrochem Solid State Lett 6: A34–A36.

Souza DCS, Pralong V, Jacobson AJ, Nazar LF (2002) A reversible solid-state crystalline transformation in a metal phosphide induced by redox chemistry. Science 296, 2012–2015.

Tarascon JM, Armand M (2001) Issues and challenges facing rechargeable lithium batteries. Nature, 414: 359–367.

Thackeray MM, David WIF, Bruce PG, Goodenough JB (1983) Lithium insertion into manganese spinels. Mat Res Bull 18: 461–472.

Thackeray MM, Kang SH, Johnson CS, Vaughey JT, Benedek R, Hackney SA (2007) Li_2MnO_3-stabilized $LiMO_2$ (M = Mn, Ni, Co) electrodes for lithium-ion batteries. J Mater Chem 17: 3112–3125.

Vadivel Murugan A, Muraliganth T, Manthiram A (2008), Rapid, size-controlled, microwave-solvothermal synthesis of phospho-olivine nanorods and their coating with a mixed conducting polymer for lithium ion batteries. Electrochem Commun 10: 903–906.

Vaughey JT, Kepler KD, Benedek R, Thackeray MM (1999) NiAs- versus zinc-blende-type intermetallic insertion electrodes for lithium batteries: lithium extraction from Li_2CuSn. Electrochem Comm 1: 517–521.

Venkatasetty HV (1984) Lithium battery technology. John Wiley, New York.

Venkatraman S, Shin Y, Manthiram A (2003) Phase relationships and structural and chemical stabilities of charged $Li_{1-x}CoO_{2-\delta}$ and $Li_{1-x}Ni_{0.85}Co_{0.15}O_{2-\delta}$. Electrochem Solid State Lett 6: A9–A12.

Wakihara M, Yamamoto O (ed) (1998) Lithium ion batteries: fundamentals and performance. Wiley-VCH, Weinheim.

Wen Z, Wang Q, Zhang Q, Li J (2007) In-situ growth of mesoporous SnO_2 on multiwalled carbon nanotubes: a novel composite with porous-tube structure as anode for lithium batteries. Adv Funct Mater 17: 2772–2778.

Winter M, Besenhard JO, Spahr ME, Novák P (1998), Insertion electrode materials for rechargeable lithium batteries. Adv Mater 10: 725–763.

Whittingham MS, Jacobson AJ (1982) Intercalation chemistry. Academic Press, New York.

Wu Y, Vadivel Murugan A, Manthiram A (2008) Surface modification of high capacity layered $Li[Li_{0.2}Mn_{0.54}Ni_{0.13}Co_{0.13}]O_2$ cathodes by $AlPO_4$. J Electrochem Soc 155: A635–A641.

Zhang WM, Hu JS, Guo YG, Zheng SF, Zhong LS (2008) Tin-nanoparticles encapsulated in elastic hollow carbon spheres for high performance anode material in lithium ion batteries. Adv Mater 20: 1160–1165.

Zheng T, Xue JS, Dahn JR (1996) Lithium insertion in hydrogen-containing carbonaceous materials. Chem Mater 8: 389–393.

Zhong Q, Banakdarpour A, Zhang M, Gao Y, Dahn JR (1997) Synthesis and electrochemistry of $LiNi_xMn_{2-x}O_4$. J Electrochem Soc 144: 205–213.

Part V
Selected Applications of Energy Harvesting Systems

Chapter 15
Feasibility of an Implantable, Stimulated Muscle-Powered Piezoelectric Generator as a Power Source for Implanted Medical Devices

B.E. Lewandowski, K. L. Kilgore, and K.J. Gustafson

Abstract A piezoelectric energy generator that is driven by stimulated muscle and is implantable into the human body is under development for use as a self-replenishing power source for implanted electronic medical devices. The generator concept includes connecting a piezoelectric stack generator in series with a muscle tendon unit. The motor nerve is electrically activated causing muscle contraction force to strain the piezoelectric material resulting in charge generation that is stored in a load capacitor. Some of the generated charge is used to power the nerve stimulations and the excess is used to power an implanted device. The generator concept is based on the hypothesis that more electrical power can be converted from stimulated muscle contractions than is needed for the stimulations, a physiological phenomenon that to our knowledge has not previously been utilized. Such a generator is a potential solution to some of the limitations of power systems currently used with implanted devices.

15.1 Introduction

Implanted electronic medical devices provide beneficial therapies and increase the quality of life of many patients. In particular, functional electrical stimulation (FES) devices, also referred to as neural prostheses, restore some neurological function in spinal cord injured (SCI) patients. There are approximately 11,000 new cases of SCI each year in the United States (NSCISC, 2006), resulting in various degrees of impairment of the many functions humans take for granted. Motor function for reaching and grasping objects, interacting with computers and other machines or appliances, bending, and walking can be impaired along with involuntary functions such as respiration and bladder control. FES devices use electrical current pulses to artificially stimulate nerves in patterns that result in muscle contractions that allow these various functions to be restored to some extent (Bhadra et al., 2001;

B.E. Lewandowski (✉)
NASA Glenn Research Center, Bioscience and Technology Branch, Cleveland, OH 44135, USA; Department of Biomedical Engineering, Case Western Reserve University, Cleveland, OH 44106, USA
e-mail: beth.lewandowski@nasa.gov

S. Priya, D.J. Inman (eds.), *Energy Harvesting Technologies*,
DOI 10.1007/978-0-387-76464-1_15 © Springer Science+Business Media, LLC 2009

Bhadra and Peckham, 1997; Creasey, 1993; Glenn et al., 1986; Keith et al., 1989). FES devices improve the quality of life of persons with SCI by allowing them to perform activities of daily living independently and in some cases return to work. FES devices are implanted into the body and require electrical power for operation. Power is obtained from either batteries implanted along with the device or from an external transcutaneous power source.

The majority of spinal cord injuries occur in young adults between the age of 16 and 30 (NSCISC, 2006). Therefore, the timeframe over which the FES device is needed can be quite long, potentially 50 years or longer, as life expectancy for SCI patients with less severe injuries are only slightly less than people without SCI (NSCISC, 2006). Batteries that are implanted with an electronic device need to be replaced when they are depleted. Battery depletion may occur several times over the lifetime of the device, requiring surgery each time battery replacement is needed. Replacement of implanted batteries requires frequent, costly surgeries with increased risk of complications. Documented clinical experience with pacemakers and implanted defibrillators highlight the limitations of implanted batteries. The mean time to when pacemaker battery replacement is needed is 8 years after the initial implantation (Kindermann et al., 2001). Depletion of implanted batteries is the reason for more than 70% of pacemaker replacement surgeries and the complication rate after a replacement surgery is three times greater than for the initial implant (Deharo and Djiane, 2005). For implanted defibrillators, patients require battery replacement surgery 3–4 years after initial implantation with costs up to $10,000 (Vorperian et al., 1999).

Transcutaneous energy transfer systems provide high levels of power to neural prostheses through radio frequency energy transfer between external and internal coils. These systems require bulky external equipment including a coil fixed to the chest, a coil driver power supply and wire leads between the driver and the coil (Puers and Vandevoorde, 2001). The external equipment can be damaged and the wires can tangle, it is burdensome to carry, cosmetically unappealing and it is unable to be used in the shower or in a rehabilitation pool. Small movements of the external or internal coils out of alignment will reduce the efficiency of the power transferred and large misalignments can severely reduce the power transferred, resulting in situations of device malfunction (Ozeki et al., 2003; Puers and Vandevoorde, 2001). The resistance inherent in the internal coil causes the coil to heat during operation. The heat that is generated has the potential to cause tissue necrosis or an inflammatory reaction (Puers and Vandevoorde, 2001). So while implanted medical devices are very beneficial to patients, there is room for improvement in how power is supplied to the devices. A totally implanted generator driven by a physiological process resulting in a replenishable and sustainable source of power is an attractive solution to the limitations of the power sources currently used to power FES devices.

15.2 Generator Driven by Muscle Power

Our generator will be driven from the force and power associated with the physiological process of muscle contraction. Muscle contraction is initiated through natural or artificial electrical stimulation of the motor nerve, resulting in an action potential traveling the length of the nerve. When the action potential reaches the nerve ending acetylcholine is released. This causes acetylcholine-gated channels on the muscle fibers to open, allowing sodium ions to flow through. The increase in sodium ions within the muscle fibers causes an action potential to be generated and propagated throughout the muscle fiber. The muscle fiber action potential causes the sarcoplasmic reticulum to release calcium ions. These calcium ions play a role in activating the attraction between the actin and myosin filaments within the muscle fiber, which is what causes the contractile forces of the muscle to occur. Prior to the attraction between the actin and the myosin filaments, the chemical energy available from adenosine triphosphate (ATP) is utilized by the myosin filaments to cause a conformational change in a portion of the filaments. The conformational change allows the myosin filament to be in the correct position for interaction with the actin filament and to have the energy needed for the muscle to produce contractile forces and mechanical power (Guyton and Hall, 2000).

There is a large body of literature available reporting the force characteristics of muscles when they are artificially stimulated with current pulses (Ding et al., 2007, 2005, 2000; Griffin et al., 2002; Gustafson et al., 2006; Karu et al., 1995; Lertmanorat et al., 2006; Parmiggiani and Stein, 1981; Wexler et al., 1997). A single current pulse, ranging in amplitude from 0.5 to 1 mA, with a pulse width ranging from 10 to 500 μs, applied to a motor nerve will cause the muscle to produce a single burst of force, referred to as a twitch. Figure 15.1A provides an illustration of a twitch force burst with a generalized amplitude and time scale. The amplitude of the force burst depends on the size of the muscle and its duration depends on the muscle fiber type. When a train of current pulses are applied to the motor nerve multiple force bursts result. At lower frequencies (1–30 Hz), the force bursts will look like individual twitches repeated at regular intervals, at higher frequencies (30–50 Hz) the force bursts occur more quickly resulting in each subsequent force burst in the train adding to the one before it, as shown in Fig. 15.1B. When current pulse trains are applied to the motor nerve at the frequencies greater than approximately 50 Hz, the force bursts fuse together to form one large force burst

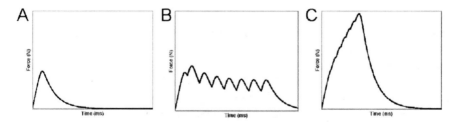

Fig. 15.1 Generalized example of muscle force production resulting from current pulse application to the motor nerve. (**A**) A single pulse of current produces a low-amplitude burst of force. (**B**) A train of current pulses applied at a mid-range frequency (30–50 Hz) results in addition of the force bursts but they are un-fused. (**C**) A train of current pulses applied at high frequencies (>50 Hz) results in a fused, high-amplitude force burst

(Fig. 15.1C). As the frequency continues to increase, the amplitude of this force burst will increase to a maximum level, after which increases in the frequency will no longer result in increases in the force. An estimate of the maximum contraction forces can be found by multiplying the muscle's physiological cross-sectional area by a conversion factor of $35\,N\,cm^{-2}$ (Guyton and. Hall, 2000). The force produced by a twitch contraction is approximately 10–30% of the maximal contractive force of the muscle. The physiological cross-sectional area of the muscles of the limbs and trunk of the human body range from 0.2 to $230\,cm^2$ (Fukunaga et al., 1992; Marras et al., 2001; Maurel, 1998; Pierrynowski, 1995) and therefore have the capacity to produce maximal forces of 8–8000 N.

While the muscles have the capability to produce enormous amounts of force, they are unable to sustain this force production for very long due to muscle fatigue. The sustainable mechanical output power of a muscle is a function of the force produced by the muscle, the distance traveled by the muscle fibers during contraction and the contraction rate. As an example, a study experimentally quantifying the sustained output power of muscle used a muscle contraction force over the range of 10–30 N, a change in muscle length of 1–3 cm and a contraction rate of 30–60 contractions per minute (Gustafson et al., 2006). The results of this study and others found a conservative estimate of the sustained output power of stimulated, conditioned muscle producing isotonic maximal contractions to be 1 mW/g (Araki et al., 1995; Gustafson et al., 2006; Trumble et al., 1997). The mass of the muscles of the limbs and trunk of the human body range from 3 to 814 g (Pierrynowski, 1995) and therefore have the capacity to produce up to approximately 800 mW of mechanical power. If this muscle power is generated from electrical stimulation, an estimate of the range of electrical stimulation power needed to produce this amount of muscle output power is 0.05–6 μW. The high end of this range is based on 1 mA, 500 μs current pulses, applied at 50 Hz for 250 ms per contraction at a rate of 1 contraction per second, assuming a 1 kΩ impedance. The low end of the range is based on single current pulse of 500 μA for 200 μs through a resistance of 1 kΩ, operating at 1 Hz.

When comparing the mechanical output power to the electrical power necessary for motor nerve stimulation, we see that muscle acts as a power amplifier. Just a small amount of electrical power initiates the chemical reaction that converts the chemical energy within the muscle to mechanical power. Our generator will exploit this power amplification characteristic of muscle, a physiological phenomenon that, to our knowledge, has not been previously utilized. An illustration of the fundamental concept of our implantable generator is shown in Fig. 15.2. A generator that converts mechanical energy to electrical energy is connected in series with a muscle-tendon unit and bone. Repetitive stimulation of the nerve innervating the muscle results in repetitive muscle contractions that are used to drive the generator. The generated power is stored in energy storage circuitry. A portion of the generated output power will be used to power the nerve stimulator and the remaining power will be available to power the targeted application. Existing conversion techniques will be used to convert the mechanical power of the muscle to electrical power.

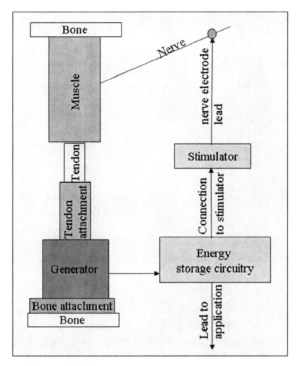

Fig. 15.2 The generator concept includes a piezoelectric stack connected in series with a muscle-tendon unit and bone. Stimulation of the motor nerve results in repetitive muscle contractions that strain the piezoelectric material producing charge that is stored in load circuitry. A portion of the generated output power will be used to power the nerve stimulator and the remaining power will power the targeted application

15.3 Selection of Mechanical-to-Electrical Conversion Method

The most common method for converting mechanical-to-electrical energy is an electromagnetic induction generator. Electromagnetic induction generators convert kinetic energy to electricity through the movement of a magnet through a coil, or visa versa. A simple example of a linear electromagnetic generator is shown in Fig. 15.3A. A force pushes the magnet through the coil, the spring reverses the motion and pushes it back through the coil, resulting in the magnet moving through the coil in a back and forth motion. The open-circuit voltage generated by the linear electromagnetic generator (V_{Mag}) is:

$$V_{\text{Mag}} = NBA_{\text{Coil}}v \qquad (15.1)$$

where N is the number of turns of the coil, B is the magnetic strength of the magnet, A_{Coil} is the cross-sectional area of the coil, and v is the velocity of the magnet as it travels through the coil (Roundy et al., 2004). While this method is appropriate for many different applications, it is not an appropriate application for an implantable,

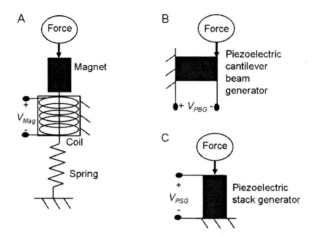

Fig. 15.3 Simple illustrations of mechanical-to-electrical energy conversion methods. (**A**) Linear electromagnetic generator. (**B**) Piezoelectric cantilever beam generator. (**C**) Piezoelectric stack generator

muscle driven generator. The reason for this is that the voltage produced in the coil is dependent on velocity. More voltage, and ultimately power, is available from the system the faster the magnet moves relative to the coil. As can be seen from the paragraphs above, fast, vibratory movement is not what the muscle produces. It produces large amounts of force, but with small displacements. To avoid fatiguing the muscle, contraction repetition rates must be kept at 1 Hz or less. The velocity could potentially be increased with the use of a spring or other mechanical device, but implantation complications will arise with such a design. When a device of any type is implanted into the body, the body's response is to encapsulate it with fibrous growth. Previous attempts at chronically implanting power-generating devices that require movement for operation have resulted in reductions in the efficiency of the generator due to the fibrous growth. For example, after 12 weeks of operation a 65% reduction in output pressure was found with a device used to convert muscle power to pneumatic pressure for cardiac assist (Mizuhara et al., 1996).

The method that will be used to convert mechanical-to-electrical energy in our system is through the use of piezoelectric material, which has a unique property where charge is generated when the material is strained by an external stress. There are two popular types of piezoelectric generators, cantilever beam generators and stack generators, as shown in Fig. 15.2B and C. Force is repetitively applied at the tip of a cantilever beam made of piezoelectric material in the cantilever beam piezoelectric generator. The resulting open-circuit voltage of this generator (V_{PBG}) is:

$$V_{PBG} = \frac{3}{4} \frac{g_{31} L F}{W t} \quad (15.2)$$

where g_{31} is the piezoelectric constant of the material for the case when the force is applied perpendicular to the direction in which the material is poled, L is the length of the piezoelectric beam, W is the width of the piezoelectric beam, t is the thickness of the beam, and F is the force application (http://www.americanpiezo.com/piezo_ theory/chart2.html#, 2005). The displacement of the beam depends on the length of the beam and the elasticity of the material. An estimate of the displacement for our application would be in the millimeter range. While this is less displacement than in the case of the electromagnetic generator, this still would most likely suffer from a decrease in efficiency after implantation, as was seen in a generator developed for powering a pacemaker. A mass was placed on the end of a single piezoelectric cantilever beam ceramic wafer ($2 \times 5 \times 1$ cm) and packaged – in a box for attachment to the heart. When 80 bpm mechanical pulses shook the box, the piezoelectric material was vibrated at 6.5 Hz, resulting in 160 μW of power (Ko, 1966; 1980). When chronically driven by actual canine heart contractions, the efficiency of the generator was reduced to a sustained output power of 30 μW (Ko, 1980). The reduction in efficiency was due to the reaction of the tissue to wall-off the generator thus minimizing force transfer to the generator.

A piezoelectric stack generator is made up of many thin layers of piezoelectric material mechanically connected in series and electrically connected in parallel. As shown in Fig. 15.1C, a compressive force applied to the stack will result in an open-circuit voltage (V_{PSG}):

$$V_{PSG} = \frac{g_{33} t F}{A} \qquad (15.3)$$

where g_{33} is the piezoelectric constant of the material for the case in which the force is applied in the same direction the material is poled, A is the cross-sectional area of the piezoelectric material, t is the thickness of the individual layers of the stack, and F is a compressive force applied to the stack (http://www.americanpiezo.com/piezo_theory/chart2.html#, 2005; Roundy et al., 2004). The displacement of the piezoelectric stack generator is in the micrometer range for force applications in the range possible from muscles. This amount of movement should be undetectable by the surrounding tissue. This minimal excursion of the piezoelectric material dictates that isometric muscle contractions be used. While shortening muscle contractions and long excursions of a power-generating device is the ideal scenario for power production by muscle and may initially appear to be more advantageous, in the long term the efficiency of such a device will decrease. In contrast, the efficiency of our conversion device should not decrease due to tissue encapsulation.

15.4 Properties of Piezoelectric Material Relevant to the Generator System

An illustration of a circuit representation of a piezoelectric generator and a simple load circuit that can be used to harness the charge developed from strained piezoelectric material is shown in Fig. 15.4. A piezoelectric generator can be electrically represented as a voltage source (V_p) in series with a capacitance (C_p). A simple load

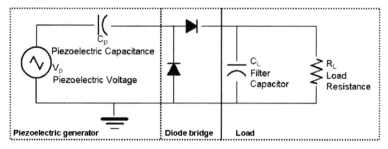

Fig. 15.4 Circuit representation of a piezoelectric generator and a load circuit

circuit includes a diode bridge, a load capacitor (C_L) and a load resistor (R_L). V_p depends on the type of piezoelectric generator used and is described by equations equating the piezoelectric voltage to force such as in Eqs. (15.2) and (15.3). C_p is described by the equation:

$$C_p = \frac{n E_r E_o A}{t} \tag{15.4}$$

where n is the number of layers of piezoelectric material, E_r is the dielectric constant of the piezoelectric material, E_o is the dielectric constant of free space $= 8.9\,\mathrm{pFm^{-1}}$, and A and t are as defined above (Cobbold, 1974). The diode bridge rectifies the piezoelectric voltage and the charge generated by the piezoelectric generator is stored in the load capacitor. The charging time increases and the leakage current decreases as the size of the load capacitor increases. The load resistor is matched to the impedance of the piezoelectric generator for maximum power conversion:

$$R_L = \frac{1}{f C_p} \tag{15.5}$$

where f is the frequency of force application (Ottman et al., 2003). Calculation of the output power of the generator is achieved using the steady-state voltage (V_{Lss}) across the load resistor with the equation:

$$P_{out} = \frac{V_{Lss}^2}{R_L} \tag{15.6}$$

When the impedances are matched, the steady-state output voltage will equal to one-half of the peak piezoelectric voltage, neglecting the voltage drop in the diodes:

$$V_{Lss} = \frac{V_{pm}}{2} = \frac{g_{33} t}{2A} F_m \tag{15.7}$$

F_m is the peak amplitude of the input force pulse and V_{pm} is the peak piezoelectric voltage. Substituting Eqs. (15.2) into (15.4) and Eqs. (15.4) and (15.5) into

(Eq. 15.3) result in the following equation for the average optimal output power $(P_{out\ opt})$ in terms of the system parameters:

$$P_{out\ opt} = \frac{V_{pm}^2 f C_p}{4} = \frac{g_{33}^2 F_m^2 tn E_r E_o f}{4A} \tag{15.8}$$

From the above equations, it is evident that the output power of the generator is dependent on the material properties of the piezoelectric generator. Ceramic and polymer materials are the two main classes of material from which commercially available piezoelectric material is made. Variations in the composition of materials within these two classes have resulted in many different types of materials with different piezoelectric properties. Table 15.1 lists some different types of ceramic and polymer piezoelectric materials and their piezoelectric and dielectric material properties. The output power of the generator is dependent on the square of the voltage generated by the piezoelectric material, its capacitance, and the frequency at which the generator is driven. The piezoelectric voltage is dependent on the piezoelectric constant of the material and the input force and the capacitance is dependent on the dielectric constant of the material. While for maximum power all of these variables should be maximized, through inspection of Table 15.1, we see that a tradeoff exists since there is an inverse relationship between the piezoelectric constants and the dielectric constant. The polymer materials have much higher piezoelectric constants than the ceramic materials, but also have a much lower dielectric constant.

PVDF is typically manufactured into a thin film. From the values in Table 15.1 and the above equations, one can see that the capacitance of PVDF thin films will be very small in low-frequency applications. This will cause the impedance to be large and difficult to match in the load circuit. This is not well suited for use in a muscle-driven generator since the frequency of muscle contractions needs to be kept low to avoid fatigue. In addition, application of large forces will result in the production of extremely large piezoelectric voltages. Since the forces available from the muscle are high, a method would be needed to step down the generated voltage to a usable level. This would add more complexity and sources of loss

Table 15.1 Properties of some example piezoelectric materials

Material type	g_{31} $(V\ mN^{-1})$	g_{33} $(V\ mN^{-1})$	E_r
Ceramic			
Lead magnesium niobate–lead titanate	−0.024	0.043	4629
Lead zirconate titanate	−0.0095	0.013	5400
Lead metaniobate	−0.007	0.032	270
Barium titanate	−0.005	0.013	1250
Bismuth titanate	−0.004	0.017	120
Polymer			
Polyvinylidene fluoride	0.216	N/A	12.5
Copolymer of polyvinylidene fluoride	0.162	N/A	7.5

to the system. However, others have successfully used PVDF generators in their low-power biological applications. The feasibility of a PVDF piezoelectric material mounted on a simply supported beam for use as a bone strain sensor and telemeter has been assessed. The prototype produced sub-microwatts of power (Elvin et al., 2001). In another application, PVDF piezoelectric material was rolled into a tube and connected between two ribs in a canine. The rib displacement during breathing produced a strain on the piezoelectric material, which produced 17 μW of power for a microprocessor-controlled insulin delivery pump application (Hausler and Stein, 1988). While PDVF was suitable for these low power applications, a ceramic piezo-electric stack was selected for our application in an effort to achieve greater output power. A ceramic stack will produce an operating voltage in a useable range in response to the force levels that will be produced by the muscle and it has a larger capacitance so that the load impedance can be more easily matched at the low-operating frequencies.

15.5 Predicted Output Power of Generator

Further development and analysis of our generator is reported in more depth elsewhere (Lewandowski et al., 2007), but is summarized here. A software model of the generator concept was developed to aid in the design of our system and simulations were performed to predict the output power of the generator. To verify the accuracy of the software model, simulated output power was compared to output power from experimental tests. In the experimental tests, a material testing system (MTS) (MTS Systems Corporation, Eden Prairie, MN) was used to apply 250 ms compressive, triangular force pulses, with peak values of 25, 50, 100, 150, 200, and 250 N, to a PZT stack (part number T18-H5-104, Piezo Systems, Inc. Cambridge, MA) at 1 Hz for 240 s. The output power of the generator was calculated from the recorded steady-state load voltage. A mathematical representation of muscle force was used as input to the software model. The dimensions and material properties of the stack were entered into the software model and simulations were run using the same input force conditions as in the experimental tests to find the predicted output power. An example of a 100 N input force is shown in Fig. 15.5A. Figure 15.5B shows the resulting load voltage from the simulations and the experimental tests. There is a good comparison between the simulated and the experimental output power (Fig. 15.5C), verifying the accuracy of the software model.

With the accuracy of the software model confirmed, an analysis was completed to determine the affect of the system parameters on the output power. The system parameters were varied within their physical and physiological constraints. The result of the analysis was size and material property specifications for a piezoelectric stack best suited for our application. We identified that the cross-sectional area (A) of the stack should be minimized and the length (L) (the number of layers times the thickness of one layer) should be maximized. The length of our generator will essentially replace the tendon of the muscle and will depend on the space available for implantation.

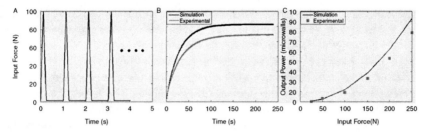

Fig. 15.5 Approximation of the force produced by muscle twitches used as input to the software model. This force was also applied to a prototype stack with an MTS machine in experimental trials. (**B**) Simulated and experimental load voltage of the generator system. (**C**) Simulated and experimental output power of the generator as a function of peak input force

A relationship for E_r and g_{33} was determined from data sheets for several commercially available piezoelectric materials. Values for g_{33} and E_r which balanced the tradeoff between these values were identified. The input force (F) will depend on the size of the muscle used to drive the generator and the frequency (f) will depend on the rate at which muscle contractions can be sustained without fatigue.

Using parameter values resulting from our analysis, Fig. 15.6 shows the predicted output power of the generator over a range of input force. As stated above, the power necessary to stimulate the motor nerve will range from 0.05 to 6 μW. In addition to the power necessary for stimulation, power will also be needed for circuitry to control and deliver the stimulation parameters. We estimate that a simple controller designed with commercially available electronics will consume approximately 2–8 μW of power. Therefore, the total input power for our design is

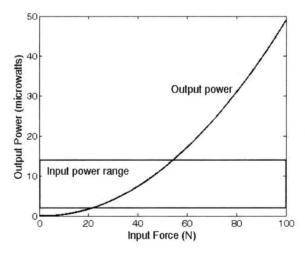

Fig. 15.6 Predicted output power of generator system with system parameters best suited for this application over a range of input force. Also included is the range of input power necessary for motor nerve stimulation

the controller power plus the stimulation power, or approximately $2–14\,\mu W$. This range is highlighted in the graph in Fig. 15.6 so that a comparison between the predicted output power and the required input power can be made. We can see that a force of approximately 55 N is needed to produce $14\,\mu W$ of power. Therefore, we predict that a muscle that can produce a sustained force greater than 55 N will produce power in excess of that needed for stimulation. The larger the force the muscle can produce, the more excess power there will be for the target application. Additional increases in power can occur as improvements in the material properties of piezoelectric generators are realized.

15.6 Steps Towards Reduction to Practice

The software model of the piezoelectric generator indicated that it should be feasible to convert mechanical power from muscle into electrical power in excess of that needed for motor nerve stimulation. The next step toward practical application of this concept is to demonstrate generator operation in acute animal studies. The purpose of the acute animal tests is to demonstrate that it is possible to charge the generator using physiological muscle contractions. Stimulation and load circuitry will be designed and built to assess the ability to generate electrical power from a muscle in excess of that needed for motor nerve stimulation. Following the success of acute animal studies, chronic animal studies will be needed to prove long-term operation of the generator and to address issues such as a permanent tendon and bone attachment strategy and the biocompatibility of the materials used in the system.

As stated above, the generator will be attached between a muscle-tendon unit and bone. Some preliminary work has been completed on a mechanical device designed to hold the piezoelectric generator in place and to convert the tensile force produced by stimulated muscle contractions into a compressive force that is applied to the piezoelectric material (Lewandowski et al., 2007). The MTS testing demonstrated that the enclosure converting tension into compression produced very little loss in the force transmitted to the piezoelectric generator. Attachment to the tendon will likely be with artificial tendon. Artificial tendon is a commercially available product (for example, CardioEnergetics, Inc., Cincinnati, OH) that has been used for other purposes. For example, Trumble et al. (2002) found that connecting their ventricular assist device to artificial tendon made of polyester fibers and incorporating it into the natural tendon was more stable than connecting the device directly to natural tendon. Additionally, it is likely that existing bone attachment strategies developed for orthopedic prosthetics can be used to attach our generator to a bone in close proximity to the tendon. Ideally, the muscle–tendon–generator system will be attached across the length of a single long bone so that no limb movement is produced during electrical stimulation of the muscle.

Since the piezoelectric generator is intended to be chronically implanted inside the body, it must be safe to the tissue, i.e., biocompatible. There are two strategies for achieving biocompatibility. One alternative is to identify a biocompatible piezoelectric material which is safe for direct contact with the body. The second alternative is to encapsulate the entire piezoelectric generator in a biocompatible material, which is the approach commonly taken with most implantable electronic devices. The following paragraphs contain a summary of studies investigating biocompatible piezoelectric material and encapsulation of piezoelectric material.

There are some types of piezoelectric material that may be biocompatible. Barium titanate ($BaTiO_3$) is a material that can be made to have piezoelectric properties and studies have provided evidence that it is biocompatible. It was implanted subcutaneously in the backs of rabbits for 20 weeks and into the canine femurs for up to 99 days. In both cases, the histology results showed no evidence of inflammation or foreign body reactions and only a thin fibrous capsule surrounded the implant (Park et al., 1981, 1977). The purpose of the studies was to assess the feasibility of using $BaTiO_3$ as a material at bone implant interfaces since its piezoelectric properties could be used as a current supply to help with the bone implant healing process. Further studies showed tissue growth into the cylinders of porous $BaTiO_3$ implanted into canine femurs after 16 days and a mature, healthy bone implant interface after 86 days (Park et al., 1980). Polarized $BaTiO_3$ was also implanted into the jaw bone of dogs for up to 12 weeks and showed an improvement in osteogenesis over non-polarized ceramic implants (Feng et al., 1997). Another application in which piezoelectric material is in direct contact with biological tissue is an electrode for stimulation or sensing such as in a pacemaker. Piezoelectric materials suitable for implantation have less attractive piezoelectric properties, resulting in large current consumption (Ljungstrom et al., 2003). This has led researchers to develop biocompatible piezoelectric materials with improved piezoelectric properties consisting of different combinations of Na, K, and NbO_3 (Ljungstrom et al., 2003).

Another strategy for maintaining biocompatibility includes coating the entire generator, electrical circuitry and leads with a soft encapsulate such as silastic or enclosing it in a titanium case. The key design consideration for the piezoelectric generator is that the externally applied tendon tension must be transmitted through the encapsulation material to the piezoelectric component with a minimum of loss. A few studies have reported the use of PZT in encapsulated medical devices that have been chronically implanted into humans and experimental animals. Piezoelectric material has been used as a transducer in implantable hearing aids (Zenner et al., 2000, 2004), as ultrasound transducers for cardiac dimensional and functional analysis (Olsen et al., 1984) and as actuators to produce current to aid in bone fracture healing (Cochran et al., 1988). In these cases, the piezoelectric material has been encapsulated in a hermetically sealed package made of biocompatible material such as titanium or Teflon or sealed in epoxy. In these studies, the function of the active piezoelectric element was not affected due to the encapsulation.

15.7 Conclusion

The conclusion of this investigation is that theoretically it is possible to generate electrical power in excess of the amount necessary for motor nerve stimulation from stimulated muscle contractions, using a piezoelectric stack generator. Such a generator has the potential to replace or augment power systems currently used to power neural prostheses, reducing or eliminating some of the limitations of the power systems currently used with these devices.

References

K. Araki, T. Nakatani, K. Toda, Y. Taenaka, E. Tatsumi, T. Masuzawa, Y. Baba, A. Yagura, Y. Wakisaka, K. Eya, Takano H, and Koga Y "Power of the fatigue resistant in situ latissimus dorsi muscle," *ASAIO J.*, vol. 41, no. 3, p. M768–M771, July, 1995.

N. Bhadra, K. L. Kilgore, and P. H. Peckham, "Implanted stimulators for restoration of function in spinal cord injury," *Med. Eng Phys.*, vol. 23, no. 1, pp. 19–28, Jan, 2001.

N. Bhadra and P. H. Peckham, "Peripheral nerve stimulation for restoration of motor function," *J. Clin. Neurophysiol.*, vol. 14, no. 5, pp. 378–393, Sept, 1997.

R. S. Cobbold, *Transducers for Biomedical Measurements: Principles and Applications*. New York: John Wiley & Sons, Inc., 1974, p. 486.

G. V. Cochran, M. P. Kadaba, and V. R. Palmieri, "External ultrasound can generate microampere direct currents in vivo from implanted piezoelectric materials," *J. Orthop. Res.*, vol. 6, no. 1, pp. 145–147, 1988.

G. H. Creasey, "Electrical stimulation of sacral roots for micturition after spinal cord injury," *Urol. Clin. North Am.*, vol. 20, no. 3, pp. 505–515, Aug, 1993.

J. C. Deharo and P. Djiane, "Pacemaker longevity. Replacement of the device," *Ann. Cardiol. Angeiol. (Paris)*, vol. 54, no. 1, pp. 26–31, Jan, 2005.

J. Ding, L. W. Chou, T. M. Kesar, S. C. Lee, T. E. Johnston, A. S. Wexler, and S. A. Binder-Macleod, "Mathematical model that predicts the force-intensity and force-frequency relationships after spinal cord injuries," *Muscle Nerve*, vol. 36, no. 2, pp. 214–222, Aug, 2007.

J. Ding, S. C. Lee, T. E. Johnston, A. S. Wexler, W. B. Scott, and S. A. Binder-Macleod, "Mathematical model that predicts isometric muscle forces for individuals with spinal cord injuries," *Muscle Nerve*, vol. 31, no. 6, pp. 702–712, June, 2005.

J. Ding, A. S. Wexler, and S. A. Binder-Macleod, "A predictive model of fatigue in human skeletal muscles," *J. Appl. Physiol*, vol. 89, no. 4, pp. 1322–1332, Oct, 2000.

N. Elvin, A.A. Elvin, and M. Spector, "A self-powered mechancial strain energy sensor," *Smart Mater. Struct.*, vol. 10, pp. 293–299, 2001.

J. Feng, H. Yuan, and X. Zhang, "Promotion of osteogenesis by a piezoelectric biological ceramic," *Biomaterials*, vol. 18, no. 23, pp. 1531–1534, Dec, 1997.

T. Fukunaga, R. R. Roy, F. G. Shellock, J. A. Hodgson, M. K. Day, P. L. Lee, H. Kwong-Fu, and V. R. Edgerton, "Physiological cross-sectional area of human leg muscles based on magnetic resonance imaging," *J. Orthop. Res.*, vol. 10, no. 6, pp. 928–934, Nov, 1992.

W. W. Glenn, M. L. Phelps, J. A. Elefteriades, B. Dentz, and J. F. Hogan, "Twenty years of experience in phrenic nerve stimulation to pace the diaphragm," *Pacing Clin. Electrophysiol.*, vol. 9, no. 6 Pt 1, pp. 780–784, Nov, 1986.

L. Griffin, S. Godfrey, and C. K. Thomas, "Stimulation pattern that maximizes force in paralyzed and control whole thenar muscles," *J. Neurophysiol.*, vol. 87, no. 5, pp. 2271–2278, May, 2002.

K. J. Gustafson, S. M. Marinache, G. D. Egrie, and S. H. Reichenbach, "Models of metabolic utilization predict limiting conditions for sustained power from conditioned skeletal muscle," *Ann. Biomed. Eng*, vol. 34, no. 5, pp. 790–798, May, 2006.

A. C. Guyton and J. E. Hall, *Textbook of Medical Physiology*. Philadelphia, PA: Elsevier/Saunders, 2000, p. 968.

E. Hausler and L. Stein, "Implantable physiological power supply with PVDF film," in *Medical Applications of Piezoelectric Polymers*. Galletti P. M., De Rossi D. E., and De Reggi A. S., (Eds.) New York: Gordon and Breach Science Publishers, 1988, pp. 259–264.

Z. Z. Karu, W. K. Durfee, and A. M. Barzilai, "Reducing muscle fatigue in FES applications by stimulating with N-let pulse trains," *IEEE Trans. Biomed. Eng*, vol. 42, no. 8, pp. 809–817, Aug, 1995.

M. W. Keith, P. H. Peckham, G. B. Thrope, K. C. Stroh, B. Smith, J. R. Buckett, K. L. Kilgore, and J. W. Jatich, "Implantable functional neuromuscular stimulation in the tetraplegic hand," *J. Hand Surg. [Am.]*, vol. 14, no. 3, pp. 524–530, May, 1989.

M. Kindermann, B. Schwaab, M. Berg, and G. Frohlig, "Longevity of dual chamber pacemakers: device and patient related determinants," *Pacing Clin. Electrophysiol.*, vol. 24, no. 5, pp. 810–815, May, 2001.

Ko W. H., "Piezoelectric energy converter for electronic implants," *Proc. Ann. Conf. on Eng. in Med. E. Biol.* 8, 1966, p. 67.

Ko W.H., "Power sources for implant telemetry and stimulation systems," in *A Handbook on Biotelemetry and Radio Tracking*. Amlaner C.J. and MacDonald D., Eds. Elmsford, NY: Pergamon Press, Inc., 1980, pp. 225–245.

Z. Lertmanorat, K. J. Gustafson, and D. M. Durand, "Electrode array for reversing the recruitment order of peripheral nerve stimulation: experimental studies," *Ann. Biomed. Eng*, vol. 34, no. 1, pp. 152–160, Jan, 2006.

B. E. Lewandowski, K. L. Kilgore, and K. J. Gustafson, "Design considerations for an implantable, muscle powered piezoelectric system for generating electrical power," *Ann. Biomed. Eng*, vol. 35, no. 4, pp. 631–641, Apr, 2007.

K. Ljungstrom, K. Nilsson, J. Lidman, and C. Kjellman, "Medical implant with piezoelectric material in contact with body tissue," United States Patent 6,571,130, May 27, 2003.

W. S. Marras, M. J. Jorgensen, K. P. Granata, and B. Wiand, "Female and male trunk geometry: size and prediction of the spine loading trunk muscles derived from MRI," *Clin. Biomech. (Bristol., Avon.)*, vol. 16, no. 1, pp. 38–46, Jan, 2001.

W. Maurel, "3D Modeling of the human upper limb including the biomechancis of joints, muscles and soft tissues." Ph.D. Ecole Polytechnique Federale de Lausanne, 1998.

H. Mizuhara, T. Oda, T. Koshiji, T. Ikeda, K. Nishimura, S. Nomoto, K. Matsuda, N. Tsutsui, K. Kanda, and T. Ban, "A compressive type skeletal muscle pump as a biomechanical energy source," *ASAIO J.*, vol. 42, no. 5, p. M637–M641, Sep, 1996.

"Modes of vibration for common piezoelectric ceramic shapes, http://www. americanpiezo.com/piezo_theory/chart2.html#," 2005.

NSCISC "Spinal Cord Injury: Facts and Figures at a Glance from the National Spinal Cord Injury Statistical Center (NSCISC), http://www.spinalcord. uab.edu/show.asp?durki=21446," 2006.

C. O. Olsen, S. J. Abert, D. D. Glower, J. A. Spratt, G. S. Tyson, J. W. Davis, and J. S. Rankin, "A hermetically sealed cardiac dimension transducer for long-term animal implantation," *Am. J. Physiol*, vol. 247, no. 5 Pt 2, p. H857–H860, Nov, 1984.

G. K. Ottman, H. F. Hofmann, and G. A. Lesieutre, "Optimized piezoelectric energy harvesting circuit using stepdown converter in discontinuous conduction mode," *IEEE Trans. Power Electron.*, vol. 18, pp. 696–703, 2003.

T. Ozeki, T. Chinzei, Y. Abe, I. Saito, T. Isoyama, S. Mochizuki, M. Ishimaru, K. Takiura, A. Baba, T. Toyama, and K. Imachi, "Functions for detecting malposition of transcutaneous energy transmission coils," *ASAIO J.*, vol. 49, no. 4, pp. 469–474, July, 2003.

J. B. Park, B. J. Kelly, G. H. Kenner, A. F. von Recum, M. F. Grether, and W. W. Coffeen, "Piezoelectric ceramic implants: in vivo results," *J. Biomed. Mater. Res.*, vol. 15, no. 1, pp. 103–110, Jan, 1981.

J. B. Park, G. H. Kenner, S. D. Brown, and J. K. Scott, "Mechanical property changes of barium titanate (ceramic) after in vivo and in vitro aging," *Biomater. Med. Devices Artif. Organs*, vol. 5, no. 3, pp. 267–276, 1977.

J. B. Park, A. F. von Recum, G. H. Kenner, B. J. Kelly, W. W. Coffeen, and M. F. Grether, "Piezoelectric ceramic implants: a feasibility study," *J. Biomed. Mater. Res.*, vol. 14, no. 3, pp. 269–277, May, 1980.

F. Parmiggiani and R. B. Stein, "Nonlinear summation of contractions in cat muscles. II. Later facilitation and stiffness changes," *J. Gen. Physiol*, vol. 78, no. 3, pp. 295–311, Sep, 1981.

M. R. Pierrynowski, "Analytic representation of muscle line of action and geometry," in *Three-Dimensional Analysis of Human Movement*. P. Allard, I. A. F. Stokes, and Blanchi J. P., Eds. Champaign, IL: Human Kinetics, 1995, pp. 215–256.

R. Puers and G. Vandevoorde, "Recent progress on transcutaneous energy transfer for total artificial heart systems," *Artif. Organs*, vol. 25, no. 5, pp. 400–405, May, 2001.

S. Roundy, P. K. Wright, and J. M. Rabaey, *Energy scavenging for wireless sensor networks*. Norwell, MA: Kluwer Academic Publishers, 2004.

D. R. Trumble, W. A. LaFramboise, C. Duan, and J. A. Magovern, "Functional properties of conditioned skeletal muscle: implications for muscle-powered cardiac assist," *Am. J. Physiol*, vol. 273, no. 2 Pt 1, p. C588–C597, Aug, 1997.

D. R. Trumble, D. B. Melvin, and J. A. Magovern, "Method for anchoring biomechanical implants to muscle tendon and chest wall," *ASAIO J.*, vol. 48, no. 1, pp. 62–70, Jan, 2002.

V. R. Vorperian, S. Lawrence, and K. Chlebowski, "Replacing abdominally implanted defibrillators: effect of procedure setting on cost," *Pacing Clin. Electrophysiol.*, vol. 22, no. 5, pp. 698–705, May, 1999.

A. S. Wexler, J. Ding, and S. A. Binder-Macleod, "A mathematical model that predicts skeletal muscle force," *IEEE Trans. Biomed. Eng*, vol. 44, no. 5, pp. 337–348, May, 1997.

H. P. Zenner, H. Leysieffer, M. Maassen, R. Lehner, T. Lenarz, J. Baumann, S. Keiner, P. K. Plinkert, and J. T. McElveen, Jr., "Human studies of a piezoelectric transducer and a microphone for a totally implantable electronic hearing device," *Am. J. Otol.*, vol. 21, no. 2, pp. 196–204, Mar, 2000.

H. P. Zenner, A. Limberger, J. W. Baumann, G. Reischl, I. M. Zalaman, P. S. Mauz, R. W. Sweetow, P. K. Plinkert, R. Zimmermann, I. Baumann, M. H. De, H. Leysieffer, and M. M. Maassen, "Phase III results with a totally implantable piezoelectric middle ear implant: speech audiometry, spatial hearing and psychosocial adjustment," *Acta Otolaryngol.*, vol. 124, no. 2, pp. 155–164, Mar, 2004.

Chapter 16
Piezoelectric Energy Harvesting for Bio-MEMS Applications

William W. Clark and Changki Mo

Abstract This chapter presents an analysis of piezoelectric unimorph diaphragm energy harvesters as a potential tool for generating electrical energy for implantable biomedical devices. First the chapter discusses current and developing biomedical devices that require energy, and the need for capture of energy from the environment of the implant. Next, a general discussion of piezoelectric harvesters is presented, and a case is made for the use of 31 mode diaphragm harvesters for conversion of energy from blood pressure variations within the body. The chapter then presents derivations of available electrical energy for unimorph diaphragm harvesters, starting with general boundary conditions, and then proceeding to simply supported and clamped conditions of various piezoelectric and electrode coverage. Using these analytical results, the chapter ends with by presenting a brief set of numerical results illustrating the amount of power that could be harvested for a particular size of device, and how that power may be used as a source for a given implanted medical device. In summary, it is shown that the harvester could potencially provide enough power to operate a 10 mW device at reasonable intermittent rates. The relationships provided here may enable other optimal designs to be realized, for medical or for many other applications.

16.1 Introduction

Implantable biotechnology has undergone rapid advances during the past two decades. Of particular interest in this chapter is the technology that has benefited from down-scaling in size and power requirements of electronics and from down-scaling in the size of mechanical and chemical sensors and actuators due to the advances in micro-electro-mechanical systems (MEMS). A host of implantable devices are either realities today, or are currently being investigated (Pelletier, 2004), including cochlear implants, artificial retinas, electrical neuro-stimulators, enhanced

W.W. Clark (✉)
University of Pittsburgh, Pittsburgh, PA 15261, USA
e-mail: wclark@pitt.edu

S. Priya, D.J. Inman (eds.), *Energy Harvesting Technologies*,
DOI 10.1007/978-0-387-76464-1_16 © Springer Science+Business Media, LLC 2009

inter-device telemetry, automated wireless alarm signaling, advanced sensors, and drug-delivery systems, advanced sensors and so-called "laboratories-on-a-chip."

While small size opens the door to previously unattainable applications, there are other factors, such as packaging for structural soundness and harsh environments, and power requirements, which may limit the feasibility of some applications. Of primary interest to this chapter is the considerable energy that is required over the lifetime of many implantable devices, which can dictate the useful lifetime of many implants due to either the longevity of the power source or the ability to recharge the batteries (Soykan, 2002). This chapter addresses the energy issue by considering the amount of power that can be reasonably harvested from within the human body. Specifically addressed it the question of how much energy can be obtained from a blood pressure source driving a piezoelectric diaphragm harvester.

The number of publications on energy harvesting topics has exploded in recent years, however, those focused specifically on biomechanical sources have been limited. One of the earliest and most-cited studies is by Starner (1996) in which human body power sources (e.g., chest motions during breathing and footsteps during walking) were quantified for wearable computer applications. This paper has spurred many additional studies, but they have primarily focused on harvesters that are external to the body and which extract energy from biomechanical motions. For example, another classic early study on human-powered energy harvesting by Kymissis et al. (1998) evaluated energy harvesting from heel strikes during walking. Niu et al. (2004) analyzed joint motions as sources of power and reported some of the largest available estimates on the amount of power that could be generated.

Obviously, implanted devices require implantable power sources. Efforts in this area are much rarer. Starner (1996; Starner and Paradiso, 2004) suggested power availability from internal sources (e.g., blood pressure), but only recently have works appeared that have begun to analytically and experimentally quantify the amount of power available from within the body. Suzuki et al. (2002, 2003) have suggested a charging device composed of a small generator and a permanent-magnet coupling, as well as a hybrid device using magnetic and ultrasonic energies through magnetostriction and piezoelectric vibrators. Platt et al. (2005) carried out an experimental study of energy harvesting in an artificial knee joint after total knee replacement surgery.

Clark and Ramsay (2000) outlined the basic relationships for assessing the amount of power that can be generated from a volume of piezoelectric material, and then used those relationships as a guide to design a harvester for a biomedical implant that utilizes blood pressure as a source of energy. In this chapter, we follow up on that original work, following the same theme of converting blood pressure energy to electrical energy by way of a diaphragm structure. The focus is not membrane stretching, as described in that paper, but is on a bending diaphragm, so the 31-mode of the piezoelectric material is still utilized.

Section 16.2 will review the general conversion principles for a volume of piezoelectric material, as done in the original paper, and then a detailed model will be developed in which the available electrical energy for conversion is derived for a number of different boundary and electrode conditions. There are a significant number of previous works on unimorph piezoelectric circular plates as actuators (Liu et al. 2003; Mo et al. 2006; Prasad et al., 2002), but not as energy harvesters. The fully covered PZT circular diaphragm with clamped edge was developed and tested in previous work (Kim et al. 2005a, b), and the simply supported case was shown by Mo et al. (2007). This chapter follows a similar modeling approach as was done in the previous works, but does so by presenting a uniform modeling approach for arbitrary boundary conditions. Finally, the chapter ends with an analysis of the amount of power available to operate a given bio-MEMS application.

16.2 General Expression for Harvesting Energy with Piezoelectric Device

There are many factors that govern the total electrical power output of a piezoelectric device. In this section, we outline the piezoelectric effect for mechanical-to-electrical energy conversion, identifying the variables that govern the overall electrical power output. The expressions for output voltage, current, and power derived in this chapter will be referred to as "available" quantities, which are the voltage, charge per time, and power that appear on the electrodes of the piezoelectric material. These quantities do not necessarily reflect the harvested voltage, current, or power, since the amount of energy that is actually stored or used by the load will depend greatly on the electrical properties of the harvesting circuit or the load. Available power, however, does provide a good means for designing the harvesting device itself, and for comparing one design to another, since it is independent of the attached circuit. An ideal harvesting circuit would transfer all of the available power to the storage device or load.

When used as a motor, a piezoelectric element undergoes a change in dimensions when it is stressed electrically, or as a generator it produces an electric charge when stressed mechanically. We will make use of the generator behavior of the piezoelement. The relationship between mechanical input and electrical response depends on the piezoelectric properties of the material, the size, and shape of the element, and the direction of mechanical excitation and electrical response.

Three orthogonal axes, 1, 2, and 3, are typically used to identify the directions within a piezoelectric element. The 3-axis is conventionally taken to be parallel to the direction of polarization of the material, established during manufacture. Two common generator modes of operation of a piezoelectric element are shown in Fig. 16.1. The "33" mode, Fig. 16.1a, implies that a tensile or compressive mechanical force is applied along the 3-axis (in the direction of polarization) on the same surface that charge is collected (charge is collected on electrodes placed on opposite sides of the element along the 3-axis). The "31" mode, shown in Fig. 16.1b, implies that the force is applied along the 1-axis (perpendicular to the direction of polarization) and charge is collected on the same surface as before (Jaffe et al., 1971; IEEE Std., 1978).

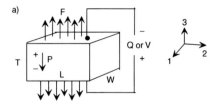

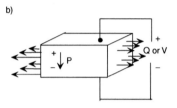

Fig. 16.1 Schematic of force/charge relationship on piezogenerator elements; element in (a) 33-mode, (b) 31-mode

For a given applied force, the electrical energy that can be extracted from a piezo-electric element is

$$\text{Energy} = VQ \tag{16.1}$$

where V and Q are the charge and voltage on the element, respectively. For the element in 33-mode, the charge and voltage can be written as

$$Q = Fd_{33}, \qquad V = \frac{Fg_{33}T}{LW} \tag{16.2}$$

and for the 31-mode

$$Q = \frac{Fd_{31}L}{T}, \qquad V = \frac{Fg_{31}}{W} \tag{16.3}$$

In Eqs. (16.2) and (16.3), F is the applied force L, T, and W are the element's length, width, and thickness (as shown in Fig. 16.1), and the d_{ij} and g_{ij} coefficients are the strain and voltage constants of the material, respectively. For one application of the mechanical stress, the electrical energy available for extraction from the piezoelement is determined by Eq. (16.1), or by the product of the charge and voltage terms of Eqs. (16.2) and (16.3). In this analysis, we assume that one can obtain an average power output of the element assuming the mechanical stress is applied repeatedly during some time period. Assuming a frequency of application of the stress of f (Hz), the available electrical power output, P, of the element is

$$P = VQf \tag{16.4}$$

Table 16.1 Typical constants for a common piezoelectric material, PZT-5H

	d_{33}	d_{31}	g_{33}	g_{31}	Young's modulus
PZT-5H	380×10^{-12}	-260×10^{-12}	12.5×10^{-3}	8.0×10^{-3}	$\sim 6.2 \times 10^{10}$
Units	$(C/m^2)/(N/m^2)$	$(C/m^2)/(N/m^2)$	$(V/m)/(N/m^2)$	$(V/m)/(N/m^2)$	N/m^2

Table 16.2 Unit power factors for different PZT-5H modes (When the numbers shown are multiplied by the coefficients at the bottom of each column, the product will be in units of Watts)

	Fixed coefficients: g, d (other terms variable)		Fixed coefficients: $g, d, L = W \equiv 0.01$ m		Fixed coefficients: $g, d, L = W = 0.01$ m, $T = 127 \times 10^{-6}$ m	
	33-mode	31-mode	33-mode	31-mode	33-mode	31-mode
Unit Power Factor	4.75×10^{-12}	2.08×10^{-12}	4.75×10^{-16}	2.08×10^{-16}	6.03×10^{-20}	2.645×10^{-20}
Multiplier	$\sigma^2 TWLf$	$\sigma^2 TWLf$	$\sigma^2 Tf$	$\sigma^2 Tf$	$\sigma^2 f$	$\sigma^2 f$

For each mode this is

$$33\text{-mode}: \quad P = \sigma^2 g_{33} d_{33} TWLf \tag{16.5}$$

$$31\text{-mode}: \quad P = \sigma^2 g_{31} d_{31} TWLf \tag{16.6}$$

Equations (16.5) and (16.6) show the influence of each variable on the power output of a piezoelectric element. With an eye toward designing a power generator, we will discuss each variable.

First, the strain and voltage constants, d and g, are dependent on the choice of material. Table 16.1 shows typical values for these constants for a commercially available piezoelectric material PZT-5H. Based on the products of d- and g-constants in Table 16.1, it would seem that the 33-mode is the better choice for energy generation, given the order of magnitude difference between the modes.

To illustrate this, Table 16.2 shows what can be called "unit power factors" for the different materials and modes. The first two columns show the influence of the d- and g-constants, with no particular choice for the other variables. The 33-mode holds an approximate order of magnitude advantage in terms of the material constants only.

If a unit square (1×1 cm) area of material is considered, then the unit power factors are as given in the third and fourth columns. Finally, considering a typical thickness for these materials (127 μm), the unit power factors are as shown in the fifth and sixth columns of Table 16.2. From these numbers, the 33-mode advantage is maintained.

Now the final two coefficients, applied mechanical stress, σ, and frequency of application, f, can be considered. It is important to maximize these two variables in the generator, since power output is proportional to the square of applied stress, and linearly proportional to the frequency of application of that stress (with no consideration given for the electrical dynamics of the circuit). While it is possible to introduce frequency multiplication devices in a harvester, it is assumed that this will change the absolute energy generated in the devices but not the relative amount of energy between devices. Here, the source is taken to be the heart beat, so a frequency of 1 Hz is assumed.

The final parameter to consider in Eqs. (16.5) and (16.6) is the mechanical stress. The mode of operation of the piezoelectric element is again important. In the 33-mode, the mechanical stress is applied along the three-axis as shown in Fig. 16.1. Unless a mechanical advantage is used then the stress will be directly proportional to the applied force on the piezoelectric element. Achieving a high-power factor using the 33-mode can be difficult to do, given the relatively low forces observed in biomedical applications. For example, normal blood pressure variations are $5300 \, N/m^2$ (or 40 mmHg), which in a 33-mode application is the stress imparted to the piezoelectric material. In order to increase power output, then, one can increase the thickness of the active material, achieving a linear increase in power. It was shown in Clark and Ramsay (2000) that since $5300 \, N/m^2$ is well below the limit of operating stress in these materials, one could achieve much higher generated power by increasing the active stress. A way to do this is to operate the material in the 31-mode, such as in a stretching or a bending diaphragm subjected to the same blood pressure. Since power relates to the square of the stress, this enables a quadratic as opposed to a linear increase in generated power. For that reason, this chapter focuses on diaphragm-type harvesters that undergo bending due to pressure loading.

In the following sections, the equations for the available power from various pressure-driven unimorph diaphragm structures are derived (bimorph structures would follow the same approach). The general approach is to derive expressions for the energy in the structure (both the mechanical strain energy and the coupled electrical energy) due to application of a pressure load. The available energy for harvesting is that portion of the energy that is converted to electricity.

16.3 Unimorph Diaphragm in Bending

This section derives the expressions for generated electrical current and voltage for a unimorph piezoelectric plate with general boundary conditions. The solution used follows the restrictions that there is no shear strain and no stress along the thickness direction (z-dir) and that the deflection is smaller than the thickness of the plate. An energy method will be used to describe the total energy in a pressure-loaded piezoelectric plate, and to ultimately calculate the electrical charge generated from a pressure source for arbitrary boundary conditions. Once the general expression is obtained, specific boundary conditions will be applied and example calculations will be made for certain cases.

The structure to be studied is shown (in cross section) in Fig. 16.2, and includes a piezoelectric layer bonded to a substrate. The piezoelectric layer is assumed to

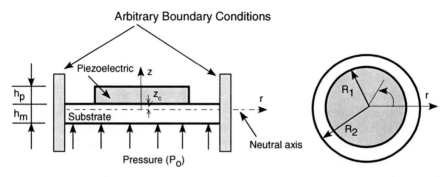

Fig. 16.2 Cross section of unimorph PZT circular plate with arbitrary boundary conditions

be covered on the top and bottom with an electrode. Neither the electrode nor the bonding agent is assumed to have any effect on the model.

At this point, several generalities are included in the derivation, as depicted in the figure. First, the boundary conditions are shown to be arbitrary, so that both clamped and simply supported boundaries can be studied. Second, the piezoelectric layer is shown to have arbitrary radius, R_1, which allows the piezoelectric layer to be of arbitrary size, including full coverage of the substrate where $R_1 = R_2$.

The general approach taken for deriving the generated electrical energy is to first derive the total energy in the structure from an applied load, and then to extract the coupled energy. For the plate structure, the energy for small volumes of the piezoelectric and substrate materials can be described as

$$dU_p = \frac{1}{2}e_r\sigma_{rp} + \frac{1}{2}e_\theta\sigma_{\theta p} + \frac{1}{2}D_3E_3$$
$$dU_m = \frac{1}{2}e_r\sigma_{rm} + \frac{1}{2}e_\theta\sigma_{\theta m}$$

(16.7)

where subscripts p and m denote piezoelectric and substrate (metal) layers and subscripts r and θ denote radial and circumferential directions, respectively. The symbol e is strain and σ is stress, D_3 is charge density on the piezoelectric material and E_3 is electric field strength. Following conventions used in the piezoelectric constitutive equations, the radial (r) direction is taken to be 1, the angular (θ) direction is 2, and the direction perpendicular to the surface (z) is 3. The subscripts r and θ are used instead of 1 and 2 for the stress and strain, respectively.

In order to distinguish between inner and outer regions, since layers or electrodes will be different, two different sets of energy expressions are defined. For the inner region ($r < R_1$)

$$dU_p^{(i)} = \frac{1}{2}e_r^{(i)}\sigma_{rp}^{(i)} + \frac{1}{2}e_\theta^{(i)}e_{\theta p}^{(i)} + \frac{1}{2}D_3E_3$$
$$dU_m^{(i)} = \frac{1}{2}e_r^{(i)}\sigma_{rm}^{(i)} + \frac{1}{2}e_\theta^{(i)}\sigma_{\theta m}^{(i)}$$

(16.8)

and for the outer region ($R_1 \leq r \leq R_2$)

$$dU_m^{(o)} = \frac{1}{2}e_r^{(o)}\sigma_{rm}^{(o)} + \frac{1}{2}e_\theta^{(o)}\sigma_{\theta m}^{(o)} \tag{16.9}$$

which includes only the substrate material since there is no piezoelectric material in this region. It is clear from Eqs. (16.7), (16.8) and (16.9) that the substrate has only strain energy (purely mechanical), while the piezoelectric material can possibly have electrical energy (depicted by the last term which represents stored energy in its capacitance due to applied electric field) and coupled energy that represents either strain energy due to applied electric field or, in the energy harvesting case, electrical energy due to applied mechanical load. By integrating Eqs. (16.7) and (16.9) over the total volume of the structure, the total energy and therefore the generated (coupled) electrical energy can be found as

$$U = \int_0^{R_1} \int_0^{2\pi} \left(\int_0^{h_p} dU_p^{(i)} \, dz + \int_{-h_m}^0 dU_m^{(i)} \, dz \right) r \, d\theta \, dr$$

$$+ \int_{R_1}^{R_2} \int_0^{2\pi} \left(\int_0^{h_p} dU_p^{(o)} \, dz + \int_{-h_m}^0 dU_m^{(o)} \, dz \right) r \, d\theta \, dr \tag{16.10}$$

where h_p and h_m are the thickness of the respective layers, as shown in Fig. 16.2.

In general, the total energy is expressed in three specific terms as

$$U = U_{P_0^2}(P_0^2) + U_{P_0 E_3}(P_0 V) + U_{E_3^2}(V^2) \tag{16.11}$$

where the first term represents the mechanical (strain) energy, the third term represents the electrical energy (stored energy in the capacitance of the piezoelectric material due to externally applied voltage, which does not exist in this problem), and the second term is the coupled electro-mechanical energy from mechanical deformation or applied voltage, which in this study will be isolated as the generated energy available for harvesting. One can further analyze this term by differentiating with respect to voltage to get the generated charge; by obtaining the effective capacitance of the device from the relation $Q = CV$; and finally using the capacitance relationship to find the generated voltage. These analyses will be done for each of the structures of interest in the following sections.

To carry out the integration in Eq. (16.10), stresses and strains in Eqs. (16.8) and (16.9) must be found in terms of the applied load. The piezoelectric constitutive equations (IEEE Std., 1978) can be written for the plate as

$$e_r^{(i)} = s_{11}^E(\sigma_{rp}^{(i)} - v\sigma_{\theta p}^{(i)}) - d_{31}E_3$$

$$e_\theta^{(i)} = s_{11}^E(\sigma_{\theta p}^{(i)} - v\sigma_{rp}^{(i)}) - d_{31}E_3 \qquad (16.12)$$

$$D_3 = -d_{31}(\sigma_r^{(i)} + \sigma_\theta^{(i)}) + \varepsilon_{33}^T E_3$$

where s_{11}^E is the piezoelectric material's elastic compliance at constant electric field, v is its Poisson ratio $(-s_{12}^E/s_{11}^E)$, ε_{33}^T is the permittivity of the piezoelectric material at constant stress, and d_{31} is the piezoelectric constant. The poling direction is assumed to be downward in Fig. 16.2. Using Eq. (16.12), stress can be described in terms of strain and electric field. For the piezoelectric layer, the stress is

$$\sigma_r^{(i)} = \frac{1}{s_{11}^E(1 - v^2)}[e_r^{(i)} + ve_\theta^{(i)} + (1 + v)d_{31}E_3]$$

$$\sigma_{\theta p}^{(i)} = \frac{1}{s_{11}^E(1 - v^2)}[ve_r^{(i)} + e_\theta^{(i)} + (1 + v)d_{31}E_3] \qquad (16.13)$$

and for the non-piezoelectric layer

$$\sigma_{rm} = \frac{1}{s_m(1 - v^2)}(e_r + ve_\theta)$$

$$\sigma_{\theta m} = \frac{1}{s_m(1 - v^2)}(ve_r + e_\theta) \qquad (16.14)$$

where s_m is the elastic compliance of the substrate material and Eq. (16.14) is the same in the inner and outer regions. By substituting Eqs. (16.12), (16.13) and (16.14) into Eqs. (16.8) and (16.9), the differential energy expressions will be written in terms of strain and electrical field only as

$$dU_p = \frac{1}{2s_{11}^E(1 - v^2)}(e_r^2 + e_\theta^2 + 2e_r e_\theta v) - \frac{1}{2}\left(\frac{2d_{31}^2}{s_{11}^E(1 - v^2)} + \varepsilon_{33}^T\right)E_3^2$$

$$dU_m = \frac{1}{2s_m(1 - v^2)}(e_r^2 + e_\theta^2 + 2e_r e_\theta v) \qquad (16.15)$$

The following derivation relates strain to the applied pressure. The moments at any point in the plate can be found by integrating the stresses over the height of the layers (over the dimension, z). For the inner region this is

$$M_r^{(i)} = \int_0^{h_p} \sigma_{rp}^{(i)}(z - z_c)dz + \int_{-h_m}^0 \sigma_{rm}^{(i)}(z - z_c)dz$$

$$M_\theta^{(i)} = \int_0^{h_p} \sigma_{\theta p}^{(i)}(z - z_c)dz + \int_{-h_m}^0 \sigma_{\theta m}^{(i)}(z - z_c)dz \qquad (16.16)$$

and for the outer region

$$M_r^{(o)} = \int_{-h_m}^{0} \sigma_{rm}^{(o)}(z - z_c)dz$$

$$M_\theta^{(o)} = \int_{-h_m}^{0} \sigma_{\theta m}^{(o)}(z - z_c)dz$$

(16.17)

Substituting the stress terms from Eqs. (16.13) and (16.14) these become

$$M_r^{(i)} = \int_0^{h_p} \left(\frac{e_r^{(i)} + ve_\theta^{(i)}}{(1 - v^2)s_{11}^E} + \frac{(1 + v)d_{31}E_3}{(1 - v^2)s_{11}^E} \right)(z - z_c)dz + \int_{-h_m}^{0} \frac{e_r^{(i)} + ve_\theta^{(i)}}{(1 - v^2)s_m}(z - z_c)dz$$

$$M_\theta^{(i)} = \int_0^{h_p} \left(\frac{ve_r^{(i)} + e_\theta^{(i)}}{(1 - v^2)s_{11}^E} + \frac{(1 + v)d_{31}E_3}{(1 - v^2)s_{11}^E} \right)(z - z_c)dz + \int_{-h_m}^{0} \frac{ve_r^{(i)} + e_\theta^{(i)}}{(1 - v^2)s_m}(z - z_c)dz$$

$$M_r^{(o)} = \int_{-h_m}^{0} \frac{e_r^{(o)} + ve_\theta^{(o)}}{s_m(1 - v^2)}(z - z_c)dz$$

$$M_\theta^{(o)} = \int_{-h_m}^{0} \frac{ve_r^{(o)} + e_\theta^{(o)}}{s_m(1 - v^2)}(z - z_c)dz$$

(16.18)

An alternate way to obtain the moments in the plate is to use the deformation. The general equations governing the transverse deflection, bending, and shear of a circular plate subjected to a uniformly distributed constant pressure load P_0 are given by (Vinson, 1974)

$$\nabla^4 W = \frac{1}{r}\frac{\partial}{\partial r}\left(r\frac{\partial}{\partial r}\left(\frac{1}{r}\frac{\partial}{\partial r}\left(r\frac{\partial W}{\partial r} \right) \right) \right) = \frac{P_0}{D}$$

(16.19)

$$M_r = -D\left(\frac{\partial^2 W}{\partial r^2} + \frac{v}{r}\frac{\partial W}{\partial r} \right)$$

$$M_\theta = -D\left(\frac{1}{r}\frac{\partial W}{\partial r} + v\frac{\partial^2 W}{\partial r^2} \right)$$

(16.20)

$$Q_r = -D\frac{\partial}{\partial r}\left(\frac{1}{r}\frac{\partial}{\partial r}\left(r\frac{\partial W}{\partial r} \right) \right)$$

(16.21)

where W is the deflection of the plate in the z-direction, P_0 is the uniform applied pressure, and D is the flexural rigidity of the plate.

Equating Eqs. (16.18) and (16.20) results in

$$-D_c\left(\frac{\partial^2 W}{\partial r^2} + \frac{v}{r}\frac{\partial W}{\partial r}\right) = \int_0^{h_p}\left(\frac{e_r^{(i)} + v e_\theta^{(i)}}{(1-v^2)s_{11}^E} + \frac{(1+v)d_{31}E_3}{(1-v^2)s_{11}^E}\right)(z - z_c)dz$$

$$+ \int_{-h_m}^0 \frac{e_r^{(i)} + v e_\theta^{(i)}}{(1-v^2)s_m}(z - z_c)dz$$

$$-D_c\left(\frac{1}{r}\frac{\partial W}{\partial r} + v\frac{\partial^2 W}{\partial r^2}\right) = \int_0^{h_p}\left(\frac{v e_r^{(i)} + e_\theta^{(i)}}{(1-v^2)s_{11}^E} + \frac{(1+v)d_{31}E_3}{(1-v^2)s_{11}^E}\right)(z - z_c)dz$$

$$+ \int_{-h_m}^0 \frac{v e_r^{(i)} + e_\theta^{(i)}}{(1-v^2)s_m}(z - z_c)dz$$

$$-D_m\left(\frac{\partial^2 W}{\partial r^2} + \frac{v}{r}\frac{\partial W}{\partial r}\right) = \int_{-h_m}^0 \frac{e_r^{(o)} + v e_\theta^{(o)}}{s_m(1-v^2)}(z - z_c)dz$$

$$-D_m\left(\frac{1}{r}\frac{\partial W}{\partial r} + v\frac{\partial^2 W}{\partial r^2}\right) = \int_{-h_m}^0 \frac{v e_r^{(o)} + e_\theta^{(o)}}{s_m(1-v^2)}(z - z_c)dz \tag{16.22}$$

where D_c and D_m are the flexural rigidity of the inner region and the outer region of the plate, respectively.

One can solve for the strain in the r and θ directions as functions of deflection and generated electric field for both inner and outer regions as

$$e_r^{(i)} = \frac{2s_m s_{11}^E(1-v^2)D_c}{h_m^2 s_{11}^E - h_p^2 s_m + 2z_c(h_p s_m + h_m s_{11}^E)}\frac{\partial^2 W}{\partial r^2}$$

$$+ \frac{s_m h_p d_{31}(h_p - 2z_c)}{h_m^2 s_{11}^E - h_p^2 s_m + 2z_c(h_p s_m + h_m s_{11}^E)}E_3$$

$$e_\theta^{(i)} = \frac{2s_m s_{11}^E(1-v^2)D_c}{h_m^2 s_{11}^E - h_p^2 s_m + 2z_c(h_p s_m + h_m s_{11}^E)}\frac{\partial W}{r\partial r}$$

$$+ \frac{s_m h_p d_{31}(h_p - 2z_c)}{h_m^2 s_{11}^E - h_p^2 s_m + 2z_c(h_p s_m + h_m s_{11}^E)}E_3$$

$$e_r^{(0)} = \frac{2s_m(1-v^2)D_m}{h_m(h_m+2z_c)} \frac{\partial^2 W}{\partial r^2}$$

$$e_\theta^{(0)} = \frac{2s_m(1-v^2)D_m}{h_m(h_m+2z_c)} \frac{\partial W}{r\partial r} \qquad (16.23)$$

The final differential energy expressions then become

$$dU_p^{(i)} = \frac{2(1-v^2)s_m^2 s_{11}^E D_c^2}{B_{11}^2} \left(\left(\frac{\partial^2 W}{\partial r^2}\right)^2 + \frac{2v}{r}\left(\frac{\partial^2 W}{\partial r^2}\right)\left(\frac{\partial W}{\partial r}\right) + \frac{1}{r^2}\left(\frac{\partial W}{\partial r}\right)^2 \right)$$

$$+ \frac{2s_m^2 h_p d_{31}(1+v)D_c(h_p-2z_c)}{B_{11}^2} \left(\frac{\partial^2 W}{\partial r^2} + \frac{\partial W}{r\partial r} \right) E_3$$

$$+ \frac{\varepsilon_{33}^T}{2} \left(1 - \left(1 - \frac{s_m^2 h_p^2(h_p-2z_c)^2}{B_{11}^2} \right) \frac{2K_{31}^2}{(1-v)} \right) E_3^2$$

$$dU_m^{(i)} = \frac{2(1-v^2)s_m s_{11}^E D_c^2}{B_{11}^2} \left(\left(\frac{\partial^2 W}{\partial r^2}\right)^2 + \frac{2v}{r}\left(\frac{\partial^2 W}{\partial r^2}\right)\left(\frac{\partial W}{\partial r}\right) + \frac{1}{r^2}\left(\frac{\partial W}{\partial r}\right)^2 \right)$$

$$+ \frac{2s_m s_{11}^E h_p d_{31}(1+v)D_c(h_p-2z_c)}{B_{11}^2} \left(\frac{\partial^2 W}{\partial r^2} + \frac{\partial W}{r\partial r} \right) E_3$$

$$+ \frac{\varepsilon_{33}^T s_{11}^E h_p^2(h_p-2z_c)^2}{B_{11}^2} \frac{K_{31}^2}{(1-v)} E_3^2$$

$$dU_m^{(0)} = \frac{2(1-v^2)D_m^2}{h_m^2(h_m+2z_c)^2} \left(\left(\frac{\partial^2 W}{\partial r^2}\right)^2 + \frac{2v}{r}\left(\frac{\partial^2 W}{\partial r^2}\right)\left(\frac{\partial W}{\partial r}\right) + \frac{1}{r^2}\left(\frac{\partial W}{\partial r}\right)^2 \right)$$

$$(16.24)$$

where

$$B_{11} = h_m^2 s_{11}^E - h_p^2 s_m + 2z_c(h_p s_m + h_m s_{11}^E).$$

Note that these expressions are for any circular plate with pressure loading. One needs only to solve for the actual deflection of the plate for the specific boundary conditions and then integrate the energy over the total volume of the structure.

16.3.1 Simply Supported Unimorph Diaphragm that Is Partially Covered with Piezoelectric Material

In this section, the energy is solved for the case in which the boundaries of the composite plate are simply-supported. The general solutions of Eq. (16.19) for this case include eight constants for the inner and outer regions. Boundary conditions consist of a finite value at $r = 0$ (which force two constants to vanish) and two edge conditions (Mo et al., 2006; Prasad et al., 2002).

$$\left.\frac{dW_r}{dr}\right|_{r=0} < \infty, \quad W_r|_{r=R_2} = 0, \quad M_r|_{r=R_2} = 0 \qquad (16.25)$$

The matching conditions at the interface (at $r = R_1$) for the other four constants are

$$W_r^{(i)}\big|_{r=R_1} = W_r^{(o)}\big|_{r=R_1}$$

$$\left.\frac{dW_r^{(i)}}{dr}\right|_{r=R_1} = \left.\frac{dW_r^{(o)}}{dr}\right|_{r=R_1} \qquad (16.26)$$

$$Q_r^{(i)}\big|_{r=R_1} = Q_r^{(o)}\big|_{r=R_1}$$

$$M_r^{(i)}\big|_{r=R_1} = M_r^{(o)}\big|_{r=R_1}$$

where the superscripts (i) and (o) indicate inner and outer regions of the plate, respectively. The general solutions of the transverse displacement for the piezoelectric composite plate are then given by

$$W = \begin{cases} C_1 + C_2 r^2 + \frac{P_0 r^4}{64 D_c} & \text{when } r \leq R_1 \\ C_3 + C_4 \ln(r) + C_5 r^2 + \frac{P_0 r^4}{64 D_m} & \text{when } R_1 < r \leq R_2 \end{cases} \qquad (16.27)$$

where

$$C_1 = \frac{P_o}{64} \frac{R_1^6}{D_1 D_2 C_0} \left(\left((v-1) + \left((3-v) + 4(1+v)\ln\left(\frac{R_2}{R_1}\right) + 12\ln(R_1) \right) \right. \right.$$

$$\left(\frac{R_2}{R_1}\right)^2 - \left((v+7) + 12\ln(R_2) + 4v\ln\left(\frac{R_2}{R_1}\right) \right) \times \left(\frac{R_2}{R_1}\right)^4 + (v+5)$$

$$\left.\left(\frac{R_2}{R_1}\right)^6\right) D_1^2 + \left(2(1-v) - \left(4(v+1)\ln\left(\frac{R_2}{R_1}\right) \right) - 2(v-2) \right) \left(\frac{R_2}{R_1}\right)^2$$

$$+ \left(4(v+3)\ln\left(\frac{R_2}{R_1}\right) + (v+7)\left(\frac{R_2}{R_1}\right)^4 - (v-5)\left(\frac{R_2}{R_1}\right)^6 \right) D_1 D_2$$

$$\left. + (v-1)\left(1 - \left(\frac{R_2}{R_1}\right)^2 \right) D_2^2 \right)$$

$$C_0 = (-R_2^2 v + R_1^2 v - R_1^2 - R_2^2) D_1 + (R_2^2 v - R_1^2 v + R_1^2 - R_2^2) D_2$$

$$C_2 = -\frac{P_o}{32}\frac{R_1^4}{(1+v)D_1 C_0}\left(\left((1-v^2)\left(\left(\frac{R_2}{R_1}\right)^2 - 1\right) - (1+2v)\left(\frac{R_2}{R_1}\right)^4\right)\right.$$

$$\left.-(1-v^2)\left(\left(\frac{R_2}{R_1}\right)^2 - 1\right)D_2\right)$$

$$C_3 = -\frac{P_o}{64}\frac{R_1^4 R_2^2}{D_2 C_0}\left((v+1)\left(2(1-v)+4(1+v)\ln\left(\frac{R_2}{R_1}\right) - ((1-v)\right.\right.$$

$$\left.+ 4(3+v)\ln(R_2))\left(\frac{R_2}{R_1}\right)^2 + (v+5)\left(\frac{R_2}{R_1}\right)^4\right)D_1 + (2(v^2-1)-4(v+1)^2$$

$$\ln(R_2) + \left((1-v^2)+4(1+v)(3+v)\ln(R_2)\right)\left(\frac{R_2}{R_1}\right)^2 - (v-1)(5+v)$$

$$\left(\frac{R_2}{R_1}\right)^4\right)D_2)$$

$$C_4 = \frac{P_o}{16}\frac{R_1^2 R_2^2}{D_2 C_0}(-R_2^2 v + vR_1^2 - 3R_2^2 + R_1^2)(D_1 - D_2)$$

$$C_5 = -\frac{P_o}{32}\frac{(-R_2^4 v + vR_1^4 - 3R_2^4 - R_1^4)}{(1+v)D_2 C_0}((1+v)D_1 - (v-1)D_2)$$

The flexural rigidities for the inner composite plate and for the outer annular plate are

$$D_c = \frac{1}{12}\frac{4h_p^3 s_{11}^E s_m h_m + h_p^4 s_m^2 + h_m^4 s_{11}^{E^2} + 4h_m^3 h_p s_{11}^E s_m + 6h_m^2 h_p^2 s_{11}^E s_m}{s_{11}^E s_m(1-v^2)(h_m s_{11}^E + h_p s_m)} \quad \text{and}$$

$$D_m = \frac{1}{12}\frac{h_m^2}{(1-v^2)}$$

By substituting Eq. (16.27) into Eq. (16.22), one can solve for the strains, which can then be used in Eqs. (16.8)–(16.14) to find the total energy in the composite diaphragm structure.

The total energy can be found as

$$U = U_{P_0^2}P_0^2 + \frac{\varepsilon_{33}^T \pi R_1^2 h_p}{2}\left(1 - \frac{2}{(1-v)}\frac{B_{35}}{B_{31}}K_{31}^2\right)E_3^2$$

$$-\frac{3\pi R_2^3 R_1^2 d_{31} s_{11}^E s_m h_m S_h S_f B_{33}h_p(h_m + h_p)}{2}\frac{}{B_{31}B_{34}}E_3 P_0$$

(16.28)

where $U_{P_0^2}$ is long and complex coefficient.

Substituting for the electric field E_3 with V/h_p, where V is applied voltage and differentiating with respect to voltage (V) produces the expression for the electric charge

$$Q = \frac{\varepsilon_{33}^T \pi R_1^2}{h_p} \left(1 - \frac{2}{(1-v)} \frac{B_{35}}{B_{31}} K_{31}^2\right) V - \frac{3}{2} \frac{\pi R_2^3 R_1^2 d_{31} s_{11}^E s_m h_m S_h S_f B_{33}(h_m + h_p)}{B_{31} B_{34}} P_0 \quad (16.29)$$

The generated charge can be found (by differentiating the coupled energy term with respect to voltage) to be

$$Q_{gen} = -\frac{3}{2} \frac{\pi R_2^3 R_1^2 d_{31} s_{11}^E s_m h_m S_h S_f B_{33}(h_m + h_p)}{B_{31} B_{34}} P_0 \quad (16.30)$$

where

$$B_{33} = h_p{}^4 s_m{}^2 + 6 s_{11}^E s_m h_m{}^2 h_p{}^2 + 4 s_{11}^E s_m h_m h_p \left(h_p{}^2 + h_m{}^2\right) + h_m{}^4 s_{11}^{E2}$$

$$\begin{aligned} B_{34} = {} & R_2^2 (s_{11}^E s_m (6 h_m{}^2 h_p{}^2 + 4 h_p{}^3 h_m + 5 h_m{}^3 h_p + 3 v h_m{}^3 h_p + 4 v h_p{}^3 h_m \\ & + 6 v h_m{}^2 h_p{}^2) + v h_p{}^4 s_m + h_p{}^4 s_m{}^2 + 2 h_m{}^4 s_{11}^{E2}) \\ & - R_1^2 h_p s_m (v-1) \times (h_p{}^3 s_m + 4 h_p{}^2 h_m s_{11}^E + 3 h_m{}^3 s_{11}^E + 6 h_m{}^2 h_p s_{11}^E) \end{aligned}$$

$$\begin{aligned} B_{35} = {} & 6 s_{11}^E s_m{}^2 h_m{}^2 h_p{}^3 R_1^2 + 2 s_{11}^E R_2 h_p{}^4 R_1 h_m s_m{}^2 + s_{11}^{E3} R_2^2 h_m{}^5 + s_{11}^{E2} R_2^2 s_m h_p{}^3 h_m{}^2 \\ & + 2 s_{11}^{E2} R_2 R_1 h_p h_m{}^4 s_m + h_p{}^5 R_1^2 s_m{}^3 \\ & + 3 s_{11}^E s_m{}^2 h_p{}^4 R_1^2 h_m + 4 s_{11}^E s_m{}^2 R_1^2 h_p{}^2 h_m{}^3 \end{aligned}$$

$S_f = 3 R_2{}^2 + v R_2{}^2 - R_1{}^2 - v R_1{}^2$ and K_{31} is the electro-mechanical coupling coefficient and is defined as $K_{31} = d_{31}^2 / \sqrt{\varepsilon_{33}^T s_{11}^E}$.

From the relation $Q = CV$, open-circuit capacitance is

$$C_{free} = \frac{\varepsilon_{33}^T \pi R_1^2}{h_p} \left(1 - \frac{2}{(1-v)} \frac{B_{35}}{B_{31}} K_{31}^2\right) \quad (16.31)$$

and the voltage that appears on the electrodes is

$$V_{gen} = \frac{Q_{gen}}{C_{free}} = -\frac{3}{2} \frac{\pi R_2^3 d_{31} s_{11}^E s_m h_m h_p S_h S_f B_{33}(h_m + h_p)}{\varepsilon_{33}^T B_{31} B_{34} \left(1 - \frac{2}{(1-v)} \frac{B_{35}}{B_{31}} K_{31}^2\right)} P_0 \quad (16.32)$$

Thus, the electrical energy generated by external pressure P_0 is

$$U_{gen} = \frac{1}{2} Q_{gen} \cdot V_{gen} = \frac{9}{8} \frac{\pi R_2^6 R_1^2 d_{31}^2 s_{11}^{E2} s_m^2 h_m^2 h_p S_h^2 S_f^2 (h_m + h_p)^2}{\varepsilon_{33}^T (B_{31} B_{34})^2 \left(1 - \frac{2}{(1-v)} \frac{B_{35}}{B_{31}} K_{31}^2\right)} P_0^2 \quad (16.33)$$

16.3.2 *Clamped Unimorph Diaphragm that is Partially Covered with Piezoelectric Material*

In this section, the derivation is repeated for the plate with the clamped edge. The general solutions of the transverse displacement for the piezoelectric composite plate are given by

$$W = \begin{cases} C_6 + C_7 r^2 + C_8 D_m r^4 & \text{when } r \le R_1 \\ C_9 + C_{10} r^2 + C_8 r^4 & \text{when } R_1 < r \le R_2 \end{cases} \tag{16.34}$$

where

$$C_6 = \left(\left((D_m - D_c) R_1^4 + D_c R_2^4 \right) C_8 + 4 D_c R_2^2 R_1^2 \left((1+v) \left((D_m - D_c) R_2^2 \right. \right. \right.$$
$$\left. \left. \left. + D_c R_1^2 \right) - 2 D_m v R_1^2 \right) \ln \left(\frac{R_2}{R_1} \right) \right)$$

$$C_7 = -2 D_m P_0 \left((D_m - D_c) \left((1+v) R_1^4 + (1-v) R_1^2 R_2^2 \right) + 2 D_c R_2^4 \right)$$

$$C_8 = P_0 \left((1+v) \left((D_m - D_c) R_1^2 + D_c R_2^2 \right) + (1-v) D_m R_2^2 \right)$$

$$C_9 = P_0 R_2^2 \left((1+v) \left((D_m - D_c) \left(2 R_1^2 - R_2^2 \right) R_1^2 - D_c R_2^4 \right) + (1-v) D_m R_2^4 \right.$$
$$\left. + 4(1-v)(D_m - D_c)(R_2^2 - R_1^2) R_1^2 \ln \left(\frac{R_2}{R_1} \right) \right)$$

$$C_{10} = -2 P_0 \left((1+v) \left((D_m - D_c) R_1^4 + D_c R_2^4 \right) + (1-v) D_m R_2^4 \right)$$

The electric charge can be expressed as

$$Q = \frac{\pi R_1^2 \varepsilon_{33}^T}{h_p} \left(1 - \frac{2}{(1-v)} \frac{B_{35}}{B_{31}} K_{31}^2 \right) V - \frac{3}{2} \frac{\pi R_1^2 R_2^3 d_{31} s_{11}^E s_m h_m S_h B_{33}(h_m + h_p)(1+v)(R_2^2 - R_1^2)}{B_{31} B_{36}} P_0 \tag{16.35}$$

where

$$B_{31} = B_{35} + 3 s_{11}^{E^2} s_m R_2^2 h_p h_m^2 (h_p + h_m)^2$$

$$B_{36} = h_p s_m (1+v)(R_2^2 - R_1^2)(4 h_p^2 h_m s_{11}^E + h_p^3 s_m + 3 h_m^3 s_{11}^E + 6 h_m^2 h_p s_{11}^E)$$
$$+ R_2^2 h_m^3 s_{11}^E (h_p s_m + h_m s_{11}^E)$$

and $S_h = s_{11}^E h_m R_2 + s_m h_p R_1$.

The generated charge can then be found to be

$$Q_{\text{gen}} = -\frac{3}{2} \frac{\pi R_1^2 R_2^3 d_{31} s_{11}^E s_m h_m S_h B_{33}(h_m + h_p)(1+v)(R_2^2 - R_1^2)}{B_{31} B_{36}} P_0 \tag{16.36}$$

Noting that open-circuit capacitance is similar to Eq. (16.31), the voltage that appears on the electrodes is

$$V_{gen} = \frac{Q_{gen}}{C_{free}} = -\frac{3}{2} \frac{R_2^3 d_{31} s_{11}^E s_m h_m h_p S_h B_{33}(h_m + h_p)(1 + v)(R_2^2 - R_1^2)}{\varepsilon_{33}^T B_{31} B_{36} \left(1 - \frac{2}{1-v} \frac{B_{35}}{B_{31}} K_{31}^2\right)} P_0 \quad (16.37)$$

The electrical energy generated by the pressure P_0 is

$$U_{gen} = \frac{1}{2} Q_{gen} \cdot V_{gen} = \frac{9}{8} \frac{\pi R_1^2 R_2^6 d_{31}^2 s_{11}^{E^2} s_m^2 h_m^2 h_p s_h^2 B_{33}^2 (h_m + h_p)^2 (1 + v)^2 (R_2^2 - R_2^2)^2}{\varepsilon_{33}^T (B_{31} B_{36})^2 (1 - \frac{2}{(1-v)} \frac{B_{35}}{B_{31}} K_{31}^2)} P_0 \quad (16.38)$$

16.3.2.1 Clamped Unimorph Diaphragm with Fully Covered Electrode (Unmodified Case)

The first case to be considered is the case where not only does the piezoelectric material extend to the boundaries, but also the electrode fully covers the piezo-electric material with no breaks (the so-called unmodified case). Equations (16.19), (16.20) and (16.21) can be applied to this case, however only four constants are required in the solution (the matching conditions at $r = R_1$ are not needed since there is no discontinuity in the composite diaphragm). Boundary conditions consist of a finite value at $r = 0$ (which force two constants to vanish) and two clamped edge conditions (Mo et al., 2006; Prasad et al., 2006),

$$W_r|_{r=R_2} = 0, \quad \frac{\partial W_r}{\partial r}\bigg|_{r=R_2} = 0 \quad (16.39)$$

The deflection solution can be found to be

$$W_r = \frac{P_0(R_2^4 - 2R_2^2 r^2 + r^4)}{64D} \quad (16.40)$$

The flexural rigidity, D, of the composite plate can be derived as

$$D = \frac{1}{12} \frac{4h_p^3 h_m s_{11}^E s_m + h_p^4 s_m^2 + h_m^4 s_{11}^{E^2} + 4h_m^3 h_p s_{11}^E s_m + 6h_m^2 h_p^2 s_{11}^E s_m}{s_{11}^E s_m (1 - v^2)(h_m s_{11}^E + h_p s_m)} \quad (16.41)$$

In solving the energy, there is no outer region, so only the inner region equations need to be applied. Substituting Eq. (16.41) into Eq. (16.22) and solving for strain, and then substituting this result into the energy expression (Eq. 16.10) produces the following energy for the diaphragm:

$$U = \int_0^{R_2} \int_0^{2\pi} \left(\int_0^{h_p} dU_p dz + \int_{-h_m}^0 dU_m dz \right) r d\theta dr$$

$$= \frac{\pi R_2^6 s_{11}^E s_m (h_p s_m + h_m s_{11}^E)(1 - \nu^2)}{32 B_{31}} P_o^2$$

$$+ \frac{\varepsilon_{33}^T \pi R_2^2 h_p}{2} \left[1 - \frac{2}{(1 - \nu)} \left(1 - \frac{3 s_{11}^{E^2} s_m h_m^2 h_p \left(h_p + h_m \right)^2}{(h_p s_m + h_m s_{11}^E) B_{31}} \right) K_{31}^2 \right] E_3^2$$

(16.42)

Electric field is assumed to be constant throughout the piezoelectric structure. After integrating the total energy, the electric field (E_3) can be replaced in Eq. (16.42) by V/h_p to give

$$U = \frac{\pi R_2^6 s_{11}^E s_m (h_p s_m + h_m s_{11}^E)(1 - \nu^2)}{32 B_{31}} P_o^2$$

$$+ \frac{\varepsilon_{33}^T \pi R_2^2 h_p}{2} \left[1 - \frac{2}{(1 - \nu)} \left(1 - \frac{3 s_{11}^{E^2} s_m h_m^2 h_p (h_p + h_m)^2}{(h_p s_m + h_m s_{11}^E) B_{31}} \right) K_{31}^2 \right] \left(\frac{V}{h_p} \right)^2$$

(16.43)

Since Eq. (16.43) has no coupled term, we can conclude that this model predicts no net electrical energy conversion from the applied pressure on the unmodified clamped plate. The result of no net energy generation is not intuitive. Presumably relaxing the assumption of bending strain only, (no strain of the neutral axis) or allowing elastic boundary conditions would change this result. Nonetheless, this model does offer some interesting insights for different electrode patterns (which are supported by experimental results, (Kim et al., 2005b)).

Given the assumption of pure-bending strain in the unimorph, a physical explanation for the prediction of zero net electrical energy generation is as follows. It is clear that even though there is no net electrical energy generated from uniform applied pressure, there is stress in the plate. In fact, locally there is substantial stress in the piezoelectric layer due to the applied pressure (Fig. 16.3), but some of it results in positive charge (regions in which the piezoelectric layer is in tension) and some results in negative charge (compressive regions). When integrated (Eq. 16.42),

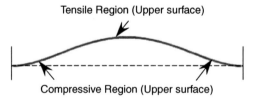

Fig. 16.3 Deflection and stress of clamped plate

the positive contributions cancel the negative contributions, so the net generated electrical energy for the clamped boundary condition using this model is zero.

To extend the above idea, the partial energy of the plate was calculated and compared to Eq. (16.43) for an inner volume defined by a radius r_1 where $0 \leq r_1 \leq R_2$, where r_1 is an arbitrary constant. The energy for this volume is

$$
\begin{aligned}
U =& \int_0^{\tau_1} \int_0^{2\pi} \left(\int_0^{h_p} dU_p dz + \int_{-h_m}^0 dU_m dz \right) r d\theta dr \\
=& \frac{r_1^2 \pi \varepsilon_{33}^T}{2h_p} \left(1 - \frac{2}{1-v} \left(1 - \frac{3s_{11}^{E2} s_m h_p h_m^2 (h_p + h_m)^2}{(s_m h_p + s_{11}^E h_m) B_{31}} \right) K_{31}^2 \right) V^2 \\
&- \frac{3 \, \pi r_1^2 d_{31} s_{11}^E s_m h_m (1+v)(h_p + h_m)(R_2^2 - r_1^2)}{2} \frac{P_0 V}{B_{22}} \\
&+ \frac{\pi r_1^2 s_{11}^E s_m (s_m h_p + s_{11}^E h_m)(1-v^2)(3v(r_1^4 + R_2^4) - 6r_1^2 R_2^2(1+v) + 5r_1^4 + 3R_2^4)}{64 B_{31}}
\end{aligned}
$$

$$(16.44)$$

This equation shows that the mechanical–electrical-coupling term (the term containing $P_0 V$) approaches zero when r_1 goes to R_2; and when $r_1 = R_2$, Eqs. (16.44) and (16.42) are identical. By inspecting the coupled energy term in Eq. (16.44), the coupled energy has the following relation:

$$ U_{\text{coupled}} \propto r_1^2 \left(R_2^2 - r_1^2 \right) \tag{16.45} $$

Therefore, the maximum coupled energy for this inner plate volume can be obtained when Eq. (16.45) is maximum, which occurs when $r_1 = \sqrt{2}/2 = 0.707 R_2$. Since the plate is fully covered with electrode, the outer region is also generating energy, albeit with the opposite charge. So if the charge in the outer region were to be reversed and combined with that of the inner region, the total energy would be maximized if the regions were separated at a radius of $r_1 = 0.707 R_2$. This separation and inversion of sign is defined as "regrouping" of the electrodes (Kim et al., 2005a).

16.3.2.2 Clamped Unimorph Plate with Segmented Electrode (Regrouped Case)

In this analysis, the electrodes on the piezoelectric layer of the plate will be divided into two areas corresponding to the regions of positive and negative stress (and thus positive and negative charge), as shown in Fig. 16.4. The electric field directions of these inner and outer regions will be opposite to each other as a result of the stress distribution from constant pressure. By separately collecting the charge on these regions (e.g., by reversing the leads that attach the respective electrodes), one will get a net addition of charge as opposed to the cancelation that occurs when the electrode is fully covered.

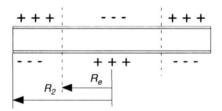

Fig. 16.4 Cross section of the regrouped PZT layer showing the reversal of polarity on the charge that would be observed after regrouping of the electrodes

The division of positive and negative charge (or stress) occurs at a radius R_e, so two new inner and outer regions are defined. Now the inner region is defined by $r < R_e$ and the other is the outer region by $R_e \leq r \leq R_2$. The inner region defined here behaves exactly the same as that in earlier sections. The outer region, however, has two differences. The first difference is that the outer region will use the opposite electric field direction, so its constitutive equations have a sign change. The constitutive equations for the inner region then are (same as Eq. 16.12)

$$e_r^{(i)} = s_{11}^E \left(\sigma_r^{(i)} - v\sigma_\theta^{(i)} \right) - d_{31} E_3$$

$$e_\theta^{(i)} = s_{11}^E \left(\sigma_\theta^{(i)} - v\sigma_r^{(i)} \right) - d_{31} E_3 \qquad (16.46)$$

$$D_3 = -d_{31} \left(\sigma_r^{(i)} + \sigma_\theta^{(i)} \right) + \varepsilon_{33}^T E_3$$

and for the outer region

$$e_r^{(o)} = s_{11}^E \left(\sigma_r^{(o)} - v\sigma_\theta^{(o)} \right) + d_{31} E_3$$

$$e_\theta^{(o)} = s_{11}^E \left(\sigma_\theta^{(o)} - v\sigma_r^{(o)} \right) + d_{31} E_3 \qquad (16.47)$$

$$D_3 = -d_{31} \left(\sigma_r^{(o)} + \sigma_\theta^{(o)} \right) + \varepsilon_{33}^T E_3$$

The second difference is that the outer region in this case now includes both piezoelectric and substrate layers. The differential energy for the inner region is the same as given in Eq. (16.8), but the outer region has an additional term for the piezoelectric layer. Eq. (16.9) becomes

$$dU_p^{(o)} = \frac{1}{2} e_r^{(o)} \sigma_{rp}^{(o)} + \frac{1}{2} e_\theta^{(o)} \sigma_{\theta p}^{(o)} - \frac{1}{2} d_{31} \left(\sigma_{rp}^{(o)} + \sigma_{\theta p}^{(o)} \right) E_3 + \frac{1}{2} \varepsilon_{33}^T E_3^2$$

$$dU_m^{(o)} = \frac{1}{2} e_r^{(o)} \sigma_{rm}^{(o)} + \frac{1}{2} e_\theta^{(o)} \sigma_{\theta m}^{(o)} \qquad (16.48)$$

For the piezoelectric layer, the stresses in the inner and outer regions are

$$\sigma_{rp}^{(i)} = \frac{1}{s_{11}^E(1-v^2)} \left[e_r^{(i)} + ve_\theta^{(i)} + (1+v)d_{31}E_3 \right]$$

$$\sigma_{\theta p}^{(i)} = \frac{1}{s_{11}^E(1-v^2)} \left[ve_r^{(i)} + e_\theta^{(i)} + (1+v)d_{31}E_3 \right]$$

$$\sigma_{rp}^{(o)} = \frac{1}{s_{11}^E(1-v^2)} \left[e_r^{(o)} + ve_\theta^{(o)} + (1+v)d_{31}E_3 \right] \tag{16.49}$$

$$\sigma_{\theta p}^{(o)} = \frac{1}{s_{11}^E(1-v^2)} \left[ve_r^{(o)} + e_\theta^{(o)} + (1+v)d_{31}E_3 \right]$$

and for the non-piezoelectric layer

$$\sigma_{rm} = \frac{1}{s_m(1-v^2)}(e_r + ve_\theta)$$

$$\sigma_{\theta m} = \frac{1}{s_m(1-v^2)}(ve_r + e_\theta) \tag{16.50}$$

where once again there is no need to distinguish between inner and outer regions of stress and strain in the substrate material.

Given the deformation, W, the strain in Eq. (16.22) can be found and then substituted into the preceding equations to finally find the differential energy. The boundary conditions for this case are the same as for the unmodified case (Eq. 16.39) and the deflection solution is also the same (Eq. 16.40), so the same procedure as above can be followed, with the exception that the integration of energy must be separated into inner and outer regions of the diaphragm to account for the differences in stress and strain.

$$U = \int_0^{R_e} \int_0^{2\pi} \left(\int_0^{h_p} dU_p^{(i)} dz + \int_{-h_m}^0 dU_m dz \right) rd\theta dr + \int_{R_e}^{R_2} \int_0^{2\pi} \left(\int_0^{h_p} dU_p^{(o)} dz \right.$$

$$\left. + \int_{-h_m}^0 dU_m dz \right) rd\theta dr = \frac{\pi R_2^6 s_{11}^E s_m (h_p s_m + h_m s_{11}^E)(1-v^2)}{32 B_{31}} P_o^2$$

$$- \frac{3}{2} \frac{\pi R_e^2 d_{31} s_{11}^E s_m h_m (1+v)(h_p + h_m)(R_2^2 - R_e^2)}{B_{31}} P_o V$$

$$+ \frac{s_{33}^T \pi R_2^2}{h_p} \left[\frac{1}{2} - \frac{1}{(1-v)} \left(1 - \frac{3s_{11}^{E^2} s_m h_m^2 h_p (h_p + h_m)^2}{(h_p z_m + h_m s_{11}^E) B_{31}} \right) K_{31}^2 \right] \left(\frac{V}{h_p} \right)^2 \tag{16.51}$$

The first integral term in Eq. (16.8) is for the inner region, so the appropriate stress and strain terms are used in forming $dU_p^{(i)}$, and the second integral term is for the outer region. This energy equation can be reduced to the unmodified diaphragm energy, Eq. (16.43), by substituting $R_e = R_2$. The generated charge from applied pressure can be found to be

$$Q_{\text{Gen}} = \frac{\varepsilon_{33}^T \pi R_2^2}{h_p} \left[1 - \frac{2}{(1-v)} \left(1 - \frac{3s_{11}^{E^2} s_m h_m^2 h_p (h_p + h_m)^2}{(h_p s_m + h_m s_{11}^E) B_{31}} \right) K_{31}^2 \right] V$$
$$- \frac{3}{2} \frac{\pi R_e^2 d_{31} s_{11}^E s_m h_m (1+v)(h_p + h_m)(R_2^2 - R_e^2)}{B_{31}} P_0$$
$$\tag{16.52}$$

The open-circuit capacitance is

$$C_{\text{free}} = \frac{\varepsilon_{33}^T \pi R_2^2}{h_p} \left[1 - \frac{2}{(1-v)} \left(\frac{3s_{11}^{E^2} s_m h_m^2 h_p (h_p + h_m)^2}{(h_p s_m + h_m s_{11}^E) B_{31}} \right) K_{31}^2 \right] \tag{16.53}$$

and the voltage that appears on the electrodes is

$$V_{\text{gen}} = \frac{Q_{\text{Gen}}}{C_{\text{free}}} = -\frac{3R_e^2 d_{31} s_{11}^E s_m h_m h_p (1+v)(h_p + h_m)(R_2^2 - R_e^2)}{2B_{31}\varepsilon_{33}^T R_2^2 \left[1 - \frac{2}{1-v} \left(1 - \frac{3s_{11}^{E^2} s_m h_m^2 h_p (h_p + h_m)^2}{(h_p s_m + h_m s_{11}^E) B_{31}} \right) K_{31}^2 \right]} P_0 \tag{16.54}$$

In this case, the generated voltage and charge are non-zero and they can be maximized with the proper choice or R_e.

$$U_{\text{gen}} = \frac{1}{2} Q_{\text{Gen}} \cdot V_{\text{Gen}} = \frac{9\pi R_e^4 d_{31}^2 s_{11}^{E^2} s_m^2 S_h h_m^2 h_p (1-v)(1+v)^2 (h_p + h_m)^2 (R_2^2 - R_e^2)}{8B_{31}\varepsilon_{33}^T R_2^2 \left((1-v) S_h B_{31} - 2(S_h B_{31} - 3s_{11}^{E^2} s_m h_m^2 h_p (h_p + h_m)^2 K_{31}^2) \right)} P_0^2$$
$$\tag{16.55}$$

16.4 Simulation Results and Analysis

A simple design study to investigate the performance of power generation based on analysis in the preceding sections is presented here. In this analysis, PZT-5H is adopted as the piezoelectric material in a circular plate to examine energy harvesting performance using the energy Eqs. (16.33), (16.38), and (16.55), assuming a uniform pressure load. It is assumed that one can obtain an average power output of the device assuming the mechanical stress is applied repeatedly during some time period. A 40 mmHg (5330 Pa) uniform pressure with 1 Hz frequency is assumed to be applied to the harvesting devices. Material properties used in this analysis include $d_{31} = -320 \times 10^{-12}$, Poisson ratio of 0.3, relative permittivity equal to 3800, and mechanical compliance of $s_{11}{}^E = 1.61 \times 10^{-11}$. The properties used in this analysis are based on the PZT-5H of Piezo Systems, Inc.

It was shown in the earlier works that there is an optimal radius ratio (R_1/R_2) of the piezoelectric layer to the substrate or an optimal regrouped radius ratio (R_e/R_2), where optimal is defined as maximum electric power output (Kim et al., 2005a,b and Mo et al., 2007). This ratio varies depending on the types of piezoelectric

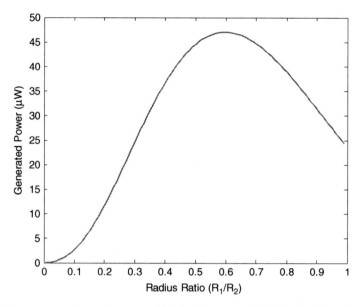

Fig. 16.5 Power-generating performance of the simply supported PZT circular plate with partially covered electrode

and substrate materials, and it shifts according to the thickness ratio (h_p/h_m) of these layers.

Figure 16.5 shows the power-generating performance of the simply supported PZT diaphragm harvester with a partially covered electrode. In this simulation, the radius of substrate (R_2) is fixed to 1.27 cm and the optimal thickness ratio of PZT to aluminum substrate is used. This device can generate maximum power at the ratio of $R_1/R_2 = 0.6$.

The generating performance of the clamped PZT plate with partially covered electrode is shown in Fig. 16.6. The generated power is reduced when compared with the simply supported case. For this case, maximum power can be generated at the ratio of $R_1/R_2 = 0.87$. It is also noted that the generated power is zero as R_1/R_2 goes to 1, which is the fully covered electrode or unmodified case as described in Section 16.3.2.1.

The generated power is zero for the unmodified case with the clamped edge, but it can generate positive power when the electrode is regrouped. Figure. 16.7 shows power-generating performance of the clamped PZT circular plate with fully covered regrouped electrode. Maximum power occurs at the ratio of $R_e/R_2 = 0.7$ as described in Eq. (16.45).

The various designs described so far are assumed to be exposed on one side to the periodic blood pressure, and on the other side to some constant pressure chamber. The output power is calculated assuming an excitation frequency of 1 Hz (60 beats per minute) and a pressure of 5333 N/m^2 (or 40 mmHg). According to theoretical analysis and numerical simulation, the available power results for

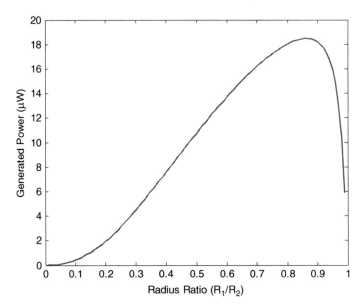

Fig. 16.6 Power-generating performance of the clamped PZT circular plate with partially covered electrode

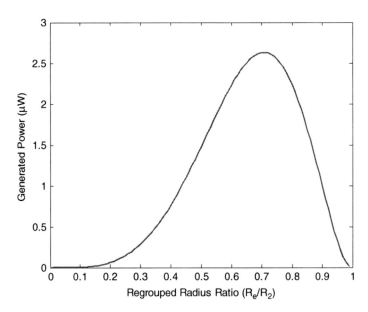

Fig. 16.7 Power-generating performance of the clamped PZT circular plate with fully covered regrouped electrode

Table 16.3 Power generation for continuous and intermittent usage by blood pressure

Harvesters	Design* specification	Generated power for continuous use (μW)	Required time for intermittent use of 10 mW
Plate (127 μm PZT)	ss, pce	47.0	4 min.
	cl,pce	18.5	9 min.
	cl,fcr	2.6	1 h 4 min.

*ss: simply supported, pce: partially covered electrode, cl: clamped, fcr: fully covered and regrouped electrode

the best case are listed in Table 16.3. It is noted that the effective loss due to the necessary rectification and regulation electronics is neglected from all calculations. The power outputs listed in Table 16.3 may not be high enough to continuously power implantable devices such as a DNA chip, which requires 10 mW for operation and telecommunication (Cain et al., 2001, Sohn et al., 2005). However, calculation of usage (or effectively a duty cycle), which represents the time for energy consumption related to the time for energy harvesting, shows the feasibility of providing intermittent power instead of continuous power. The usage is expressed as required harvesting time for the intermittent use of 10 mW in Table 16.3. These results show that for even a relatively high-power (10 mW) consuming implantable device, it is theoretically possible to generate sufficient amounts of power with a simple blood pressure harvester to operate the device at reasonable intermittent times.

16.5 Conclusions

This chapter has presented a brief analysis of 33- versus 31-mode harvesters for implantable devices, and has made an argument for the case of using the 31-mode because of the potential for generating high stresses, and therefore relatively higher power, with the mechanical advantage of the 31-mode for the relatively low excitation forces within the body. The chapter then proceeded to present a derivation for the available electrical energy for a general pressure-loaded piezoelectric unimorph diaphragm structure. Using this framework, the expressions of generated energy were derived for specific boundary conditions, piezoelectric coverage conditions, and electrode conditions. While the complete population of possible configurations was not exhausted, enough cases were presented to provide an understanding of the key features to consider when designing such a harvester. Using the resulting relationships, the chapter then calculated the generated power for a specific geometry (a 1.27 cm diameter diaphragm). It was shown that while the device would not be able to continuously power, say a 10 mW implanted sensor device, it could potentially provide enough power to operate the device at reasonable intermittent rates. The relationships provided here may enable other optimal designs to be realized.

References

Cain JT, Clark WW, Ulinski D, Mickle MH (2001) Energy harvesting for DNA gene sifting and sorting. International Journal of Parallel and Distributed Systems and Networks, 4(5): 140–149.

Clark WW, Ramsay MJ (2000) Smart material transducers as power sources for MEMS devices. International Symposium on Smart Structures and Microsystems, Hong Kong.

IEEE std. 176 (1978) IEEE Standards on Piezoelectricity, The Institute of Electrical and Electronics Engineers.

Jaffe B, Cook R, Jaffe H (1971) Piezoelectric Ceramics, Academic Press, New York.

Kim S, Clark WW, Wang QM (2005a) Piezoelectric energy harvesting using a bimorph circular plate: analysis. Journal of Intelligent Material Systems and Structures 16(10): 847–854.

Kim S, Clark WW, Wang QM (2005b) Piezoelectric energy harvesting using a bimorph circular plate: experimental study. Journal of Intelligent Material Systems and Structures 16(10): 855–864.

Kymissis J, Kendall C, Paradiso J, Gershenfeld N (1998) Parasitic power harvesting in shoes. Second IEEE International Conference on Wearable Computing.

Liu C, Cui T, Zhou Z (2003) Modal Analysis of a Unimorph Piezoelectrical Transducer, Microsystem Technologies 9, pp. 474–479, Springer-Verlag.

Mo C, Wright R, Slaughter WS, Clark WW (2006) Behavior of a unimorph circular piezoelectric actuator. Smart Materials and Structures 15:1094–1102.

Mo C, Radziemski LJ, Clark WW (2007) Analysis of PMN-PT and PZT circular diaphragm energy harvesters for use in implantable medical devices. Proceedings of the SPIE Smart Structures and Materials & Nondestructive Evaluation and Health Monitoring Conference, Paper No. 6525–06.

Niu P, Chapman P, Riemer R, Zhang X (2004) Evaluation of motions and actuation methods for biomechanical energy harvesting. 35th Annual IEEE Power Electronics Specialists Conference 2100–2106.

Pelletier F (2004) Inner Space. MPT-medicare technology, Wicklow, pp. 32–34.

Platt SR, Farritor S, Garvin K, Haider H (2005) The Use of Piezoelectric Ceramics for Electric Power Generation Within Orthopedic Implants. IEEE/ASME Transactions on Mechatronics 10(4):455–461.

Prasad SAN, Sankar BV, Cattafesta LN, Horowitz S, Gallas Q, Sheplak M (2002) Two-port electroacoustic model of an axisymmetric piezoelectric composite plate, AIAA-2002-1365, Denver, Colorado.

Sohn JW, Choi SB, Lee DY (2005) An investigation on piezoelectric energy harvesting for MEMS power sources. Proceedings of the IMechE Part C: Journal of Mechanical Engineering Science 219:429–436.

Soykan O (2002) Power sources for implantable medical devices. Business Briefing: Medical Device Manufacturing & Technology 76–79.

Starner T (1996) Human-powered wearable computing IBM SYSTEMS Journal 35:618.

Starner T, Paradiso JA (2004) Human generated power for mobile electronics. In Piguet C (ed) Low Power Electronics Design (Piguet, C. (ed.)) CRC Press, Boca Raton.

Suzuki S, Katane T, Saotome H, Saito O (2002) Electric power-generating system using magnetic coupling for deeply implanted medical electronic devices. IEEE Trans. on Magnetics 38(5):3006–3008.

Suzuki S, Katane T, Saotome H, Saito O (2003) Fundamental study of an electric power transmission system for implanted medical devices using magnetic and ultrasonic energy. Journal of Artificial Organs 6:145–148.

Vinson JR (1974) Structural Mechanics: The Behavior of Plates and Shells. John Wiley & Sons New York.

Chapter 17
Harvesting Energy from the Straps of a Backpack Using Piezoelectric Materials

Henry A. Sodano

Abstract Over the past few decades the use of portable and wearable electronics has grown steadily. These devices are becoming increasingly more powerful; however, the gains that have been made in the device performance have resulted in the need for significantly higher power to operate the electronics. This issue has been further complicated due to the stagnate growth of battery technology over the past decade. In order to increase the life of these electronics, researchers have begun investigating methods of generating energy from ambient sources such that the life of the electronics can be prolonged. Recent developments in the field have led to the design of a number of mechanisms that can be used to generate electrical energy, from a variety of sources including thermal, solar, strain, inertia, etc. Many of these energy sources are available for use with humans, but their use must be carefully considered such that parasitic effects that could disrupt the user's gait or endurance are avoided. These issues have arisen from previous attempts to integrate power harvesting mechanisms into a shoe such that the energy released during a heal strike could be harvested. This chapter will present research into a novel energy harvesting backpack that can generate electrical energy from the differential forces between the wearer and the pack. The goal of this system is to make the energy harvesting device transparent to the wearer such that his or her endurance and dexterity is not compromised, therefore to preserve the performance of the backpack and user, the design of the pack will be held as close to existing systems as possible.

17.1 Introduction

The advances in low-power electronics, wireless technology, and wearable-computing devices have led to an ever increasing amount of electronics carried by a person. While these devices increase our ability to communicate they can also be cumbersome and require the use of electrochemical batteries to supply power to each

H.A. Sodano (✉)
Department of Mechanical Engineering – Engineering Mechanics, Michigan Technological University, Houghton, MI 49931-1295, USA
e-mail: Henry.Sodano@asu.edu

S. Priya, D.J. Inman (eds.), *Energy Harvesting Technologies*,
DOI 10.1007/978-0-387-76464-1_17 © Springer Science+Business Media, LLC 2009

device. In the case of emergency personnel, field-based environmental researchers, or backcountry sport enthusiasts, these devices also result in substantial loads. These loads are greatly increased due to the need to carry heavy electrochemical batteries as the energy source for each device. Additional complications occur due to stagnant battery technology which has not progressed along with the increasing power demands of current electronics, as shown in Fig. 17.1. Furthermore, since each battery only contains a finite lifespan additional backup batteries must be carried to ensure that the devise can maintain functionality.

Power harvesting is the act of converting ambient energy into electrical energy that can then be used to power other devices. Recent developments in the field have led to the design of a number of mechanisms that can be used to generate electrical energy from a variety of sources, including thermal, solar, strain, inertia, and so on. Many of these energy sources are available from humans, but their integration must be carefully considered such that parasitic effects that could disrupt the user's gait or endurance are avoided. Using the power harvested, it is desirable to construct the system such that it can be used to provide a direct energy source to the electronics, as a means of supplementing the electrochemical battery to increase its life, or to recharge the battery.

Several studies have been performed to investigate the energy available from various sources of human power. Perhaps, the earliest of these was published by Starner (1996), who examined the energy available from the leg motion of a human and surveyed other human sources of mechanical energy, including blood pressure, breathing, and typing. The author claimed that 8.4 W of useable power could be achieved from a PZT device mounted in a shoe to harvest the force generated during walking. While the extraction of energy from walking motion can generate high levels of power, the device typically interferes with the user's gait resulting in a

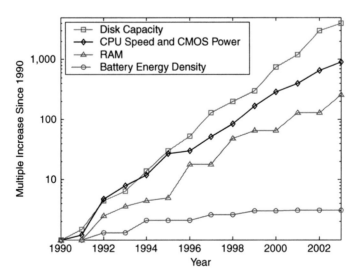

Fig. 17.1 Advances in computer and battery technology since 1990. (Data from Paradiso and Starner, 2005.)

reduction of the wearer's endurance. Several studies have been performed to investigate the power available from the force generated between the heel and the shoe. Kymissis et al. (1998) examined the three different devices that could be built into a shoe to harvest excess energy and generate electrical power parasitically while walking. The devices that were considered included a "Thunder" actuator constructed of piezoceramic composite material located in the heel, a rotary magnetic generator also located under the heel, and a multilayer PVDF foil laminate patch located in the sole of the shoe. The Thunder actuator was developed by NASA and has a rainbow (arch) configuration that allows the impact of the heel to be translated into bending strain for electrical power generation. The electromagnetic generator used the pressure of the heel to spin a flywheel and rotary generator to extract the power from the pressure of the heel during walking. The last device used was a laminate of piezofilm, or "stave", which was used to harness the energy lost during the bending of the sole. In order to compare the performance of the three methods, a working prototype was constructed for each and its performance was measured. The peak powers were observed to approach 20 mW for the PVDF stave, 80 mW for the PZT unimorph, and the shoe-mounted rotary generator averaged to about 250 mW. For a full review of power harvesting using piezoelectric materials, see Sodano et al. (2004a), Anton and Sodano (2007).

More recently, SRI International built a dielectric elastomer generator that was designed to replace the sole of a soldier's boot (Kornbluh et al., 2002). The study configured the dielectric elastomer materials such that when the heel pressed down it ballooned between a set of holes built into the frame, thus increasing the strain applied to the material. The device was capable of generating 800 mW of power per shoe when walking at a pace of two steps per second. While the system was demonstrated to effectively generate large power levels, it requires a substantial bias voltage and a switching circuit to pull energy from the material and maximize charge. This can make the energy harvesting system difficult to implement.

Several studies have also investigated the storage of electrical energy generated by a power-harvesting device. Umeda et al. (1997) investigated the characteristics of energy storage by a piezogenerator with a bridge rectifier and capacitor. Their study used a small piezoelectric bender as the energy source and varied several parameters to determine the effect on energy storage. Kymissis et al. (1998) also investigated the storage of energy in a capacitor and developed a circuit that used a capacitor to accumulate the electrical energy along with various other components to regulate the charging and discharging cycle of the capacitor. The circuit was found to function well in their application, but the capacitor charged and discharged very quickly resulting in only intermitted power output, as shown in Fig. 17.2. Later Sodano et al. (2005a,b) performed a series of studies investigating the use of an electrochemical battery as the energy storage device. The authors showed that a piezoelectric patch could recharge a small nickel metal hydride battery in a few hours when excited with the level of energy available on an automobile engine.

Previous studies have demonstrated that it is difficult to obtain electrical energy from a shoe without disrupting the wearer's gait or endurance. To avoid these issues, researchers have begun to look into obtaining electrical energy from the differential forces between a human and backpack that occur during walking. Rome et al. (2005)

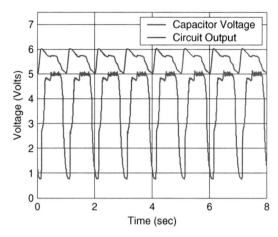

Fig. 17.2 Plot of the capacitor's voltage and the corresponding voltage output for a commonly used power-harvesting circuit (Sodano et al. 2005a)

investigated the design of a backpack that could convert mechanical energy from the vertical movement of carried loads to electricity. The study designed the backpack such that a linear bearing and a set of springs suspended the load relative to a frame and shoulder harness. This configuration allows the load to move vertically relative to the frame. This relative motion was then converted to electrical energy using a rotary electric generator with a rack and pinion, as shown in Fig. 17.3. This system was demonstrated to generate a maximum power of approximately 7.37 W. However, the authors indicate through analysis of the O_2 intake and CO_2 produced by the wearer that the motion of the pack increased the energy expended by 19.1 W or about a 3.2% increase over the energy expelled without the harvesting device (Kuo, 2005). While the backpack does generate significant power levels, the additional degree of freedom provided to the load could impair the user's dexterity and lead to increased fatigue.

While many of these systems are compatible with the energy present around an emergency worker or soldier, they typically do not generate sufficient energy, are cumbersome, or interfere with the gait of the wearer. Thus, the focus of this research is to design the system such that the power-harvesting backpack provides no additional stress or load to the wearer over that of a conventional backpack. The research effort presented here will utilize two systems to harvest the dynamic energy from a backpack, namely, the piezoelectric polymer polyvinylidene fluoride (PVDF) and mechanically amplified stack actuators. Piezoelectric materials function in such a way that an applied electric potential forms a mechanical strain and an applied strain results in the formation of an electrical charge.

The PVDF bulk material is widely available, and is low cost; however, it requires processing to obtain piezoelectric properties. In order to make the backpack as close to a typical design as possible, the fabric straps are to be replaced with a PVDF polymer strap. As the soldier walks with the backpack, the differential forces between wearer and backpack would be transferred to the polymer straps which then convert

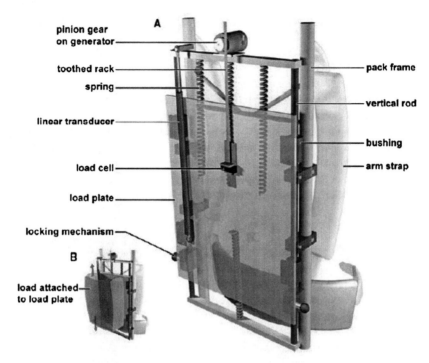

Fig. 17.3 Schematic showing the energy harvesting backpack developed by Rome et al. (2005)

the applied force to electrical energy. PVDF polymer is a high-density polymer and has an elastic modulus approximately equal to that of nylon or PVC, which makes it well suited for this application because a 100 N load on a 100 μm thick strap will only result in a 0.6% strain of the strap. This indicates that the strap would perform very similarly to the traditional strap in this application. This level of strain does not pose an issue for the PVDF material; however, typical electrodes consist of a solid metallic film that is applied to the surface of the polymer using sputter coating and cannot withstand high levels of cyclic strain or the high shear stress associated with the proposed design. Therefore, the use of a PVDF polymer strap necessitates the application of an advanced electrode that can withstand the intended environment. To overcome this issue, a nanostructured electrode has been fabricated using NanoSonic, Inc.'s proprietary self-assembly process. This electrode design provides the required robustness and durability such that the functionality of the pack can be guaranteed in the harsh conditions experienced during outdoor activities.

The second energy harvesting system presented here will use piezoelectric stack actuators for energy harvesting. Stack actuators have seen little use for power-harvesting applications due to their high stiffness which makes straining (power proportional to strain) the material difficult under typical ambient vibration levels. This issue can be alleviated using the mechanically amplified stack actuator which employs a simple kinematic design to transform the low force in the strap to a high force at the piezoelectric stack. This design can be effectively used to significantly reduce the system stiffness to a level appropriate for use in energy harvesting applications.

This chapter will first describe the development of a piezoelectric polymer-based strap followed by a study into the use of mechanically amplified stack actuators. A theoretical model will be first developed to predict the power generated by the piezoelectric strap followed by experimental testing to validate its accuracy. Following the validation of the model, simulations will be performed using actual loading data measured from the straps of a backpack during walking to predict the energy available from the backpack. The model will then be modified to describe the amplified piezoelectric stack and testing will again be performed to validate its accuracy on this system. The results will show that both systems could be used as a power supply for low-power electronics or sensors.

17.2 Model of Power-Harvesting System

In order to predict the energy generated by a strap of piezoelectric material subjected to a dynamic tension, the piezoelectric constitutive equations are used in coordination with a single degree of freedom model. When defining the constitutive equations, it is typical that the poling direction of the strap be defined as the 3 direction and loading be in the -1 direction for this application, as shown in Fig. 17.4. The linear constitutive equations for piezoelectric materials are defined as

$$
\begin{aligned}
S_1(t) &= s_{11}^E T_1(t) + d_{31} E_3(t) \\
D_3(t) &= d_{31} T_1(t) + \varepsilon_{33}^T E_3(t)
\end{aligned}
\tag{17.1}
$$

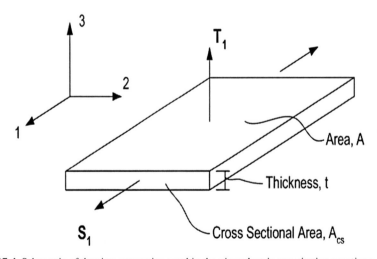

Fig. 17.4 Schematic of the sign convention used in the piezoelectric constitutive equations

where $S_1(t)$ is the strain, s_{11} is the compliance, $T_1(t)$ is the stress, d_{31} is the piezo-electric coupling coefficient, $E_3(t)$ is the electric field, $D_3(t)$ is the electric displacement, ε_{33} is the dielectric permittivity, and the subscripts represent the direction of each property. The first equation defines the mechanical response of the material, while the second equation defines the electrical response. Because our system is not operating over a wide range of frequencies and assuming the strap is subjected to a known tension (neglecting bending, see Sodano et al. (2004b) for general power harvesting models) from the experimental characterization, the constitutive equations can be simplified using the following relationships

$$S_1(t) = \frac{x(t)}{L} \quad T_1(t) = \frac{F(t)}{A_{cs}} \quad E_1(t) = \frac{v(t)}{t} \quad D_3(t) = \frac{Q_3(t)}{A} \tag{17.2}$$

where $x(t)$ is the displacement, $Q(t)$ is the charge, L is the length of the strap, t is the material thickness, A_{cs} is the cross-sectional area, A is the surface area, $F(t)$ is the applied force to a single strap, and $v(t)$ is the voltage. Because the system will include a number of straps mechanically in parallel to strengthen the strap while maintaining a desired capacitance, the number of straps, n, is introduced into the equations. This term affects the capacitance as well but has been omitted from the equations because the variation is dependent on the electrical connection between each strap, which can be modified. Substitution of these terms allows the constitutive equations to be written as

$$x(t) = \frac{L}{nA_{cs}Y}F(t) + \frac{d_{31}L}{t}v(t) \tag{17.3}$$

$$Q(t) = \frac{d_{31}L}{t}F(t) + \frac{\varepsilon_{33}^T A}{t}v(t). \tag{17.4}$$

where Y is the modulus of elasticity. With the constitutive equations in a convenient form they can be used with the single degree of freedom model shown in Fig. 17.5 to define the dynamics of the strap. The mass shown in Fig. 17.5 only represents the effective strap mass because the strap tension is directly measured using a load cell, therefore capturing the gravitational and inertial effects of the pack load. Solving the force balance in Fig. 17.5, substituting the piezoelectric force in Eq. (17.3), and realizing $v(t) = Q(t)/C$ allows the mechanical response of the system to be written as

$$M\ddot{x} = -C_p\dot{x} - F_p(t) + F(t)$$

$$\Rightarrow M\ddot{x} + C_p\dot{x} + \frac{YnA_{cs}}{L}x(t) - \frac{d_{31}YnA_{cs}}{t}C^{-1}Q(t) = F(t) \tag{17.5}$$

where M is the mass, C_p is the damping of the PVDF material, C is the capacitance defined as $C = \varepsilon_{33}A/t$, $F_P(t)$ represents the stiffness and actuation force of the

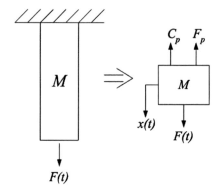

Fig. 17.5 Single degree of freedom representation of the piezoelectric strap

piezoelectric polymer, and $F(t)$ is the dynamic tension or force in the strap measured using a load cell. The electrical response can then be defined by solving for $v(t)$ in Eq. (17.4) and substituting the longitudinal stiffness $k = EA/L$ to give

$$-v(t) - C^{-1}\frac{d_{31}AY}{t}x(t) + C^{-1}Q(t) = 0. \tag{17.6}$$

Once the dynamics of the system have been coupled to the electrical response of the piezoelectric material, the electrical boundary conditions can be included by defining the load resistance R as

$$v(t) = -R\dot{Q}(t) \tag{17.7}$$

Substituting Eq. (17.7) into Eqs. (17.5) and (17.6) gives the coupled electro-mechanical response of the power-harvesting system as

$$M\ddot{x} + C_p\dot{x} + \frac{YnA_{cs}}{L}x(t) - \frac{d_{31}YnA_{cs}}{t}C^{-1}Q(t) = F(t)$$

$$R\dot{Q}(t) - C^{-1}\frac{d_{31}AY}{t}x(t) + C^{-1}Q(t) = 0 \tag{17.8}$$

where $i(t) = \dot{Q}(t)$, and the voltage output of the system across the load resistance is defined by the $v(t) = R\dot{Q}(t)$ term. The power output can then be calculated assuming that the load impedance will be the same as the source defined as

$$Z = \frac{1}{j\omega C} = \frac{t}{j\omega\varepsilon_{33}^T A} \tag{17.9}$$

where ω is the walking frequency and j defines the complex impedance of the capacitive piezoelectric material. Using the defined impedance (here a resistance between the electrodes) of Eq. (17.9), the power can then be defined as

$$P(t) = \frac{v(t)^2}{R}. \tag{17.10}$$

With the equation defined above, the electric response of the strap material when subjected to a known dynamic tension can be determined.

17.2.1 Experimental Testing of Piezoelectric Strap

In order to identify the level of energy available from a backpack instrumented with a PVDF polymer strap, the dynamic tension resulting from walking with a 220 N (50lb) load was identified. The varying tension was measured using a load cell integrated into the top and bottom of the backpack strap to allow direct measurement of the tension, as shown in Figure 17.6. A 220 N (50lb) steel plate was placed inside the pack to act as the load and the pack was worn during walking on a treadmill at speeds ranging from 2-3mph (0.9–1.3 m/s). The results of these tests showed that the load from pack was not evenly distributed through the strap. The resulting force measured in the top and bottom sections of the strap are shown in Figure 17.7. These results indicate that the load is very close to evenly dispersed over the top and bottom straps of the pack. The mean load in the top strap was determined to be 84 N (18.81 lbs) and 83 N (18.67 lbs) measured in the bottom strap. The data obtained from these tests can be directly used along with a teoritical model to predict the power output from the alternating force in the PVDF strap.

After characterizing the loading in the backpack straps, a set of PVDF straps were fabricated for experimental characterization and validation of the model. The flexibility of the harness design requires a series of equally flexible electrodes which can withstand the rigors of everyday use. To be successful in this application, the electrodes must possess four distinct properties. The first is the flexibility and conformability of the electrode to avoid failure and prevent interference with the

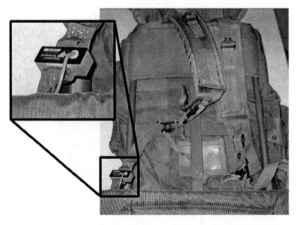

Fig. 17.6 Instrumented backpack used to experimentally determine the loading in the straps

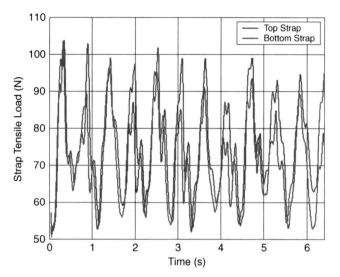

Fig. 17.7 Resulting load applied to the top and bottom straps of the pack

mobility of the wearer. The second is the resistance properties of the electrode as a function of strain and high-cycle loading. The third feature corresponds to the adhesion and durability properties of the electrode, while the fourth is the patterning capabilities of the electroding method.

Typically, PVDF sensors use a thin-metallized film for electrodes of either aluminum which cannot withstand high-strain and cyclic loading or a silver ink which has poor adhesion properties. Therefore, the sensor developed here uses a self-assembly method which offers the ability to produce highly uniform coatings and electrodes which range from nano- to macro-scale. The electrostatic self-assembly (ESA) process is shown in Fig. 17.8, and consists of the simple soaking of a chosen substrate into alternate aqueous solutions containing anionic and cationic materials resulting in covalently bonded layers with a nearly perfect molecular order of the individual monolayers. The design of the individual precursor molecules, and control of the order of the multiple molecular layers

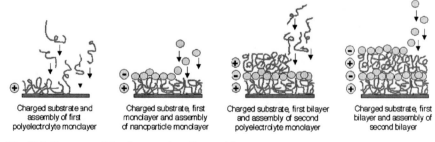

| Charged substrate and assembly of first polyelectrolyte monolayer | Charged substrate, first monolayer and assembly of nanoparticle monolayer | Charged substrate, first bilayer and assembly of second polyelectrolyte monolayer | Charged substrate, first bilayer and assembly of second bilayer |

Fig. 17.8 Summary of the electrostatic self-assembly process

through the thickness of the film, allows control over the macroscopic properties. For this specific application, NanoSonic's proprietary nanocomposite Metal Rubber^{TM} offers performance capabilities that can be tailored to meet the necessary requirements for this power-harvesting system. Metal Rubber^{TM} has the ability to undergo strains of 1000% while maintaining conductivity, and return to its original shape and conductivity when released. The modulus properties can be tailored from less than 0.1 MPa to greater than 500 MPa and the resistive properties can be tuned to provide near constant conductivity for strains less than 5%. NanoSonic has also demonstrated the ability to fabricate complex electrode patterns using their self-assembly process.

Several 20.32 × 24.4 cm samples of the unmetallized 28 and 52 μm films were self-assembled with Metal Rubber^{TM} using systematic variations in the initial chemical treatment of the samples to prepare them for the ESA process. This initial treatment corresponds to the first step in the ESA summary shown in Fig. 17.8. Electrode thickness is estimated to be ~100 nm based on the number of cycles used in the ESA process. The electrode resistance was measured along each edge of the samples, and along the diagonal. The surface resistance slowly increases over 4 s when probes are applied to the electrode, rising from 0 to 5.8 Ω where it stabilizes. This behavior was also seen in the metallized samples purchased from MSI and occurs due to the samples capacitance. The samples were then poled using a 71 MV/m field on the 28 μm films and a field of 36 MV/m on the 52 μm films. Two separate field values were used due to limitations in the power supply used for poling. However, both samples exhibited equally high coupling.

Once the PVDF samples had been fabricated, experimental testing was performed to characterize their use for energy harvesting in a backpack strap and to validate the accuracy of the model such that predictions on the available power output from the entire backpack harness could be made. Testing was performed using a material testing system (MTS) with one to four straps, such that the energy output could be identified for various strength straps and the effect of the electrical connection between the samples could be identified. This configuration was used such that a controlled tension equivalent to the measurements made during testing of the backpack could be applied to the sample, while the energy output was measured. The MTS machine with the PVDF samples is shown in Fig. 17.9. Since the straps did not have any protective coating applied, multiple straps were electrically separated using spacer blocks. These spacers had electrical contacts to complete the circuit with the piezoelectric and allowed for wiring in series and parallel configurations. A preload of approximately 40 N was applied to the straps to simulate the static weight in the backpack, while a 20 N sine wave with frequency of 5 Hz was applied to simulate the alternating load in the backpack. This load was chosen because it is representative of the force found during testing of the loaded backpack. The MTS load cell was fixed to the stationary clamp to avoid its inertia affecting the dynamics of the system. Tests of two or more straps were run using load control; however, single strap tests required position control because the PVDF is too compliant for the MTS to auto-tune its controller.

Fig. 17.9 Experimental setup used to apply a controlled tension to the strap, while the power output is measured

For each test, the capacitance of the piezoelectric straps was measured and a resistive load with matched impedance was applied across the piezoelectric strap. The resistance was matched for each configuration taking into account the input impedance of the oscilloscope and attached probe. The voltage output was measured using a Tektronix (model TDS 2002) digital oscilloscope with a 10× probe to increase the voltage range of the oscilloscope. The oscilloscope was set to record a time history of the voltage output of the piezoelectric and the tensile force on the strap during the test. These two datasets were used in modeling the system to validate its accuracy. Tables 17.1 and 17.2 summarize the strap properties and test configurations performed. Because the force applied to the strap is directly measured

Table 17.1 Mechanical and electrical properties of the PVDF materials

Material property	Symbol	PVDF thickness	
		28 μm	52 μm
Elastic modulus	E	4 GPa	5 GPa
Piezoelectric coupling	d_{31}	25 pC/N	27 pC/N
Permittivity	E_{33}	110 pF/m	110 pF/m
Strap width	W	21.7 mm	21.7 mm
Strap active length	L	180 mm	180 mm
Strap mass	M	0.33 g	0.60 g

Table 17.2 Capacitance and impedance of the PVDF strips

Thickness	No. of straps	Series		Parallel	
		Capacitance (nF)	Resistance (MΩ)	Capacitance (nF)	Resistance (MΩ)
28 μm	1	19.5	1.65	19.5	1.65
	2	9.6	3.20	39.0	0.83
	3	7.0	4.70	60.0	0.57
	4	5.2	6.60	75.0	0.43
52 μm	1	10.5	0.78	10.5	0.78
	2	5.0	6.10	21.0	1.46
	3	3.4	8.90	32.0	1.00
	4	2.6	10.00	41.0	0.77

using a load cell, the mass in the back can be neglected and only the effective mass of the strap needs to be used in Eq. (17.8).

17.2.2 Results and Model Validation

Experimental tests were performed with one, two, three, and four straps with both parallel and series configurations. The data collected from each of the test scenarios were used to validate the accuracy of the theoretical model such that a prediction of the total power output from a complete backpack could be identified. The first set of tests performed was for the straps wired in series. Connecting the piezoelectric straps in this way increases voltage output, decreases capacitance, and increases impedance. However, since the testing was performed using load control, the voltage output remained fairly consistent even when the number of straps was altered. This is due to a decrease in strain per strap as the number of straps was increased for the same load. Figure 17.10 shows the comparison between the voltage and the corresponding power for the experimental and simulated data for one and four strips, respectively, in parallel and series configurations. The plots show that the model accurately predicts the voltage output and the system dynamics. The predicted power amplitude for a single strap is slightly lower than the experimental data because the model underpredicts power for single strap tests. This may result from overstraining the piezoelectric strip in the test stand leading to a nonlinear stiffness and higher strain than the model would predict under the same load. Figure 17.10c depicts the voltage and power output of four straps wired in parallel. In this configuration, the voltage output is decreased while the current output is increased, resulting in approximately the same amount of power as the four strip test wired in series. Again, the model accurately predicts the voltage and power output.

Figure 17.11a–d shows plots of the mean power against the number of straps for the 28 μm (Fig. 17.11a and b) and 52 μm (Fig. 17.11c and d) thicknesses in series (Fig. 17.11a and c) and parallel (Fig. 17.11b and d) wiring configurations. From these figures, it can be seen that the model accurately predicts the power output over each of the 14 configurations tested. The plots display the decreasing

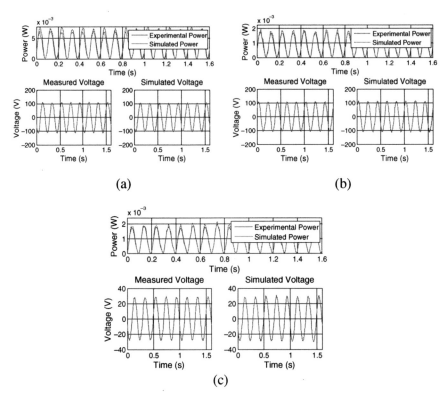

Fig. 17.10 Measured and simulated voltage and power output, (**a**) one strap, (**b**) four straps in series, and (**c**) four straps in parallel

power output as the number of straps is increased, which results due to less strain applied to each strap. The apparent dip in the mean power output for the 28 μm, three strap configuration results because the load impedance was not tuned to the piezoimpedance as well as in the other runs. This was because of the difficulty in creating an appropriate resistance while using the 10× probe and oscilloscope. However, the model captures the impedance mismatch, thus demonstrating that it can be accurately used to predict the cases in which the load electronics may not be tuned to the piezoelectric's impedance. The test of a single, 52 μm strap also did not have impedance matched exactly, resulting in a lower than expected power output. Table 17.3 lists the mean power output and the percent difference between the experimental and simulated results. The model is capable of predicting power output within 13% of the actual value for all cases.

17.2.3 Backpack Power Prediction

Once the model had been validated using the results of the experimental tests, it was used to predict the power that could be generated by a backpack-containing piezo-

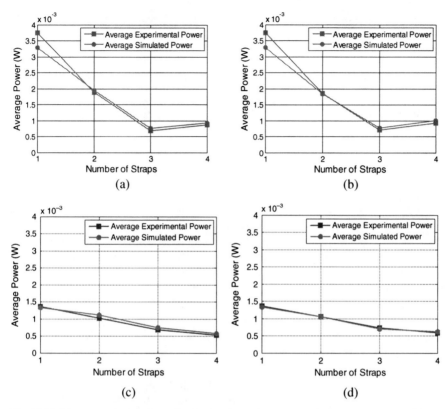

Fig. 17.11 Mean power output for each case tested, (**a**) 28 μm series, (**b**) 28 μm parallel, (**c**) 52 μm series, and (**d**) 52 μm parallel

electric straps. The power output was theoretically predicted rather than experimentally identified, because the MTS machine could not support the length of PVDF strap required. Table 17.4 lists the backpack strap parameters used to estimate the power output from a loaded pack. In order to generate the highest level of power

Table 17.3 Mean power output for each case experimentally tested and the error in the predicted output

Thickness	Number of straps	Mean power in series (mW)			Mean power in parallel (mW)		
		Exp.	Sim.	Percent difference	Exp.	Sim.	Percent difference
28 μm	1	3.75	3.28	12.6			
	2	1.89	1.94	2.8	1.85	1.83	0.8
	3	0.68	0.76	10.4	0.71	0.78	10.2
	4	0.87	0.94	7.6	0.93	1.01	9.3
52 μm	1	1.36	1.33	2.3			
	2	1.02	1.12	9.1	1.05	1.06	1.1
	3	0.68	0.75	9.9	0.73	0.71	2.5
	4	0.53	0.57	7.5	0.60	0.62	4.9

Table 17.4 Mechanical and electrical properties and dimensions used to simulate a PVDF backpack harness

| Material property | Symbol | PVDF thickness | |
		28 μm	52 μm
Elastic modulus	E	4.0 GPa	5.0 GPa
Piezoelectric coupling	d_{31}	25 pC/N	27 pC/N
Permittivity	ε_{33}	110 pF/m	110 pF/m
Strap width	W	51 mm	51 mm
Strap active length (top)	L	1.016 m	1.016 m
Strap active length (bottom)		203 mm	203 mm
Mass per strap	m	3.5 g	6.3 g

possible, the strap length must be maximized. This is achieved using a continuous strap running through the pack's frame and making a complete loop as shown in Fig. 17.12. From this table, it can be seen that the total strap length is 1.2 m. The loading applied in this simulation was identical to that identified through testing of an instrumented backpack carrying a 444 N load as previously discussed. The strap tension walking data were then used to run the simulation. Figure 17.13 shows the estimated power output for a backpack with two, 52 μm thick piezoelectric straps per backpack shoulder strap (four 52 μm piezoelectric straps total) connected electrically in parallel. From this simulation, the maximum instantaneous power

Fig. 17.12 Schematic of the backpack with piezoelectric straps

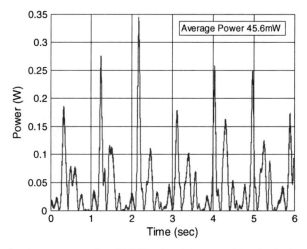

Fig. 17.13 Predicted power output for a PVDF backpack harness with a single piece of 52 μm film on each strap

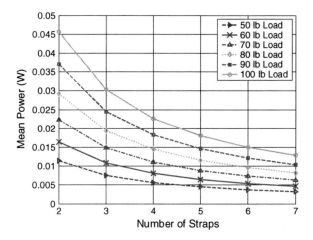

Fig. 17.14 Mean power output for the each of the simulated conditions

can be seen to be 0.345 W and an average power of 45.6 mW over the duration of the simulation. This configuration of strap thickness and number does not generate the highest power, but was found to provide enough strength to carry the simulated load.

The plot shown in Fig. 17.14 shows the power output for the two shoulder straps based on the load in the backpack and the number of piezoelectric straps per shoulder strap for a 52 μm strap thickness. Due to the high impedance of the PVDF material, the energy output is at a very large voltage, but low current. For this reason, a parallel wiring configuration is typically chosen to generate lower voltages with higher current levels. The increase in current compared with a series configuration

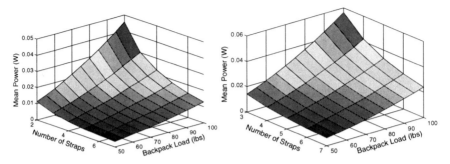

Fig. 17.15 Mean power output for several loads and number of straps: (*left*) 52 μm and (*right*) 28 μm

would be more beneficial when used for charging small batteries or supercapacitors. As is seen in the figure, the load plays a very important role in the level of power generated by the energy harvesting system. Also, fewer straps create more power because they have higher associated strains. Figure 17.15 provides a surface plot demonstrating the mean power output (continuous power) for a variety of loads and number of straps. It is anticipated that the actual power obtained from a strap of PVDF would be higher than the predictions due to bending effects. The resulting power output could certainly be used to power some small, low-power electronics or could be accumulated over the duration of the excursion leading to supplemental energy and lower number of batteries carried without increasing the mass of the pack or inducing parasitic effects on the wearer. This backpack design will lead to minimal parasitic effects therefore making it a feasible method of gathering energy from human motion.

17.3 Energy Harvesting Using a Mechanically Amplified Piezoelectric Stack

Two practical coupling modes exist in piezoelectric materials: the −31 mode and the −33 mode. In the −31 mode, a force is applied in the direction perpendicular to the poling direction, an example of which is a bending beam that is poled on its top and bottom surfaces. In the −33 mode, a force is applied in the same direction as the poling direction, such as the compression of a piezoelectric block that is poled on its top and bottom surfaces. An illustration of each mode is shown in Fig. 17.16. Conventionally, the −31 mode has been the most commonly used coupling mode (the piezoelectric polymer backpack strap used this mode); however, the −31 mode yields a lower coupling coefficient, k, than the −33 mode. This mode of operation is typically capitalized on by stacking a large number of thin piezoceramic wafers together, called the stack configuration, with the electric field applied along the length of the stack.

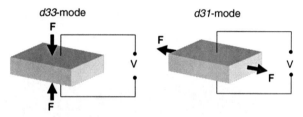

Fig. 17.16 Illustration of −33 mode and −31 mode operation for piezoelectric materials

Baker et al. (2005) have shown that for three different types of piezoelectric materials, the −31 mode has a lower coupling coefficient, k, than the −33 mode. Upon comparing a piezoelectric stack operating in the −33 mode to a cantilever beam operating in the −31 mode of equal volumes, however, it was observed that although the stack was more robust and had a higher coupling coefficient, the cantilever produced two orders of magnitude more power when subjected to the same force. This result is due to the high-mechanical stiffness in the stack configuration which makes straining of the material difficult. It was concluded that in a small force, low-vibration level environment, the −31 configuration cantilever proved most efficient, but in a high-force environment, such as a heavy manufacturing facility or in large-operating machinery, a stack configuration would be more durable and generate useful energy. This result was also presented by Roundy et al. (2003), who concluded that the resonant frequency of a system operating in the −31 mode is much lower, making the system more likely to be driven at resonance in a natural environment, thus providing more power.

While the piezoelectric stack actuator has a higher energy density than benders due to their operation in the d_{33}-direction; they have seen little use in energy harvesting applications because typical ambient vibration levels cannot effectively strain the material. This issue can be alleviated through the use of a mechanical amplification system to transform a low load/high displacement input into a high force on the stack. This amplification is typically achieved using relatively simple kinematic designs, and can be effectively designed to reduce the system stiffness to a level appropriate for use in energy harvesting applications. Figure 17.17 shows the design of the mechanical amplifier used in to convert the low forces in the strap of the pack to high forces on the stack actuator. A finite element model of the system was used to design the compliant mechanism such that it could both support the load in the pack and provide an amplification factor that would allow the device to function effectively as an energy harvesting system. The device was fabricated from 2024-T351 Aircraft grade aluminum with a modulus of elasticity of 73.1 GPa and density of 2780 kg/m³. Figure 17.18 shows the CAD model of the mechanical amplifier and the loading applied to simulate the static load in the strap. The finite element simulation was performed with Tetra4 Mesh elements with an element size of 1 mm. The resulting deformation is shown in Fig. 17.20 with respect to the original shape. By taking the ratio of the deformation of the bolt hole (0.026 mm) to that of the

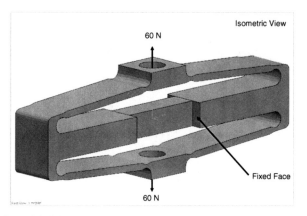

Fig. 17.17 CAD model of the mechanically amplified piezoelectric stack actuator

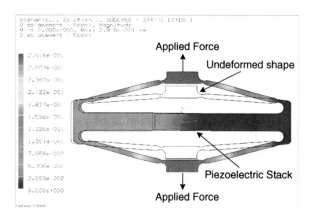

Fig. 17.18 FEA results of the mechanically amplified stack actuator; the color scale represents the displacement magnitude. The design shown provides a force magnification of 10.8

piezoelectric stack face (0.00479 mm), it can be determined that the amplification ratio for this design is 10.8.

17.3.1 Model and Experimental Validation of Energy Harvesting System

Once the amplification factor is known, a model similar to that developed in Section 17.2 can be formulated to predict the energy generated by the amplified piezoelectric actuator when subjected to a dynamic tension. Using the piezoelectric constitutive equations of Eq. (17.1) and the relationships shown in Eq. (17.2), the equations defining the electrical and mechanical responses of the material can be modified to account for the piezoelectric stack. Because a stack actuator has many layers of piezoelectric material, the number of layers, n, is introduced into

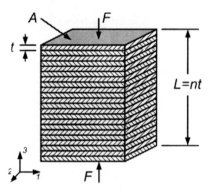

Fig. 17.19 Schematic of piezoelectric stack actuator indicating the notation used

the equations using the relationship that $L = nt$ as shown in Fig. 17.19, this results in the capacitance of the stack being defined as $C = n\varepsilon_{33}A/t$. Accounting for the affect of the multilayer stack properties, the coupled electromechanical equations for this energy harvesting system are defined as

$$M\ddot{x} + kx(t) - \frac{d_{33}kL}{t}C^{-1}Q(t) = \alpha F(t)$$

(17.11)

$$R\dot{Q}(t) - C^{-1}d_{33}nkx(t) + C^{-1}Q(t) = 0$$

where α is the amplification factor modifying the force applied to the stack, and $i(t) = \dot{Q}(t)$, and the voltage output of the system across the load resistance defined by the $v(t) = R\dot{Q}(t)$ term. Like the equations defined in Section 17.2, the power output from the stack can be identified through the voltage and load resistance.

In order to validate the model of Eq. (17.11), experimental testing was performed using a servo-hydraulic Instron load frame. The loading identified through experimental testing in Section 17.3 was used to scale the force input. As before, various input forces were used to characterize the performance of the stack amplifier such that the energy output, however, all loading was representative of the forces in the backpack strap. A preload of approximately 40 N was applied to the stack to simulate the static weight in the backpack, while various frequencies and amplitudes were applied to simulate the alternating load in the backpack and provide different conditions to test the model with. This load was chosen because it is representative of the force found during testing of the loaded backpack. The MTS load cell was fixed to the stationary clamp to avoid inertial effects corrupting the dynamics of the system. All tests were run under position control due to the inability of the MTS to compensate for the high compliance of the stack amplifier under load control.

The voltage output of the stack was measured using a dSPACE data acquisition system (model RTI1104) with a 10× probe to increase the measurable voltage

Table 17.5 Mechanical and electrical properties of the stack actuator

Material property	Symbol	
Elastic modulus	E	44 GPa
Piezoelectric coupling	d_{33}	650 pC/N
Permittivity	ε_{33}	6200 pF/m
Capacitance	C	1.59 μF
Stack area	A	25 mm^2
Stack active length	L	16 mm
Stack mass	m	2.3 g
Number of layers	n	130
Stack amplification	amp	10.9

Table 17.6 Parameters of each test configuration performed

Test number	Amplitude (mm)	Frequency (Hz)	Resistance (kΩ)
1	0.35	5	4.94
2	0.35	5	9.72
3	0.35	5	19.2
4	0.35	5	28.3
5	0.35	5	37.0
6	0.35	2.5	37.0
7	0.35	10	9.72
8	0.2	5	19.2
9	0.5	5	19.2

range. The dSPACE was set to record a time history of the voltage output of the piezoelectric, the displacement of the MTS, and the tensile force on the stack amplifier during each test. The force and time histories were then directly applied to the model allowing a verification of the models accuracy. Tables 17.5 and 17.6 summarize the piezoelectric actuator properties and test configurations performed, respectively.

17.3.2 Results and Model Validation

Experimental tests were performed with the configurations provided in Table 17.2. The first five tests were used to verify that the model could predict the power generation of the system with a load resistance that was not tuned to the impedance of the stack. Tests 6 and 7 investigated the effect of frequency on the system and the last two tests were to verify the predicted power under different input amplitudes. The data collected from each of the test scenarios were used to validate the accuracy of the theoretical model such that a prediction of the total power output from a complete backpack could be identified. Figures 17.20 and 17.21 compare the voltage and power experimentally measured and predicted for the parameters of test 2 (matched impedance) and test 4 (detuned impedance), respectively. The plots show the model

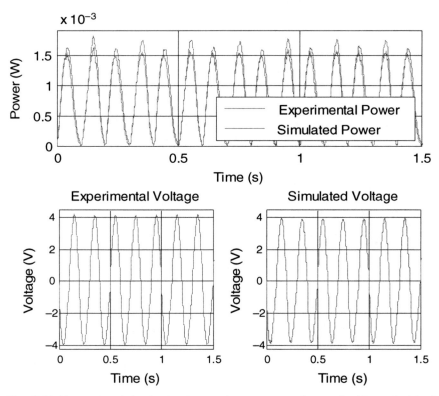

Fig. 17.20 Measured and simulated voltage and power output for test 2 with matched load impedance

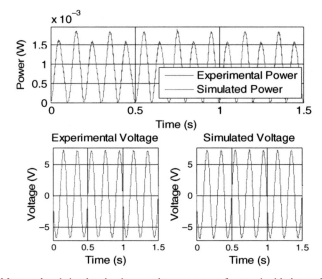

Fig. 17.21 Measured and simulated voltage and power output for test 4 with detuned impedance

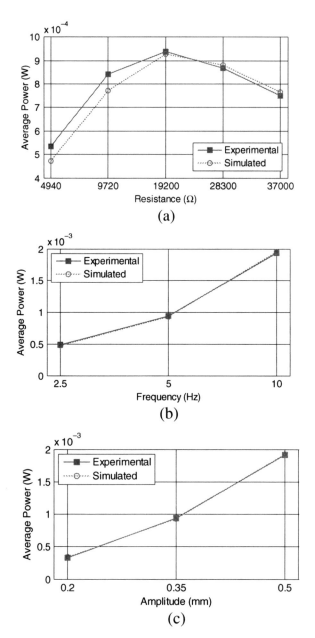

Fig. 17.22 Mean power output for each test scenario, (**a**) Impedance tuning, (**b**) varying frequency, and (**c**) varying amplitude

Table 17.7 Mean power output for each case experimentally tested and the error in the predicted output

Test number	Experimental (mW)	Simulation (mW)	Absolute difference	Percent difference
1	0.750	0.767	0.17	2.3
2	0.0535	0.473	−0.62	11.6
3	0.843	0.771	−0.73	8.8
4	0.938	0.928	−0.10	1.1
5	0.868	0.882	0.14	1.6
6	0.484	0.478	−0.06	1.2
7	1.939	1.956	0.17	0.9
8	0.329	0.337	0.08	2.5
9	1.909	1.920	0.10	0.6

accurately predicts the voltage for a tuned and detuned system. These results were also representative of the other tests performed.

Comparisons between the model and experimentally measured data for each of the three varied parameters, load impedance, frequency, and amplitude are shown in Fig. 17.22. From these figures, it can be seen that the model accurately predicts the power output for all of the test cases. The optimal energy is harvested when the load impedance is matched to the source, in this case a 19.2 kΩ. Table 17.7 provides the mean power output and the percent difference between the experimental and the simulated results for each test. The model is capable of predicting power output within 12% of the actual value for all cases, providing confidence in its prediction for the backpack loading.

Once the model had been validated using the MTS machine, the stack amplifier was connected as shown in Fig. 17.23. This setup was used to measure the force

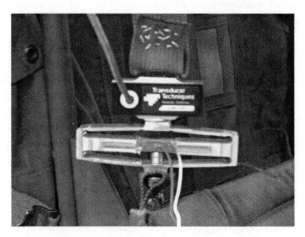

Fig. 17.23 Stack amplifier installed in series with a force transducer for simultaneous measurement of force and voltage

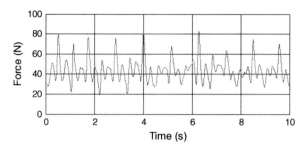

Fig. 17.24 Measured force on the stack amplifier with a 176 N (40 lb) load in the backpack

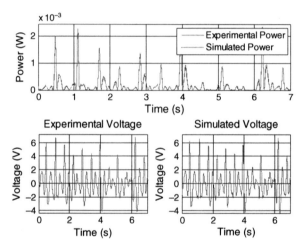

Fig. 17.25 Experimental and simulated voltage and power output for the amplified piezoelectric device in the backpack with a 176 N (40 lb) load

on the stack amplifier as well as the voltage from the stack amplifier. The stack was designed for testing in the MTS machine with loads up to 220 N (50 lb). To minimize the chance of breaking the device, only 176 N (40 lb) was loaded into the backpack in the form of dumbbells. Figure 17.24 shows the measured load. The force transducer data were fed into the model and the resulting voltage and power time histories are shown in Fig. 17.25. The average simulated power was 0.175 mW and the average measured power was 0.176 mW (0.6% difference).

17.3.3 Backpack Power Prediction

With the model validated under both sinusoidal and walking loads, it can be used to predict the level of power that could be generated by a backpack that contain various loading. The loading applied in this simulation was identical to that identified through testing of an instrumented backpack carrying a 220 N load as previously

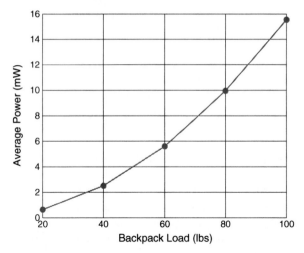

Fig. 17.26 Simulated mean power output for the amplified piezoelectric device with respect to the load in the backpack

discussed in Section 17.2.1. Using the loading data measured from the backpack, the piezoelectric stack was simulated to identify the level of energy available. The results of this simulation are shown in Fig. 17.26 for a backpack with a stack in each strap (two in total) and demonstrate that the available power increases with increasing load as would be expected. At a load of 100 lb, the continuous power output would be ~16 mW, which is sufficient to power some small, low-power electronics or could be accumulated over the duration of the excursion leading to emergency energy when needed. This backpack design will lead to minimal parasitic effects, therefore, making it a feasible method of gathering energy from human motion.

17.4 Conclusions

The past decade has seen a rapid increase in the number of wireless sensors deployed and portable electronics carried by the modern individual. Each of these devices is typically powered using a traditional electrochemical battery which can lead to issues due to their finite lifetime. To overcome this issue, the field of power harvesting has grown, which looks to convert ambient energy surrounding the system to usable electrical energy. This chapter has presented the development of a novel energy harvesting system, which generates electrical energy from the differential forces generated in the straps of a backpack during walking. Two systems were developed to generate electrical energy from this source, the first used the piezoelectric polymer PVDF to replace the nylon strap in a typical backpack and generate electrical energy, and the second used a mechanically amplified stack actuator.

A theoretical model was developed for each system and experimental testing was performed to both validate the model and determine the level of energy available through each system. To ensure the polymer strap could withstand the rigors of use in an outdoor environment and high-cyclic loading, a compliant electrode was grown on the surface using electrostatic self-assembly (ESA) processes. Once the model had been verified, simulations were performed to predict the energy available from a complete backpack. The results showed that 45.6 mW of continuous power could be obtained from the PVDF polymer system, while ~16 mW could be obtained by the amplified stack.

The development of energy harvesting systems that can generate electrical energy could lead to improved life of portable electronics or in the ideal case to power the electronics for their useful life. However, many systems that are currently being designed result in significant parasitic effects to the wearer. The backpack proposed and investigated here is designed to preserve both the wearer and the pack performance by holding the system as close to existing harnesses as possible.

References

Anton, SR and Sodano, HA (2007) A Review of Power Harvesting Using Piezoelectric Materials (2003–2006). *Smart Materials and Structures*, 16:R1–R21.

Baker J, Roundy S and Wright P (2005) Alternative geometries for increasing power density in vibration energy scavenging for wireless sensor networks. *Proceedings of the 3rd Int. Energy Conversion Engineering Conference (San Francisco, CA, August 15–18)* 2005–5617:959–970.

Kornbluh, RD, Pelrine, R, Pei, Q, Heydt, R, Stanford, S Oh, S and Eckerle, J (2002) Electroelastomers: Applications of Dielectric Elastomer Transducers for Actuation, Generation, and Smart Structures. *Smart Structures and Materials 2002: Industrial and Commercial applications of Smart Structures Technologies*, 4698:254–270.

Kuo, AD (2005) Harvesting Energy by Improving the Economy of Human Walking. *Science*, 309:1686–1687.

Kymissis, J, Kendall, C, Paradiso, J and Gershenfeld, N (1998) Parasitic Power Harvesting in Shoes. Second IEEE International Symposium on wearable Computers, October 19–20th, Pittsburg, PA, pp. 132–139.

Paradiso, JA and Starner T (2005) Energy Scavenging for Mobile and Wireless Electronics. *Pervasive Computing*, January-March, pp. 18–27.

Rome, LC, Flynn, L, Goldman, EM and Yoo, TD (2005) Generating Electricity While Walking with Loads. *Science*, 309: 1725–1728.

Roundy S, Wright PK and Rabaey J (2003) A Study of Low Level Vibrations as a Power Source for Wireless Sensor Nodes. *Computer Communications*, 26:1131–1144.

Sodano, HA, Park, G and Inman, DJ (2004a) A Review of Power Harvesting Using Piezoelectric Materials. *Shock and Vibration Digest*, 36(3):197–206.

Sodano, HA, Park, G and Inman, DJ (2004b) Estimation of Electric Charge Output for Piezoelectric Energy Harvesting. *Journal of Strain*, 40: 49–58.

Sodano, HA, Park, G and Inman, DJ (2005a) Generation and Storage of Electricity from Power Harvesting Devices. *Journal of Intelligent Material Systems and Structures* 16(1): 67–75.

Sodano, HA, Park, G, and Inman, DJ (2005b) Comparison of Piezoelectric Energy Harvesting Devices for Recharging Batteries. *Journal of Intelligent Material Systems and Structures*, 16(10):799–807.

Starner, T (1996) Human-Powered Wearable Computing. *IBM Systems Journal*, 35(3–4):618–628.

Umeda, M, Nakamura, K and Ueha, S (1997) Energy Storage Characteristics of a Piezo-Generator Using Impact Induced Vibration. *Japanese Journal of Applied Physics*, 36(5B):3146–3151.

Chapter 18
Energy Harvesting for Active RF Sensors and ID Tags

Abhiman Hande, Raj Bridgelall, and Dinesh Bhatia

Abstract This chapter highlights the importance and significance of energy harvesting in applications involving use of active RF sensors and ID tags. The chapter begins by providing a basic overview on radio frequency identification (RFID) operation, various types of RFID tags, and the need for energy harvesting, especially for active RFID tags. Unlike passive tags, active tags utilize a battery to emit rather than reflect or backscatter RF energy. Advantages of active tags include improved range and read rate in electromagnetically unfriendly environments and improved link quality. Typical applications include monitoring enterprise/supply chain assets (e.g. laptops, computers, peripherals, electronic equipment, pallets, inventory items, etc.), personnel, patients, vehicles, and containers. Although a battery can substantially improve performance, it limits maintenance-free operational life. Therefore, harvesting energy from sources such as vibration or light has been shown to address this shortcoming but these sources must be adequate, available throughout the life of the application, and highly efficient. These available energy harvesting technologies are described, and basic design procedures and components for such systems are identified. This includes three key components namely, the energy harvesting transducer, power management circuit, and energy storage device. Each component of the energy harvesting system is described and important design criteria are highlighted. Specific emphasis is placed on the design of the power management component and the available energy storage device technologies. Disadvantages of using off-the-shelf DC-DC converters and rectifiers are emphasized and possible power management solutions for solar and vibrational energy harvesting are explained. Finally, the chapter concludes by describing the future directions and scope including development of integrated multiple source energy harvesting systems on thin-film substrates.

A. Hande (✉)
Electrical Engineering Department, University of Texas at Dallas 800 W Campbell Road, Richardson, TX 75080, USA; Texas Micropower Inc., 18803 Fortson Ave, Dallas Texas, 75252 USA
e-mail: ahande@texasmicropower.com

S. Priya, D.J. Inman (eds.), *Energy Harvesting Technologies*,
DOI 10.1007/978-0-387-76464-1_18 © Springer Science+Business Media, LLC 2009

18.1 Introduction

Radio frequency (RF) sensors and tags find applications in several areas including inventory control, pallet/container tracking, identification (ID) badges and access control, fleet maintenance, equipment/personnel tracking in hospitals, parking lot access and control, car tracking in rental lots, monitoring product health in manufacturing, and so on. Potentially, one of the fastest growing RFID applications is within the retail supply chain. Major retailers need to track goods to and from their worldwide suppliers as containers are transported overseas, and through distribution centers and warehouses. They are seeking technology that provides a high return on investment, whereby manual labor and cost overhead can be eliminated through increasing degrees of automation. Figure 18.1 maps typical demands within an important functional space of RFID technologies by studying the logistics of tracking goods throughout the supply chain. Retailers want the ability to reliably and transparently track pallets, stacks of boxes, and various other forms of traded units, transport devices, and logistical units as they move through portals that are up to 4 m wide and tall. Major retailers and manufactures such as Walmart have begun to successfully deploy RFID tag technology to enable communications with high-value items that move through the supply chain. There is strong evidence that suggests RFID tag technology will soon provide the long awaited, cost-effective mechanism that will fully automate supply chain logistics (Bridgelall, 1999). Automation will deliver greatly improved efficiencies and productivity, while significantly improving product availability.

End-users plan to deploy larger scale RFID tag communications infrastructure as initial pilots validate their expected return on investment. Initially, this RFID infras-

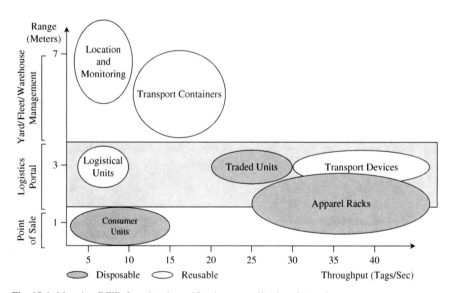

Fig. 18.1 Mapping RFID functional specification to application demands

tructure will be separate from a wireless local area network (WLAN) but in time both will become more highly integrated. As the growing demand for RFID tags continue to accelerate their cost reduction, the technology will begin to penetrate the retail point-of-sale. Multi-bit RFID tags have already begun to upgrade the existing single-bit electronic article surveillance (EAS) security tags. Pervasive deployment of wireless infrastructure that provides dual-mode WLAN and RFID tag communications will provide the necessary foundation for even larger scale deployment, and hence more substantial cost reduction. At some point, it is expected that the value proposition for tagging individual low cost items will reach an equilibrium state.

The existence of a robust and pervasive dual-mode communications infrastructure for WLAN and RFID tags will trigger numerous opportunities for applications around m-commerce. Consumers will eventually utilize PDA-size multi-technology mobile computers that incorporate both wireless network connectivity and RFID tag communications. Imagine being able to automatically sense and physically locate the exact model of a digital camcorder or TV in a showroom, then view its webpage, evaluate its performance and features, compare prices, and finally place an on-line purchase via the WLAN connection.

Most of these RF devices are traditionally passive without any energy storage device on board. However, as will be seen in the following section, there are applications that require a battery on board for better throughput and performance that consequently results in limited life and the need for replacement. This chapter will review the potential for energy harvesting (EH) for such RF sensors and ID tags. There will be a concerted focus in EH from vibrations and light. Feasible applications and potential circuits will be identified accordingly.

18.2 RFID Tags

RF sensor and ID tag products appear in the 125 kHz, 13.56 MHz, 915 MHz, and 2.45 GHz (Microwave) frequency bands, thereby providing evidence that no one technology can equally meet the demands of all application requirements. RFID tag technology falls into two broad categories: passive or active. Simply stated, active has an on board power supply (e.g., a battery), while passive relies on capturing and reusing a small portion of the wake up signal's energy to transmit its RFID tag ID back to the receiver. Passive tags have the advantage of being manufactured and sold at a much lower price point today than active tags. This is a critical element in many RFID supply chain applications requiring the tagging of millions of units. However, passive tags often struggle to provide reliable reads given the performance limitations of a technology using only a small amount of power to push its signal off metal surfaces, through layers of palletized products, and so on. In addition, passive tags sometimes struggle to provide a highly reliable signal when supply chain goods are in motion. Therefore, active tags have an innate performance advantage over passive tags when it comes to providing a consistently robust, penetrating signal. So the higher value assets (e.g., at the pallet level) often require active tags. Considering the total cost of ownership, an active tag's higher cost is offset by a lower cost reader and processor infrastructure making the cost justification easier. Containers, trucks, and trailers are the best examples of high-value items that require active tags.

The three fundamental RFID architectures in use today are *passive, battery-assisted passive* (BAP), and *active*. This section provides an overview of each type of RFID architecture.

18.2.1 Passive RFID

Passive RFID tags backscatter or reflect RF energy to the reader and they do not require batteries for operation. These tags are low cost and their construction is relatively simple. A high-performance passive RFID tag shown in Fig. 18.2 consists of a tiny integrated circuit chip, a printed antenna, and an adhesive label substrate for application to items.

Passive RFID systems have improved substantially since the introduction of first generation of ultra-high frequency (UHF) systems. Improvements in range

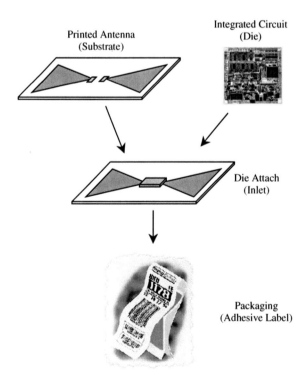

Fig. 18.2 Example of a passive RFID tag

and interoperability, multi-tag arbitration speed, and interference susceptibility were promised and delivered with the ratification of EPC Class I Generation 2 (C1G2) and the ISO 18000-6 standards (EPCglobal Inc., 2005). Although passive UHF RFID performance enhancements, cost reduction, and end-user mandates helped to improve the technology adoption rate, the level of deployment has yet to meet industry expectations. A dominant reason for the slow adoption rate of RFID is a mismatch in expectations between technological capabilities and application requirements. The passive tag's read rates, localization accuracy, interference resilience, infrastructure simplicity, and maintenance costs have not met the needs of several high-profile applications (United States Government Accountability Office (GAO), 2006; Holt, 2007).

18.2.2 Battery-Assisted Passive (BAP) RFID

BAP tags also backscatter RF energy to the reader. However, a battery is used to improve the tag's receiving sensitivity, avoid remote power transfer from the reader, and improve omnidirectional range (Bridgelall, 2002). It will be seen that utilizing a battery for a tag to transmit power instead of reflecting or backscattering power will provide a relatively large magnitude of improvement in omnidirectional range especially with small antennas. BAP tags are also sometimes called *semi-passive* tags.

18.2.3 Active RFID

A tag that utilizes the battery to emit rather than reflect or *backscatter* RF energy is often called an *active* tag. Although a battery can substantially improve performance, it will also limit maintenance-free operational life. Harvesting energy from sources such as vibration or light has been shown to address this shortcoming, but these sources must be adequate and available throughout the life of the application (Hande, 2007).

Active tags substantially improve range and read rate in electromagnetically unfriendly environments. Greater range improves the link quality in most applications but also exacerbates the problems of RF interference and position determination. Multi-path interference causes signal nulls and signal reflections. This makes it difficult to pinpoint a tag's location and provide for directionality of movement, even when using sophisticated computational techniques (Assad, 2007). UHF RFID systems, whether active, passive or semi-passive, propagate a signal in the *far-field* that can also generate interference for nearby readers and tags (Leong et al., 2006). Some active systems such as Wi-Fi, ZigBee, and Bluetooth incorporate complex protocols to reduce the impact from multi-path interference, but the trade-off is increased power consumption (IEEE Std. 802.11; IEEE Std. 802.15.4; IEEE Std.

Table 18.1 Comparing key RFID parameters to other wireless standards

Parameters	SRD		WLAN		WWAN	
	RF Sensors	Bluetooth	802.11b	802.11a	GSM/ GPRS	IS2000
Multiple Access	FHSS/TDMA	FHSS/TDMA	FDM/CSMA/CA	FDM/CSMA/CA	FDMA/TDMA	CDMA
Frequency Band	125 kHz–2.5 GHz	2402–2484 MHz	2402–2484 MHz	5.2–5.8 GHz	800–2000 MHz	1885–2200 MHz
Data Rate	1–200 kbps	1 Mbps	1–11 Mbps	1–54 Mbps	14–115 kbps	100–2000 Mbps
RF Modulation	ASK, mFSK, mPSK	FSK	DQPSK	nPSK/QAM-n OFDM	GMSK	QPSK (DL) BPSK (UL)
Transmission Power	1 mW–4 W	1 mW–100 mW	100 mW	100 mW	2 W	600 mW
Typical Range	20′ (passive) 400′ (active)	30 ft–100 ft	400 ft	200 ft	20 miles	12 miles

802.15.1-2002). Complex air-interface protocols will require larger and more expensive batteries than a simpler protocol.

Table 18.1 provides a brief comparison of the key RFID tag system parameters and how they compare with those of other wireless communication systems with which they must coexist. It can be observed that an RFID tag is simply another short-range device (SRD) amongst other popular wireless entities, and may interfere with the operation of other systems sharing nearby channels. This issue will become more pronounced as RFID technology becomes pervasive for item identification and tracking (Bridgelall, 2003).

Equipment manufacturers, end-users, and standards organizations are progressively addressing the traditional barriers to significant RFID deployment. These are no different from the initial barriers to significant deployment of cellular telephony and IEEE 802.11 WLAN networks. A recent acceleration in the performance and cost reduction trends for RFID tags is being witnessed. Commercially available passive RFID systems now operate robustly at a distance of 20 ft, and provide hundreds of tags per second throughput. This range is expected to reach 50 ft within the next 2 years. Manufacturers have recently demonstrated novel manufacturing processes to provide end users soon with low-cost RFID tags (Alien Technology Corporation).

In summary, passive and active RFID capabilities are related, but they are distinctly different technologies which should be matched to the application's technical and economic requirements. Automatically identifying personnel, assets, and vehicles are "active" applications and the cornerstones of automated visibility, security, and quality improvements in the enterprise. In the supply chain, for example, total visibility, security, and quality can only be attained by utilizing both. Figure 18.3

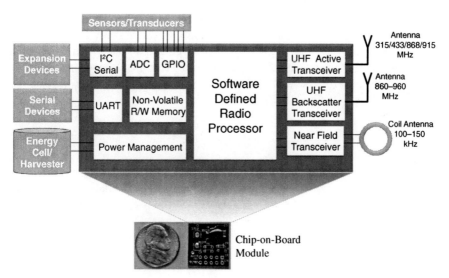

Fig. 18.3 Enterprise Dot^{TM} architecture

shows one such architecture of the Enterprise Dot™ tag (Axcess International, Inc.) that is a multi-mode hybrid architecture capable of operating both as a passive and as an active RFID tag (Bridgelall, 2008). The EH module forms an integral part of the architecture to ensure near perpetual operation.

18.3 RFID Operation and Power Transfer

Backscatter tags transmit data by "reflecting" the continuous wave (CW) energy received from a reader. The rate of change between energy absorption and reflection states encodes the bits. For example, the reader will interpret the time between magnitude and/or phase changes of the reflected energy as information bits. Hence, this type of communications is sometimes referred to as modulated backscatter transmission. Passive backscatter tags harvest energy from the reader to power their circuits. It is also possible for tags to harvest energy from other sources such as vibration and light. BAP tags utilize a battery to power their circuits instead of harvesting RF energy.

Given regulatory power and bandwidth limitations, the link margin between the reader and the backscatter tag is constrained by either tag receiver sensitivity or reader receiver sensitivity. The higher the tag receiver sensitivity, the weaker the signal, and the further away it can reliably decode reader commands. The reader's transmitted power diminishes at the rate of $(1/r^2)$ in the unobstructed far-field, where r is the separation distance between the reader and tag antennas. The tag receives power as given by Eq. (18.1),

$$P_{\text{tag\_rx}} = \psi_r(1/r^2) \qquad (18.1)$$

The parameter ψ_r is proportional to the reader transmitted power and the carrier wavelength (Bridgelall, 1999). This equation is accurate only in the far-field. The near-field/far-field boundary is typically defined as a distance from the antenna where the wave impedance quickly approaches the free-space impedance value of 377 Ω (Krause, 1950). The near-field radius is a function of the carrier wavelength, λ and is given by equation (18.2),

$$R_{\text{NF}} = \lambda/(2\pi) \qquad (18.2)$$

For most practical far-field UHF RFID systems, this distance is less than 1 ft. Upon decoding the reader command, the tag backscatter modulates a response. Some of the energy incident on the tag's antenna will be lost to impedance boundary absorption. Therefore, the propagating reader signal is further weakened before it is reflected from the tag. In addition to these absorption losses, the signal undergoes a return path loss that is identical to the forward link path loss. Therefore, the total backscatter signal loss will be a function of $(1/r^4)$ as given by Eq. (18.3),

$$P_{\text{reader\_rx}} = \Psi_t(1/r^4) \qquad (18.3)$$

The parameter Ψ_t includes losses from energy absorbed by the tag (Bridgelall, 2002). A tag reader can decode the reflected signal if it has adequate receiver sensitivity. For most commercially available systems, passive tag range is typically limited by tag receiver sensitivity rather than reader receiver sensitivity. This is due to a design trade-off between optimizing for maximum RF EH and optimizing for maximum receiver sensitivity. Since RF EH is not a requirement for semi-passive tags, their designs can be optimized for maximum receiver sensitivity. Consequently, the reader receiver sensitivity typically becomes the limiting factor for semi-passive RFID range. Reader receiver sensitivity is, in turn, limited by its architecture, which is constrained by the RADAR backscatter problem of isolating a transmitted signal from a co-located receiver (Skolnik, 1980).

These technical constraints in receiver sensitivity adversely limit the far-field backscatter RFID link margin. Tags are generally required to operate in electromagnetically unfriendly environments consisting of liquids and metals. Liquids absorb RF energy. Metals contribute to multi-path reflections that can cause signal cancelation when reflected signal components combine at the receiver in anti-phase. While the free-space path loss for backscatter RFID systems is of $(1/r^4)$, an RF absorptive and reflective environment can contribute an even greater signal loss of $(1/r^{12})$ (Smith, 1998).

Active far-field tags transmit a signal by emitting energy transformed from its power source rather than reflecting incident energy from the reader. A typical active tag can transmit orders of magnitude more power than a backscatter tag will reflect at the same distance. For example, a 4 W ($\sim$36 dBm) signal transmitted from a passive RFID reader at 915 MHz will become approximately 3 μW (-25.36 dBm) after traveling a distance of 100 ft in free-space. This is with respect to the maximum signal level legally allowed in North America for transmission in the unlicensed 915 MHz frequency band (U.S. Code of Federal Regulations (CFR)). A backscatter tag, whether battery assisted or not, reflects a portion of this signal to the passive tag reader. A low-power active tag in the same band will typically transmit a signal at 1 mW. The radiated signal power from the active tag is over 300 times greater than that of the backscatter signal power reflected from a passive tag. Thus, a 915 MHz reader antenna will receive an active tag signal from 100 ft away that is over 300 times greater. A stronger signal improves a tag's chances of being received in any application.

Although this improved link margin can result in higher read rates, the stronger signal can also degrade tag localization accuracy. Multi-path interference adversely affects far-field passive, semi-passive, and active RFID systems in the same manner. Zone control and tag directionality are compromised when signals reflect uncontrollably in a highly cluttered environment. Attempts to increase range by radiating a stronger signal only exacerbates this problem (Skolnik, 1980).

18.4 Battery Life

Although a battery adds initial material cost over passive solutions, the savings accumulated from "soft" benefits could compensate for this added expense over the life of the tagged item. For many applications, these soft benefits include improved data capture reliability, greater asset visibility, ease of deployment, and negligible upfront business process adjustments. In general, the longer the battery life, the lower the amortized cost of an active tag over the life of a tagged item.

Given a tag's battery, its useful life is a function of both the tag's sleep mode power consumption and activation duty cycle. Power consumption depends on the electronic architecture, anti-collision protocol, and power management schemes. The architecture complexity is primarily determined from the wireless standard that a tag is designed to support. For example, Wi-Fi tags must be compatible with the IEEE 802.11 series of standards that were developed for high-speed wireless internet applications. ISO 18000-7 tags are designed for operation within a less complex narrow-band channel at 433.92 MHz (ISO 18000-7, 2004). In general, a lower complexity protocol will require fewer transistors to implement the hardware and also fewer processor instructions to complete the information exchange, thereby consuming less power.

Thin-film and printed battery technologies promise to dramatically lower the cost of active RFID technologies because they will be compatible with high-volume roll-to-roll electronic manufacturing assembly processes. However, these emerging battery technologies cannot yet provide the energy densities of standard "coin-cell" or cylindrical batteries (Starner, 2003). Low-power consumption architectures that implement low-complexity RFID protocols can best leverage this early stage low-capacity battery technology to provide many months of maintenance-free operation.

18.5 Operational Characteristics of RF Sensors and ID Tags

Active RFID solutions will continue their cost reduction trend as chipsets become more widely available. Given similar costs and form-factors, chip-set differentiation will tend to focus on omnidirectional range, multi-tag arbitration speed, and battery life. It has been discovered that even when a design achieves the optimum balance between range, speed, and battery life for one application, it is not necessarily the solution that is preferred across all applications. Therefore, a solution that can adapt to an application's requirements is desirable. Domain knowledge indicates that an architecture based on software definable radio (SDR) techniques can meet the majority of these needs most of the time.

RFID and wireless technologies tend to utilize a single mode of operation and are consequently limited in their application scope. The chart of Fig. 18.4 summarizes the four fundamental RFID or wireless sensor categories and their operational characteristics in terms of key benefits and deficiencies. Each type utilizes one of several standards indicated. There are also numerous variations in operational characteristics within each category. For example, active far-field devices are available for both narrow-band and wide-band channels. Narrow-band RFID devices utilize tens of kilohertz of spectrum and typically operate in the globally available 433.92 MHz band. Wide-band RFID typically utilizes tens of megahertz of spectrum and requires spread spectrum, orthogonal-frequency-division-multiplexing (OFDM), or ultra-

Near-Field		Far-Field	
Active e.g. RuBee (IEEE P1902.1), NFC (ISO 18092)		e.g. Wi-Fi (IEEE 802.11), UHF RFID (ISO18000-7)	
Benefits • Robust link around dense RF media • Magnetic field zone control • Simple narrow-band protocols maximize battery life	**Deficiencies** • Range limited to antenna loop diameter • Multi-tag arbitration speed limited by data rate	**Benefits** • Long range from RF propagation and higher transmit power • High multi-tag arbitration rates possible due to larger bandwidth & data-rate	**Deficiencies** • Some bands require spread spectrum and complex multiple access protocols; leads to higher power consumption • Poor zone control
Backscatter e.g. HF RFID (ISO 14443), LF RFID (ISO 14223-1)		e.g. UHF RFID (ISO18000-6 & EPC)	
Benefits • Excellent zone control • Robust near-field energy harvesting for passive HF/LF RFID • Robust media penetration	**Deficiencies** • Backscatter reader sensitivity and loop antenna diameter limits practical range to within one meter • Multi-tag arbitration limited by bandwidth and data rate	**Benefits** • Tens of meters of range for passive tags • Longer range for semi-passive tags; limited primarily by reader sensitivity • High multi-tag arbitration rate • Longer battery life	**Deficiencies** • Poor zone control • Poor RF media penetration • High orientation sensitivity due to weaker backscatter and multi-path propagation

Fig. 18.4 Operational characteristics of RFID and wireless sensors

wide band modulation techniques in order to co-exist and share the spectrum. Depending on the power consumption requirements, any of the technologies in each of these four fundamental categories can harvest energy or utilize a battery. The energy source may be utilized to power only the logic and sensing circuits or to also radiate energy for communications.

Near-field backscatter tags use a low-frequency (LF) magnetic field ranging from 100 to 150 kHz initiated from a near-field reader or generator. Once the preamble is recognized and the protocol identified, the RF sensor or ID tag can respond with synchronous backscatter or load modulation using the appropriate sub-carrier modulation type. Such digital modulation procedures include amplitude shift keying (ASK), frequency shift keying (FSK), or phase-shift keying (PSK), and the appropriate symbol-encoding scheme, for example, Manchester or Miller. Alternatively, one of several proximity access control protocols can be instantiated by vectoring to the appropriate instruction sequence within on board non-volatile memory.

Far-field backscatter tags include UHF carriers ranging from 860 to 960 MHz using ASK modulation. Most tags use the EPC C1G2 (ISO18000-6c) protocol. Far-field active tags utilize a UHF carrier within the 433.92 MHz band. "Class-IV" bi-directional protocol based on noise adaptive multiple access techniques can be used (US Patent #6034603, 2000). Such devices may also be utilized for communications with other wireless devices in a wireless sensor network (US Patent #7005985, 2006). Applications include asset protection and security. For example, a belt worn tag may continuously query and monitor for the presence of specific items within a specified communications range.

18.6 Why EH Is Important?

The widespread need for RF sensors and tags is evident from the proliferation of short-range wireless standards such as Bluetooth, ZigBee, Wi-Fi, and RFID. Sensor nodes for wireless personal area network (WPAN) that utilize smart mesh network protocols are traditionally designed to optimize for low cost and low power. Each wireless node is intended for deployment in large quantities at remote locations to sense critical data and relay its measurements to other network nodes for monitoring and control purposes. In this new era of high security and vigilance, wireless sensors are being deployed at major events so as to predict catastrophes such as acts of terrorism (Axcess International Inc., 2007). Another application combines the data from vehicular sensors, such as wheel hub-odometers, tire pressure sensors, and asset tags to improve safety, reliability, and reduce fleet maintenance costs. As these types of applications proliferate, it will become necessary to deploy RF sensors in hard to reach places. Once thousands of low-cost RF sensors are deployed for any given application, replacing batteries will become an impractical task. Therefore, self-sufficient devices that can operate for an indefinite period of time will be required. The availability of small, rugged, low-cost, and self-powered wireless sensors and tags will revolutionize the supply chain industry and open up numerous applications in both the military and the commercial sectors. Many foresee these embedded system devices in applications ranging from industrial automation to home networking (Rabaey et al., 2000; Gates, 2002; Hitachi, 2003). RFID devices are used for personnel (e.g., badges for unattended access control, work-force management, personnel safety, and regulatory compliances) and asset (laptops, medical equipment, pallets, and so on) tracking will require energy harvesting solutions for extended life. Other applications include temperature and light monitoring in remote locations, sensing chemicals in traffic-congested areas, measuring tire pressure, and monitoring acceleration in automobiles.

Active RF sensors and ID tags have fairly moderate densities and consume power of the order of several megawatts, while receiving and transmitting data packets but they are designed to draw a few microwatts in their sleep state. These devices can form dense ad hoc networks transmitting data from 1 to 10 m indoors and as high as 100–150 m outdoors. In fact, for indoor communication over distances greater than 10 m, the energy-to-transmit data rapidly dominate the system (Rabaey et al., 2002).

The issue of powering these systems becomes critical when one considers the prohibitive cost of wiring power to them or replacing their batteries. Obviously, such devices have to be small in size so that they can conveniently be placed in remote locations. This places a severe restriction on their life if alkaline or similar batteries are used to power them. To make matters worse, battery technology has not improved in terms of energy density and size over the last decade, especially for low-power mobile applications such as sensor networks (Paradiso and Starner, 2005). While an effort is being made to improve the energy density of batteries, additional energy resources need to be investigated to increase the life of these devices. Exploiting renewable energy resources in the device's environment offer a power source limited by the device's physical survival rather than an adjunct energy source. EHs true legacy dates to the water wheel and windmill, and credible approaches that harvest energy from waste heat or vibrations have been around for many decades (Starner, 1996; Starner and Paradiso, 2004). Nonetheless, the field has encountered renewed interest as low-power electronics, wireless standards, and miniaturization conspire to populate the world with RF tags, sensor networks, and mobile devices (Roundy et al., 2004; Joseph and Srivastava, 2005).

18.7 EH Technologies and Related Work

There are several sources of energy that can be used to power RF sensors and ID tags. Table 18.2 compares the power generation potential of some of the typical EH modalities which include ambient radiation (Yeatman, 2004), temperature gradients (Stevens, 1999), light (Schmidhuber and Hebling, 2001), and vibrations (Shearwood and Yates, 1997; Amirtharajah and Chandrakasan, 2004; Meninger et al., 1999, 2001; Roundy et al., 2003; Min et al., 2002; Chandrakasan et al., 1998; Glynne-Jones et al., 2001; Ottman et al., 2003). Among these sources of energy, solar EH through photovoltaic conversion and vibrational energy through piezoelectric elements provide relatively higher power densities, which makes them the modalities of choice. However, the design of both EH modules involve complex tradeoffs due to the interaction of several factors such as the characteristics of the energy sources, chemistry, and capacity of the energy storage device(s) used, power supply requirements, and power management features of the embedded system, and application behavior. It is, therefore, essential to thoroughly understand and judiciously exploit these factors in order to maximize energy efficiency of the harvesting modules. Moreover, the power output from both these sources is highly nonlinear in nature and depends on a variety of factors.

There have been a few concerted efforts towards developing EH circuits to power RF sensor nodes. Among others, the UCLA-Center for Embedded Networked Sens-

Table 18.2 Power densities of energy harvesting technologies

Energy Scavenging Source	Power Density (μW/cm^3)	Information Source
Solar (Outdoors)	15,000 – Direct Sun 150 – Cloudy Day	Commonly Available
Solar (Indoors)	6 – Office Desk	Experiments
Vibrations	100–200	Roundy et. al.
Acoustic Noise	0.003 at 75 dB 0.96 at 100 dB	Theory
Daily Temp. Variation	10	Theory
Temp. Gradient	15 at 10° Celsius	Stordeur & Stark, 1997
Piezo Shoe Inserts	330	Starner 1996

Source: Reprinted from Roundy et al. (2003), with permission from Elsevier.

ing (CENS) (Joseph and Srivastava, 2005; Raghunathan et al., 2002, 2005; Kansal et al., 2004) and UC Berkeley (Roundy et al., 2003, 2004), in particular, have investigated solar EH and piezoelectric vibration EH, respectively. Several publications are available that document the strategies used and power output available from the prototypes. The CENS group investigated solar energy harvesting to energize a variety of motes, including MICA2, MICAz, Telos, and Intel's Stargate mote (Crossbow Technologies, 2008). The Heliomote platform in particular uses outdoor solar energy to power MICA2 motes with rechargeable NiMH batteries serving as the energy storage device. The EH setup has useful features such as battery overcharge and undercharge protection, regulated $3\,V_{DC}$ output, intermediate battery status transmission to base node, and environmental aware routing. As seen from Table 18.2, it is clear that although solar energy is the most efficient natural energy source available for outdoor applications, for indoor applications it is important to note that the efficiency of photovoltaic cells is very low. Typically, the indoor lighting intensity found in hospitals and offices is less than $10\,W/m^2$ when compared with $100–1000\,W/m^2$ outdoors. Monocrystalline solar cells have an efficiency of less than 1–3% under typical indoor lighting conditions (Randall and Jacot, 2002). However, in spite of such poor efficiencies, these cells still have a power density of at least $0.5–1\,mW/cm^2$ under indoor $1–5\,W/m^2$ light intensity conditions, which is higher than their nearest EH competitor. Also, amorphous (flexible) solar cells have been found to have slightly higher efficiencies of 3–7% under indoor conditions (Randall and Jacot, 2002). Similarly, publications by UC Berkeley (Roundy et al.) have focused on circuits that harvest energy from piezoelectric vibration generators. Detailed power output data with respect to vibration frequency have been documented.

A system-level block diagram of an RF sensor operating on a natural source of energy is shown in Fig. 18.5. Typically, such a device consists of an 8/16 bit microcontroller with adequate resources to operate its kernels. The microcontroller manages power to the sensors and data acquisition elements, as well as responds to commands (e.g., from the reader). For example, an Axcess International RFID

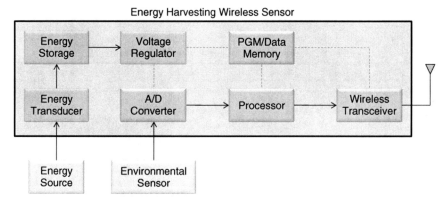

Fig. 18.5 Block diagram of an energy harvesting RF sensor

tag uses a 16 bit TI MSP430 microcontroller with necessary coils and power amplifier circuit (Axcess International Inc., 2008). Similarly, ZigBee-based RF sensors such as the TI EZ430-RF2500 use an MSP430 microcontroller integrated with an IEEE 802.15.4 compliant, ZigBee ready CC2500 transceiver (TI EZ430-RF2500, 2008). Typically, primary non-rechargeable batteries are used to provide a 3 V_{DC} output voltage for the device's microcontroller, sensor board, and transceiver. The device consumes about 15–20 mA during full operation and about 6–10 μA during the dormant or sleep mode.

18.8 EH Design Considerations

Several key issues have to be addressed before embarking upon designing EH circuits for RF sensors and tags. First, the natural source of energy has to be harvested by an EH system and second, there has to be an efficient means of storing this energy (Rahimi et al., 2003). The system must also effectively route the stored energy to the RF sensor. This means that there has to be an intelligent power management strategy in place. Obviously, this strategy must be efficient and should serve to lengthen the life of the energy storage devices.

18.8.1 Energy Storage Technologies

Perhaps, the most complex (and crucial) design decision involves the energy storage mechanism. The two practical choices available for energy storage are batteries and electrochemical double layer capacitors – also known as ultra-capacitors. Batteries are a relatively mature technology and have a higher energy density (more capacity for a given volume/weight) than ultra-capacitors. Four types of rechargeable batteries are commonly used: nickel cadmium (NiCd), nickel metal hydride (NiMH), lithium–ion (Li-Ion), and sealed lead acid (SLA). Of these, SLA and NiCD batteries are not preferred because the former has a relatively low-energy density and the latter suffers from temporary capacity loss caused by shallow discharge cycles. The choice between NiMH and Li–ion batteries involves several tradeoffs. Li–ion batteries are much more efficient than NiMH batteries, have a longer cycle lifetime, and involve a lower rate of self-discharge. However, they are more expensive, even after accounting for their increased cycle life (Berndt, 1997; Linden, 1984). An additional consideration is battery aging due to charge–discharge cycles. For example, NiMH batteries (when subjected to repeated 100% discharge) yield a lifetime of about 500 cycles, at which point the battery will deliver around 80% of its rated capacity. The residual capacity is significantly higher if the battery is subjected to shallow discharge cycles only. At the rate of one discharge cycle per day, the battery will last for several years before its capacity becomes zero.

Several other battery-related factors, which are usually insignificant for conventional mobile devices, also play a role due to the nature of the target system/application. First, the battery non-ideality termed as rate capacity effect is non-existent since the system's current draw (few tens of milliampere) is an order of magnitude less than the rated current of most present day batteries. Second, the operating temperature of the batteries will vary, leading to changes in battery characteristics. For example, battery self-discharge rate approximately doubles with every 10 °C increase in ambient temperature (Raghunathan et al., 2005). Thus, the choice of battery chemistry for a harvesting system depends on its power usage, recharging current, and the specific point on the cost-efficiency tradeoff curve that a designer chooses.

Ultra-capacitors have a higher power density than batteries and have traditionally been used to handle short duration power surges. Batteries are not as robust as ultra-capacitors in terms of depth of discharge, and they tend to lose capacity when exposed to outdoor temperatures (Raghunathan et al., 2005). Ultra-capacitors are rated for several thousand charge–discharge cycles when compared with a few hundred cycles for Li–ion batteries. Battery cycle life is severely reduced when exposed to cold temperatures as well (Berndt, 1997). On the other hand, ultra-capacitors possess robust characteristics over a wide temperature range (Burke, 2000). Recently, such capacitors have been explored for energy storage, since they are more efficient than batteries and offer higher lifetime in terms of charge–discharge cycles. However, they suffer from severe leakage and are expensive.

Therefore, the choice of energy storage device will depend on a variety of factors including peak power requirements, cycle life, energy storage capacity, cost, and form factor. It is also possible to implement a tiered energy storage mechanism using an ultra-capacitor and a battery, to make use of the advantages of both these devices, although there might be some overhead due to the energy storage overhead in the power management module.

18.8.2 Energy Requirements and Power Management Issues

The core of any EH system is the harvesting circuit, which draws power from the EH transducer (e.g., solar cells, piezoelectric vibration generator, and so on), manages energy storage, and routes power to the target system. The most important consideration in the design of this circuit is to maximize energy efficiency, enhance device reliability, and lengthen the life of the RF sensor. Figure 18.6 shows a

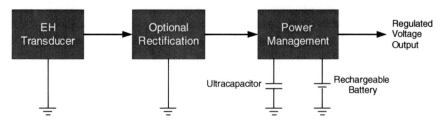

Fig. 18.6 EH system for an RF sensor

simplified diagram of the EH system. The power management circuit is responsible for both charging the energy storage device as well as routing energy to the sensor. The RF sensor's microcontroller will be responsible for effectively implementing these tasks.

In the case of solar EH systems, it is important to track an optimal operating point that yields maximal power output. The harvesting circuit should ensure operation at (or near) this maximal power point, which is done by clamping the output terminals of the solar panel to a fixed voltage. Since the maximal power point changes slightly with the time of day (i.e., as the incident radiation changes), a maximal power point tracker (MPPT) circuit is necessary to continuously track and operate at the optimal point (Raghunathan et al., 2005). This forms an essential portion of the power management module. The use of a switch-mode DC–DC converter is therefore warranted to assist in achieving optimal power flow. The DC–DC converter provides regulated supply voltage to the embedded system.

In case of systems with a piezoelectric element, depending on the frequency of excitation an AC voltage is generated. This voltage needs to be rectified and therefore such systems will require a rectifier circuit with a low-pass filter (Ottman et al., 2003). A DC–DC converter is used for maximum power transfer and forms an essential part of the power management module (Chi-ying et al., 2005; Shengwen et al., 2005). The choice of DC–DC converter depends on the operating voltage range of the energy storage device as well as the supply voltage required by the target system. If the supply voltage falls within the voltage range of the energy storage device, a buck-boost converter is required, since the rectified voltage will have to be increased or decreased depending on its state. However, if the supply voltage falls outside the battery's voltage range, either a boost converter or a buck converter is sufficient.

To enable harvesting aware power management decisions, the harvesting module must have energy measurement capabilities. The power management scheme can contain a low-power monitor IC that will track the status of the energy storage banks. The target system microcontroller would be able to query the harvesting module for the instantaneous power that is available. The terminal voltage of the energy storage device would provide information on their respective energy status. With this information in hand, it is possible for the RF sensor microcontroller to learn the power availability pattern from the EH transducer, and build and train a power macro-model that provides information about future power arrival (Raghunathan et al., 2002; Kansal et al., 2004; Pedram and Rabaey, 2002; Min et al., 2000; Shah and Rabaey, 2002).

18.8.3 Vibrational Energy Harvesting

Mechanically excited piezoelectric elements possess their own distinct characteristics. A vibrating piezoelectric device differs from a typical electrical power source in that it has a capacitive rather than inductive source impedance, and maybe driven by mechanical vibrations of varying amplitudes (Kymissis et al., 1998; Shenck and Paradiso, 2001; Smalser, 1997; Glynne-Jones et al., 2001). Piezoelectric devices such as Quickpack QP20 W from Mide Technology Corp. are available as a two-layer device that generates an AC voltage when vibrated in a direction perpendicular to its mid-plane (Mide Technology Corp., 2008). Therefore, EH circuits that use such devices as sources of energy require a rectifier and low-pass filter to obtain a DC voltage for the embedded system. These components must have low power loss to ensure high-system efficiency. The magnitude of current, and hence the optimal rectifier voltage may not be constant as it depends on the vibration level exciting the piezoelectric element. This creates the need for flexibility in the circuit, i.e., the ability to adjust its average input impedance to achieve maximum power transfer. This means such EH setups will need a DC–DC converter between the rectifier and the energy storage mechanism (Ottman et al., 2002).

Previous studies have characterized the performance of the three important types of vibration energy transducers: piezoelectric, electrostatic, and electro-magnetic. Energy is produced by the vibration work done to change strain, capacitance, and inductance in the three types of transducers, respectively, shown in Fig. 18.7.

Piezoelectric generators work by straining a piezoelectric material to produce charge separation across the material. For most approaches, the amount of power generated is proportional to the g-force acceleration experienced by an oscillating mass, as well as the oscillating frequency.

Figure 18.8 shows the two operating modes (d_{33} and d_{31}) in which the piezoelectric material can be used to generate power. In d_{33} mode, both the mechanical stress and the output voltage act in the three directions. In d_{31} mode, the stress acts in the one direction and voltage acts in the three directions. Operation in the d_{31}

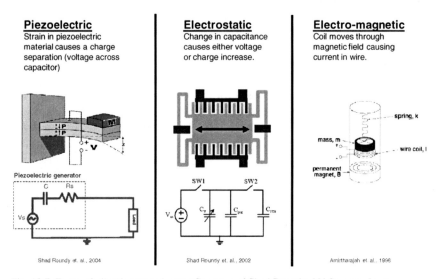

Fig. 18.7 Types of vibration transducers (Courtesy of Shad Roundy, LV Sensors, Inc.)

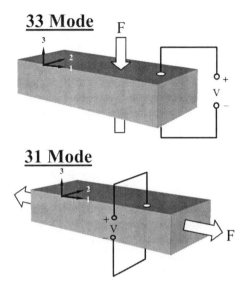

Fig. 18.8 Piezoelectric conversion modes (Reprinted from Roundy et al. (2003), with permission from Elsevier)

mode leads to the use of thin-bending elements such as bimorphs, in which two separate sheets are bonded together, sometimes with a center shim in between them. The electro-mechanical coupling for d_{31} mode is lower than for d_{33} mode. However, d_{31} systems can produce larger strains with smaller input forces. Also, the resonant frequency is much lower. An immense mass would be required in order to design a piezoelectric converter operating in d_{33} mode with a resonant frequency below 60 Hz. Therefore, this mode of operation is most suitable for powering RF sensors and ID tags. Experiments show that a practical generator of 1 cm^3 is capable of producing more than 300 μW of power from an oscillating mass that experiences 0.25 g of acceleration at 120 Hertz (Roundy and Wright, 2004).

Electrostatic generators can utilize a micro-electro-mechanical structure (MEMS) to modulate the capacitance between two surface areas so as to amplify a reference voltage across another fixed value capacitance. The input voltage to the structure is amplified by the mechanical work done against the electrostatic forces between the two plates. It is important to note that electrostatic generators require an initial voltage source to create the initial charge separation for amplification. These converters are not capable of converting as much power per unit volume. However, they are still attractive because they offer more potential for integration with microsystems.

There are several types of devices that can transform vibrations into capacitance changes. UC Berkeley developed three types of MEMS structures and evaluated two types referred to as in-plane overlap and in-plane gap (Roundy et al., 2002). The study concluded that the former required large spring deflections and high Q-factors to produce maximum power, while the latter has an optimum spring deflection and can accommodate lower Q-factors. The Q-factor is of significant interest. A high

Q-factor will be more vibration–frequency selective, while a lower Q-factor can accommodate a wider range of vibration frequencies. Therefore, it might become necessary to invent a resonant adaptive structure. A significant portion of any initial design involves carefully characterizing the acceleration profile of vibration sources for the intended application. This determines the design parameters and optimization criteria for a vibration transducer that can deliver maximum energy.

Electro-magnetic generators work by moving a coil through a magnetic field (or a permanent magnet through a coil) to induce an electric current in the coil. Existing commercial products such as the crank radio and the shaker flashlight works on this principle. These types of generators produce very small voltages of less than 1 V. This low voltage increases the difficulty of designing power efficient DC–DC converters.

These studies indicate that piezoelectric cantilever structures operating in d_{31} mode seem to be a good choice for powering active RF sensors and ID tags. Note that such tags typically draw about 6–10 μA during sleep mode and about 15–20 mA for short periods of about 20 msec when operational (Dot™, 2008). Therefore, a continuous EH source providing at least 100–200 μA hr should be more than sufficient to power these tags.

In order to harvest energy from vibrations, the first step involves obtaining the vibration spectrum from the source or structure. It is important to determine the range of frequencies at which maximum force (acceleration) occurs. For example, vibrational transducers for vehicle monitoring RFID tags that are placed on the windshield will generate optimal power if they are designed to resonate at the windshield vibrational frequency (Hande et al., 2008). A suitable accelerometer can be used for measuring the acceleration profile. Next, the frequency spectrum is obtained by running a fast Fourier transform on the data. Figure 18.9 shows one such dataset for a 2003 Honda Civic under city-driving conditions. These data are used to define the frequency band for transducer design. It is conceivable that multiple bimorphs can be integrated within this frequency spectrum to improve transducer

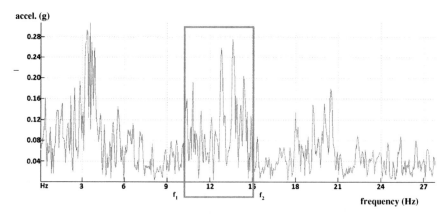

Fig. 18.9 Frequency spectrum of automobile windshield vibrations

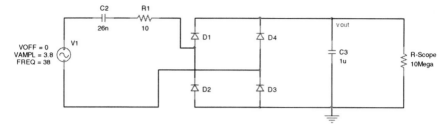

Fig. 18.10 Equivalent model of piezoelectric generator

energy output. Typical values of frequency and acceleration for this application can be seen to reside in the range of 10–15 Hz and 0.1–0.3 g, respectively.

An appropriate piezoelectric bimorph can be designed with adequate specifications (e.g., width, length, thickness, tip mass, and so on) that provides the necessary output power at the desired resonance frequency. Since, this application requires resonant frequency (f_0) to be in the 10–15 Hz, it might be necessary to use a large tip mass to lower f_0. Other applications such as industrial health monitoring might need transducers with f_0 at 60 Hz and consequently, lower tip mass.

It is important to model the piezoelectric transducer at f_0 using a typical model such as an AC voltage source with series capacitance (C_2) and resistance (R_1) as shown in Fig. 18.10 (Lefeuvre et al., 2007). This model works well under resonant operating conditions where the transducer output impedance (Z_0) is inversely proportional to f_0 and given by:

$$Z_0 = 1/(2\pi f_0 C) \tag{18.4}$$

Such a model also provides information on maximum power that can be obtained at matching load impedance. Z_0 is required for designing a DC–DC converter with average input impedance (Z_{in}) equal to Z_0 for maximum power transfer. For example, Fig. 18.11 shows that maximum power of 16 μW was obtained at 38 Hz resonant frequency with a 220 kΩ load for a PZT-5 based cantilever size of 0.3 cm^3 at 0.5 g excitation. This would, therefore, require the DC–DC converter to be designed for $Z_{in} \approx 220$ kΩ.

18.8.4 Solar Energy Harvesting

Solar cells have vastly differing characteristics from batteries. For example, the V-I characteristics of a single 3.75 in. × 2.5 in. 4-4.0-100 solar panel from Solar World, Inc., under indoor fluorescent lighting conditions are shown in Fig. 18.12 (Hande et al., 2006). Similar curves can be observed under outdoor light conditions as well. The V–I curve shows that two parameters, the open-circuit voltage (V_{OC}) and the short-circuit current (I_{SC}), influence the panel operation. These form the x- and y-intercepts of the V–I curve, respectively. From Fig. 18.12, it is clear that

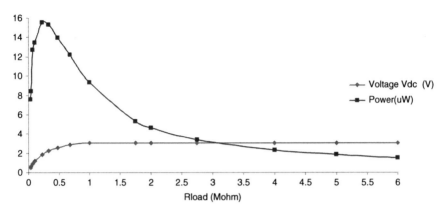

Fig. 18.11 Output power and voltage versus load resistance of a PZT-5-based piezoelectric cantilever

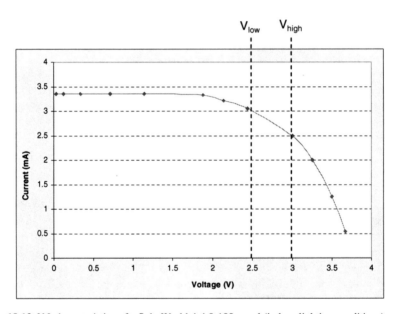

Fig. 18.12 V-I characteristics of a SolarWorld 4-4.0-100 panel (indoor lighting conditions)

a solar panel behaves as a voltage limited current source (as opposed to a battery which is a voltage source). Also, there exists an optimal operating point at which the power extracted from the panel is maximized. Finally, as the amount of incident solar radiation decreases (increases), the value of I_{SC} also decreases (increases). However, V_{OC} remains almost constant (Panasonic Corporation, 1998). Due to its current source-like behavior, it is difficult to power the target system directly from the solar panel, since the supply voltage would depend on the time varying load impedance (Voigt et al., 2003; Krikke, 2005). A power management scheme is therefore required to provide regulated power to store energy in devices such as batteries and ultra-capacitors.

Efficient solar energy harvesting involves monitoring the maximum power point tracking (MPPT). For example, Fig. 18.12 indicates a 7.5 mW window wherein maximum power is obtained. This voltage window is fairly constant for different levels of solar light intensity. Circuits need to be designed to optimally track this window so that maximum power is transferred to the storage device. Researchers at UCLA have developed an analog approach to implement this task for outdoor energy harvesting applications. The circuit consists of comparators and references, and is active all the time (Raghunathan et al., 2005).

Figure 18.13 shows a typical method wherein a DC–DC buck converter is used for the energy conversion process. Here, voltage across the input capacitor is monitored until it reaches V_{high}. At this point, the MOSFET switch is turned on to route energy to the storage device. When voltage across the input capacitor reduces to V_{low}, the converter is turned off, thus realizing maximum power transfer and consequently, higher efficiency.

A similar approach has been adopted by researchers at UT Dallas for indoor solar energy harvesting from overhead fluorescent lights (Hande et al., 2006). This setup was designed for an application, wherein RF sensors were used to monitor

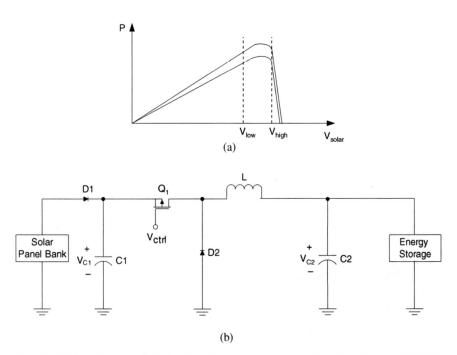

Fig. 18.13 (**a**) Online control for tracking PV curve variations with incident light. (**b**) DC–DC buck converter

patient vital sign data from a remote location in hospital premises. It consisted of an adequate number of solar panels to charge ultra-capacitors that were used as energy storage devices which in turn supply regulated power to the RF node. The core of the energy harvesting module consisted of a power management circuit, which draws power from the solar panels and manages energy storage and power routing to the RF node. The power management circuit provided regulated power to the RF node and also simultaneously stored energy in ultra-capacitors. It ensured node operation at the optimal MPPT point for maximum power transfer. Figure 18.14 shows the router node with its EH circuit. The node harvests energy from indoor overhead 34 W fluorescent lights. An adequate number of Solar World 4-4.0-100 monocrystalline solar panels rated at an open-circuit voltage (OCV) of 4 V_{DC} and short-circuit current of 100 mA under 1000 W/m^2 light conditions were used in series–parallel to support charging of the ultra-capacitor bank. Later, higher efficiency Solar World PowerFilm amorphous cells rated at an OCV of 4.8 V_{DC} and short-circuit current of 100 mA were used. The ultra-capacitor bank is comprised of two Maxwell PC5-5 ultra-capacitors. Both overcharge and undercharge protection circuits are implemented along with the programmable regulated output voltage for operating RF nodes with different voltage requirements. The power management circuit provides an efficiency of 82% which bodes fairly well for most applications.

RF sensors and tags will vary their power consumption depending on the task performed at any given moment in time. For example, during data logging, the tag must activate appropriate sensors and convert an analog value (voltage or current) into a digital value. The analog-to-digital converter will require some amount of power during this process. The processor must then store the digital value in memory. Also, when queried, the tag must transmit data to the reader. Each of these processes require a different level of energy consumption. A DC–DC converter must allow optimum power transfer while adapting to the variable power demand.

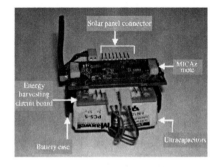 

Fig. 18.14 Indoor solar energy harvesting sensor node

18.9 Relevant Circuits and Systems

Figure 18.15 shows the typical architecture for an EH system. The EH transducer output can be an AC or DC voltage. In the case of piezoelectric vibrational transducers, an AC voltage is generated. Therefore, in this case, the first stage involves design of an AC–DC rectifier. This is not required for solar EH systems. The second stage involves DC–DC converter design.

18.9.1 AC–DC Rectifier

Several different strategies are available for rectifier design. The most basic, using pn junction diodes as a bridge rectifier, is relatively straightforward. However, each silicon diode requires a 0.6–0.7 V_{DC} voltage drop to be overcome before conduction begins. This forward voltage drop causes power loss during conduction and consequently lower efficiency. Alternatively, germanium diodes have a relatively lower 0.3 V_{DC} forward voltage drop. But this is still fairly large when input power flow from the piezoelectric cantilever structure is of the order of only hundreds of microwatts. Low leakage (<200 nA) Schottky diodes have lesser forward voltage drop ($\approx$0.2 V_{DC}) and work well (Lefeuvre et al., 2007). Also, synchronous rectifiers can be used instead of Schottky diodes to improve efficiency. Here, the body diode of a MOSFET is used instead of a discrete pn junction diode. The transistor is turned on when the body diode begins conduction. Due to the extremely low R_{ds}(on) (few milliohms) specifications of MOSFETs, the voltage drop across the body diode during conduction is negligible, consequently yielding very low power loss and high efficiency. Han et al. have tried using two diode rectifiers with comparators to aid in transistor turn-on (Lefeuvre et al., 2007). Similarly, Guo et al. have developed

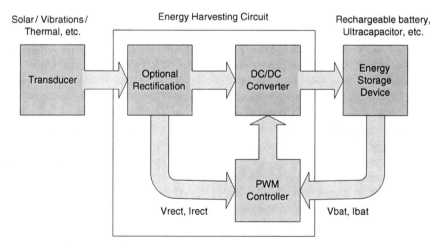

Fig. 18.15 Energy harvesting architecture

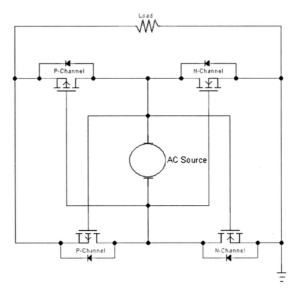

Fig. 18.16 Synchronous rectifier

high efficiency bridge rectifiers with comparators using 0.35 μm CMOS technology (Guo and Lee, 2007). Figure 18.16 shows such a circuit where the diagonally opposite leg diodes turn on during each half cycle using the source AC voltage. Use of zero-threshold MOSFETs that are available from Advanced Linear Devices might make the system even more efficient (Zero Threshold MOSFETs, 2008).

18.9.2 DC–DC Switch-Mode Converters

The second stage involves design of a high-efficiency DC–DC converter to provide regulated DC voltage at the output. The overall efficiency of the system depends on efficiency of the rectifier in conjunction with the DC–DC converter. There are several DC–DC converter topologies to choose from. Typically, non-isolated switch-mode DC–DC converters seem to be the energy conversion device of choice due to relative ease in matching the transducer impedance. The three most basic converter topologies: buck (step-down), boost (step-up), and buck-boost (step-up/step-down) can all be used to achieve impedance matching. Alternatively, charge pumps or switched capacitor converters are also feasible in certain cases due to ease of integration on an integrated circuit (IC). These converters have the advantage of consisting of only capacitors and MOSFETs (no inductor). However, from a maximum power transfer standpoint, their input resistance cannot be easily tuned to match the output impedance of the transducer. Low dropout (LDO) regulators are not considered as candidates since they suffer from poor efficiency and can only step down the voltage (no step-up possible).

18.9.2.1 Buck Converter

Ottman et al. and Lesieutre et al. have shown that it is feasible to use a DC–DC
buck converter operating in the discontinuous conduction mode (DCM) along with
a pulse-charging circuit to maximize harvested power from piezoelectric elements
(Ottman et al., 2002, 2003). Continuous conduction mode (CCM) operation causes
reduced efficiency because feedback circuits need to be employed to measure output
voltage and current causing additional power loss. The optimal duty cycle of the
converter is dependent on output filter inductance, switching frequency, the piezoelectric element's output capacitance, and the frequency of mechanical excitation.
Kim et al. have also shown that such a converter can significantly reduce the matching impedance of the circuit (Kim et al., 2007). This is very important because unlike
typical voltage sources, many EH transducers (such as piezoelectric elements) are
capacitive in nature with high output impedances, while typical loads such as batteries have an impedance of the order of few hundred ohms. Without appropriate
impedance matching, there can be significant loading issues on the voltage source
element causing inefficient power transfer.

A typical DC–DC buck converter with a MOSFET input switch is shown in
Fig. 18.17. The freewheeling diode (D_1) can be replaced with a synchronous rectifier for higher efficiency. It has been shown that maximum energy harvesting efficiency is obtained at the optimal duty cycle (D) given by:

$$D = \sqrt{(4\omega L C_p f_s / \pi)} \tag{18.5}$$

where

(a) ω is the resonant frequency of the mechanical host structure,
(b) L is the output filter inductance,
(c) C_p is the piezoelectric element capacitance, and
(d) f_s is the switching frequency.

This equation only holds good for large input-to-output converter voltage difference ($V_{rect} \gg V_{bat}$). High efficiencies (≈ 65–70%) were observed at high levels

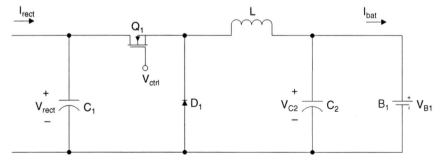

Fig. 18.17 DC–DC buck converter with battery load

of input excitation (OCV $> 45\ V_{DC}$). Another version had an efficiency of about 74–78% (Kim et al., 2007).

18.9.2.2 Buck-Boost Converter

The buck-boost converter circuit works for input voltages higher and lower than its output voltage. This allows optimization of different EH transducers with energy storage cells of different voltages. Figure 18.18 shows the schematic of a typical buck-boost converter. Under DCM, the average input resistance (R_{in}) is given by (Lefeuvre et al., 2007):

$$R_{in} = (2Lf_s)/(D^2) \tag{18.6}$$

This indicates that for a constant switching frequency (f_s), inductance (L), and duty cycle (D), R_{in} is constant and does not depend on the output battery voltage (V_{B1}) and current (I_{bat}). Unlike the buck converter, it is not necessary to have a high input-to-output converter voltage differential. The converter can also operate in CCM with R_{in} given by (Lefeuvre et al., 2007):

$$R_{in} = ((1 - D)/D)^2(V_{bat}/I_{bat}) \tag{18.7}$$

Here, R_{in} is not constant and is necessary to incorporate feedback circuits to measure V_{B1} and I_{bat}, and adjust D to obtain desired R_{in} for appropriate impedance matching. The constraint to enable DCM mode of operations is as follows:

$$V_{rect} < V_{bat}(1 - D)/D \tag{18.8}$$

Each of these converters typically supply DC power to an energy storage element at their output.

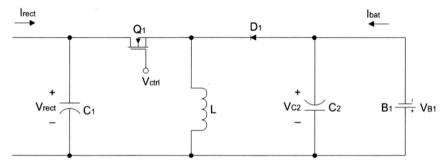

Fig. 18.18 DC–DC buck-boost converter with battery load

18.10 Future Directions and Scope

The future directions seem to suggest integration of EH transducer, circuits and systems, energy storage devices, power management, and tags on flexible thin-film substrates thus realizing smart active labels. Figure 18.19 shows the concept for a vibrational EH smart label. Similarly, one can envision amorphous solar cells forming one of the layers or a combination of solar and vibrational EH layers.

For energy from vibrations, piezoelectric MEMS-based transducers seem to be the obvious choice. Although these systems are not commercially available, researchers in China and MIT have designed such cantilevers. The MIT system consists of a composite micro-cantilever beam with a PZT thin film layer and electrode layer operating in d_{33} mode (Sood, 2003). A single piezoelectric micro-power generator (PMPG) device could deliver about 1 µW at 2.36 V_{DC} with energy density of 0.74 mW hr/cm^2. The second generation PMPG could provide 0.173 mW at 3 V_{DC} with 1 g excitation at 155.5 Hz (Xia et al., 2006). Investigators from Micro/Nano Fabrication Technology, Shanghai, China have shown that at the resonant frequency of about 608 Hz, a MEMS-based generator prototype can output about 0.89 V AC peak–peak voltage output with output power of 2.16 mW (Fang et al., 2006).

Flexible amorphous silicon and other nano-based materials are available to harvest energy from light. Amorphous silicon seems to be the current choice, but these are relatively inefficient under outdoor conditions and expensive. However, higher efficiency amorphous panels are available from vendors such as PowerFilm, Inc. (http://www.powerfilmsolar.com/). Alternatively, other nano-based compounds such as solar cells made out of CIGS (copper indium gallium selenide) are being pursued by many companies, including Nanosolar (http://www.nanosolar.com/). Other companies such as Konarka have developed nano-enabled polymer photovoltaic materials that are lightweight, flexible and more versatile than traditional solar materials (http://www.konarka.com/).

Thin-film solid-state energy storage device research and production are also making considerable progress. Infinite power solutions (IPS) offer Lite Star batteries operating with standard 4.2 V_{DC} constant voltage chargers that recharge to 90% of capacity in few minutes (http://www.infinitepowersolutions.com/). These devices can operate for thousands of charge–discharge cycles with no memory effect and

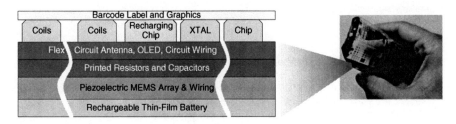

Fig. 18.19 Smart active label

minimal charge loss over time. The IPS LS101 version is rated at 0.7 mA hr/in.[2] (110 μm thickness) and the second generation (LS201) will be designed with higher capacity. Similarly, Cymbet Corporation has two versions of EnerChip solid-state batteries that have 12, 50, and 85 μA hr capacity and require 4.1 V_{DC} charging circuit (http://www.cymbet.com/content/products.asp). These devices are surface mount chips and rated for several thousand charge–discharge cycles.

Progress made by the above technologies indicate good feasibility towards development of smart active labels. Obviously, initially each layer of the tag will have to be developed individually with roll-to-roll manufacturing and then integrated to realize the smart active label. Such devices will reduce the form factor and weight of existing active sensors used in applications such as asset and personnel monitoring, and open the door to new applications such as smart bandages, labels, stamps, and so on.

18.11 Conclusions

The design of EH circuits for active RF sensors and ID tags is very important primarily because of the rapid development of this market and secondly, due to the limited life of battery and consequently, replacement. This, in particular, adds cost and limits deployment of these devices to high-value asset and limited life applications. Although the advantages of these devices are enormous in terms of communication range, throughput, and reliability, the cost of battery, and its replacement do hamper adoption rates. Feasible EH from vibrations and solar energy can extend battery life and at the bare minimum replenish idle energy consumption, which is a dominant factor in battery life reduction. Design of new RF sensor fabrication (e.g., smart labels) and device technology (e.g., communication protocols) are steadily decreasing the energy requirements and therefore, smart power management and EH techniques allow a realistic possibility for perpetual node life.

References

Alien Technology Corporation, Fluidic Self Assembly Manufacturing Process, http://www.alientechnology.com/products/index.php.

R. Amirtharajah and A. Chandrakasan, "Self-Powered Signal Processing Using Vibration-Based Power Generation", IEEE Journal of Solid-State Circuits, Vol. 33, No. 5, pp. 687–694, April 2004.

M. A. Assad, "A Real-Time Laboratory Test Bed for Evaluating Localization Performance of Wi-Fi RFID Technologies", M.S. Thesis, Worchester Polytechnic Institute, **Worcester**, MA, May 4th, 2007.

Axcess International Inc., "AXCESS Awarded $2 Million Contract for Integrated RFID and Sensor Port Security Solution in Barbados", Press Release, March 2007.

Axcess International Inc., 2008, http://www.axcessinc.com/products/tags.html.

D. Berndt, "Maintenance-Free Batteries", 2nd Ed., Research Studies Press Ltd., Taunton, England, 1997.

R. Bridgelall, "UHF Tags – The Answer to The Retail Supply Chain Prayers?," RF Innovations Magazine, August 1999.

R. Bridgelall, "Bluetooth/802.11 Protocol Adaptation for RFID Tags," 4th European Wireless Conference, Florence, Italy, February 28th, 2002.

R. Bridgelall, "Enabling Mobile Commerce Through Pervasive Communications with Ubiquitous RF Tags", IEEE Wireless Communications and Networking Conference, pp. 2041–2046, New Orleans, LA, March 2003.

R. Bridgelall, "Introducing a Micro-Wireless Architecture for Business Activity Sensing", IEEE RFID Conference (to appear), Las Vegas, NV, April 2008.

A. Burke, "Ultracapacitors: Why, How, and Where is the Technology", Journal of Power Sources, Vol. 91, pp. 37–50, 2000.

A. Chandrakasan, R. Amirtharajah, R. Goodman, and W. Rabiner, "Trends in Low Power Digital Signal Processing", 1998 International Symposium on Circuits and Systems, pp. 604–607, 1998.

T. Chi-ying, S. Hui, K. Wing-Hung, and S. Feng, "Ultra-low Voltage Power Management and Computation Methodology for Energy Harvesting Applications", Symposium on VLSI Circuits, pp. 316–319, June 2005.

Crossbow Technologies, http://www.xbow.com/Home/HomePage.aspx, 2008.

Cymbet Corporation, EnerChip products, http://www.cymbet.com/content/products.asp.

"Dot^{TM}", Axcess International Inc., http://www.axcessinc.com/products/dot.html, 2008.

EPCglobal Inc., "EPC^{TM} Radio-Frequency Identification Protocols Class-1 Generation-2 UHF RFID Protocol for Communications at 860 MHz – 960 MHz," Specification version 1.1.0, December 17th, 2005.

H. Fang, J. Liu, Z. Xu, L. Dong, L. Wong, D. Chen, B. Cai, and Y. Liu, "Fabrication and Performance of MEMS-based Piezoelectric Power Generator for Vibration Energy Harvesting", Microelectronics Journal, Vol. 37, No. 11, pp. 1280–1284, 2006.

B. Gates, "The Disappearing Computer", The Economist, Special Issue: The World in 2003, pp. 99, December 2002.

P. Glynne-Jones, S. Beeby, E. James, and N. White, "The Modelling of a Piezoelectric Vibration Powered Generator for Microsystems", Transducers 01/Eurosensors XV, June 2001.

P. Glynne-Jones, S. P. Beeby, and N. M. White, "Towards a Piezoelectric Vibration-Powered Microgenerator", IEEE Science Measurement and Technology, Vol. 148, pp. 68–72, March 2001.

S. Guo and H. Lee, "An Efficiency-Enhanced Integrated CMOS Rectifier with Comparator-Controlled Switches for Transcutaneous Powered Implants", 2007 IEEE Custom Integrated Circuits Conference, San Jose, CA, September 2007.

J. Han, A. Von Jouanne, T. Le, K. Mayaram, and T. Fiez, "Novel Power Conditioning Circuits for Piezoelectric Micro Power Generators", 19th IEEE Applied Power Electronics Conference, pp. 1541–1546, 2004.

A. Hande, "Energy Harvesting for Mobile and Wireless Electronics", IEEE Dallas Section Meeting, April 2007.

A. Hande, T. Polk, W. Walker, and D. Bhatia, "Indoor Solar Energy Harvesting for Sensor Network Router Nodes", accepted for publication, Journal of Microprocessors and Microsystems – Special Issue on Sensor Systems, December 2006.

A. Hande, A. Rajasekaran, R. Bridgelall, P. Shah, and S. Priya, "High Efficiency Vibrational Energy Harvester for Active RFID Tags", Second Annual nanoPower Forum (to appear), Irvine, CA, June 2–4 2008.

Hitachi, "Hitachi Unveils Smallest RFID Chip", RFiD Journal, March 2003.

M. Holt, "Does Asset Tracking Live Up To The Hype?", Health IT World News, March 20th, 2007.

IEEE Std. 802.11; Specification is a bases for Wi-Fi^{TM}.

IEEE Std. 802.15.4; Specification is a bases for ZigBee.

IEEE Std. 802.15.1-2002; Specification is a bases for Bluetooth.

Infinite Power Solutions, http://www.infinitepowersolutions.com/

ISO 18000-7, "Parameters for Active Air Interface Communications at 433 MHz", 2004.

A. Joseph and M. Srivastava, "Energy Harvesting Projects", IEEE Pervasive Computing, Vol. 4, No. 1, pp. 69–71, January–March, 2005.

A. Kansal, D. Potter, and M. Srivastava, "Performance Aware Tasking for Environmentally Powered Sensor Networks", ACM International Conference on Measurement and Modeling of Computer Systems, pp. 223–234, 2004.

H. Kim, S. Priya, H. Stephanou, and K. Uchino, "Consideration of Impedance Matching Techniques for Efficient Piezoelectric Energy Harvesting", IEEE Transactions on Ultrasonics, Ferroelectrics, and Frequency Control, Vol. 54, No. 9, pp. 1851–1859, September 2007.

Konarka Technologies Inc., http://www.konarka.com/

J. Krause, "Antennas", McGraw-Hill, New York, NY, 1950.

J. Krikke, "Sunrise for Energy Harvesting Projects", IEEE Pervasive Computing, Vol. 4, No. 1, pp. 4–5, January–March, 2005.

J. Kymissis, C. Kendall, J. Paradiso, and N. Gerhenfeld, "Parasitic Power Harvesting in Shoes", 2nd International Symposium Wearable Computing Conference, pp. 132–139, Pittsburgh, PA, Oct. 19–20, 1998.

E. Lefeuvre, D. Audigier, C. Richard, and D. Guyomar, "Buck-Boost Converter for Sensorless Power Optimization of Piezoelectric Energy Harvester", IEEE Transactions on Power Electronics, Vol. 22, No. 5, September 2007.

K. S. Leong, M. L. Ng, P. H. Cole, "Positioning Analysis of Multiple Antennas in Dense RFID Reader Environment", Applications and the Internet Workshops, January 23rd, 2006.

D. Linden, "Handbook of Batteries and Fuel Cells", McGraw-Hill, New York, 1984.

S. Meninger, J. Mur-Miranda, R. Amirtharajah, A. Chandrakasan, and J. Lang, "Vibration-to-Electric-Energy Conversion", IEEE Transactions on Very Large Scale Integration (VLSI) Systems, Vol. 9, No. 1, pp. 64–75, February 2001.

S. Meninger, et al., "Vibration-to-electric Energy Conversion", ACM/IEEE International Symposium on Low Power Electronics and Design, pp. 48–53, 1999.

Mide Technology Corp., "Quickpack Piezoelectric Actuator", http://www.mide.com/products/qp/qp_catalog.php, 2008.

R. Min, M. Bhardwaj, S. Cho, N. Ickes, E. Shih, A. Sinha, A. Wang, and A. Chandrakasan, "Energy-Centric Enabling Technologies for Wireless Sensor Networks", IEEE Wireless Communications, Vol. 9, No. 4, pp. 28–39, August 2002.

R. Min, et al., "An Architecture for a Power-Aware Distributed Microsensor Node", IEEE Workshop on Signal Processing Systems, pp. 581–590, 2000.

Nanosolar Inc., http://www.nanosolar.com/

G. Ottman, H. Hofmann, A. Bhatt, and G. Lesieutre, "Adaptive Piezoelectric Energy Harvesting Circuit for Wireless Remote Power Supply", IEEE Transactions on Power Electronics, Vol. 17, No. 5, pp. 669–676, September 2002.

G. Ottman, H. Hofmann, and G. Lesieutre, "Optimized Piezoelectric Energy Harvesting Circuit Using Step-Down Converter in Discontinuous Conduction Mode", IEEE Transactions on Power Electronics, Vol. 18, No. 2, pp. 696–703, March 2003.

Panasonic Corporation, "Panasonic Solar Cells Technical Handbook", http://www.solarbotics.net/library/datasheets/sunceram.pdf, 1998.

J. Paradiso and T. Starner, "Energy Scavenging for Mobile and Wireless Electronics", IEEE Pervasive Computing, Vol. 4, No. 1, pp. 18–27, January–March, 2005.

M. Pedram and J. Rabaey, "Power Aware Design Methodologies", Kluwer Academic Publishers, Boston, MA, 2002.

PowerFilm Inc., http://www.powerfilmsolar.com/

J. Rabaey, M. Ammer, J. da Silva, D. Patel, and S. Roundy, "PicoRadio Supports Ad Hoc Ultra-Low Power Wireless Networking", IEEE Computer, Vol. 33, No. 7, pp.42–48, July 2000.

J. Rabaey, M. Ammer, T. Karalar, S. Li, B. Otis, M. Sheets, and T. Tuan, "PicoRadios for Wireless Sensor Networks: The Next Challenge in Ultra-Low-Power Design", International Solid-State Circuits Conference, San Francisco, CA, February 2002.

V. Raghunathan, A. Kansal, J. Hsu, J. Friedman, and M. Srivastava, "Design Considerations for Solar Energy Harvesting Wireless Embedded Systems", Fourth International Symposium on Information Processing in Sensor Networks, pp. 457–462, April 2005.

V. Raghunathan, C. Schurgers, S. Park, and M. Srivastava, "Energy Aware Wireless Microsensor Networks", IEEE Signal Processing Magazine, Vol. 19, No. 2, pp. 40–50, March 2002.

M. Rahimi, H. Shah, G. Sukhatme, J. Heideman, and D. Estrin, "Studying the Feasibility of Energy Scavenging in a Mobile Sensor Network", IEEE International Conference on Robotics and Automation, pp. 19–24, 2003.

J.F. Randall and J. Jacot, "The Performance and Modelling of 8 Photovoltaic Materials Under Variable Light Intensity and Spectra", World Renewable Energy Conference VII, Cologne, Germany, 2002.

S. Roundy, P. Wright, and J. Rabaey, "A Study of Low Level Vibrations as a Power Source for Wireless Sensor Nodes", Journal of Computer Communications, Vol. 26, pp. 1131–1144, July 2003.

S. Roundy, P. Wright, and J. Rabaey, "Energy Scavenging for Wireless Sensor Networks with Special Focus on Vibrations", ISBN 1-4020-7663-0, Kluwer Academic Publishers, Dordrecht, 2004.

S. Roundy and P. K. Wright, "A Piezoelectric Vibration Based Generator for Wireless Electronics", Smart Materials and Structures, Vol. 13, pp. 1131–1142, 2004.

S. Roundy, P. K. Wright, and K. S. J. Pister, "Micro-Electrostatic Vibration-to-Electricity Converters", ASME International Mechanical Engineering Congress & Exposition IMECE2002, November 2002.

H. Schmidhuber and C. Hebling, "First Experiences and Measurements with a Solar Powered Personal Digital Assistant (PDA)", 17th European Photovoltaic Solar Energy Conference, pp. 658–662, ETA-Florence and WIP-Munich, 2001.

R. C. Shah and J. M. Rabaey, "Energy Aware Routing for Low Energy Ad Hoc Sensor Networks", IEEE Wireless Comm. and Networking Conference, pp. 350–355, 2002.

C. Shearwood and R. Yates, "Development of an Electromagnetic Micro-generator", Electronics Letters, Vol. 33, No.22, pp. 1883–1884, October 1997.

N. Shenck and J. A. Paradiso, "Energy Scavenging with Shoe-mounted Piezoelectrics", IEEE Micro, Vol. 21, pp. 30–42, May–June 2001.

X. Shengwen, K. Ngo, T. Nishida, C. Gyo-Bum, and A. Sharma, "Converter and Controller for Micro-power Energy Harvesting", 20th Annual IEEE Applied Power Electronics Conference and Exposition, pp. 226–230, March 2005.

M. Skolnik, "Introduction to RADAR Systems", MMcGraw-Hill, New York, NY, 1980.

P. Smalser, "Power Transfer of Piezoelectric Generated Energy", U.S. Patent, 5 703 474, 1997.

A. A. Smith Jr., "Radio Frequency Principles and Applications", IEEE Press, New York, 1998.

R. Sood, "Piezoelectric Micro Power Generator: A MEMS-base Energy Scavenger", M.S. Thesis, Massachusetts Institute of Technology, Cambridge, MA, September 2003.

T. Starner, "Human-powered Wearable Computing" IBM Systems Journal, Vol. 35. No. 3–4, 1996.

T. Starner, "Powerful Change Part 1: Batteries and Possible Alternatives for the Mobile Market", IEEE Pervasive Computing Mobile and Ubiquitous Systems, Vol. 2, No. 4, pp. 86–88, October 2003.

T. Starner and J.A. Paradiso, "Human-Generated Power for Mobile Electronics", Low-Power Electronics Design, C. Piguet, ed., Chapter 45, pp. 1–35, CRC Press, **Boca** Raton, FL, 2004.

J. Stevens, "Optimized Thermal Design of Small ΔT Thermoelectric Generators", 34th Intersociety Energy Conversion Engineering Conference, paper 1999-01-2564, Society of Automotive Engineers, 1999.

TI EZ430-RF2500, http://focus.ti.com/docs/toolsw/folders/print/ez430-rf2500.html, 2008.

United States Government Accountability Office (GAO), US Border Security US-VISIT, Report No. 07–248, December 2006.

US Code of Federal Regulations (CFR), Title 47, Chapter 1, Part 15: Radio-frequency devices, U.S. Federal Communications Commission.

US Patent #6034603, "Radio Tag System and Method with Improved Tag Interference Avoidance," March 7th, 2000.

US Patent #7005985, "Radio Frequency Identification System and Method", February 28th, 2006.

T. Voigt, H. Ritter, and J. Schiller, "Utilizing Solar Power in Wireless Sensor Networks", IEEE Conference on Local Computer Networks, 2003.

R. Xia, C. Farm, W. Choi, and S. Kim, "Self-Powered Wireless Sensor System using MEMS Piezo-electric Micro Power Generator", IEEE Sensors Conference, October 2006.

E. Yeatman, "Advances in Power Sources for Wireless Sensor Nodes", International Workshop Wearable and Implantable Body Sensor Networks, pp. 20–21, Imperial College, 2004, http://www.doc.ic.ac.uk/vip/bsn_2004/program/index.html

"Zero Threshold MOSFETs", Advanced Linear Devices Inc., http://www.aldinc.com/ald_przerothreshold.htm, 2008.

Chapter 19
Powering Wireless SHM Sensor Nodes through Energy Harvesting

Gyuhae Park, Kevin M. Farinholt, Charles R. Farrar, Tajana Rosing, and Michael D. Todd

Abstract The concept of wireless sensor nodes and sensor networks has been widely investigated for various applications, including the field of structural health monitoring (SHM). However, the ability to power sensors, on board processing, and telemetry components is a significant challenge in many applications. Several energy harvesting techniques have been proposed and studied to solve such problems. This chapter summarizes recent advances and research issues in energy harvesting relevant to the embedded wireless sensing networks, in particular SHM applications. A brief introduction of SHM is first presented and the concept of energy harvesting for embedded sensing systems is addressed with respect to various sensing modalities used for SHM and their respective power requirements. The power optimization strategies for embedded sensing networks are then summarized, followed by several example studies of energy harvesting as it has been applied to SHM embedded sensing systems. The paper concludes by defining some future research directions that are aimed at transitioning the concept of energy harvesting for embedded sensing systems from laboratory research to field-deployed engineering prototypes.

19.1 Introduction

Structural health monitoring (SHM) is the process of detecting damage in aerospace, civil, and mechanical infrastructure. The goal of SHM is to improve the safety, reliability, and/or ownership costs of engineering systems by autonomously monitoring the conditions of structures and detecting damage before it reaches a critical state. To achieve this goal, technology is being developed to replace qualitative visual inspection and time-based maintenance procedures with more quantifiable and automated condition-based damage assessment processes. The authors believe that all approaches to SHM, as well as all traditional non-destructive evaluation procedures can be cast in the context of a statistical pattern recognition problem

G. Park (✉)
The Engineering Institute, Los Alamos National Laboratory, Los Alamos, NM 87545, USA
e-mail: gpark@lanl.gov

S. Priya, D.J. Inman (eds.), *Energy Harvesting Technologies*,
DOI 10.1007/978-0-387-76464-1_19 © Springer Science+Business Media, LLC 2009

(Farrar et al., 2001; Sohn et al., 2004; Doebling et al., 1996). Solutions to this problem require the four steps of (1) operational evaluation, (2) data acquisition, (3) feature extraction, and (4) statistical modeling for feature classification. Inherent in parts 2–4 of this paradigm are the processes of data normalization, data compression, and data fusion.

A major concern with any embedded wireless sensing networks is their long-term reliability and sources of power. If the only way to provide power is by direct connections, then the need for wireless protocols is eliminated, as the cabled power link can also be used for the transmission of data. However, if one elects to use a wireless network, the development of micro-power generators is a key factor for the deployment of the network. A possible solution to the problem of localized power generation is technologies that enable harvesting ambient energy to power the instrumentation. Although energy harvesting for large-scale alternative energy generation using wind turbines and solar cells is mature technology, the development of energy harvesting technology on a scale appropriate for small, low-power, embedded sensing systems is still in the developmental stages, particularly when applied to SHM sensing systems. Given these reasons, the amount of research devoted to energy harvesting has been rapidly increasing, and the SHM and sensing network community have investigated the energy harvesters as an alternative power source for the next generation of embedded sensing systems.

This chapter summarizes the state-of-the-art in energy harvesting as it has been applied to SHM embedded sensing systems. First, various existing and emerging sensing modalities used for SHM and their respective power requirements are summarized followed by a discussion of power optimization strategies. This chapter also addresses current SHM energy harvesting applications and system integration issues. It should be, however, noted that much of the technologies discussed herein is applicable to powering any type of low-power embedded sensing system regardless of the application.

19.2 Sensing System Design for SHM

For efficient SHM, one must establish an appropriate sensor network that can adequately observe changes in the system dynamics caused by damage and manage these data for suitable signal processing, feature extraction, and classification. The goal of any SHM sensor network is to make the sensor reading as directly correlated with, and as sensitive to, damage as possible. At the same time, one also strives to make the sensors as independent as possible from all other sources of environmental and operational variability, and, in fact, independent from each other (in an information sense) to provide maximal data with minimal sensor array outlay. To meet best these goals, the following design parameters must be defined, as much as possible, *a priori*: types of data to be acquired; sensor types, number, and locations; bandwidth, sensitivity, and dynamic range; data acquisition/telemetry/storage system; power requirements; sampling intervals; processor/memory requirements; and excitation source needs (for active sensing).

With these design parameters and issues in mind, the sensing systems for SHM that have evolved to date consist of some or all of the following components: transducers that convert changes in the field variable of interest to changes in an electrical signal; actuators that can be used to apply a prescribed input to the system; analog-to-digital (A/D) and digital-to-analog (D/A) converters; signal conditioning; power; telemetry; processing capability; and memory for data storage.

19.3 Current SHM Sensor Modalities

The sensing component (transducer) refers to the transduction mechanism that converts a physical field (such as acceleration) into an electronically measurable form (usually an electrical potential difference). If the sensing system involves actuation, then the opposite is required, i.e., a voltage command is converted into a physical field (usually displacement). The most common measurements currently made for SHM purposes are, in order of use: acceleration, strain, Lamb wave, and electrical impedance.

The use of accelerometers and strain gauges by far the most common approach used in SHM applications today, primarily because of the relative maturity and commercial availability of associated hardware. The energy consumed by these devices themselves is very small because of their passive nature, but the centralized multiplexing, amplification, and signal conditioning units required to obtain usable data can often have power requirements that approach a few Watts. There is a considerable recent work suggesting the use of micro-electromechanical systems (MEMS) accelerometers for SHM applications, but to date this type of accelerometer has seen little actual use in SHM applications.

Most wave propagation approaches to SHM make use of piezoelectric patches as both sensors and actuators. Arrays of these devices can be configured to sequentially induce local motion at various locations on the structure, and the same array can also be used to measure the response to these excitations. In this mode, the sensor–actuator pairs interrogate a structure in a manner analogous to traditional pitch–catch or pulse-echo ultrasonic inspection. Alternatively, many researchers have measured the electrical impedance across a piezoelectric patch as an indictor of damage (Park et al., 2003). It has been shown that this electrical impedance is related to the local mechanical impedance of the structure, with the assumption that the mechanical impedance will be altered by damage.

In the passive sensing mode, piezoelectric transducers would consume much less energy, compared with accelerometers or strain gauges, because they do not require any electrical peripherals such as signal conditioning and amplification units, which are typically embedded and required for piezoelectric accelerometers to operate. However, this low-power consumption characteristic will be modified if one needs to use charge amplifiers or voltage follower circuits to improve the signal-to-noise ratio depending on applications or frequency range of interest. When used in an active sensing mode, a D/A converter and a waveform generator are also needed along with higher speed A/D converters, additional memory, and possibly multiplexers in order to control and manage a network of piezoelectric transducers. These extra components will inherently demand more energy.

19.4 Energy Optimization Strategies Associated with Sensing Systems

This section will discuss strategies to minimize the energy demands of a sensor network in an effort to efficiently power these sensing systems.

Embedded system design is characterized by a tradeoff between a need for good performance and a low-power consumption. The proliferation of wireless sensing devices has stressed even more the need for energy minimization as the battery capacity has improved very slowly (by a factor of 2–4 over the last 30 years), while the computational demands have drastically increased over the same time frame, as shown in Fig. 19.1.

Since the introduction of wireless computing, the demands on the battery lifetime have grown even more. In fact, in most of today's embedded sensing devices, the wireless connectivity consumes a large fraction of the overall energy consumption. Figure 19.2 shows a power consumption breakdown for a small sensor node (top of the figure) and a larger embedded device based on a strong ARM processor (200 MHz) coupled with a wireless local area network for communication. On small sensor nodes, as much as 90% of the overall system power consumption can go to wireless communication, while on the larger devices, such as the one shown at the bottom of the Fig. 19.2, the wireless takes approximately 50% of the overall

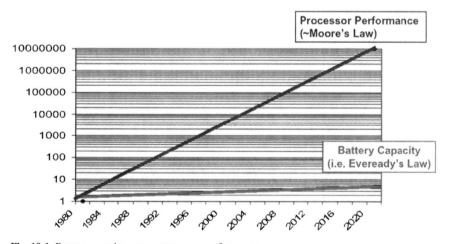

Fig. 19.1 Battery capacity versus processor performance

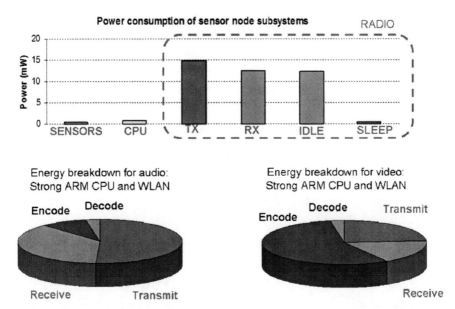

Fig. 19.2 Power consumption of two different embedded system designs (Source: Sensors Tutorial, 7th Annual International Conference on Mobile Computing and Networks)

power budget. In both cases, the second most power-hungry device is the processor. Therefore, in order to achieve the desired energy efficiency, both optimization of both computing and communication energy consumption are critically important.

Better low-power circuit design techniques have helped to lower the power consumption (Chandrakasan and Brodersen, 1995; Rabaey and Pedram, 1996; Nabel and Mermet, 1997). On the other hand, managing power dissipation at higher levels can considerably decrease energy requirements and thus increase battery lifetime and lower packaging and cooling costs (Ellis, 1999; Benini, and De Micheli, 1997). Two different approaches for lowering the power consumption at the system level have been proposed: dynamic voltage scaling, primarily targeted at the processing elements, and dynamic power management, which can be applied to all system components. The rest of this section provides an overview of state-of-the-art dynamic power management and dynamic voltage scaling algorithms that can be used to reduce the power consumption of both processing and communication in wireless sensing devices.

19.4.1 Dynamic Voltage Scaling

Embedded sensing systems are designed to be able to deliver peak performance when needed, but in most cases, their components operate at utilizations less than 100%. One way of lowering the power consumption is by slowing down the execution, and, when appropriate, also lowering the component's voltage of operation.

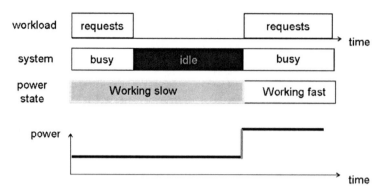

Fig. 19.3 Dynamic voltage scaling on a single processor

This power reduction is done with *dynamic Voltage Scaling (DVS)* algorithms.

$$P_{\text{dyn}} \propto f V_{\text{dd}}^2 \tag{19.1}$$

$$f \propto (V_{\text{dd}} - V_{\text{threshold}})^2 / V_{\text{dd}} \tag{19.2}$$

The primary motivation comes from the observation that dynamic power consumption, P_{dyn}, is directly proportional to the frequency of operation, f, and the square of the supply voltage, V_{dd}^2 (see Eq. 19.1). Frequency, in turn, is a linear function of V_{dd} (see Eq. 19.2), so decreasing the voltage results in a cubic decrease in the power consumption. Clearly, decreasing the voltage also lowers the frequency of operation, which, in turn, lowers the performance of the design.

Figure 19.3 shows the effect of DVS on power and performance of a processor. Instead of having longer idle period, the central processing unit (CPU) is slowed down to the point where it completes the task in time for the arrival of the next processing request, while at the same time saving quite a bit of energy. DVS algorithms are typically implemented at the level of an operating system's (OS) scheduler. There have been a number of voltage scaling techniques proposed for real-time systems. Early work typically assumed that the tasks run at their worst case execution time, while the later research work relaxes this assumption and suggest a number of heuristics for the prediction of task execution time. A more detailed overview on various DVS algorithms can be found in Kim and Simunic Rosing (2006).

19.4.2 Dynamic Power Management

In contrast to DVS, system-level *dynamic power management (DPM)* decreases the energy consumption by selectively placing idle components into lower power states. DVS can only be applied to CPUs, while DPM can be used to reduce the energy consumption of wireless communication, CPUs, and all other components that have low-power states. While slowing down the CPU with DVS can provide quite a bit

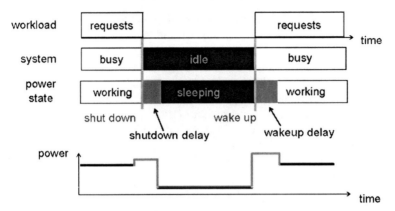

Fig. 19.4 Dynamic power management for a single device

of power savings, applying DPM typically increases the savings by at least a factor of 10, and in many systems by significantly more than that. On the other hand, changing processor speed happens relatively quickly, while the transitions in and out of sleep states can be quite costly in terms of both energy and performance. Figure 19.4 shows both power and performance overheads incurred during the transition. At a minimum the device needs to stay in the low-power state for long enough (defined as the break even time, T_{BE}) to recuperate the cost of transitioning. The break even time, as defined in Eq. (19.3), is a function of the power consumption in the active state, P_{on}, the amount of power consumed in the low-power state, P_{sleep}, and the cost of the transition in terms of both time, T_{tr}, and power, P_{pr}.

$$T_{BE} = T_{tr} + T_{tr}\frac{P_{tr} - P_{on}}{P_{on} - P_{sleep}} \tag{19.3}$$

If it were possible to predict ahead of time the exact length of each idle period, then the ideal power management policy would place a device in the sleep state only when an idle period would be longer than the break even time. Unfortunately, in most real systems, such perfect prediction of idle periods is not possible. As a result, one of the primary tasks, DPM algorithms have to predict when the idle period will be long enough to amortize the cost of transition to a low-power state, and to select the state to transition to. Three classes of policies can be defined – time-out-based, predictive, and stochastic. Time-out policy is implemented in the most operating systems. The drawback of this policy is that it wastes power while waiting for the time-out to expire. Predictive policies developed for interactive terminals (Srivastava et al., 1996; Hwang and Wu, 1997) force the transition to a low-power state as soon as a component becomes idle if the predictor estimates that the idle period will last long enough. An incorrect estimate can cause both performance and energy penalties. Both time-out and predictive policies are heuristic in nature, and thus do not guarantee optimal results. In contrast, approaches based on stochastic models can guarantee optimal results. Stochastic models use distributions to describe the times between arrivals of user requests (*interarrival times*), the length of time it takes for a device to service a user's request, and the time it takes for the device to transition between its power states. The optimality of stochastic approaches depends on the accuracy of the system model and the algorithm used to compute the solution.

Finally, much recent work has looked at combining DVS and DPM into a single power management implementation. Shorter idle periods are more amiable to DVS, while longer ones are more appropriate for DPM. Thus, a combination of the two approaches is needed for the most optimal results. It should also be pointed out that the studies in the current SHM sensing hardware development (Spencer et al., 2004; Lynch, and Loh, 2006) have not yet fully incorporated the power-awareness design described in this section.

19.5 Applications of Energy Harvesting to SHM

Although the energy harvesting techniques are still in a development stage, several conceptual designs for applications into SHM have been proposed. Elvin et al. (2002) proposed a self-powered damage detection sensor using piezoelectric patches. A network of self-powered strain energy sensors were embedded inside a structure, and a moving cart capable of applying a time-varying dynamic load was driven over the structure. The piezoelectric harvesters convert this applied load into electricity and provide a power for sensors in order to measure the strain and to send the results to the moving cart, as shown in Fig. 19.5.

James et al. (2004) also proposed a prototype of self-powered system for condition monitoring applications. The devices, using a low-power accelerometer as a sensor, are powered by a vibration-based electromagnetic generator, which provides a constant power of 2.5 mW. However, the systems are not equipped with a local computing capability and only send out the direct sensor readings.

Discenzo et al. (2006) developed a prototype self-powered sensor node that performs sensing, local processing, and telemeters the result to a central node for pump

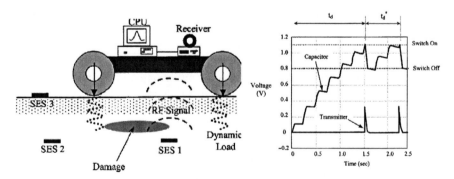

Fig. 19.5 Implementation of self-powered sensors for damage detection (Source: Elvin et al., 2002)

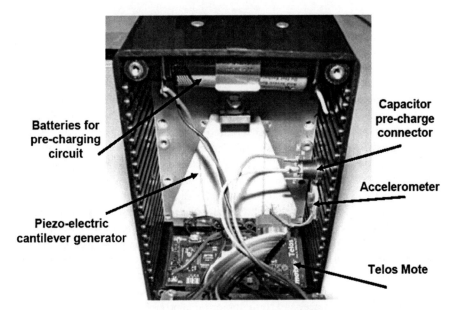

Capacitor pre-charge connector

Batteries for pre-charging circuit

Accelerometer

Piezo-electric cantilever generator

Telos Mote

Fig. 19.6 Self-powered sensor node (Source: Discenzo et al. (2006) reprinted with permission of Sound and Vibration Magazine)

condition monitoring applications, shown in Fig. 19.6. A wireless mote system was integrated with a piezoelectric energy harvesting technique. The device was mounted on an oil pump, and a cantilever piezoelectric beam tuned to the excitation frequency was embedded with the sensor node to extracted energy from the pump vibration. The maximum power output of 40 mW was achieved.

Pfeifer et al. (2001) investigated the development of self-powered sensor tags that can be used to monitor the health of a structure. A micro-controller was powered by a piezoelectric patch (7.5 × 5 cm). Once powered, the micro-controller operates the sensor array, performs the local computing, and saves the results of computation into a RFID tag. By storing the data in nonvolatile memory, the data can be retrieved by a mobile host, even if the sensor node does not have enough power to operate. In a laboratory setting, the piezoelectric harvester can deliver enough energy to the micro-controller for 17 s of operation.

Inman and Grisso (2006) propose an integrated autonomous sensor node that contains the elements of energy harvesting from ambient vibration and temperature gradients, a battery charging circuit, local computing and memory, active sensors, and wireless transmission. These elements could be autonomous, self-contained, and unobtrusive when compared with the system being monitored.

Mascarenas et al. (2007) experimentally investigated the radio frequency (RF) wireless energy transmission as a power source for wireless SHM sensor nodes. They experimentally demonstrated that the delivered RF energy could be stored and used to successfully operate the radio, which is the largest power consumer in a SHM sensor node. This work was based on the new and efficient SHM sensing network, whereby the electric power and interrogation commands are wirelessly provided by a mobile agent (Farrar et al., 2006; Todd et al., 2007). This approach involves using an unmanned mobile host node to generate an RF signal near sensors that have been embedded on the structure. The sensors measure the desired response at critical areas on the structure and transmit the signal back to the mobile host again via the RF communications. This research takes traditional sensing networks to the next level, as the mobile hosts (such as UAV), will fly to known critical infrastructure based on a GPS locator, deliver required power, and then begin to perform an inspection without human intervention. The mobile hosts will search for the sensors on the structure and gather critical data needed to perform the structural health evaluation, which schematically described in Fig. 19.7. This integrated technology will be directly applicable to rapid structural condition assessment of buildings and bridges after an earthquake. It should be emphasized that this technology can be hybrid in that the sensor node is still equipped with energy harvesting devices and the mobile host would provide additional energy if the energy harvesting device is not able to provide enough power to operate the sensor nodes. Even the energy harvesting device provides sufficient power, the mobile agent can wirelessly trigger the sensor nodes, collect the information, and/or provide computational resources, which significantly relax the power and computation demand at the sensor node level. Their recent work shows that 0.1 F supercapacitor can be charged in less than 30 s, which provides enough power to operated an SHM active-sensor node (Farinholt, in press).

Ha and Chang (2005) assessed the suitability and efficiency of energy harvesting techniques for an SHM system based on the network of piezoelectric sensors and actuators. They concluded that total power requirement of the piezoelectric Lamb-wave-based SHM far exceeds the current energy harvesting capability. However, they suggested that the passive sensing system, which uses passive acoustic

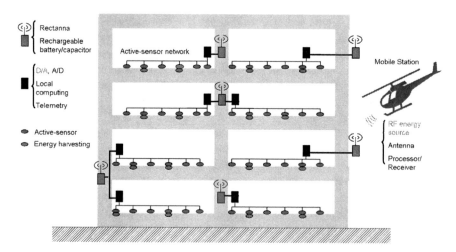

Fig. 19.7 A new sensing network that includes wireless energy transmission and energy harvesting and that is interrogated by an unmanned robotic vehicle

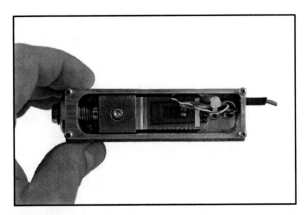

Fig. 19.8 A prototype of sensor node with piezoelectric energy harvester. (Source: image courtesy Microstrain, Inc. (http://www.microstrain.com/) reprinted with permission)

emission and detects an accidental impact event, would be a good candidate for energy harvesting technology because of the low power requirement and very low duty-cycle.

Although vibration-based energy harvesters are still under the development stage, several commercial solutions are available. Most research efforts are still in proof-of-concept demonstrations in a laboratory setting. Microstrain, Inc. (http://www.microstrain.com/) developed a prototype of piezoelectric-based energy harvester, shown in Fig. 19.8. The sensor node is equipped with temperature and humidity sensors with wireless telemetry. It is claimed that the piezoelectric harvester can produce up to 2.7 mW of instant power at 57 Hz vibration. Perpetuum, Inc. (http://perpetuum.co.uk/) commercialized electromagnetic energy converters, which are capable of generating up to 3.3 V and 5 mW of instant power under the 100 mg vibration. The operating frequencies could be tuned in the range of 47–100 Hz. Ferro Solution, Inc. (http://www. Ferrosi.com/) also produced electromagnetic generators that have a 9.3 mW power capability with 100 mg input vibration.

Energy harvesting is slowly coming into full view of the SHM and the more general sensing network communities. With the continual advances in wireless sensor/actuator technology, improved signal processing technique, and the continued development of power efficient electronics, energy harvesting will continue to attract the attentions of researchers and field engineers. However, it should be emphasized that a tremendous research effort is still required to convert, optimize, and accumulate the necessary amount of energy to power such electronics.

19.6 Future Research Needs and Challenges

While it is noted that there is tremendous research into the development of energy harvesting schemes for large-scale alternative sources such as wind turbines and solar cells and that these large-scale systems have made the transition from research to commercial products, energy harvesting for embedded sensing systems is still in its infancy. Also, there is no clearly defined design process to develop such energy harvesting for embedded sensing systems. Therefore, some future research areas for energy harvesting will be outlined in order to transition the current state-of-the-art to full-scale deployment in the current practice of embedded sensing networks.

As identified, the major limitations facing researchers in the field of energy harvesting revolve around the fact that the energy generated by harvesting devices is far too small to directly power most electronics. Therefore, the efficient and innovative methods of storing electric energy are the key technologies that will allow energy harvesting to become a source of power for electronics and wireless sensors. It should also be emphasized that, when using any storage medium, the duty cycle of the application must be considered, as this factor drastically changes the design parameters and associated electronics. It is necessary to match the duty cycle to the time required to store enough energy until it is needed by electronics.

Research on energy harvesting materials has focused mainly on determining the extent of power capable of being generated rather than investigating applications and uses of the harvested energy. The practical applications for energy harvesting systems, such as wireless self-powered SHM sensing networks, must be clearly identified with the emphasis on power management issues. Application-specific, design-oriented approaches are needed to help with the practical use of these technologies. It is also suggested that the biggest roadblock for using energy harvesting devices is the lack of clear design guidelines that help determine how to characterize the ambient energy, what circuits and storage devices are best for a given application, and what strategies are best to integrate the harvesting devices into embedded sensor units. Developing such guideline demands substantial research efforts to define the key parameters and predictive models affecting efficient energy harvesting.

Reliability is an essential requirement for any energy sources. Because many vibration-based harvesters are designed to operate at their resonances, the systems will be inherently unstable after the long operation cycles. Also, any energy sources for field use should be able to withstand harsh environmental conditions. The reliability and robustness must be proved before the energy harvesting techniques can be used in practice.

Few studies addressed the integrated use of available energy harvesting devices. Each energy harvesting scheme needs to be compared precisely to the other methods and, if necessary, integrated together to maximize the energy generation under a given environmental condition. To realize this integration, a general standard should be established to address the technical capabilities of each energy source for system integrators so that they can easily assemble components for final design.

The goal of maximizing the amount of the harvested energy involves several factors, including electronics optimization, characterization of the available ambient energy, selection and configuration of energy harvesting materials, integration with storage mechanisms, along with the power-optimization and power-awareness design. Few studies have addressed these issues in an integrated manner from the multidisciplinary engineering perspective. Finally, it has been identified that, although several energy harvesting devices are developed and fabricated as a prototype, the performance of these techniques in real operational environments needs to be verified and validated.

19.7 Conclusion

This chapter presents the state-of-the-art in energy harvesting as it has been applied to SHM embedded sensing systems. Various existing and emerging sensing modalities used for SHM and their respective power requirements were first summarized and several sensor network power optimization strategies were discussed. This chapter also summarizes the current energy harvesting applications to SHM or other embedded sensing systems. Some future research directions and possible technology demonstrations are also discussed in order to transition the concept of energy harvesting for embedded sensing systems from laboratory research to field-deployed engineering prototypes.

References

Benini, L. and De Micheli, G., 1997, *Dynamic power management: design techniques and CAD tools*, Kluwer, Dordrecht.

Chandrakasan, A. and Brodersen, R. 1995, *Low power digital CMOS design*, Kluwer, Dordrecht.

Discenzo, F.M., Chung, D. and Loparo, K.A., 2006, "Pump condition monitoring using self-powered wireless sensors," *Sound and Vibration*, **40**(5), pp. 12–15.

Doebling, S.W., Farrar, C.R., Prime M.B. and Shevitz, D., 1996 "Damage identification and health monitoring of structural and mechanical systems from changes in their vibration characteristics: a literature review," Los Alamos National Laboratory report LA-13070-MS.

Ellis, C., 1999, "The case for higher-level power management," *7th IEEE Workshop on Hot Topics in Operating Systems*, pp. 162–167.

Elvin, N., Elvin, A. and Choi, D.H., 2002, "A self-powered damage detection sensor," *Journal of Strain Analysis*, **38**, pp. 115–124.

Farinholt, K.M., Park, G., Farrar, C.R. in press. "Energy harvesting and wireless energy transmission for SHM sensor nodes," *Encyclopedia of Structural Health Monitoring*, John Wiley & Sons Ltd.

Farrar, C.R., Doebling S.W., and Nix, D.A., 2001, "Vibration-based structural damage identification," *Philosophical Transactions of the Royal Society: Mathematical, Physical & Engineering Sciences*, **359**(1778) pp. 131–149.

Farrar, C.R., Park, G., Allen, D.W., Todd, M.D., 2006. "Sensor network paradigms for structural health monitoring," *Structural Control and Health Monitoring*, **13**(1), pp. 210–225.

Ha, S. and Chang, F.K., 2005, "Review of energy harvesting methodologies for potential SHM applications, *Proceedings of 2005 International Workshop on Structural Health Monitoring*, pp. 1451–1460.

Hwang, C.-H. and Wu, A., 1997, "A predictive system shutdown method for energy saving of event-driven computation," *International. Conference on Computer Aided Design*, pp. 28–32.

Inman, D.J. and Grisso, B.L., 2006, "Towards autonomous sensing," *Proceedings of SPIE*, 6174, pp. T1740–T1749.

James, E.P., Tudor, M.J., Beeby, S.P., Harris, N.R., Glynne-Jones, P., Ross, J.N. and White, N.M., 2004. "An investigation of self-powered systems for condition monitoring applications," *Sensors & Actuators*, **110**, pp. 171–176.

Kim, J. and Simunic Rosing, T., 2006, "Power-aware resource management techniques for low-power embedded systems," in *Handbook of real-time embedded systems* S. H. Son, I. Lee and J. Y-T Leung, (Editors), Taylor-Francis Group LLC, New York.

Lynch, J.P. and Loh, K.J., 2006, "A summary review of wireless sensors and sensor networks for structural health monitoring," *The Shock and Vibration Digest*, **38**(2), pp. 91–128.

Mascarenas, D.L., Todd, M.D., Park, G., Farrar, C.R. 2007, "Development of an impedance-based wireless sensor node for structural health monitoring," *Smart Materials and Structures*, **16**, 2137–2145.

Nabel, W. and Mermet, J. (Editors), 1997, *Lower power design in deep submicron electronics*, Kluwer, Dordrecht.

Park, G., Sohn, H., Farrar, C.R. and Inman, D.J., 2003, "Overview of piezoelectric impedance-based health monitoring and path forward," *Shock and Vibration Digest*, **35**(6) pp. 451–463.

Pfeifer, K.B., Leming, S.K. and Rumpf, A.N., 2001, *Embedded self-powered micro sensors for monitoring the surety of critical buildings and infrastructures*, Sandia Report, SAND2001-3619, Sandia National Laboratory.

Rabaey, J. and Pedram, M. (Editors), 1996, *Low power design methodologies*, Kluwer, Dordrecht.

Sohn, H., Farrar, C.R., Hemez, F.M., Shunk, D.D., Stinemates, D.W. and Nadler, B.R., 2004, "A review of structural health monitoring literature from 1996–2001," Los Alamos National Laboratory report LA-13976-MS.

Spencer, B.F., Ruiz-Sandoval, M.E. and Kurata, N., 2004, "Smart sensing technology: opportunities and challenges," *Structural Control and Health Monitoring*, **11**(4) pp. 349–368.

Srivastava, M.B., Chandrakasan, A.P. and Brodersen, R.W., 1996, "Predictive system shutdown and other architectural techniques for energy efficient programmable computation," *IEEE Transactions on VLSI Systems*, **4**(1) pp. 42–55.

Todd, M.D., Mascarenas, D.L., Flynn, E.B., Rosing, T., Lee, B., Musiani, D., Dasgupta, S., Kpotufe, S., Hsu, D., Gupta, R., Park, G., Overly, T., Nothnagel, M., Farrar, C.R., 2007. "A different approach to sensor networking for SHM: remote powering and interrogation with unmanned Arial vehicles," *Proceedings of 6th International Workshop on Structural Health Monitoring*, September 11–13, Stanford, CA.

Appendix A
First Draft of Standard on Vibration Energy Harvesting

At the Second Annual Energy Harvesting Workshop held on January 30–31, 2007 in Fort Worth, TX, a committee was formed consisting members from academia, industry, and federal labs. This committee was assigned the task of compiling current practices used to characterize the vibration energy harvesting devices and come up with a metric which can allow the comparison of all prototype harvesters. This first draft of standard is just a start and will be discussed in detail at the Fourth Annual Energy Harvesting Workshop to be held on January 28–29, 2009 in Virginia Tech (http://www.cpe.vt.edu/ehw). Following is the list of the committee members:

Committee Members (alphabetical order)

Bob O'Neil, Morgan Electroceramics
Brad Mitchell, Boeing
Chris Ludlow, Mide Technology Corporation
Dan Inman, Virginia Tech
Farhad Mohammadi, Advanced Cerametrics, Inc.
J. K. Huang, Ferro Solutions
Jan Kunzmann, Smart Material
John Blottman, Naval Undersea Warfare Center
M. G. Prasad, Stevens Institute of Technology
Robert O'Handley, MIT/Ferro Solutions
Roger Richards, Naval Undersea Warfare Center
Shashank Priya, Virginia Tech

Energy harvesting: Energy recovery from freely available environmental resources. Primarily, the selection of the energy harvester when compared with the other alternatives such as battery depends on two main factors: cost effectiveness and reliability. Another goal for energy harvesters has been to recharge the batteries in existing applications.

In the recent years, several energy harvesting approaches have been proposed using photovoltaic, thermoelectric, electromagnetic, piezoelectric, and capacitive schemes. This first draft of standard addresses the issues related with the vibration energy harvesting which primarily utilizes electromagnetic, piezoelectric, and

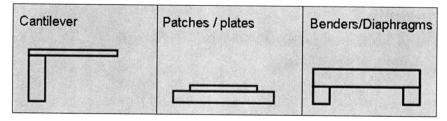

Fig. A.1 Common structures utilized for harvesting mechanical energy using piezoelectric transducers

capacitive schemes. Most means of vibration energy harvesting (VEH) are based on a mechanically *resonant* device that is, hopefully, matched to the *vibration spectrum* of source. There are elements of various devices (inductive, piezo benders, magneto-electric) that are common, because they are all based on the forced, damped harmonic oscillator. Figure A.1 shows some of the common configurations used for piezoelectric harvesters consisting of cantilever, patches, and diaphragms.

It is also possible to make *non-resonant* VEH devices. These fall into at least two categories: (1) mechanical systems that have *zero restoring force* (therefore $f_r = 0$) such as the shaker flashlights in which the freely moving proof mass (the permanent magnet) essentially remains stationary while the flashlight case is shaken and (2) systems designed to harvest *impact* or impulse forces.

A.1 Potential Vibration Sources for Energy Harvesting

Following is a list of vibration sources classified according to their elastic stiffness and Table A.1 lists the sources according to the surrounding.

– Stiff structures which make a movement by their own (ships, containers, mobile devices, housings of fans, escalators and elevators in public places, appliances, refrigerator, bridges, automobiles, building structures, and trains)

Table A.1 Sources of energy available in the surrounding which are/can be tapped for generating electricity

Human body	Vehicles	Structures	Industrial	Environment
Breathing, blood pressure, exhalation, body heat	Aircraft, UAV, helicopter, automobiles, trains	Bridges, roads, tunnels, farm house structures	Motors, compressors, chillers, pumps, fans	Wind
Walking, arm motion, finger motion, jogging, swimming, eating, talking	Tires, tracks, peddles, brakes, shock absorbers, turbines	Control-switch, HVAC systems, ducts, cleaners, etc.	Conveyors, cutting and dicing, vibrating mach.	Ocean currents, acoustic waves,

- Elastic structures which show an elastic deformation of their walls (rotor blades, wind mill blades, aircraft wings, pumps, motors, HVAC Ducts, and rotorcraft)
- Soft structures with very low-elastic modulus and high-deformation ratios (different textiles, leather, rubber membranes, and piping with internal fluid flow)

A.2 Parameters Required to Describe the Source

(i) The *source of vibrations* should be described to clarify the extent to which the source is diminished or degraded by harvesting some of its energy. Sources undiminished by the VEH would be earth tremors or heavy machine vibrations while the sources diminished to some extent by VEH would include small machines or an energy harvester attached to the body.

(ii) The acceleration values for vibration source should be reported as peak-to-peak g level. The preferred unit for acceleration is in m/s^2 described in terms of "g" where $1\,g = 9.8\,m/s^2$. Acceleration can be further categorized as low (less than $10\,mg$), mid (10–$100\,mg$), and high (above $100\,mg$).

(iii) The median frequency for vibration source should be reported in unit of Hertz. Frequency can be categorized as low (less than $10\,Hz$), mid (10–$120\,Hz$), and high (above $120\,Hz$).

(iv) For non-resonant systems, the external vibration source needs to be described (by two of the four parameters *force, displacement, velocity, and acceleration*) and the resulting motion of the VEH defined. For the impact systems, the impulse, $\int F(t)dt$, must be defined as best as possible and its frequency of occurrence or duty cycle specified.

A.3 Theoretical Models Used to Describe the Vibration Energy Harvesting

A.3.1 Williams-Yates Model

The differential equation of motion describing the system in terms of the housing vibration ($y(t) = Y_0 \cos \omega t$) and relative motion of mass ($z(t)$) is given as (Williams et al., 2001):

$$m\ddot{z}(t) + d\dot{z}(t) + kz(t) = -m\ddot{y}(t) \tag{A.1}$$

where m is the seismic mass, d is the damping constant, and k is the spring constant. The total power dissipated in the damper under sinusoidal excitation was found to be given as:

$$P(\omega) = \frac{m\zeta Y_o^2 \left(\frac{\omega}{\omega_n}\right)^3 \omega^3}{\left[1 - \left(\frac{\omega}{\omega_n}\right)^2\right]^2 + \left[2\zeta\left(\frac{\omega}{\omega_n}\right)\right]^2} \tag{A.2}$$

where $\omega_n^2 = k/m$ is the system resonant frequency and $\zeta = d/2\sqrt{mk}$ is the damping ratio. If the vibration spectrum is known beforehand than the device can be tuned to operate at the resonance frequency of the system, in which case the maximum power that can be generated is given as:

$$P_{\text{max}} = \frac{m Y_o^2 \omega_n^3}{4\zeta} \tag{A.3}$$

A.3.2 Erturk–Inman Model

The previous model was proposed for *electromagnetic* energy harvesters using the magnet-coil arrangement as proposed by Williams and Yates. The mechanism of *piezoelectric* transduction is relatively complicated, where the mechanical energy dissipation due to electrical power generation is not in the form of viscous damping (unlike the case in the Williams–Yates model) (Erturk and Inman, 2008a). The coupled distributed parameter solution of a piezoelectric energy harvester under the base excitation (Fig. A.1) was given for unimorph (Erturk and Inman, 2008b) and bimorph (Erturk and Inman, 2008c) cantilevers (i.e., for cantilevers with one or two piezoceramic layers).

As summarized in Chapter 2 by Erturk and Inman, for the unimorph cantilever configuration (Fig. 2.7a), the coupled beam equation can be expressed based on the Euler–Bernoulli beam theory as

$$YI\frac{\partial^4 w_{\text{rel}}(x,t)}{\partial x^4} + c_s I\frac{\partial^5 w_{\text{rel}}(x,t)}{\partial x^4 \partial t} + c_a\frac{\partial w_{\text{rel}}(x,t)}{\partial t} + m\frac{\partial^2 w_{\text{rel}}(x,t)}{\partial t^2} + \vartheta v(t)$$
$$\times \left[\frac{d\delta(x)}{dx} - \frac{d\delta(x-L)}{dx}\right] = -[m + M_t\delta(x-L)]\frac{\partial^2 w_{\text{b}}(x,t)}{\partial t^2}, \tag{A.4}$$

where $w_{\text{b}}(x,t)$ is the effective base displacement with a translational and small rotational component, $w_{\text{rel}}(x,t)$ is the transverse displacement response of the harvester beam relative to its vibrating base, $v(t)$ is the voltage response across the resistive load, YI is the bending stiffness, m is the mass per unit length, c_a is the external viscous damping term (due to air or the respective surrounding fluid), $c_s I$ is the internal strain rate (or Kelvin-Voigt) damping term,), ϑ is the piezoelectric coupling term, M_t is the tip (proof) mass (if one is attached) and $\delta(x)$ is the Dirac delta function.

If the electrode pair covering the piezoceramic layer operates into a circuit with a resistive load only, the circuit equation can be obtained as

$$\frac{\varepsilon_{33}^S bL}{h_p} \frac{dv(t)}{dt} + \frac{v(t)}{R_1} = -\int_{x=0}^{L} e_{31} h_{pc} b \frac{\partial^3 w_{rel}(x,t)}{\partial x^2 \partial t} dx, \qquad (A.5)$$

where ε_{33}^S is the permittivity at constant strain, b is the electrode width, L is the electrode length, h_p is the thickness of the piezoceramic layer, R_1 is the load resistance, e_{31} is the piezoelectric constant and h_{pc} is the distance from the neutral axis to the center of the piezoceramic layer.

The coupled voltage response to harmonic base excitation at steady state is

$$v(t) = \frac{\sum_{r=1}^{\infty} \frac{j\omega\varphi_r F_r}{\omega_r^2 - \omega^2 + j2\zeta_r\omega_r\omega}}{\frac{1}{R_1} + j\omega C_p + \sum_{r=1}^{\infty} \frac{j\omega\varphi_r \chi_r}{\omega_r^2 - \omega^2 + j2\zeta_r\omega_r\omega}} e^{j\omega t}, \qquad (A.6)$$

where ω is the excitation frequency, j is the unit imaginary number, $C_p = \varepsilon_{33}^S bL/h_p$ is the internal capacitance of the piezoceramic layer, φ_r is the forward modal coupling term, χ_r is the backward modal coupling term, ω_r and ζ_r are the undamped natural frequency and the mechanical damping ratio of the rth mode, respectively, and F_r is the modal mechanical forcing term. Thus, $|v(t)|$ is the *peak* voltage amplitude and $|v(t)|/\sqrt{2}$ is the *root mean square* voltage. Consequently, the peak power amplitude is $|v(t)|^2/R_1$ and the average power amplitude is $|v(t)|^2/2R_1$.

For modal excitations (i.e., $\omega \cong \omega_r$), the voltage expression given by Eq. (A.6) can be simplified drastically to give the single-mode (reduced) voltage response, $\hat{v}(t)$:

$$\hat{v}(t) = \frac{j\omega R_1 \varphi_r F_r e^{j\omega t}}{(1 + j\omega R_1 C_p)(\omega_r^2 - \omega^2 + j2\zeta_r\omega_r\omega) + j\omega R_1 \varphi_r \chi_r}, \qquad (A.7)$$

which yields the following peak power amplitude expression from $|\hat{v}(t)|^2/R_1$:

$$|\hat{P}(t)| = \frac{R_1(\omega\varphi_r F_r)^2}{\left[\omega_r^2 - \omega^2(1 + 2\zeta_r\omega_r R_1 C_p)\right]^2 + \left[2\zeta_r\omega_r\omega + \omega R_1\left[C_p(\omega_r^2 - \omega^2) + \varphi_r \chi_r\right]\right]^2}. \qquad (A.8)$$

A.4 Characterization of Vibration Energy Harvester

(i) Describe whether the VEH is uniaxial, biaxial or omnidirectional in its response.
(ii) In order to make the comparison of various types of vibration energy harvesters, a frequency of 60 Hz and acceleration of 1g is being set as the benchmark.

Table A.2 Compilation of the power density and optimum operating condition for various mechanical energy harvesting devices

Power			Volume	Power density	Acceleration^2		
(μW)	f(Hz)	a (m s^{-2})	(mm^3)	(μW/mm^3)	(m^2/s^4)	Method	Mass (g)
2	80	2	125	0.0168	4	Piezoelectric	10

(iii) The direction of mounting for the optimum output should be specified. The type of mounting (using fasteners, glue based, permanent, nut-bolts, magnets) on the vibration source should be described.

(iv) The foregoing table is being given here as an example to illustrate the parameters required for VEH. This is very crucial at this stage as it provides a method for comparison of various mechanisms and designs. In Table A.2, the volume refers to the system volume which includes all the VEH components.

Using this table, a figure can be constructed in terms of energy density (power density/frequency) and square of acceleration. In the near future when the data become available, we would like to start comparing all the VEHs on this one figure (see the reference in endnote) (Kim and Priya, 2008).

(v) Describe the maximum operating temperature, range of acceleration, and life-cycles at the rated frequency.

(vi) The testing of the VEH should include following three measurements: (i) RMS power as a function of vibration frequency at fixed acceleration ($1g$) and matching load and (ii) power as a function of acceleration at fixed frequency (60 Hz) and matching load, and (iii) power as a function of load at fixed acceleration ($1g$) and frequency (60 Hz).

A.5 Characterization of the Conditioning Circuit

(i) If the VEH being described is a system such that it includes power conditioning, then the rate at which conditioned power can be delivered to a defined load should be specified, namely, DC voltage level, load impedance, and current available to that load.

(ii) It should be mentioned that how much of the harvested power is consumed by the power-conditioning circuit.

(iii) Describe the output on-time rating in X msec at Y mA.

References

C. B. Williams, C. Sherwood, M. A. Harradine, P. H. Mellor, T. S. Birch, and R. B. Yates, "Development of an electromagnetic micro-generator", *IEE Proceedings of Circuits Devices System*, **148**(6) 337–342 (2001).

A. Erturk and D.J. Inman, Issues in mathematical modeling of piezoelectric energy harvesters, *Smart Materials and Structures*, (2008a).

A. Erturk and D.J. Inman, A distributed parameter electromechanical model for cantilevered piezo-electric energy harvesters, *ASME Journal of Vibration and Acoustics*, **130**, 041002 (2008b).

A. Erturk and D.J. Inman, An experimentally validated bimorph cantilever model for piezoelectric energy harvesting from base excitations, *Smart Materials and Structures*, (2008c).

H. Kim and S. Priya, "Piezoelectric microgenerator – current status, challenges, and applications", Proceedings of the ISAF, Santa Fe, NM (2008).

Index

Breinigsville, PA USA
04 December 2009
228544BV00011B/3/P

9 780387 764634